Basic Mathematics:
A Review

SECOND EDITION

Jack Barker
James Rogers
James Van Dyke
Department of Mathematics
Portland Community College
Portland, Oregon

Saunders College Publishing
Harcourt Brace Jovanovich College Publishers
Fort Worth   Philadelphia   San Diego
New York   Orlando   Austin   San Antonio
Toronto   Montreal   London   Sydney   Tokyo

Text Typeface: 10/12 Times Roman
Compositor: Progressive Typographers
Acquisitions Editor: Leslie Hawke
Project Editor: Patrice L. Smith
Art Directors: Richard L. Moore and Carol C. Bleistine
Art/Design Assistant: Virginia A. Bollard
Text Design: Emily Harste
Cover Design: Lawrence R. Didona
New Text Artwork: Philakorp
Production Manager: Tim Frelick
Assistant Production Manager: Maureen Iannuzzi

Cover credit: Houston reflections. Copyright © Harvey Lloyd, The Image Bank

BASIC MATHEMATICS: A REVIEW. 2/e

ISBN 0-03-071588-1

Printed in the United States of America.

Library of Congress catalog card number 84-051162.

1234   032   9 8 7 6 5 4

Harcourt Brace Jovanovich, Inc.
The Dryden Press
Saunders College Publishing

To our parents:
Floyd and Atha Barker
Frank and Louise Rogers
George and Ethel Van Dyke

PREFACE

OBJECTIVE

The objective of the second edition of *Basic Mathematics: A Review* is to provide a write-in text for students who need to review the skills of basic mathematics as a prerequisite for business mathematics, advanced mathematics, or technical mathematics.

APPLICATION

A new feature of the second edition is the prominent attention given to the *applications* of the objectives of each section. The application serves as a motivation for the mastery of the section. The solution of the application immediately follows the *Examples* and is reinforced in the *Exercises* with similar problems. These applications serve as a bridge between the mathematics classroom and the fields of business, shop mathematics, health science, consumer mathematics, and others.

VOCABULARY

As in the first edition, this write-in text is a *review text* that allows a student to quickly recall and practice a skill, as opposed to one that offers an in-depth presentation of the topic. *Modularized material* allows an instructor to choose sections of material for use in a classroom situation. *Individualized units* are keyed to:

1. a chapter *Pre-Test* to determine which objectives the student needs to study
2. a chapter *Post-Test* that may serve as either a chapter review or a chapter evaluation

HOW AND WHY

The material is divided into *twelve* major units with objectives explicitly stated. The first three chapters are devoted to arithmetic. Chapter Four covers denominate numbers and measurement. Chapters Five through Eleven review algebra, including a treatment of ratio and percent. (Parts of the ratio and percent chapter could follow Chapter Four). The review of percent from the algebraic viewpoint has been quite successful in classroom and mathematics laboratory use. The final chapter is an introduction to trigonometry.

Each section begins with an objective, which is followed by an application drawn from the world of experience or related to the section in the text. The applications demonstrate the usefulness of the skill and serve to motivate the student's interest in the topic.

Each skill expected of the student is stated in the form of one or more objectives.

At the suggestion of reviewers, in this edition many explanatory paragraphs have been edited and some topics have been reordered so that the study of the material has been further simplified.

The technical vocabulary to be used is defined, described, or used in context. The text is written with a minimum of rigor, and the use of formal mathematical vocabulary is limited as much as possible. For example, there are no objectives stated in terms of the associative, commutative, or distributive laws.

An explanation of how the process is performed and why it works is included. Emphasis is placed on the "how," since this is a review text. The student is "talked" through the process, and most sections end with a simple rule for quick recall of the procedure.

An instructor's manual, which includes a set of prepared tests and a test bank, is also available.

EXAMPLES

Several examples are worked out in detail for each skill to be practiced. In this way students can follow the process step by step. This allows them to become comfortable with the process before trying it on their own.

WARM UPS

Most sections include warm-up exercises. The authors find these useful as mental computations and as class exercises to check on the students' understanding of the section before beginning practice that usually requires pencil and paper. Answers to each section's warm-up exercises precede the next section or follow the chapter Post-Test.

EXERCISES

The second edition of *Basic Mathematics: A Review* differs from the previous edition in that the more than 3000 exercises (approximately 15% are applications) are usually divided into four categories. A group can be used for more oral exercises in class so that the instructor can check on how well the skill has been learned. These are followed by group **B**. This is a set of more difficult problems to be used by the student to gain proficiency in using the skill. This is followed by group **C** problems, which offer a challenge to the better student; where deemed applicable by the instructor, they may be used to provide calculator drill. The final group (group **D**) consists of applications of the skill drawn from experiences available in today's world. These applications are placed in most sections so the student can study them as they relate to the skill being practiced. This avoids a section devoted entirely to applications with no apparent linkage to the preceding skills. The applications, when placed in this manner, serve to promote the transfer of learning to the area in which the student will be applying the mathematical skill.

TIMETABLE

The text can be used in a variety of classroom situations, depending upon the needs of the students. Four such possibilities follow:

1. A three-quarter or two-semester course for students reviewing the basic skills prior to enrolling in a technical mathematics program. In such a course all of the topics would normally be covered.
2. A two-quarter or one-semester course for vocational students. Such a course would cover Chapters One through Six and Eight. This omits the more advanced topics of algebra and the unit on trigonometry.
3. A one-quarter or one-semester course for students preparing for business mathematics or business courses. This course would cover Chapters One through Four and Eight, along with some selected objectives in algebra.
4. A one-quarter or one-semester course as a review prior to beginning a college algebra program. This course would cover Chapters Five through Seven and Nine through Twelve.

The authors are appreciative of their wives, Mary Barker, Elinore Rogers, and Carol Van Dyke, without whose unfailing patience the work would never have been completed. Special thanks are due Leslie Hawke and Patrice Smith for their suggestions and support during the development of the second edition of this text.

Acknowledgment is given for the excellent contributions to the text by the following reviewers: Brayton Danner, Lincoln College; Margarite Dennis, Howard University; Christine Gregory, Gainesville Junior College; Eutilia Maher, University of Minnesota–Duluth; Gus Pekara, South Oklahoma City Junior College; Hazel Small, Gaston College; Ian Walton, Mission College; and Ray Treadway, Bennett College.

Jack Barker
James Rogers
James Van Dyke

CONTENTS

1

Whole Numbers

1. _____

2. _____

3. _____

4. _____

5. _____

6. _____

7. _____

8. _____

9. _____

10. _____

11. _____

12. _____

13. _____

14. _____

15. _____

16. _____

17. _____

18. _____

19. _____

20. _____

21. _____

22. _____

23. _____

24. _____

25. _____

Chapter 1
PRE-TEST

The problems in the following pre-test are a sample of the material in the chapter. You may already know how to work some of these. If so, this will allow you to spend less time on those parts. As a result, you will have more time to give to the sections that you find more difficult or that are new to you. The answers are in the back of the text.

Perform the indicated operations.

1. **(Obj. 3)** *Add:* $34 + 678 + 9 + 22$

2. **(Obj. 3)** *Add:* $3052 + 607 + 40 + 4070$

3. **(Obj. 3)** *Add:*
$$\begin{array}{r} 289 \\ 55 \\ 308 \\ 1008 \\ \underline{3} \end{array}$$

4. **(Obj. 4)** *Subtract:* $\begin{array}{r} 3060 \\ \underline{828} \end{array}$

5. **(Obj. 4)** *Subtract:* $8444 - 655$

6. **(Obj. 4)** *Subtract:* $1033 - 707$

7. **(Obj. 5)** *Multiply:* $\begin{array}{r} 451 \\ \underline{39} \end{array}$

8. **(Obj. 5)** *Multiply:* $(7804)(47)$

9. **(Obj. 6)** *Divide:* $34\overline{)10404}$

10. **(Obj. 6)** *Divide:* $3996 \div 12$

11. **(Obj. 6)** *Divide:* $108\overline{)1735}$

12. **(Obj. 9)** Find the value of 14^3.

13. **(Obj. 10)** Do the indicated operations: $(16 \div 8)5 + 34$

14. **(Obj. 10)** Do the indicated operations: $6^2 - 29 + 3^4$

15. **(Obj. 10)** Do the indicated operations: $13 + 2(14) \div 4 - 6$

16. **(Obj. 11)** List all the factors of 371.

17. **(Obj. 13)** Is 127 a prime or a composite number?

18. **(Obj. 14)** Write the prime factorization of 385.

19. **(Obj. 14)** Write the prime factorization of 513.

20. **(Obj. 15)** What is the LCM of 8, 14, and 28?

21. **(Obj. 15)** What is the LCM of 9, 12, 18, and 20?

22. **(Obj. 6)** If a stock clerk can stock a shelf of cans of vegetables in 25 minutes, how many shelves can she stock in two and a half hours?

23. **(Obj. 15)** What number is the least common denominator (the LCM of all the denominators) of the fractions

$$\frac{17}{18}, \frac{23}{45}, \text{ and } \frac{43}{60}?$$

24. **(Obj. 10)** On Saturday, the chemistry lab (which has 75 stations) is open for makeup labs. If there are 13 first term students, 8 second term students, and 17 third term students making up experiments, how many stations are empty?

25. **(Obj. 5)** A Day 'n Night grocer orders 18 cases of 6-packs of diet cola. If there are four 6-packs in a case, how many cans of cola will there be?

1.1 Whole Numbers: Place Value and Word Names

OBJECTIVES

1. **Write numeral form from word names.**
2. **Write word names from numeral form.**

APPLICATION

The purchasing agent for the Upjohn Corporation received a telephone bid of twenty-three thousand eighty-one dollars as the price of a new printing press. What is the numeral form of the bid?

VOCABULARY

The *digits* are 0, 1, 2, 3, 4, 5, 6, 7, 8, and 9. These are the symbols (numerals) that name the whole numbers from zero through nine. Numbers larger than nine are written in *numeral form* (492) by placing the digits (4, 9, and 2) in a certain order according to the standard *place value*.

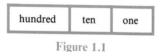

HUNDRED'S PLACE   TEN's PLACE   ONE'S PLACE

Words, spoken or written, that represent numbers are called *word names* (four hundred ninety-two).

HOW AND WHY

In our number system (called the Hindu-Arabic system), the digits are the only symbols used besides commas. This system is also called a base ten (decimal) system. From right to left, the first three place value names are one, ten, and hundred.

hundred	ten	one

Figure 1.1

For instance, in 573,   3 has place value one   (1)
                         7 has place value ten   (10)
                         5 has place value hundred   (100)

Continuing to the left, the digits are grouped in threes. The first five groups are named (from right to left) unit, thousand, million, billion, and trillion. The group on the extreme left may contain one, two, or three digits, while all other groups *must* contain three digits. Within each group the names are the same.

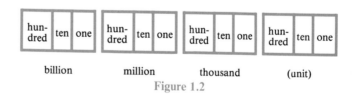

billion   million   thousand   (unit)

Figure 1.2

The place value of any position is one, ten, or hundred, followed by the group name.

WORD NAME

Any whole number can be written by writing the word name for the number represented by the digits in a group of three followed by the group name. The "units" group name is usually not read or written in naming a number.

So 34,573 is written

	3	4

5	7	3

thousand   (unit)

thirty-four thousand, five hundred seventy-three

NUMERAL FORM

The numeral form of a number can be written from the word name by writing the digits in each group of three and replacing the group names with commas.

EXAMPLES

a. Consider the number whose numeral form is 17,698,453.

Numeral form:	17,	698,	453
Word name of number in each group:	seventeen,	six hundred ninety-eight,	four hundred fifty-three
Group name:	million,	thousand,	(unit)

Word name: Seventeen million, six hundred ninety-eight thousand, four hundred fifty-three.

b. Reversing the process, we have:

Word name:

Two million, two hundred thirty-seven thousand, five hundred sixty-four.

group name   group name

(The group name "unit" is understood.)

Numeral form: 2,   2 3 7,   5 6 4

million   thousand   (unit)

APPLICATION SOLUTION

The numeral form for twenty-three thousand eighty-one is

            23,   081   or 23,081.

WARM UPS

Read (or write) the word names of these numbers.

1. 7852   **2.** 7802   **3.** 7082

4. 7820   **5.** 78502   **6.** 875,000

Read (or write) the numeral form of these numbers.

7. five hundred two

8. five hundred ninety-two

9. five thousand seventy-three

10. thirty-five thousand seventy-one

11. eighteen thousand one hundred

12. nine hundred seven thousand sixteen

Exercises 1.1

A

Write the word names of these numbers.

1. 540   **2.** 504   **3.** 5042

Write the numeral form of these numbers.

4. three hundred six   **5.** three thousand six

6. three thousand sixty

B

Write the word names of these numbers.

7. 2502   **8.** 2520   **9.** 25,200   **10.** 252,000

Write the numeral form of these numbers.

11. two hundred forty-three

12. two thousand thirty-four

13. two hundred forty-three thousand

14. two hundred three thousand forty-seven

C

Write the word names of these numbers.

15. 78,087,780   **16.** 502,520,052   **17.** 50,050,500

18. 13,131,013   **19.** 5,722,044,910

Write the numeral form of these numbers.

20. four million, two thousand, seven hundred

21. four hundred six million, two hundred forty-two thousand, seven hundred thirteen

22. three hundred fifteen million, five hundred seventy-two

23. six million, six hundred six

24. fifty-four billion, fifty-five million, fifty-six thousand, fifty seven

ANSWERS

1. _____
2. _____
3. _____
4. _____
5. _____
6. _____
7. _____
8. _____
9. _____
10. _____
11. _____
12. _____
13. _____
14. _____
15. _____
16. _____
17. _____
18. _____
19. _____
20. _____
21. _____
22. _____
23. _____
24. _____

25. _____

26. _____

27. _____

28. _____

29. _____

30. _____

D

25. A buyer for the Mason Brick Company gave a telephone bid of seventy-eight thousand sixty-three dollars for the construction materials for an apartment building. What is the numeral form of the bid?

26. A wholesale garment firm gave a telephone offer of two thousand sixty dollars to supply a retail chain with designer tee-shirts. What is the numeral form of the offer?

27. Michelle is paid 15 cents per mile when she uses her car for company business. She must report the mileage she has driven her auto in both numeral and word form. If she drives 785 miles during the month of June, what is the word name she should write for the mileage?

28. Stan uses a State Department car to go to a series of meetings. He must report the mileage he has driven in both numeral and word form. If he drove 1041 miles, what is the word name he should write for the mileage?

29. The world population during 1976 exceeded four billion. Write the number "four billion" in numeral form.

30. Write the name of the year 2001 in word form.

ANSWERS TO WARM UPS (1.1)

1. seven thousand eight hundred fifty-two

2. seven thousand eight hundred two

3. seven thousand eighty-two

4. seven thousand eight hundred twenty

5. seventy-eight thousand five hundred two

6. eight hundred seventy-five thousand

7. 502   **8.** 592   **9.** 5073   **10.** 35,071

11. 18,100   **12.** 907,016

1.2 Addition and Subtraction of Whole Numbers

OBJECTIVES

3. **Find the sum of two or more whole numbers.**
4. **Find the difference of two whole numbers.**

APPLICATION

Firebird Auto is offering a $366 rebate on all new cars priced above $6500. What is the cost of a car priced at $6785?

VOCABULARY

A natural number is any number from the group 1, 2, 3, 4, 5, . . . Notice that 0 is a whole number but not a natural number.

Words used with addition and subtraction problems are:

$$
\left.\begin{matrix} 16 \\ 23 \\ \underline{7} \end{matrix}\right\} \text{ADDENDS} \qquad \begin{matrix} 35 \leftarrow \text{MINUEND} \\ \underline{18} \leftarrow \text{SUBTRAHEND} \\ 17 \leftarrow \text{DIFFERENCE} \end{matrix}
$$
$$
46 \leftarrow \text{SUM}
$$

HOW AND WHY

To perform addition and subtraction, there are certain facts that must be memorized. The facts are commonly listed in an "addition table," which shows the sums of all possible combinations of two single-digit numbers. It is assumed from this point on that you know these facts.

ADDITION

To add two or more whole numbers:

1. **Write the numbers in columns so that place values are lined up.**
2. **Add each column, starting with the ones.**
3. **If the sum is more than nine, write the ones digit and carry the tens digit.**

To subtract 2 from 8 asks us to find the number that, when added to 2, gives a sum of 8. That is, $8 - 2$ wants to know $2 + ? = 8$. Since $2 + 6 = 8$, then $8 - 2 = 6$.

What is $47 - 15$? Since $15 + 32 = 47$, this tells us that $47 - 15 = 32$. There is a second subtraction problem associated with $15 + 32 = 47$. Do you see that this tells us that $47 - 32 = 15$?

SUBTRACTION

To subtract whole numbers:

1. **Write the numbers in columns so that place values are lined up.**
2. **Subtract each column, starting with the ones.**
3. **When the column cannot be subtracted, rename by borrowing.**

EXAMPLES

a. *Add:* $1682 + 4491 + 7629$

First, list the numbers in columns so the place values are lined up.

<table>
<tr><td>1 2 1</td><td>*Think:*</td><td></td></tr>
<tr><td>1682</td><td>ones column $2 + 1 + 9 = 12,$</td><td>*carry the "1" to the*
tens column</td></tr>
<tr><td>4491</td><td></td><td></td></tr>
<tr><td>7629</td><td>tens column $1 + 8 + 9 + 2 = 20,$</td><td>*carry the "2" to the*
hundreds
column</td></tr>
<tr><td>13802</td><td></td><td></td></tr>
<tr><td></td><td>hundreds column $2 + 6 + 4 + 6 = 18,$</td><td>*carry the "1" to the*
thousands
column</td></tr>
<tr><td></td><td>thousands column
   $1 + 1 + 4 + 7 = 13;$</td><td>*since there are no*
digits in ten
thousands column,
no carry is
necessary; record
the answer.</td></tr>
</table>

b. *Add:* $48 + 232 + 4 + 2834$

We write the problem vertically, making sure to line up the place values correctly.

```
 1 1 1
   48
  232
    4
 2834
 3118
```

c. *Subtract:* $752 - 295$

```
        752      Think:
        295      We must do two regroupings. We cannot subtract 5 ones from 2
  6  14          ones, and we cannot subtract 9 tens from 5 tens. Therefore, we
      4   12     borrow 1 ten (10 ones) from the tens to regroup with the ones (for
  7   5   2      a total of 12 ones) and then borrow 1 hundred (which is 10 tens)
  2   9   5      from the hundreds to regroup with the tens (for a total of 14 tens).
  4   5   7
```

The answer is checked by adding 295 and 457.

Check:
```
 295
 457
 752
```

d. **Calculator example**

$378 + 4091 + 52 = ?$

ENTER	DISPLAY
378	378.
+	378.
4091	4091.
+	4469.
52	52.
=	4521.

The sum is 4521.

APPLICATION SOLUTION

Since the car costs over $6500, we subtract the value of the rebate to find the cost.

6785
 366
6419

So the car costs $6419.

WARM UPS

Add:

1. $49 + 6$ **2.** $44 + 23$ **3.** $18 + 7 + 12$

4. $25 + 13 + 5$ **5.** $14 + 76$ **6.** $482 + 16$

Subtract:

7. $48 - 13$ **8.** $57 - 9$ **9.** $98 - 55$

10. $156 - 9$ **11.** $476 - 221$ **12.** $295 - 36$

Exercises 1.2

A

Add:

1. $28 + 5$ **2.** $124 + 7$ **3.** $130 + 29$

Subtract:

4. $39 - 7$ **5.** $37 - 9$ **6.** $73 - 14$

B

Add:

7. $75 + 94 + 80 + 3 + 27 + 30$ **8.** $419 + 638 + 215 + 482$

9. $491 + 3092 + 38 + 97 + 621$ **10.** $20000 + 60904 + 9130 + 72$

Subtract:

11. $274 - 181$ **12.** $517 - 399$

13. $2184 - 998$ **14.** $7002 - 3654$

C

Add:

15. $4629 + 1083 + 2194 + 4083 + 6219$

16. $1684 + 73 + 2891 + 400 + 5004$

17. $48 + 171 + 104 + 1956$

18. $28004 + 78 + 987 + 1435$

19. $13 + 8271 + 4082 + 16752$

Subtract:

20. $2704 - 809$ **21.** $40000 - 18214$

22. $7985 - 2006$ **23.** $46709 - 30510$

24. $4125 - 687$

ANSWERS

1. _____
2. _____
3. _____
4. _____
5. _____
6. _____
7. _____
8. _____
9. _____
10. _____
11. _____
12. _____
13. _____
14. _____
15. _____
16. _____
17. _____
18. _____
19. _____
20. _____
21. _____
22. _____
23. _____
24. _____

ANSWERS

25. _____

26. _____

27. _____

28. _____

29. _____

30. _____

25. Turibo Auto Company is offering a $297 rebate on all new cars priced above $5500. What is the cost of a car priced at $6225?

26. The Cookade Appliance firm is offering a $29 rebate on any of their food processors that retail $225 or more. What is the cost of a food processor that has a price tag of $334?

27. If there are 58275 tickets available for a baseball game, and all but 2863 tickets have been sold, how many tickets have been sold?

28. On Monday, Mr. Nguyen drove his car 182 miles, on Tuesday 364 miles, on Wednesday 235 miles, on Thursday 317 miles and on Friday 293 miles. What was the total mileage driven in the five days?

29. If the price of a used Trans-State is $597 more than the price of a Turibo, what is the price of the Trans-State if the Turibo costs $2987?

30. If the mileage (odometer reading) on a used Vee-Ten is 3478 miles less than mileage on a used Combi, what is the mileage on the Combi if the mileage on the Vee-Ten is 73,445?

1.3 Multiplication and Division of Whole Numbers

OBJECTIVES

5. Find the product of two whole numbers.
6. Divide two whole numbers.

APPLICATION

If the Mighty Good Dog Food Company puts a dozen cans of dog food in a case, how many cases and how many extra cans will there be in an order of 1069 cans?

VOCABULARY

There are several ways to show a multiplication problem. Here are most of them, using 3 times 17 as an example.

3×17 $3 \cdot 17$ $\begin{array}{r} 17 \\ \times 3 \end{array}$ *Used mostly in arithmetic*

$(3)(17)$ $3(17)$ $(3)17$ *Used in arithmetic and algebra*

$3 * 17$ *Used with computers*

Here are some of the ways to show a division problem, using 72 divided by 6 as an example.

$72 \div 6$ $6\overline{)72}$ *Used mostly in arithmetic*

$\dfrac{72}{6}$ *Used in arithmtic and algebra*

$72/6$ *Used with computers*

Words used with multiplication and division problems are:

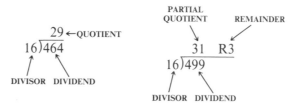

HOW AND WHY

Multiplication is often thought of as a short method for addition. $6 \cdot 8$ or 6×8 or $(6)(8)$ asks for the sum of six eights. You can now find the sum, $8 + 8 + 8 + 8 + 8 + 8 = 48$, or $6 \cdot 8 = 48$.

MULTIPLICATION

To perform multiplication without repeated addition, there are certain facts that should be memorized. These facts are commonly listed in a "multiplication table," which shows the products of all possible combinations of two single-digit numbers. It is assumed from this point on that you know these facts.

Using the multiplication facts along with the concept of place value, the product can be found as follows:

```
  629
   46
   54      6 times 9
  120      6 times 20
 3600      6 times 600
  360     40 times 9
  800     40 times 20
24000     40 times 600
28934
```

As you get better at this method of multiplying, you might want to take some shortcuts. The usual shortcut is to find the sum (mentally) of all the numbers multiplied by 6 and the sum of all the numbers multiplied by 40. This is shown below:

```
 1 3
 1 5
 629
  46
3774
25160
28934
```

DIVISION

$(24)(?) = 288$ is the same as $288 \div 24 = ?$ From this idea, division is referred to as the inverse of multiplication or as a process of undoing multiplication.

Recall that multiplication can be interpreted as summing up a number of groups. Division then is the process of discovering how many groups were summed.

The most common difficulty is that the maximum number of ones, tens, hundreds or whatever must be guessed correctly at each step of the way. This process is shown in the following example:

```
       745
 23)17135
    16100
     1035
      920
      115
      115
        0
```

Think: One group of 1000 twenty-threes is 23000, which is more than 17135; so there are no groups of 1000 twenty-threes. Try 7 one hundred groups of twenty-three or $700 \times 23 = 16100$. Subtract this from 17135. Since the difference is less than $100 \times 23 = 2300$, 7 is the maximum number of 100 groups. Now try 4 ten groups of twenty-three or $40 \times 23 = 920$. Subtract 920 from 1035. Since the difference is less than $10 \times 23 = 230$, 4 is the maximum number of ten groups. Continuing, try 5 groups of 23 or $5 \times 23 = 115$. Subtract 115 from 115. Since the difference is 0, the division is complete. Checking, $(745)(23) = 17135$, so the answer is correct.

Not all division problems involving whole numbers have a divisor (factor) that is summed a whole number of times. In

```
      3
 12)41
    36
     5
```

we see that 41 contains 3 twelves and 5 toward the next group of twelve. The answer is now written as 3 remainder 5.

The word "remainder" is usually abbreviated "R," and then the result is

$$41 \div 12 = 3 \, R \, 5$$

A check can be made by:

Divisor $\times$ partial quotient + remainder = dividend

So, $(12)(3) + 5 = 41$.

EXAMPLES

a. $0 \cdot 37 = 0$

b. $38 \cdot 74$

$$
\begin{array}{r}
74 \\
\underline{38} \\
592 \\
\underline{2220} \\
2812
\end{array}
$$

c. $4207 \div 7$

$$
\begin{array}{r}
601 \\
7\overline{)4207} \\
\underline{4200} \\
7 \\
\underline{0} \\
7 \\
\underline{7} \\
0
\end{array}
$$

(Note that a "0" must be placed above the tens digit if the place values of the 6 and 1 are to be correct.)

Check: $(7)(601) = 4207$

d. $1697 \div 55$

$$
\begin{array}{r}
30 \text{ R } 47 \\
55\overline{)1697} \\
\underline{1650} \\
47
\end{array}
$$

Check:

$$
\begin{array}{r}
55 \\
\underline{30} \\
00 \\
\underline{165} \\
1650 \\
\underline{47} \\
1697
\end{array}
$$

e. Calculator example

$346 \cdot 76 = ?$

ENTER	DISPLAY
346	346.
$\times$	346.
76	76.
$=$	26296.

The product is 26296.

f. Calculator example

$56\overline{)47432}$

ENTER	DISPLAY
47432	47432.
$\div$	47432.
56	56.
$=$	847.

The quotient is 847.

g. Calculator example

$37\overline{)6483}$

ENTER	DISPLAY
6483	6483.
÷	6483.
37	37.
=	175.2162

The partial quotient is 175.

$$175 \text{ R?}$$
$$37\overline{)6483}$$

To find the remainder, first find the product (175)(37).

ENTER	DISPLAY
37	37.
×	37.
175	175.
=	6475.

Then subtract the product from the original dividend.

ENTER	DISPLAY
6483	6483.
−	6483.
6475	6475.
=	8.

The answer is 175 R 8.

APPLICATION SOLUTION

Since each case must contain 12 cans of dog food, we divide to find out how many cases we need.

$$\begin{array}{r} 89 \\ 12\overline{)1069} \\ \underline{1068} \\ 1 \end{array}$$

The 1069 cans will fill 89 cases of dog food, with 1 can extra.

WARM UPS

Multiply:

1. 8×11 **2.** $2 \cdot 17$ **3.** $3(22)$

4. $(8)12$ **5.** 9×14 **6.** $14 \cdot 5$

Divide:

7. $7\overline{)56}$ **8.** $9\overline{)72}$ **9.** $3\overline{)123}$

10. $2\overline{)39}$ **11.** $5\overline{)87}$ **12.** $7\overline{)142}$

Exercises 1.3

ANSWERS

1. _____

2. _____

3. _____

4. _____

5. _____

6. _____

7. _____

8. _____

9. _____

10. _____

11. _____

12. _____

13. _____

14. _____

15. _____

16. _____

17. _____

18. _____

19. _____

20. _____

21. _____

22. _____

23. _____

24. _____

A

Multiply:

1. $\begin{array}{r} 122 \\ \underline{4} \end{array}$

2. $\begin{array}{r} 18 \\ \underline{3} \end{array}$

3. $\begin{array}{r} 64 \\ \underline{0} \end{array}$

Divide:

4. $7\overline{)714}$

5. $7\overline{)78}$

6. $7\overline{)359}$

B

Multiply:

7. 86(9)

8. (25)(16)

9. $26 \cdot 723$

10. 49(837)

Divide:

11. $6\overline{)372}$

12. $8\overline{)256}$

13. $1273 \div 33$

14. $2614 \div 68$

C

Multiply:

15. $\begin{array}{r} 756 \\ \underline{359} \end{array}$

16. $\begin{array}{r} 4009 \\ \underline{908} \end{array}$

17. $\begin{array}{r} 6125 \\ \underline{2234} \end{array}$

18. 505(213)

19. 79(8007)

Divide:

20. $47\overline{)14476}$

21. $18\overline{)2880}$

22. $76\overline{)44080}$

23. $408\overline{)126075}$

24. $97051 \div 507$

25. _____

26. _____

27. _____

28. _____

29. _____

30. _____

D

25. The Purrite Cat Food Corp. puts 24 cans of cat food in a case. How many cases and how many extra cans will there be if there are 2595 cans to be packed?

26. The Good Food Grocery Store ordered 234 cases of pears. The shipper packs 12 cans per case. How many cans of pears will the store expect to get?

27. The Good Food Grocery Store must pay 7 dollars for each of the 234 cases ordered in problem 26. What is the total cost of the pears?

28. Ellie ordered three hundred twenty-two 8-packs of Hot'nCola for sale at her store. How many bottles will there be?

29. The Night and Day Market orders 17 cases of single serve cans of orange juice. If each case contains 4 six-pack cartons of juice, how many single serve cans will they receive?

30. Norm's estate totaled $27,835. It is to be shared equally by his five nephews. How much money will each nephew receive?

1.4 Exponents

OBJECTIVES

7. Name the exponent when a number is written in exponential form.
8. Name the base when a number is written in exponential form.
9. Find the value of expressions written in exponential form.

APPLICATION

A Congressional committee proposes to increase the national debt by 13×10^9 dollars. Write this amount in numeral form.

VOCABULARY

Factors are numbers used as multipliers. In the expression 6×5 or $6 \cdot 5$, 6 and 5 are factors.

3^4 is an exponential form of a number, where 4 is the *exponent* and 3 is the *base*. The exponent shows how many times the base is used as a factor. The *power* or *value* is the result of the multiplication.

A *natural* or *counting* number is a number used in the ordinary process of counting (1, 2, 3, 4, 5, and so on). *Whole numbers* include both counting numbers and the number zero (0, 1, 2, 3, 4, . . .).

HOW AND WHY

The exponential form of writing a number is often used to write certain multiplication problems in a shorter form.

Any whole number can be used as an exponent.

> **To find the value of an expression with an exponent:**
>
> 1. **If the exponent is 1, the value is the base number.**
> 2. **If the exponent is larger than 1, use the base as a factor as many times as shown by the exponent.**

For example, $3^4 = 3 \cdot 3 \cdot 3 \cdot 3$. The value of 3^4 is 81, since $3 \cdot 3 \cdot 3 \cdot 3 = 9 \cdot 3 \cdot 3 = 27 \cdot 3 = 81$, and 81 is sometimes referred to as the fourth power of three.

$$\text{BASE} \rightarrow 3^{\overset{\text{EXPONENT}}{4}} = 81 \leftarrow \text{VALUE}$$

If 1 is used as the exponent, the number named is equal to the base. That is, $7^1 = 7$.

If 0 is used as the exponent, the base cannot be 0.

By definition, any natural number with 0 for the exponent is 1. For example, $2^0 = 1$, $11^0 = 1$, and $6^0 = 1$.

> $x^1 = x$, if x is any number.
> $x^0 = 1$, if x is not equal to 0 ($x \neq 0$).

EXAMPLES

a. 2^3 The exponent is 3.
The base is 2. 2^3 means $2 \cdot 2 \cdot 2$
The value is 8. $2^3 = 8$

b. 8^1 The exponent is 1.
The base is 8. 8^1 means 8
The value is 8. $8^1 = 8$

c. 3^5 The exponent is 5.
The base is 3. 3^5 means $3 \cdot 3 \cdot 3 \cdot 3 \cdot 3$
The value is 243. $3^5 = 243$

d. 7^0 The exponent is 0.
The base is 7. 7^0 means 1
The value is 1. $7^0 = 1$

e. 10^4 means $10 \cdot 10 \cdot 10 \cdot 10$
$10^4 = 10000$

f. **Calculator example**

Find the value of 13^5.

ENTER	DISPLAY
13	13.
$\times$	13.
$\times$	13.
$=$	169.
$=$	2197.
$=$	28561.
$=$	371293.*

If your calculator has an x^y key,

ENTER	DISPLAY
13	13.
x^y	13.
5	5.
$=$	371293.

APPLICATION SOLUTION

The numeral form of the number 13×10^9 can be found by multiplying

(13)(10)(10)(10)(10)(10)(10)(10)(10)(10)

or   13,000,000,000 dollars (thirteen billion dollars)

WARM UPS

Name the exponent of each expression.

1. 7^3 **2.** 3^7 **3.** 14^5

Name the base of each expression.

4. 7^6 **5.** 13^4 **6.** 5^{12}

Find the value of each expression.

7. 7^2 **8.** 1^5 **9.** 3^3

10. 6^0 **11.** 11^2 **12.** 10^3

* If your calculator does not give this display, see the instruction manual for the calculator. (Some calculators use a "constant" key instead.)

Exercises 1.4

A

Name the exponent of each expression.

1. 6^{14} **2.** 14^8

Name the base of each expression.

3. 16^2 **4.** 173^3

Find the value of each expression.

5. 2^4 **6.** 2^0

B

Name the exponent of each expression.

7. 56^{10} **8.** 33^{22}

Name the base of each expression.

9. 1^{77} **10.** 45^2

Find the value of each expression.

11. 12^2 **12.** 5^3

13. 10^5 **14.** 6^3

C

Find the value of each expression

15. 3^5 **16.** 7^3 **17.** 10^6

18. 2^6 **19.** 4^4 **20.** 99^0

21. 10^8 **22.** 11^3 **23.** 18^1

24. 10^{10}

ANSWERS

1. _____
2. _____
3. _____
4. _____
5. _____
6. _____
7. _____
8. _____
9. _____
10. _____
11. _____
12. _____
13. _____
14. _____
15. _____
16. _____
17. _____
18. _____
19. _____
20. _____
21. _____
22. _____
23. _____
24. _____

25. _____

25. One of the President's advisors recommends Federal savings to reduce the national debt by 4×10^8 dollars. Write this amount in numeral form.

26. _____

26. A western state had a fiscal surplus of 16×10^7 dollars for a given year. Write this number of dollars in both numeral and word form.

27. _____

27. The distance from earth to the nearest star (Alpha Centauri) is approximately 255×10^{11} miles. Write this distance in numeral form.

28. _____

28. The distance from earth to the sun is approximately 93×10^6 miles. Write this distance in numeral form.

29. _____

29. The average number of red blood cells in the human adult is approximately 2×10^{12}. Write this number in numeral form.

30. _____

30. At the rate of 3 miles per hour, it would take approximately 13×10^5 days to walk from the earth to the sun. Write this number in numeral form. (How many years would it take? See Section 4.3.)

1.5 Order of Operations (Problems With Two or More Operations)

OBJECTIVE

10. Perform any combination of operations (addition, subtraction, multiplication, and/or division) on whole numbers, in the conventional order.

APPLICATION

The cash price for a used car is $2795. If Mr. Macy pays $500 down and makes 30 payments of $99 each, the full price he pays is given by $P = 500 + 99(30)$. Is this more or less than the cash price? How much is the difference?

VOCABULARY

Parentheses ( ) and brackets [ ] are used in mathematics as grouping symbols. These symbols indicate that the operations inside are to be performed first. Thus,

$$(24 \div 4) \div 2 = 6 \div 2 = 3$$

and

$$24 \div (4 \div 2) = 24 \div 2 = 12$$

Other grouping symbols that are used are braces { } and bars $-$. The bar is used, most often, in division.

$$\frac{7 + 3}{2} = \frac{10}{2} = 5$$

HOW AND WHY

If we are asked to work the problem $2 \cdot 3 + 4$, it is possible to get two answers. If we do the multiplication first and then the addition, we get $2 \cdot 3 + 4 = 6 + 4 = 10$. If we do the addition first and then the multiplication, we have $2 \cdot 3 + 4 = 2 \cdot 7 = 14$.

If we are asked to work the problem $16 \div 2 + 6$, again it is possible to get two answers. That is, we could work it as follows: $16 \div 2 + 6 = 8 + 6 = 14$ or $16 \div 2 + 6 = 16 \div 8 = 2$.

The order in which the operations are performed is important. Because of this, an agreement has been made. It can be stated as follows:

In an exercise with more than one operation:

1. **Do the operations inside grouping symbols first, in the order of steps 2 to 4.**
2. **Work from left to right doing only exponentiations (powers) as you come to them.**
3. **Go back to the left and work to the right doing only MULTIPLICATIONS and DIVISIONS as you come to them.**
4. **Go back to the left and work to the right doing only ADDITIONS and SUBTRACTIONS as you come to them.**

Neither multiplication nor division takes preference over the other. They are performed in the order in which they appear. The same is true with regard to addition and subtraction.

EXAMPLES

Do the indicated operations:

a. $7 \cdot 9 + 6 \cdot 2$

$63 + 12$ *Multiply*

75 *Add*

b. $25 - 6 \div 3 + 8 \cdot 4$

$25 - 2 + 32$ *Multiply and divide in the order they come*

55 *Add and subtract in the order they come*

c. $(4 + 7 + 9) \div (8 - 3)$

$20 \div 5$ *Perform operations inside parentheses (part 1 of rule)*

4 *Divide*

d. $7^2 + 8^3$

$49 + 512$ *Compute powers first (part 2 of rule)*

561 *Add*

e. $17^2 - (39 - 5^2)$

$17^2 - (39 - 25)$ *First, do operations inside parentheses; of these, powers are done first (part 2 of rule)*

$17^2 - (14)$ *Do subtraction inside parentheses*

$289 - 14$ *Go back to the left and do powers (part 2 of rule)*

275 *Subtract*

f. Calculator example

$8 + 6 \cdot 5 \div 3 = ?$

If your calculator is a four-function, non-algebraic calculator, do the addition last:

ENTER	DISPLAY
6	6.
×	6.
5	5.
÷	30.
3	3.
+	10.
8	8.
=	18.

The answer is 18.

If your calculator has algebraic logic:

ENTER	DISPLAY
$\boxed{8}$	8.
$\boxed{+}$	8.
$\boxed{6}$	6.
$\boxed{\times}$	6.
$\boxed{5}$	5.
$\boxed{\div}$	30.
$\boxed{3}$	3.
$\boxed{=}$	18.

The answer is 18.

If your calculator has parentheses:

ENTER	DISPLAY
$\boxed{8}$	8.
$\boxed{+}$	8.
$\boxed{(}$	0.
$\boxed{6}$	6.
$\boxed{\times}$	6.
$\boxed{5}$	5.
$\boxed{\div}$	30.
$\boxed{3}$	3.
$\boxed{)}$	10.
$\boxed{=}$	18.

The answer is 18.

APPLICATION SOLUTION

In the formula $P = 500 + 99(30)$, the operations of addition and multiplication appear. Following steps 3 and 4 of the rule:

$P = 500 + 99(30)$ *Original formula*

$P = 500 + 2970$ *Multiply*

$P = 3470$ *Add*

So Mr. Macy would pay $3470 with the down payment and 30 more payments. This amount would be $675 more than the cash price.

WARM UPS

Do the indicated operations:

1. $5 + 6(2)$ **2.** $20 \div 4 - 2$

3. $20 - 4 \div 2$ **4.** $12 \div 6 \cdot 2$

5. $12 \cdot 6 \div 2$ **6.** $4 \cdot 7 + 12$

7. $16 \div 2 + 3 \cdot 2$ **8.** $19 - 3 \cdot 2$

9. $7 + 3^2$ **10.** $4^2 + 2^2$

11. $(8 + 3)2^3$ **12.** $2(3 + 4) - 7$

Exercises 1.5

A

Do the indicated operations.

1. $12 \cdot 3 - 5 \cdot 4$ **2.** $10^2 - 2 \cdot 20$

3. $64 \div 8^2$ **4.** $44 \div (2 \cdot 11)$

5. $60 \div 3 \cdot 4$ **6.** $28 \div 4 + 2^3 - 10$

B

Do the indicated operations.

7. $12(4 + 2) - 24 \div 12$ **8.** $(5 \cdot 2 + 3)2 - 10$

9. $120 \div (5^2 - 5)$ **10.** $22^2 - 11^2 + 12^2$

11. $22^2 - (11^2 + 12^2)$ **12.** $2[8 + (27 - 2 \cdot 12)]$

13. $41 - [2(5^2) - 7^2]$ **14.** $(8 - 1)^2 + 5(36 - 3^3)$

C

Do the indicated operations.

15. $7[(4^2 - 5)(4^2 + 6)]$ **16.** $14^2 - [48 - (19 - 5)]$

17. $5^3(18 - 12)^2$ **18.** $2[3(27 - 25)^3]^2$

19. $8^2 - (36 \div 3^2)$ **20.** $(15^2 - 12^2) \div (15 - 12)$

21. $(16^2 - 13^2) \div (16 - 13)$ **22.** $7^2(19 - 14)$

23. $7(19 - 14)^2$ **24.** $2\{119 - [(16 + 23)3] + 7\}$

D

25. The cash price for a used car is $4595. If Ms. Masley pays $800 down and makes 24 payments of $177 each, the full price she pays is given by $P = 800 + 24(177)$. How much is the difference between this and the cash price?

26. The cash price for a used car is $5500. If Mrs. Manyard pays $1150 down and makes 36 payments of $127 each, the full price she pays is given by $P = 1150 + 36(127)$. How much is the difference between this and the cash price?

ANSWERS

1. _____
2. _____
3. _____
4. _____
5. _____
6. _____
7. _____
8. _____
9. _____
10. _____
11. _____
12. _____
13. _____
14. _____
15. _____
16. _____
17. _____
18. _____
19. _____
20. _____
21. _____
22. _____
23. _____
24. _____
25. _____
26. _____

ANSWERS

27. _____

28. _____

29. _____

30. _____

31. _____

32. _____

33. _____

27. The Acme Delivery Service agrees to deliver telephone directories for a fixed fee of $100 plus $1 for every directory delivered. The directories are shipped in cartons that contain 24 directories each. What will be the delivery charge for 300 cartons of directories? [Hint: $C = 100 + (1)(24)(300)$]

28. The Rosaday Floral Store has 570 red roses to sell. They advertise a bouquet of 6 roses for $9 plus a $3 delivery charge. If all the roses were sold and delivered, what was the income from the sale?

29. On March 1 the balance in a checking account was $432. During the month a check was written for $63, three checks for $18 and a check for $89. There was one deposit of $22 and a service charge of $3. The balance at the end of the month can be calculated by finding the value of

$$432 - 63 - 3(18) - 89 + 22 - 3.$$

What was the balance at the end of the month?

30. Charles Mitchell is planning to carpet two rooms in his home. One has dimensions 15 feet by 15 feet, the other 25 feet by 17 feet. The number of square feet of carpet he needs can be calculated by finding the value of

$$15 \cdot 15 + 25 \cdot 17.$$

How many square feet of carpet does he need?

31. Greg wants to calculate the total resistance in a power supply circuit in a television set. The set has two resistors in parallel. If one resistor is 2400 ohms and the second is 1200 ohms, the total resistance can be calculated by finding the value of

$$R_t = 2400 \div (2400 \div 1200 + 1).$$

What is the total resistance in the circuit?

32. To check that $x = 14$ is a solution to the equation

$$125 - 6x + 4 = 45,$$

we substitute 14 for x and find the value of

$$125 - 6(14) + 4.$$

Verify that the above expression equals 45.

33. Substitute $x = 24$ in the following equation and check that it is a solution (makes the equation true).

$$\frac{x}{8} + \frac{x}{12} + 4x = 101$$

that is, $x \div 8 + x \div 12 + 4x = 101$.

1.6 Divisors and Factors

OBJECTIVES	**11. List all factors (divisors) of a given whole number.** **12. Write a whole number as the product of two factors in as many ways as possible.**

APPLICATION

A television station has 130 minutes of programming to fill. In how many ways can this time be scheduled if all programs are to be the same length and a whole number of minutes? (For example, ten programs each thirteen minutes in length, $10 \cdot 13 = 130$.)

VOCABULARY

Remember that 6 is a factor of 24 since $(6)(4) = 24$, and 6 is also called a divisor of 24 since $24 \div 6 = 4$. Another way of saying this is that 24 is *divisible* by 6.

HOW AND WHY

All the factors (or divisors) of 250 could be listed by trial and error, but we might miss a few. To make sure that we got them all, we could divide 250 by every whole number from 1 to 249, but it would take a long time. The following steps take less time.

First: Write all the counting numbers from 1 to the first number whose square is larger than 250.

1	6	11	16 *Since $16 \cdot 16 = 256$ (the square of 16), stop.*
2	7	12	
3	8	13	
4	9	14	
5	10	15	

Second: Divide each of these into 250. If it divides evenly, write the factors. If not, cross out the number.

$1 \cdot 250$	6̸	1̸1̸ 1̸6̸
$2 \cdot 125$	7̸	1̸2̸
3̸	8̸	1̸3̸
4̸	9̸	1̸4̸
$5 \cdot 50$	$10 \cdot 25$	1̸5̸

These steps give us a list of all factors (divisors) and products.

Factors of 250: 1, 2, 5, 10, 25, 50, 125, 250

250 as a product: $1 \cdot 250$, $2 \cdot 125$, $5 \cdot 50$, $10 \cdot 25$

These steps give us all of the factors because any factor larger than 16 is found as a quotient when we divide by the numbers in the list.

> To find all the factors or divisors of a given number:
>
> 1. **List the counting numbers starting with 1.**
> 2. **Stop the list when the square of the counting number is larger than the given number.**
> 3. **The numbers in the list that divide the given number (evenly) are factors of the given number. So are the quotients in these divisions.**

EXAMPLES

a. The factors of 68 can be found as follows:

$1 \cdot 68$ $4 \cdot 17$ $\not{7}$

$2 \cdot 34$ $\not{5}$ $\not{8}$

$\not{3}$ $\not{6}$ $\not{9}$ *Since* $9 \cdot 9 = 81$, *we stop here.*

The list of factors is 1, 2, 4, 17, 34, 68. The list of products is $1 \cdot 68, 2 \cdot 34$, and $4 \cdot 17$.

b. The factors of 180 are:

$1 \cdot 180$ $6 \cdot 30$ $\not{11}$

$2 \cdot 90$ $\not{7}$ $12 \cdot 15$

$3 \cdot 60$ $\not{8}$ $\not{13}$

$4 \cdot 45$ $9 \cdot 20$ $\not{14}$ *Since* $14 \cdot 14 = 196$, *we stop here.*

$5 \cdot 36$ $10 \cdot 18$

The list of factors is 1, 2, 3, 4, 5, 6, 9, 10, 12, 15, 18, 20, 30, 36, 45, 60, 90, 180. The list of products is $1 \cdot 180, 2 \cdot 90, 3 \cdot 60, 4 \cdot 45, 5 \cdot 36, 6 \cdot 30, 9 \cdot 20, 10 \cdot 18, 12 \cdot 15$.

c. The factors of 29 are:

$1 \cdot 29$ $\not{4}$

$\not{2}$ $\not{5}$

$\not{3}$ $\not{6}$

The list of factors is 1, 29. The list of products is $1 \cdot 29$.

APPLICATION SOLUTION

> Write 130 as a product in all possible ways.
>
> $1 \cdot 130$ One program (perhaps a movie) that is 130 minutes long or 130 programs that are 1 minute long
>
> $2 \cdot 65$ Two programs that are 65 minutes long or 65 programs that are 2 minutes long
>
> $\not{3}$
> $\not{4}$
>
> $5 \cdot 26$ Five programs that are 26 minutes long or 26 programs that are 5 minutes long
>
> $\not{6}$
> $\not{7}$
> $\not{8}$
> $\not{9}$
>
> $10 \cdot 13$ Ten programs that are 13 minutes long or 13 programs that are 10 minutes long
>
> $\not{11}$
> $\not{12}$ Stop

WARM UPS

Answer yes or no.

1. Is 3 a factor of 12?   **2.** Is 4 a factor of 12?

3. Is 5 a factor of 12?   **4.** Is 6 a factor of 12?

5. Is 6 a factor of 15?   **6.** Is 18 a factor of 6?

List all the factors of each number.

7. 18   **8.** 26   **9.** 24

Write each number as the product of two factors in all possible ways.

10. 24   **11.** 50   **12.** 102

Exercises 1.6

A

List all the factors of each number.

1. 25   **2.** 27   **3.** 35

Write each number as the product of two factors in all possible ways.

4. 66   **5.** 40   **6.** 45

B

List all the factors of each number.

7. 36   **8.** 80   **9.** 96   **10.** 90

Write each number as the product of two factors in all possible ways.

11. 100   **12.** 80   **13.** 96   **14.** 95

C

List all the factors of each number.

15. 128   **16.** 131   **17.** 245   **18.** 300

19. 444

Write each number as the product of two factors in all possible ways.

20. 500   **21.** 847   **22.** 720   **23.** 1311

24. 1312

ANSWERS

1. _____
2. _____
3. _____
4. _____
5. _____
6. _____
7. _____
8. _____
9. _____
10. _____
11. _____
12. _____
13. _____
14. _____
15. _____
16. _____
17. _____
18. _____
19. _____
20. _____
21. _____
22. _____
23. _____
24. _____

ANSWERS

25. _____

26. _____

27. _____

28. _____

D

25. In what ways can a television station schedule 150 minutes of time if all the programs are the same length and each runs a whole number of minutes?

26. In what ways can the TV station in Exercise 25 schedule 3 hours of time?

27. In what ways can a radio station schedule 90 minutes of news features if each is the same length and each runs a whole number of minutes?

28. In what ways can the station in Exercise 27 schedule 100 minutes of news features?

ANSWERS TO WARM UPS (1.6)

1. yes	**2.** yes
3. no	**4.** yes
5. no	**6.** no
7. 1, 2, 3, 6, 9, 18	**8.** 1, 2, 13, 26
9. 1, 2, 3, 4, 6, 8, 12, 24	**10.** $1 \cdot 24, 2 \cdot 12, 3 \cdot 8, 4 \cdot 6$
11. $1 \cdot 50, 2 \cdot 25, 5 \cdot 10$	**12.** $1 \cdot 102, 2 \cdot 51, 3 \cdot 34, 6 \cdot 17$

1.7 Primes and Composites

OBJECTIVE

13. Identify a given whole number as prime or composite.

APPLICATION

Primes can be used to find the lowest common denominator of fractions $\left(\dfrac{5}{56} + \dfrac{27}{64}\right)$ and to reduce some fractions $\left(\dfrac{456}{792}\right)$.

VOCABULARY

A *prime number* is a whole number greater than one with exactly two different factors (divisors). These two factors are the number 1 and the whole number itself. Whole numbers greater than 1 with more than two different factors are called *composite numbers.*

HOW AND WHY

Primes and composites are one useful way to classify whole numbers. The whole numbers zero (0) and one (1) are neither prime nor composite. They are in a class by themselves.

Two (2) is the first prime number ($2 = 2 \cdot 1$), since 2 and 1 are the only factors of 2.

Four (4) is a composite number ($4 = 4 \cdot 1$ and $4 = 2 \cdot 2$), since it has more than two factors. Four has the three factors 1, 2, and 4. It is possible to study each whole number and list its factors or divisors to see if it has exactly two or more than two factors.

To tell whether a number is prime or composite, make a table of possible divisors and try each one. If the number has exactly two divisors, it is prime.

To determine if 51 is prime or composite, make a table of possible divisors and try each one.

$$
\begin{array}{lll}
1 \cdot 51 & \cancel{4} & \cancel{7} \\
\cancel{2} & \cancel{5} & \cancel{8} \\
3 \cdot 17 & \cancel{6} &
\end{array}
$$

The factors of 51 are 1, 3, 17, and 51. Since it has four factors, it is a composite number.

In the following list, the first 20 prime numbers are circled. The numbers that are crossed out are composite numbers. The numbers 0 and 1 are special and in a class by themselves. Zero is the only number for which every number is a factor, and 1 is the only number that has just a single factor.

0	1	②	③	4̶	⑤	6̶	⑦	8̶	9̶	1̶0̶
⑪	1̶2̶	⑬	1̶4̶	1̶5̶	1̶6̶	⑰	1̶8̶	⑲	2̶0̶	
2̶1̶	2̶2̶	㉓	2̶4̶	2̶5̶	2̶6̶	2̶7̶	2̶8̶	㉙	3̶0̶	
㉛	3̶2̶	3̶3̶	3̶4̶	3̶5̶	3̶6̶	㊲	3̶8̶	3̶9̶	4̶0̶	
㊶	4̶2̶	㊸	4̶4̶	4̶5̶	4̶6̶	㊼	4̶8̶	4̶9̶	5̶0̶	
5̶1̶	5̶2̶	㊾	5̶4̶	5̶5̶	5̶6̶	5̶7̶	5̶8̶	㉟	6̶0̶	
㊱	6̶2̶	6̶3̶	6̶4̶	6̶5̶	6̶6̶	㊿	6̶8̶	6̶9̶	7̶0̶	
㋀										

EXAMPLES

a. 31 is a prime number since it has exactly two factors (1 and 31).

b. 101 is a prime number since it has exactly two factors (1 and 101).

c. 91 is composite since it has more than two factors (1, 7, 13, and 91).

d. Is 323 prime or composite? By division 2, 3, 5, 7, 11, and 13 are not factors. Note that we need only test prime divisors, since composite numbers include prime factors. Next try 17. Since $323 \div 17 = 19$, 323 has more than two factors (1, 17, 19, 323) and therefore is composite.

e. Is 331 prime or composite? By division 2, 3, 5, 7, 11, 13, 17, and 19 are not factors. Since 19^2 is larger than 331, it is useless to try larger primes; therefore, 331 is prime.

WARM UPS

Tell whether each number is prime or composite.

1. 7 **2.** 17 **3.** 27 **4.** 37

5. 9 **6.** 19 **7.** 29 **8.** 39

9. 11 **10.** 21

11. What is the smallest prime number?
12. What is the smallest composite number?

Exercises 1.7

A

Tell whether each number is prime or composite.

1. 12   **2.** 13   **3.** 14   **4.** 15

5. 18   **6.** 23

B

7. 27   **8.** 37   **9.** 47   **10.** 57

11. 67   **12.** 77   **13.** 87   **14.** 97

C

15. 197   **16.** 297   **17.** 397   **18.** 497

19. 597   **20.** 697   **21.** 797   **22.** 897

23. 997   **24.** 1097

1.
2.
3.
4.
5.
6.
7.
8.
9.
10.
11.
12.
13.
14.
15.
16.
17.
18.
19.
20.
21.
22.
23.
24.

1.8 Prime Factorization

OBJECTIVE

14. Write the prime factorization of a given whole number.

APPLICATION

Prime factorization will be used to reduce fractions, to find the least common multiple of numbers, and to find the lowest common denominator of fractions.

VOCABULARY

The *prime factorization* (prime factored form) of a number is the number written as a product of primes. A number is said to be *completely factored* when it is in prime factored form ($18 = 2 \cdot 3 \cdot 3$).

HOW AND WHY

To express a number as a product of primes, try to divide the number by each prime in consecutive order starting with 2. When a prime divides the number, the quotient is then divided by the same prime if possible. Continue the process until the quotient is one. The indicated product of all the primes that are divisors is the prime factorization of the whole number.

For instance:

$$
\begin{array}{r|r}
2 & 48 \\
2 & 24 \\
2 & 12 \\
2 & 6 \\
3 & 3 \\
  & 1
\end{array}
$$

$48 = 2 \cdot 2 \cdot 2 \cdot 2 \cdot 3 = 2^4 \cdot 3$

If a whole number or any of the resulting quotients is a large prime, you will recognize this when you have tried each prime whose square is smaller than the number and found no divisor. Then divide by the larger prime to complete the process. For instance:

$$
\begin{array}{r|r}
3 & 1179 \\
3 & 393 \\
131 & 131 \\
   & 1
\end{array}
$$

131 is not divisible by 2, 3, 5, 7, or 11; also, $13 \cdot 13 = 169$.
Therefore, 131 is prime.

$1179 = 3 \cdot 3 \cdot 131 = 3^2 \cdot 131$

EXAMPLES

Write the prime factorization of each of the following.

a. 42

$$2\,\lfloor 42$$
$$3\,\lfloor 21$$
$$7\,\lfloor 7$$
$$1$$

b. 848

$$2\,\lfloor 848$$
$$2\,\lfloor 424$$
$$2\,\lfloor 212$$
$$2\,\lfloor 106$$
$$53\,\lfloor 53$$
$$1$$

$$42 = 2 \cdot 3 \cdot 7$$

$$848 = 2 \cdot 2 \cdot 2 \cdot 2 \cdot 53 = 2^4 \cdot 53$$

c. 377

$$13\,\lfloor 377$$
$$29\,\lfloor 29$$
$$1$$

$$377 = 13 \cdot 29$$

WARM UPS

Give the prime factorization of each number.

1. 6   **2.** 10   **3.** 15   **4.** 21

5. 22   **6.** 23   **7.** 25   **8.** 26

9. 18   **10.** 19   **11.** 20   **12.** 24

Exercises 1.8

A

Give the prime factorization of each number.

1. 8 **2.** 9 **3.** 11 **4.** 12

5. 14 **6.** 34

B

7. 30 **8.** 36 **9.** 48 **10.** 50

11. 64 **12.** 51 **13.** 71 **14.** 91

C

15. 97 **16.** 100 **17.** 120 **18.** 150

19. 198 **20.** 325 **21.** 414 **22.** 768

23. 563 **24.** 888

1. _____
2. _____
3. _____
4. _____
5. _____
6. _____
7. _____
8. _____
9. _____
10. _____
11. _____
12. _____
13. _____
14. _____
15. _____
16. _____
17. _____
18. _____
19. _____
20. _____
21. _____
22. _____
23. _____
24. _____

1.9 Least Common Multiple

OBJECTIVE

15. **Find the least common multiple of two or more numbers.**

APPLICATION

The least common multiple of the denominators of

$$\frac{5}{6}, \frac{2}{9}, \frac{7}{12}, \text{ and } \frac{5}{18}$$

is also called the lowest common denominator of the fractions. What is the lowest common denominator of these four fractions?

VOCABULARY

A multiple of a number is the product of that number and a natural number. The multiples of 6 are $1 \cdot 6, 2 \cdot 6, 3 \cdot 6, 4 \cdot 6, 5 \cdot 6$ and so on, or 6, 12, 18, 24, ...

The *least common multiple* of two or more whole numbers is the smallest natural number that is a multiple of each of the given numbers. LCM is an abbreviation of Least Common Multiple. The LCM of 4 and 6 is 12.

HOW AND WHY

Find the LCM of 30 and 42. This is done by listing the multiples of each and finding the smallest one in both groups. This number is the LCM of 30 and 42.

Multiples of 30: 30, 60, 90, 120, 150, 180, $\boxed{210}$, 240, 270, . . .

Multiples of 42: 42, 84, 126, 168, $\boxed{210}$, 252, 294, . . .

The LCM of 30 and 42 is 210. Finding the LCM in this manner has a drawback: you might have to list hundreds of multiples. For this reason, we look for a shortcut.

To find the LCM of 12 and 18, write the prime factorization of each of them. Stack the prime factors that are the same.

PRIMES WITH
LARGEST EXPONENTS

$$12 = 2 \cdot 2 \cdot 3 \quad = 2^2 \cdot 3^1$$
$$18 = 2 \quad \bigg| \quad \cdot 3 \cdot 3 = 2^1 \cdot 3^2 \qquad\qquad 2^2, 3^2$$
$$\downarrow \quad \downarrow \quad \downarrow \quad \downarrow$$

LCM of 12 and 18 $= 2 \cdot 2 \cdot 3 \cdot 3 \ = 2^2 \cdot 3^2 = 36$

The number in each column is a prime factor of the LCM. This guarantees that the LCM is a multiple of each number and that we do not get too many prime factors. Note that the LCM is the product of the different prime factors that have the largest exponents.

What is the LCM of 16, 10, and 24?

PRIMES WITH
LARGEST EXPONENTS

$16 = 2^4$

$10 = 2^1 \cdot 5^1$

$24 = 2^3 \cdot 3^1$ $2^4, 3^1, 5^1$

$LCM = 2^4 \cdot 3^1 \cdot 5^1 = 16 \cdot 3 \cdot 5 = 240$

> **To find the Least Common Multiple (LCM) of two or more numbers:**
>
> 1. **Write each number in prime-factored form using exponents.**
> 2. **Find the product of the different factors with the largest exponents.**

EXAMPLES

a. Find the LCM of 12 and 20.

PRIMES WITH
LARGEST EXPONENTS

$12 = 2 \cdot 2 \cdot 3 = 2^2 \cdot 3^1$

$20 = 2 \cdot 2 \cdot 5 = 2^2 \cdot 5^1$ $2^2, 3^1, 5^1$

$LCM = 2^2 \cdot 3 \cdot 5 = 60$

b. What is the LCM of 18, 24, and 30?

$18 = 2 \cdot 3 \cdot 3 \quad = 2 \cdot 3^2$

$24 = 2 \cdot 2 \cdot 2 \cdot 3 = 2^3 \cdot 3$

$30 = 2 \cdot 3 \cdot 5 \quad = 2 \cdot 3 \cdot 5 \qquad 2^3, 3^2, 5$

$LCM = 2^3 \cdot 3^2 \cdot 5 \quad = 360$

APPLICATION SOLUTION

The fractions $\frac{5}{6}, \frac{2}{9}, \frac{7}{12},$ and $\frac{5}{18}$ have the denominators 6, 9, 12, and 18.

The LCM of these numbers is:

$6 = 2^1 \cdot 3^1$

$9 = \quad 3^2$

$12 = 2^2 \cdot 3^1$

$18 = 2^1 \cdot 3^2$

$LCM = 2^2 \cdot 3^2 = 36$ (36 is also the lowest common denominator)

WARM UPS

Find the LCM of each group of numbers.

1. 2, 4	**2.** 3, 4	**3.** 3, 6
4. 4, 8	**5.** 4, 6	**6.** 3, 8
7. 3, 9	**8.** 4, 9	**9.** 2, 4, 6
10. 2, 4, 8	**11.** 2, 3, 6	**12.** 2, 3, 4

Exercises 1.9

ANSWERS

1. _____

2. _____

3. _____

4. _____

5. _____

6. _____

7. _____

8. _____

9. _____

10. _____

11. _____

12. _____

13. _____

14. _____

15. _____

16. _____

17. _____

18. _____

19. _____

20. _____

21. _____

22. _____

23. _____

24. _____

A

Find the LCM of each group of numbers.

1. 3, 5   **2.** 3, 7   **3.** 4, 5   **4.** 4, 7

5. 6, 8   **6.** 6, 9

B

7. 10, 15   **8.** 15, 20   **9.** 18, 20

10. 15, 18   **11.** 3, 8, 9   **12.** 10, 12, 15

13. 75, 100   **14.** 48, 60

C

15. 36, 60   **16.** 28, 42

17. 96, 120   **18.** 15, 33, 55

19. 48, 72, 80   **20.** 75, 150, 250

21. 32, 60, 72   **22.** 21, 70, 90

23. 14, 35, 49, 91   **24.** 38, 57, 114, 171

D

Find the common denominator of each set of fractions:

25. $\dfrac{5}{12}, \dfrac{9}{16}, \dfrac{7}{24}$ **26.** $\dfrac{5}{12}, \dfrac{3}{16}, \dfrac{9}{20}$

27. $\dfrac{11}{12}, \dfrac{5}{36}, \dfrac{7}{54}$ **28.** $\dfrac{3}{14}, \dfrac{8}{21}, \dfrac{7}{8}$

29. $\dfrac{1}{4}, \dfrac{7}{10}, \dfrac{2}{15}, \dfrac{5}{12}$ **30.** $\dfrac{3}{5}, \dfrac{11}{20}, \dfrac{1}{12}, \dfrac{17}{30}$

Chapter 1
POST-TEST

1. _____

2. _____

3. _____

4. _____

5. _____

6. _____

7. _____

8. _____

9. _____

10. _____

11. _____

12. _____

13. _____

14. _____

15. _____

16. _____

17. _____

18. _____

19. _____

20. _____

21. _____

22. _____

23. _____

24. _____

25. _____

1. (Obj. 11) List all the factors of 56.

2. (Obj. 15) What is the LCM of 8, 12, and 36?

3. (Obj. 4) _Subtract:_
$$\begin{array}{r} 2009 \\ 432 \\ \hline \end{array}$$

4. (Obj. 6) _Divide:_ $24\overline{)8904}$

5. (Obj. 3) _Add:_
$$\begin{array}{r} 67 \\ 21 \\ 34 \\ 52 \\ 87 \\ 69 \\ \hline \end{array}$$

6. (Obj. 5) _Multiply:_
$$\begin{array}{r} 723 \\ 37 \\ \hline \end{array}$$

7. (Obj. 9) Find the value of 2^5.

8. (Obj. 3) _Add:_
$$\begin{array}{r} 1437 \\ 215 \\ 9213 \\ 14 \\ 30 \\ 8021 \\ \hline \end{array}$$

9. (Obj. 10) Do the indicated operations: $45 - 8 \cdot 2 \div 4 + 12$

10. (Obj. 6) _Divide:_ $48\overline{)62149}$

11. (Obj. 14) Write the prime factorization of 624.

12. (Obj. 10) Do the indicated operations: $3 \cdot 8 + 14$

13. (Obj. 3) _Add:_
$$\begin{array}{r} 45 \\ 37 \\ 21 \\ \hline \end{array}$$

14. (Obj. 9) Find the value of 5^3.

15. (Obj. 13) Is 103 a prime or a composite number?

16. (Obj. 4) _Subtract:_
$$\begin{array}{r} 2372 \\ 897 \\ \hline \end{array}$$

17. (Obj. 10) Do the indicated operations: $36 - 12 \div 4$

18. (Obj. 4) _Subtract:_
$$\begin{array}{r} 6023 \\ 2914 \\ \hline \end{array}$$

19. (Obj. 6) _Divide:_ $7\overline{)364}$

20. (Obj. 14) Write the prime factorization of 180.

21. (Obj. 5) _Multiply:_
$$\begin{array}{r} 2093 \\ 59 \\ \hline \end{array}$$

22. (Obj. 6) The distance around an automobile wheel is 45 inches. How many revolutions will it make in one mile? (1 mile = 63,360 inches)

23. (Obj. 6) An automobile traveled 792 miles and used 36 gallons of gasoline. What were the miles per gallon for the trip?

24. (Obj. 5) If the speed of sound is 1100 feet per second, how far away is a skyrocket if the sound is heard 9 seconds after the explosion is observed?

25. (Obj. 10) On a construction project 40 tons of fill gravel are to be used. If Mr. Rock has hauled four loads that weighed 12,974; 13,642; 11,921; and 14,672 pounds, how many more pounds are left to be hauled? (1 ton = 2000 pounds)

ANSWERS TO WARM UPS (1.9) **1.** 4   **2.** 12   **3.** 6   **4.** 8   **5.** 12   **6.** 24

7. 9   **8.** 36   **9.** 12   **10.** 8   **11.** 6   **12.** 12

2

**Fractions
and Mixed
Numbers**

Chapter 2
PRE-TEST

The problems in the following pre-test are a sample of the material in the chapter. You may already know how to work some of these. If so, this will allow you to spend less time on those parts. As a result, you will have more time to give to the sections that you find more difficult or that are new to you. The answers are in the back of the text.

1. **(Obj. 19)** Reduce to lowest terms: $\dfrac{36}{84}$

2. **(Obj. 20)** *Multiply:* $\dfrac{5}{18} \cdot \dfrac{6}{25}$

3. **(Obj. 20)** *Multiply:* $\dfrac{4}{5} \cdot \dfrac{15}{16} \cdot \dfrac{2}{3}$

4. **(Obj. 20)** *Multiply:* $2\dfrac{1}{3} \cdot 5\dfrac{1}{6} \cdot 1\dfrac{7}{9}$

5. **(Obj. 21)** What is the reciprocal of $6\dfrac{2}{3}$?

6. **(Obj. 22)** *Divide:* $\dfrac{3}{11} \div \dfrac{4}{5}$

7. **(Obj. 22)** *Divide:* $5\dfrac{3}{4} \div 6\dfrac{1}{2}$

8. **(Obj. 22)** *Divide:* $\dfrac{8}{9} \div 24$

9. **(Obj. 23)** Find the missing numerator: $\dfrac{3}{14} = \dfrac{?}{42}$

10. **(Obj. 24)** List the following fractions in order from smallest to largest.

$$\frac{7}{36}, \frac{1}{6}, \frac{1}{4}$$

11. **(Obj. 24)** List the following fractions in order from smallest to largest.

$$\frac{5}{9}, \frac{2}{3}, \frac{11}{18}, \frac{29}{54}$$

12. **(Obj. 25)** *Add:* $\dfrac{17}{63} + \dfrac{7}{63}$

13. **(Obj. 26)** *Add:* $\dfrac{1}{5} + \dfrac{1}{2} + \dfrac{2}{3}$

14. **(Obj. 27)** *Add:* $7\dfrac{8}{9}$

$$2\frac{2}{3}$$

$$5\frac{1}{18}$$

$$3\frac{3}{4}$$

15. _____

16. _____

17. _____

18. _____

19. _____

20. _____

21. _____

22. _____

23. _____

24. _____

25. _____

15. **(Obj. 27)** *Add:* $2\frac{7}{8} + 4\frac{2}{3} + \frac{1}{6}$

16. **(Obj. 28)** *Subtract:* $\frac{7}{9} - \frac{1}{6}$

17. **(Obj. 29)** *Subtract:* $21\frac{7}{12}$
$$9\frac{3}{4}$$
$$\overline{\phantom{9\frac{3}{4}}}$$

18. **(Obj. 29)** *Subtract:* $48 - 21\frac{5}{7}$

19. **(Obj. 29)** *Subtract:* $15\frac{5}{6}$
$$7\frac{5}{7}$$
$$\overline{\phantom{7\frac{5}{7}}}$$

20. **(Obj. 30)** Do the indicated operations. $\frac{2}{3} + \frac{1}{7} \cdot \frac{4}{3} - \frac{3}{7}$

21. **(Obj. 30)** Do the indicated operations. $\frac{3}{14} \div \frac{5}{7} + \frac{2}{7} \cdot \frac{3}{2} - \frac{1}{2}$

22. **(Obj. 27)** Four poured iron castings have weights of $31\frac{3}{8}$ lb, $40\frac{1}{2}$ lb, $45\frac{1}{8}$ lb, and $30\frac{7}{16}$ lb. What is the total weight of the iron in the four castings?

23. **(Obj. 29)** A wood beam $40\frac{3}{4}$ feet in length has $3\frac{3}{8}$ feet cut off of one end. How long is the beam now?

24. **(Obj. 20)** The cost of making an ice-cream bar is $10\frac{3}{5}$ cents. What is the cost of making 55 bars?

25. **(Obj. 22)** How many bags of sand can be made from a container of $33\frac{1}{3}$ lb of sand, if each bag is to contain $\frac{3}{4}$ lb of sand?

2.1 Fractions and Mixed Numbers

OBJECTIVES

16. Write a fraction to describe parts of units (unit regions or unit groups).
17. Change improper fractions to mixed numbers.
18. Change mixed numbers to improper fractions.

APPLICATION

If $\frac{1}{2}$ inch represents 10 feet on a scale drawing, how many feet does $2\frac{1}{2}$ inches represent?

VOCABULARY

A *fraction* is a name of a number $\left(\text{such as } \frac{6}{7}\right)$. The upper numeral (6) is the *numerator* and is a whole number. The lower numeral (7) is the *denominator* and is a natural number. That is, $\dfrac{\text{numerator}}{\text{denominator}}$.

A *proper fraction* names a number less than one $\left(\text{such as } \frac{4}{9}\right)$. Its numerator is always smaller than the denominator. An *improper fraction* names a number greater than or equal to one $\left(\text{such as } \frac{7}{6} \text{ or } \frac{9}{9}\right)$. Its numerator is equal to or larger than the denominator.

A *mixed number* is the sum of a whole number and a fraction $\left(3 + \frac{1}{2}\right)$ with the plus sign omitted $\left(3\frac{1}{2}\right)$.

HOW AND WHY

When using a rectangle to illustrate the fraction $\frac{a}{b}$, the letter b represents the number of equal parts into which the rectangle is divided. The letter a represents the number of these parts under discussion.

The following rectangle is subdivided into 7 equal parts, and 6 of the parts are shaded. The fraction $\frac{6}{7}$ represents the shaded part of Figure 2.1. The denominator (7) tells us the number of equal parts in the rectangle. The numerator (6) tells us the number of these parts under discussion. The fraction $\frac{1}{7}$ represents the unshaded part of the rectangle.

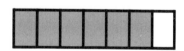

Figure 2.1

The fractions $\frac{a}{a}$ and $\frac{0}{a}$ have special significance: $\frac{a}{a} = 1$, all parts shaded, and $\frac{0}{a} = 0$, no parts shaded.

> To change an improper fraction to a mixed number, divide the numerator by the denominator. If there is a remainder, write the whole number and then write the fraction, $\dfrac{\text{remainder}}{\text{divisor}}$.

$$\frac{7}{3} = 7 \div 3 \quad \text{or} \quad 3\overline{)7} \;\; \begin{array}{r} 2 \\ \hline 7 \\ 6 \\ \hline 1 \end{array}$$

So,

$$\frac{7}{3} = 2\frac{1}{3}$$

> To change a mixed number to an improper fraction, multiply the denominator times the whole number, add the numerator, and put the sum over the denominator.

So,

$$1\frac{3}{7} = \frac{7 \cdot 1 + 3}{7} = \frac{7 + 3}{7} = \frac{10}{7}$$

EXAMPLES

These are examples of proper fractions:

a. $\dfrac{3}{5}$

b. $\dfrac{0}{3}$

   (no shaded parts)

Change each improper fraction to a mixed number.

c. $\dfrac{8}{7} = 7\overline{)8}\;\begin{array}{r}1\\\hline 8\\7\\\hline 1\end{array} = 1\dfrac{1}{7}$

d. $\dfrac{11}{3} = 3\overline{)11}\;\begin{array}{r}3\\\hline 11\\9\\\hline 2\end{array} = 3\dfrac{2}{3}$

Change each mixed number to an improper fraction.

e. $2\dfrac{4}{5} = \dfrac{5 \cdot 2 + 4}{5} = \dfrac{10 + 4}{5} = \dfrac{14}{5}$

f. $3\dfrac{1}{10} = \dfrac{10 \cdot 3 + 1}{10} = \dfrac{30 + 1}{10} = \dfrac{31}{10}$

g. $7 = 7\dfrac{0}{1} = \dfrac{1 \cdot 7 + 0}{1} = \dfrac{7}{1}$

APPLICATION SOLUTION

First change $2\frac{1}{2}$ to an improper fraction.

$$2\frac{1}{2} = \frac{2 \cdot 2 + 1}{2} = \frac{5}{2}$$

Now, if $\frac{1}{2}$ inch represents 10 ft, then $\frac{5}{2}$ inches represents 5 times as much. Since

$$5 \cdot 10 = 50$$

the $2\frac{1}{2}$ inches represents 50 ft.

WARM UPS

Write the fraction that represents the shaded part of each figure.

1. 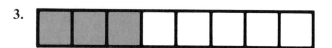   2.

3.

4.

Change to a mixed number.

5. $\dfrac{11}{4}$ 6. $\dfrac{17}{3}$ 7. $\dfrac{21}{10}$ 8. $\dfrac{15}{7}$

Change to an improper fraction.

9. $3\dfrac{1}{9}$ 10. $4\dfrac{7}{8}$ 11. $5\dfrac{1}{5}$ 12. $4\dfrac{1}{4}$

Exercises 2.1

ANSWERS

1. _____

2. _____

3. _____

4. _____

5. _____

6. _____

7. _____

8. _____

9. _____

10. _____

11. _____

12. _____

13. _____

14. _____

15. _____

A

Write the fraction that describes the shaded part of each figure.

1.

2.

Change to a mixed number.

3. $\dfrac{17}{8}$ **4.** $\dfrac{43}{4}$

Change to an improper fraction.

5. $3\dfrac{3}{4}$ **6.** $3\dfrac{2}{7}$

B

Write a fraction that describes the shaded part of each figure.

7.

8.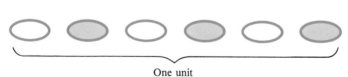

One unit

Change to a mixed number.

9. $\dfrac{87}{10}$ **10.** $\dfrac{132}{11}$ **11.** $\dfrac{133}{15}$

Change to an improper fraction.

12. $2\dfrac{0}{3}$ **13.** $5\dfrac{3}{8}$ **14.** $10\dfrac{7}{10}$

C

Write a fraction that describes the shaded part of each figure.

15.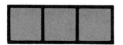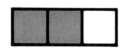

ANSWERS

16. _____

17. _____

18. _____

19. _____

20. _____

21. _____

22. _____

23. _____

24. _____

25. _____

26. _____

27. _____

28. _____

29. _____

30. _____

31. _____

32. _____

33. _____

34. _____

35. _____

36. _____

16.

Change to a mixed number.

17. $\dfrac{196}{3}$ **18.** $\dfrac{32}{23}$ **19.** $\dfrac{79}{5}$ **20.** $\dfrac{202}{3}$

Change to an improper fraction.

21. 5 **22.** $60\dfrac{5}{6}$ **23.** $16\dfrac{2}{3}$ **24.** $3\dfrac{17}{100}$

D

25. If $\dfrac{1}{3}$ inch represents 1 foot on a scale drawing, how many feet does $2\dfrac{1}{3}$ inches represent?

26. If $\dfrac{1}{8}$ inch represents 1 mile on a map, how many miles does $3\dfrac{3}{8}$ inches represent?

27. Hal's construction company must place section barriers, each $\dfrac{1}{8}$ mile long, between the two sides of a freeway. How many such sections will be needed for $24\dfrac{3}{8}$ miles of freeway?

28. A scale is marked (as usual) with a whole number at each pound. What whole number mark will be closest to a measurement of $\dfrac{50}{16}$ lb?

29. A ruler is marked (as usual) with a whole number at each centimeter. What whole number mark will be closest to a measurement of $\dfrac{87}{10}$ cm?

30. The fraction $\dfrac{373}{10}$ falls between what two whole numbers?

31. How many equal parts must be shaded in a figure of 18 unit regions to illustrate the mixed number $17\dfrac{3}{8}$?

E *Maintain Your Skills*

Add:

32. $48 + 37 + 24 + 83$ **33.** $345 + 132 + 48$ **34.** $456 + 2743 + 18$

Subtract:

35. $2856 - 1890$ **36.** $38123 - 20984$

ANSWERS TO WARM UPS (2.1)   1. $\dfrac{4}{5}$ 2. $\dfrac{1}{6}$ 3. $\dfrac{3}{8}$ 4. $\dfrac{7}{9}$ 5. $2\dfrac{3}{4}$ 6. $5\dfrac{2}{3}$

7. $2\dfrac{1}{10}$ 8. $2\dfrac{1}{7}$ 9. $\dfrac{28}{9}$ 10. $\dfrac{39}{8}$ 11. $\dfrac{26}{5}$ 12. $\dfrac{17}{4}$

2.2 Reducing Fractions

OBJECTIVE

19. Reduce a given fraction to lowest terms.

APPLICATION

On a sheet metal layout the diameter of a hole is given as $\dfrac{28}{32}$ inch. Write this diameter as a reduced fraction.

VOCABULARY

Equivalent fractions are fractions that are different names for the same number. $\dfrac{1}{4}$ and $\dfrac{2}{8}$ are equivalent fractions.

Reducing a fraction is the process of renaming it by using a smaller numerator and denominator. Reducing a fraction to *lowest terms* is writing an equivalent fraction that has the smallest possible numerator and denominator. For instance, $\dfrac{6}{12} = \dfrac{1}{2}$.

HOW AND WHY

When we compare the two units in Figure 2.2, we note that each is divided into

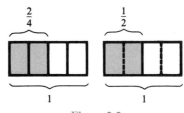

Figure 2.2

four parts. The shaded part on the left is named $\dfrac{2}{4}$, while the shaded part on the right is labeled $\dfrac{1}{2}$. It is clear that the two are the same size, and therefore we can say $\dfrac{2}{4} = \dfrac{1}{2}$.

The arithmetical way of showing that $\dfrac{2}{4} = \dfrac{1}{2}$ is to eliminate the common factors by canceling or dividing:

$$\frac{2}{4} = \frac{1 \cdot \overset{1}{\cancel{2}}}{2 \cdot \underset{1}{\cancel{2}}} = \frac{1}{2} \qquad \text{or} \qquad \frac{2}{4} = \frac{2 \div 2}{4 \div 2} = \frac{1}{2}$$

These methods work for reducing all fractions. Consider reducing $\dfrac{15}{18}$.

$$\frac{15}{18} = \frac{5 \cdot \cancel{3}^{1}}{6 \cdot \cancel{3}_{1}} = \frac{5}{6}$$ *Cancel out the common factors*

$$\frac{15}{18} = \frac{15 \div 3}{18 \div 3} = \frac{5}{6}$$ *Divide out the common factors*

When all common factors have been eliminated (canceled or divided out), the fraction is reduced to lowest terms.

$$\frac{24}{40} = \frac{12}{20} = \frac{6}{10} = \frac{3}{5}$$ *Reduced to lowest terms*

> **To reduce a fraction to lowest terms, eliminate all common factors (other than 1) in the numerator and denominator by canceling or dividing.**
>
> $$\frac{ac}{bc} = \frac{ac \div c}{bc \div c} = \frac{a}{b}$$
>
> $$\frac{a}{ab} = \frac{1 \cdot \cancel{a}^{1}}{b \cdot \cancel{a}_{1}} = \frac{1}{b}$$

If the common factors cannot be discovered easily, they can always be found by writing the numerator and denominator in prime factored form (See Example d.)

EXAMPLES

a. $\dfrac{28}{21} = \dfrac{28 \div 7}{21 \div 7} = \dfrac{4}{3}$

b. $\dfrac{16}{24} = \dfrac{8 \cdot 2}{12 \cdot 2} = \dfrac{8}{12} = \dfrac{4 \cdot 2}{6 \cdot 2} = \dfrac{4}{6} = \dfrac{2 \cdot 2}{3 \cdot 2} = \dfrac{2}{3}$ *(completely reduced or reduced to lowest terms)*

  or $\dfrac{16}{24} = \dfrac{2 \cdot 8}{3 \cdot 8} = \dfrac{2}{3}$

c. $\dfrac{16}{8} = \dfrac{\cancel{8}^{1} \cdot 2}{\cancel{8}_{1} \cdot 1} = \dfrac{2}{1} = 2$

d. $\dfrac{126}{144} = \dfrac{\cancel{2}^{1} \cdot \cancel{3}^{1} \cdot \cancel{3}^{1} \cdot 7}{\cancel{2}_{1} \cdot 2 \cdot 2 \cdot 2 \cdot \cancel{3}_{1} \cdot \cancel{3}_{1}}$ *Prime factor, since the numbers are large*

  $= \dfrac{7}{2 \cdot 2 \cdot 2}$ *Cancel out the common factors*

  $= \dfrac{7}{8}$ *Multiply*

e. $\dfrac{8}{9} = \dfrac{2 \cdot 2 \cdot 2}{3 \cdot 3} = \dfrac{8}{9}$ *No common factors other than 1*

APPLICATION SOLUTION

Reduce $\dfrac{28}{32}$ to lowest terms.

$$\dfrac{28}{32} = \dfrac{7 \cdot \overset{1}{\cancel{4}}}{8 \cdot \underset{1}{\cancel{4}}} = \dfrac{7}{8}$$

The diameter of the hole is $\dfrac{7}{8}$ inch.

WARM UPS

Reduce to lowest terms.

1. $\dfrac{12}{14}$ 2. $\dfrac{14}{24}$ 3. $\dfrac{24}{36}$ 4. $\dfrac{4}{10}$

5. $\dfrac{6}{9}$ 6. $\dfrac{8}{12}$ 7. $\dfrac{10}{25}$ 8. $\dfrac{18}{30}$

9. $\dfrac{20}{22}$ 10. $\dfrac{21}{24}$ 11. $\dfrac{30}{75}$ 12. $\dfrac{45}{75}$

Exercises 2.2

A

Reduce to lowest terms.

1. $\dfrac{65}{75}$ **2.** $\dfrac{8}{40}$ **3.** $\dfrac{10}{40}$ **4.** $\dfrac{15}{40}$

5. $\dfrac{35}{40}$ **6.** $\dfrac{21}{36}$ **7.** $\dfrac{48}{30}$

B

8. $\dfrac{16}{4}$ **9.** $\dfrac{45}{32}$ **10.** $\dfrac{64}{72}$ **11.** $\dfrac{90}{126}$

12. $\dfrac{36}{100}$ **13.** $\dfrac{121}{132}$ **14.** $\dfrac{98}{210}$

C

15. $\dfrac{268}{402}$ **16.** $\dfrac{97}{101}$ **17.** $\dfrac{98}{102}$ **18.** $\dfrac{153}{255}$

19. $\dfrac{200}{330}$ **20.** $\dfrac{546}{910}$ **21.** $\dfrac{630}{1050}$ **22.** $\dfrac{504}{1764}$

23. $\dfrac{294}{1617}$ **24.** $\dfrac{255}{480}$

D

25. On a sheet metal layout the diameter of a hole is given as $\dfrac{40}{64}$ inch. What reduced fraction represents the diameter of the hole?

26. On a drawing the length of a rectangle is given as $\dfrac{30}{32}$ inch. Write the length as a reduced fraction.

1. _____
2. _____
3. _____
4. _____
5. _____
6. _____
7. _____
8. _____
9. _____
10. _____
11. _____
12. _____
13. _____
14. _____
15. _____
16. _____
17. _____
18. _____
19. _____
20. _____
21. _____
22. _____
23. _____
24. _____
25. _____
26. _____

ANSWERS

27.

28.

29.

30.

31.

32.

33.

34.

35.

27. The number of board feet in a piece of lumber that measures 2 inches by 5 inches by 10 inches is given by

$$\text{ft. b. m.} = \frac{(2)(5)}{12} \cdot 10 = \frac{100}{12}$$

Reduce the ft. b. m. to lowest terms.

28. The volume of a small box with dimensions $\frac{2}{3}$ inch by $\frac{3}{4}$ inch by $\frac{5}{6}$ inch is given by

$$V = \left(\frac{2}{3}\right)\left(\frac{3}{4}\right)\left(\frac{5}{6}\right) = \frac{30}{72}$$

Reduce the volume to lowest terms.

29. If 12 of a total of 16 rolls of wire have been used, what fractional part of the wire is left? Reduce to lowest terms.

30. The float on a tank registers 8 ft. If the tank is full when the float registers 12 ft, what fractional part of the tank is filled? Reduce to lowest terms.

E *Maintain Your Skills*

Multiply:

| **31.** 478 | **32.** 761 | **33.** 849 |
| 379 | 208 | 963 |

Divide:

34. $58\overline{)1763200}$ **35.** $48\overline{)96048}$

2.3 Multiplication of Fractions and Mixed Numbers

OBJECTIVES

20. Find the product of numbers written as fractions or mixed numbers and reduce the product to lowest terms.
21. Find the reciprocal of a given natural number, fraction, or mixed number.

APPLICATION

Bobby drives his automobile at an average rate of 40 miles per hour for $2\dfrac{1}{2}$ hours. How many miles does he travel?

VOCABULARY

A *product* is the answer to a multiplication problem. If two fractions have a product of 1, either one is called the *reciprocal* of the other. For example, $\dfrac{2}{3}$ is the reciprocal of $\dfrac{3}{2}$.

HOW AND WHY

What is $\dfrac{1}{2}$ of $\dfrac{1}{3}$? (This is more commonly written as $\dfrac{1}{2} \cdot \dfrac{1}{3} = ?$) Refer to Figure 2.3, in which the rectangle is divided into three equal parts (thirds); the $\dfrac{1}{3}$ that we want to multiply by $\dfrac{1}{2}$ is shaded. Since we are interested in one half of this one third, let us divide each of the thirds in the rectangle into two equal parts. As a result of doing this, we see in Figure 2.4 that the rectangle is now divided into six equal parts. Since we want one half of the shaded one third, we can see that $\dfrac{1}{2}$ of $\dfrac{1}{3}$ must be $\dfrac{1}{6}$. That is, $\dfrac{1}{2} \cdot \dfrac{1}{3} = \dfrac{1}{6}$. We could also write $\left(\dfrac{1}{2}\right)\left(\dfrac{1}{3}\right) = \dfrac{1}{6}$.

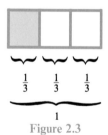

Figure 2.3   Figure 2.4

Note that we get $\dfrac{1}{6}$ if we multiply the numerators and multiply the denominators and make these products the numerator and denominator, respectively.

The product of two or more numbers written as fractions is the product of the numerators over the product of the denominators.

So

$$\frac{3}{4} \cdot \frac{5}{8} = \frac{3 \cdot 5}{4 \cdot 8} = \frac{15}{32}$$

$$\frac{5}{6} \cdot \frac{7}{9} = \frac{5 \cdot 7}{6 \cdot 9} = \frac{35}{54}$$

To multiply mixed numbers, change each mixed number to its equivalent improper fraction and multiply. (See Example e.)

The product of two or more fractions can often be found more quickly in reduced form (lowest terms) by reducing *before* performing the multiplication. Check each of the numerators and denominators to determine if there are common factors in any numerator *and* denominator. If there is a common factor in a numerator and a denominator, factor each of these whole numbers so that this common factor is seen. The multiplication is indicated by writing the product as a single fraction whose numerator and denominator are in factored form. For instance,

$$\frac{12}{35} \cdot \frac{25}{18} = \frac{2 \cdot 6 \cdot 5 \cdot 5}{7 \cdot 5 \cdot 3 \cdot 6}$$

The common factors in the numerator and denominator can then be eliminated, and the indicated multiplication can be performed. The answer will be reduced completely. So,

$$\frac{12}{35} \cdot \frac{25}{18} = \frac{2 \cdot 5}{7 \cdot 3} = \frac{10}{21}$$

If the common factors cannot be seen easily, use the prime factorization of each numerator and denominator.

The reciprocal of any fraction may be found by interchanging the numerator and denominator.

This is often called "inverting" the fraction. For instance, the reciprocal of $\frac{3}{7}$ is $\frac{7}{3}$. We check by verifying that the product is 1:

$$\frac{3}{7} \cdot \frac{7}{3} = \frac{21}{21} = 1$$

EXAMPLES

a. $\dfrac{1}{2} \cdot \dfrac{3}{4} \cdot \dfrac{5}{8} = \dfrac{1 \cdot 3 \cdot 5}{2 \cdot 4 \cdot 8} = \dfrac{15}{64}$

b. $\dfrac{5}{2} \cdot 3 = \dfrac{5}{2} \cdot \dfrac{3}{1} = \dfrac{15}{2}$

c. $\left(1\dfrac{1}{2}\right) \cdot \dfrac{3}{4} = \dfrac{3}{2} \cdot \dfrac{3}{4} = \dfrac{9}{8}$

(Note that the 8 and 12 have a common factor of 4.)

d. $\dfrac{8}{15} \cdot \dfrac{5}{12} = \dfrac{4 \cdot 2}{5 \cdot 3} \cdot \dfrac{5 \cdot 1}{4 \cdot 3} = \dfrac{4 \cdot 2 \cdot 5 \cdot 1}{5 \cdot 3 \cdot 4 \cdot 3} = \dfrac{2 \cdot 1}{3 \cdot 3} = \dfrac{2}{9}$

(5 and 15 have a common factor of 5.)

e. $\left(3\dfrac{3}{4}\right)\left(2\dfrac{2}{5}\right) = \dfrac{15}{4} \cdot \dfrac{12}{5} = \dfrac{3 \cdot 5}{4} \cdot \dfrac{4 \cdot 3}{5} = \dfrac{3 \cdot 5 \cdot 4 \cdot 3}{4 \cdot 5} = \dfrac{3 \cdot 3}{1} = 9$

f. The reciprocal of $\dfrac{7}{10}$ is $\dfrac{10}{7}$. *Check:* $\dfrac{7}{10} \cdot \dfrac{10}{7} = \dfrac{70}{70} = 1$.

g. The reciprocal of $1\dfrac{4}{5}$ is ? First change $1\dfrac{4}{5}$ to an improper fraction: $1\dfrac{4}{5} = \dfrac{9}{5}$.

The reciprocal of $1\dfrac{4}{5}$ is $\dfrac{5}{9}$. *Check:* $\dfrac{9}{5} \cdot \dfrac{5}{9} = 1$.

h. $6 \cdot 11\dfrac{2}{3} = \dfrac{\overset{2}{\cancel{6}}}{1} \cdot \dfrac{35}{\underset{1}{\cancel{3}}} = \dfrac{70}{1} = 70$

APPLICATION SOLUTION

To find the distance Bobby traveled, use the formula:
distance = (rate)(time).

Formula: $d = rt,\ r = 40$ mph, $t = 2\dfrac{1}{2}$ hours

Substitute: $d = (40)\left(2\dfrac{1}{2}\right)$

Solve: $= \dfrac{40}{1} \cdot \dfrac{5}{2}$

$= \dfrac{200}{2}$

$= 100$

Answer: So Bobby drove 100 miles.

WARM UPS

Multiply. Reduce to lowest terms.

1. $\dfrac{1}{3} \cdot \dfrac{8}{9}$ **2.** $\dfrac{2}{5} \cdot \dfrac{7}{9}$ **3.** $\dfrac{1}{2} \cdot \dfrac{1}{8}$

4. $\dfrac{1}{2} \cdot \dfrac{4}{3}$ **5.** $\dfrac{3}{4} \cdot \dfrac{4}{7}$ **6.** $\dfrac{2}{3} \cdot \dfrac{5}{6}$

7. $\dfrac{3}{7} \cdot \dfrac{14}{9}$ **8.** $\dfrac{4}{5} \cdot \dfrac{7}{8}$ **9.** $\dfrac{5}{9} \cdot \dfrac{9}{10}$

10. $\dfrac{5}{7} \cdot \dfrac{7}{8}$

11. What is the reciprocal of $\dfrac{5}{7}$?

12. What is the reciprocal of $\dfrac{8}{9}$?

Exercises 2.3

A

Multiply and reduce each answer to lowest terms.

1. $\dfrac{3}{2} \cdot \dfrac{8}{7}$ **2.** $\dfrac{5}{6} \cdot \dfrac{12}{11}$ **3.** $\dfrac{1}{8} \cdot \dfrac{12}{7}$

4. $\dfrac{1}{2} \cdot \dfrac{2}{3} \cdot \dfrac{3}{4}$

5. What is the reciprocal of $\dfrac{5}{8}$?

6. What is the reciprocal of $\dfrac{3}{10}$?

B

Multiply and reduce to lowest terms.

7. $\dfrac{7}{8} \cdot \dfrac{8}{21}$ **8.** $\dfrac{5}{18} \cdot \dfrac{2}{3} \cdot \dfrac{27}{35}$ **9.** $\dfrac{15}{8} \left(1\dfrac{3}{5}\right)$

10. $\left(7\dfrac{1}{2}\right)\left(4\dfrac{3}{4}\right)$ **11.** $\left(\dfrac{5}{8}\right)\left(\dfrac{16}{7}\right)\left(2\dfrac{3}{4}\right)$ **12.** $\dfrac{4}{6} \cdot \dfrac{9}{30}$

13. What is the reciprocal of 3?

14. What is the reciprocal of $\dfrac{5}{13}$?

C

Multiply and reduce to lowest terms.

15. $\dfrac{2}{3} \cdot \dfrac{4}{5} \cdot \dfrac{6}{7}$ **16.** $5 \cdot \dfrac{1}{4} \cdot \dfrac{8}{15}$ **17.** $1\dfrac{4}{5} \cdot \dfrac{7}{8} \cdot \dfrac{8}{21}$

18. $\dfrac{5}{6} \cdot 1\dfrac{1}{5}$ **19.** $\left(4\dfrac{1}{2}\right)\left(\dfrac{2}{9}\right)\left(\dfrac{7}{8}\right)$ **20.** $\dfrac{3}{8} \cdot \dfrac{4}{7} \cdot \dfrac{8}{3} \cdot \dfrac{14}{24}$

21. $\left(2\dfrac{1}{4}\right)\left(2\dfrac{2}{3}\right)\left(5\dfrac{1}{6}\right)$ **22.** $\left(15\dfrac{2}{3}\right)\left(7\dfrac{5}{8}\right)$

23. What is the reciprocal of $2\dfrac{2}{3}$?

24. What is the reciprocal of $6\dfrac{1}{8}$?

ANSWERS

1. _____
2. _____
3. _____
4. _____
5. _____
6. _____
7. _____
8. _____
9. _____
10. _____
11. _____
12. _____
13. _____
14. _____
15. _____
16. _____
17. _____
18. _____
19. _____
20. _____
21. _____
22. _____
23. _____
24. _____

25. _____

26. _____

27. _____

28. _____

29. _____

30. _____

31. _____

32. _____

33. _____

34. _____

35. _____

36. _____

D

25. Mary drove for $3\frac{2}{3}$ hours at 60 miles per hour. How many miles did Mary travel?

26. Joe jogged for $1\frac{1}{2}$ hours at an average speed of 8 miles per hour. How many miles did Joe jog?

27. If Tammy can make $2\frac{2}{3}$ drawings in an hour, how many can she make in $6\frac{3}{4}$ hours?

28. If a steel bar weighs $\frac{3}{4}$ pound per foot, what is the weight of a bar that is $20\frac{1}{2}$ feet in length?

29. The owner of a sporting goods store usualy sells 1200 tennis racquets a year at a price of $18 each. He decides to raise the price per racquet by $\frac{1}{8}$, which will in turn decrease his sales by $\frac{1}{6}$. What will be his net income under the new system? Is it more or less than under the old pricing system?

30. Each student in the television repair class is allowed $6\frac{3}{4}$ in. of wire solder for his work. How many inches of wire must be ordered for a class of 32 students?

31. The water pressure during a bad fire is reduced to $\frac{5}{9}$ of its original pressure at the hydrant. What is the reduced pressure if the original pressure was $73\frac{2}{3}$ lb/in²?

E   *Maintain Your Skills*

32. Name the exponent in the expression 5^4.

33. Name the base in the expression 25^7.

Find the value of each of the following expressions.

34. 8^2 **35.** 10^3 **36.** 78^0

1. $\dfrac{8}{27}$ 2. $\dfrac{14}{45}$ 3. $\dfrac{1}{16}$ 4. $\dfrac{2}{3}$ 5. $\dfrac{3}{7}$ 6. $\dfrac{5}{9}$

7. $\dfrac{2}{3}$ 8. $\dfrac{7}{10}$ 9. $\dfrac{1}{2}$ 10. $\dfrac{5}{8}$ 11. $\dfrac{7}{5}$ 12. $\dfrac{9}{8}$

2.4 Division of Fractions and Mixed Numbers

OBJECTIVE

22. Find the quotient of two numbers written as fractions, mixed numbers, or whole numbers.

APPLICATION

If each turn of a wood screw sinks it $\dfrac{3}{16}$ of an inch, how many turns are needed to sink it $1\dfrac{1}{4}$ inches?

VOCABULARY

Recall that the reciprocal of $\dfrac{3}{5}$ is $\dfrac{5}{3}$ because $\dfrac{3}{5} \cdot \dfrac{5}{3} = 1$.

HOW AND WHY

Consider $\dfrac{3}{4} \div \dfrac{1}{3} = ?$ This can be thought of as $\dfrac{1}{3} \cdot (?) = \dfrac{3}{4}$, since multiplication and division are inverse operations. Replace the "?" with $\dfrac{3}{1} \cdot \dfrac{3}{4}$ and you have the right answer.

$$\dfrac{1}{3}\left(\dfrac{3}{1} \cdot \dfrac{3}{4}\right) = \dfrac{3}{4}$$

$$1 \cdot \dfrac{3}{4} = \dfrac{3}{4} \qquad \textit{Note:} \ \dfrac{1}{3} \cdot \dfrac{3}{1} = 1$$

$$\dfrac{3}{4} = \dfrac{3}{4}$$

So, $\dfrac{3}{4} \div \dfrac{1}{3} = \dfrac{3}{4} \cdot \dfrac{3}{1} = \dfrac{9}{4}$; $\dfrac{3}{4}$ is multiplied by the reciprocal of the divisor $\dfrac{1}{3}$.

In general, to divide two fractions, change the problem to multiplication by using the reciprocal of the divisor (that is, invert the divisor and multiply). The pattern is:

$$\dfrac{d}{e} \div \dfrac{f}{g} = \dfrac{d}{e} \cdot \dfrac{g}{f} = \dfrac{dg}{ef}$$

To divide whole numbers and/or mixed numbers:

1. write them as improper fractions, and
2. divide, following the procedures for fractions.

EXAMPLES

a. $\dfrac{8}{3} \div \dfrac{4}{5} = \dfrac{8}{3} \cdot \dfrac{5}{4} = \dfrac{\cancel{4} \cdot 2 \cdot 5}{3 \cdot \cancel{4}} = \dfrac{10}{3} = 3\dfrac{1}{3}$

b. $3\dfrac{1}{3} \div 6\dfrac{7}{8} = \dfrac{10}{3} \div \dfrac{55}{8} = \dfrac{10}{3} \cdot \dfrac{8}{55} = \dfrac{\cancel{5} \cdot 2 \cdot 8}{3 \cdot \cancel{5} \cdot 11} = \dfrac{16}{33}$

c. $12\dfrac{6}{10} \div 21\dfrac{3}{5} = \dfrac{126}{10} \div \dfrac{108}{5} = \dfrac{126}{10} \cdot \dfrac{5}{108} = \dfrac{\cancel{2} \cdot \cancel{3} \cdot \cancel{3} \cdot 7 \cdot \cancel{5}}{\cancel{2} \cdot \cancel{5} \cdot 2 \cdot 2 \cdot \cancel{3} \cdot \cancel{3} \cdot 3} = \dfrac{7}{12}$

d. $4 \div 3\dfrac{1}{3} = \dfrac{4}{1} \div \dfrac{10}{3} = \dfrac{4}{1} \cdot \dfrac{3}{10} = \dfrac{\cancel{4}^{2}}{1} \cdot \dfrac{3}{\cancel{10}_{5}} = \dfrac{6}{5} = 1\dfrac{1}{5}$

APPLICATION SOLUTION

Since each turn of the screw sinks it $\dfrac{3}{16}$ inch, we need to find out how many $\dfrac{3}{16}$'s are in $1\dfrac{1}{4}$ inches. This is found by division.

$\left(1\dfrac{1}{4}\right) \div \dfrac{3}{16}$

$\dfrac{5}{4} \div \dfrac{3}{16}$

$\dfrac{5}{\cancel{4}_{1}} \cdot \dfrac{\cancel{16}^{4}}{3}$ *Multiply by the reciprocal of the divisor.*

$\dfrac{20}{3}$

So it takes $\dfrac{20}{3}$ or $6\dfrac{2}{3}$ turns to sink the screw $1\dfrac{1}{4}$ inches.

WARM UPS

Divide. Reduce answers to lowest terms and express as mixed numbers where possible.

1. $\dfrac{3}{7} \div \dfrac{1}{2}$ 　　　　2. $\dfrac{1}{2} \div \dfrac{1}{3}$ 　　　　3. $\dfrac{1}{3} \div \dfrac{1}{2}$

4. $\dfrac{2}{3} \div \dfrac{5}{7}$ 　　　　5. $\dfrac{2}{5} \div \dfrac{5}{7}$ 　　　　6. $\dfrac{1}{10} \div 3$

7. $\dfrac{3}{4} \div 1\dfrac{1}{4}$ 　　　　8. $5 \div \dfrac{5}{6}$ 　　　　9. $1\dfrac{2}{5} \div 1\dfrac{1}{10}$

10. $2\dfrac{1}{2} \div 3\dfrac{1}{2}$ 　　　　11. $\dfrac{5}{8} \div \dfrac{35}{36}$ 　　　　12. $\dfrac{9}{16} \div \dfrac{1}{48}$

Exercises 2.4

A

Divide. Reduce answers to lowest terms and express as mixed numbers where possible.

1. $7 \div \dfrac{1}{5}$ 　　　 **2.** $\dfrac{3}{2} \div \dfrac{1}{3}$ 　　　 **3.** $\dfrac{5}{7} \div \dfrac{5}{6}$ 　　　 **4.** $\dfrac{1}{4} \div \dfrac{3}{4}$

5. $\dfrac{8}{9} \div \dfrac{2}{9}$ 　　　 **6.** $\dfrac{8}{9} \div \dfrac{8}{3}$

B

7. $\dfrac{8}{9} \div \dfrac{2}{3}$ 　　　 **8.** $\dfrac{5}{8} \div 1\dfrac{3}{4}$ 　　　 **9.** $3\dfrac{7}{8} \div 7\dfrac{3}{4}$

10. $\dfrac{2}{5} \div \dfrac{3}{5}$ 　　　 **11.** $12\dfrac{2}{9} \div \dfrac{2}{3}$ 　　　 **12.** $\dfrac{7}{8} \div 3\dfrac{3}{4}$

13. $\dfrac{30}{49} \div \dfrac{20}{21}$ 　　　 **14.** $9 \div 6$

C

15. $8 \div 12$ 　　　 **16.** $\dfrac{4}{15} \div \dfrac{5}{8}$ 　　　 **17.** $26 \div 4\dfrac{5}{16}$

18. $2\dfrac{4}{9} \div 3$ 　　　 **19.** $\dfrac{51}{10} \div \dfrac{14}{15}$ 　　　 **20.** $5\dfrac{11}{15} \div 2\dfrac{4}{5}$

21. $2\dfrac{11}{15} \div 5\dfrac{4}{5}$ 　　　 **22.** $\dfrac{75}{132} \div \dfrac{245}{308}$ 　　　 **23.** $9\dfrac{5}{7} \div 11\dfrac{11}{14}$

24. $5\dfrac{1}{16} \div 7\dfrac{1}{5}$

1. _____
2. _____
3. _____
4. _____
5. _____
6. _____
7. _____
8. _____
9. _____
10. _____
11. _____
12. _____
13. _____
14. _____
15. _____
16. _____
17. _____
18. _____
19. _____
20. _____
21. _____
22. _____
23. _____
24. _____

ANSWERS

25. _____

26. _____

27. _____

28. _____

29. _____

30. _____

31. _____

32. _____

33. _____

34. _____

35. _____

36. _____

37. _____

D

25. If each turn of a metal screw sinks it $\frac{5}{12}$ inch, how many turns are needed to sink it $2\frac{1}{4}$ inches?

26. If each strike by an automatic hammer sinks a spike $\frac{3}{4}$ inch, how many strikes are needed to sink it $5\frac{3}{4}$ inches?

27. Carol has $\frac{7}{8}$ lb of cheese. If an omelet recipe calls for $\frac{1}{4}$ lb of cheese, how many omelets can she make?

28. A machinist takes $74\frac{1}{2}$ minutes to machine 5 pins. How long will it take to machine one pin?

29. A welder is to drill 13 equally spaced holes on the center line of a flat metal bar. The two end holes are to be 1 inch from each end of the bar. What will be the distance between the centers of the holes if the bar is 52 inches long?

30. As part of her job at a pet store, Becky feeds each gerbil $\frac{1}{8}$ cup of seeds each day. If the seeds come in packages of $\frac{3}{4}$ cup, how many gerbils can be fed from one package?

31. Norma has a ribbon that is $\frac{3}{4}$ of a yard long. From this ribbon, she wants to cut 6 pieces of equal length. How long will each piece be?

32. A sheet metal worker must make seven slits in a piece of metal 72 inches long. The two end slits must be 5 inches from the end of the piece of metal and the slits must be equal distances apart. What will be the distance between the centers of the slits?

E   *Maintain Your Skills*

Do the indicated operations.

33. $5(4) - 8 + 3$

34. $8[7 - 2(5 - 3)]$

35. $15(2) - 5(4)$

36. $150 - [5(6) - 5]$

37. $12 - 3 + (8)2 - 7$

ANSWERS TO WARM UPS (2.4)   **1.** $\dfrac{6}{7}$ **2.** $1\dfrac{1}{2}$ **3.** $\dfrac{2}{3}$ **4.** $\dfrac{14}{15}$ **5.** $\dfrac{14}{25}$ **6.** $\dfrac{1}{30}$

7. $\dfrac{3}{5}$ **8.** 6   **9.** $1\dfrac{3}{11}$ **10.** $\dfrac{5}{7}$ **11.** $\dfrac{9}{14}$ **12.** 27

2.5 Building Fractions

OBJECTIVE

23. Build a fraction by finding the missing numerator.

APPLICATION

To add, subtract, or compare $\dfrac{7}{16}$ and $\dfrac{3}{24}$, we must rename them with a common denominator. The common denominator is 48. Find the missing numerators.

$$\frac{7}{16} = \frac{?}{48} \qquad \frac{3}{24} = \frac{?}{48}$$

VOCABULARY

Recall that *equivalent fractions* are fractions that are different names for the same number. The fractions $\dfrac{3}{6}$ and $\dfrac{4}{8}$ are equivalent since both represent one half of a unit.

HOW AND WHY

The process of writing equivalent fractions described below is sometimes referred to as "building" fractions. It may be thought of as the opposite of the process called "reducing" fractions. It is based on the fact that multiplying a number by 1 does not change its value (multiplication property of one). We may build new fractions from any given fraction.

$$\frac{3}{5} = \frac{3}{5} \cdot 1 = \frac{3}{5} \cdot \frac{2}{2} = \frac{6}{10}$$

$$\frac{3}{5} = \frac{3}{5} \cdot 1 = \frac{3}{5} \cdot \frac{3}{3} = \frac{9}{15}$$

$$\frac{3}{5} = \frac{3}{5} \cdot 1 = \frac{3}{5} \cdot \frac{4}{4} = \frac{12}{20}$$

and so on. We may write

$$\frac{3}{5} = \frac{6}{10} = \frac{9}{15} = \frac{12}{20} = \frac{15}{25} = \frac{18}{30} = \frac{21}{35} = \cdots$$

Multiplying any fraction by $\dfrac{2}{2}, \dfrac{3}{3}, \dfrac{4}{4}$ (or any form of the number one) builds the fraction to an equivalent form but does not change its value.

To find a missing numerator such as

$$\frac{3}{5} = \frac{?}{60}$$

divide 60 by 5 to find out what fraction to multiply by.

$$60 \div 5 = 12$$

The correct multiplier is $\frac{12}{12}$.

$$\frac{3}{5} = \frac{3}{5} \cdot \frac{12}{12} = \frac{36}{60}$$

The short cut is to multiply the numerator (3) by 12.

$$\frac{3}{5} = \frac{3 \cdot 12}{60} = \frac{36}{60}$$

The fractions $\frac{3}{5}$ and $\frac{36}{60}$ are equivalent. Either fraction can be used in place of the other.

> **To find a missing numerator:**
>
> 1. **Divide the larger denominator by the smaller denominator.**
> 2. **Multiply the answer by the numerator. This product is the missing numerator.**

EXAMPLES

a. $\frac{3}{5} = \frac{?}{80}$ $\frac{3}{5} \cdot \frac{16}{16} = \frac{48}{80}$ *Note that* $80 \div 5 = 16$.

b. $\frac{2}{7} = \frac{?}{56}$ $\frac{2}{7} \cdot \frac{8}{8} = \frac{16}{56}$

c. $\frac{13}{10} = \frac{?}{120}$ $\frac{13}{10} \cdot \frac{12}{12} = \frac{156}{120}$

APPLICATION SOLUTION

> To add, subtract, or compare the fractions we find the missing numerators.
>
> $\frac{7}{16} = \frac{?}{48}$ $\frac{7}{16} \cdot \frac{3}{3} = \frac{21}{48}$
>
> $\frac{3}{24} = \frac{?}{48}$ $\frac{3}{24} \cdot \frac{2}{2} = \frac{6}{48}$
>
> So we add, subtract, or compare the fractions as $\frac{21}{48}$ and $\frac{6}{48}$.

WARM UPS

Find the missing numerators.

1. $\frac{1}{2} = \frac{}{10}$ 2. $\frac{3}{2} = \frac{}{10}$ 3. $\frac{7}{2} = \frac{}{10}$

4. $\frac{7}{2} = \frac{}{18}$ 5. $\frac{5}{8} = \frac{}{48}$ 6. $\frac{3}{4} = \frac{}{100}$

7. $\frac{7}{4} = \frac{}{16}$ 8. $\frac{}{12} = \frac{2}{3}$ 9. $\frac{3}{11} = \frac{}{66}$

10. $\frac{23}{6} = \frac{}{12}$ 11. $\frac{9}{5} = \frac{}{100}$ 12. $\frac{}{300} = \frac{7}{15}$

Exercises 2.5

ANSWERS

1. _____
2. _____
3. _____
4. _____
5. _____
6. _____
7. _____
8. _____
9. _____
10. _____
11. _____
12. _____
13. _____
14. _____
15. _____
16. _____
17. _____
18. _____
19. _____
20. _____
21. _____
22. _____
23. _____
24. _____

A

Find the missing numerators.

1. $\dfrac{5}{7} = \dfrac{}{28}$

2. $\dfrac{7}{9} = \dfrac{}{81}$

3. $\dfrac{2}{3} = \dfrac{}{300}$

4. $\dfrac{3}{8} = \dfrac{}{56}$

5. $\dfrac{3}{4} = \dfrac{}{56}$

6. $\dfrac{5}{6} = \dfrac{}{66}$

B

7. $\dfrac{3}{4} = \dfrac{}{68}$

8. $\dfrac{6}{9} = \dfrac{}{108}$

9. $\dfrac{}{90} = \dfrac{11}{15}$

10. $\dfrac{}{126} = \dfrac{19}{42}$

11. $\dfrac{16}{7} = \dfrac{}{147}$

12. $\dfrac{7}{10} = \dfrac{}{1000}$

13. $\dfrac{7}{8} = \dfrac{}{1000}$

14. $\dfrac{1}{5} = \dfrac{}{1000}$

C

15. $\dfrac{5}{14} = \dfrac{}{112}$

16. $\dfrac{5}{12} = \dfrac{}{168}$

17. $\dfrac{13}{15} = \dfrac{}{255}$

18. $\dfrac{5}{24} = \dfrac{}{360}$

19. $\dfrac{7}{30} = \dfrac{}{510}$

20. $\dfrac{17}{32} = \dfrac{}{480}$

21. $\dfrac{11}{14} = \dfrac{}{238}$

22. $\dfrac{17}{23} = \dfrac{}{1265}$

23. Write a fraction equivalent to $\dfrac{3}{4}$ that has a denominator of 28.

24. What fraction with denominator 132 is equivalent to $\dfrac{2}{3}$?

ANSWERS

25. _____

26. _____

27. _____

28. _____

29. _____

30. _____

31. _____

32. _____

D

25. To add $\frac{5}{8}$, $\frac{7}{10}$, and $\frac{3}{4}$, we need to write each as an equivalent fraction with a common denominator. Write each fraction as an equivalent fraction using 40 as the denominator.

26. To subtract $\frac{17}{20}$ and $\frac{11}{15}$, we write each as an equivalent fraction with a common denominator. Write each using 60 as a denominator.

27. To compare $\frac{5}{8}$, $\frac{7}{9}$, and $\frac{5}{6}$, we write each as an equivalent fraction with a common denominator. Write each using 72 as the denominator.

E   *Maintain Your Skills*

List all of the factors of each of the following whole numbers.

28. 462   **29.** 156   **30.** 400

Write each of the following whole numbers as a product of two factors in as many ways as possible.

31. 210   **32.** 192

2.6 Listing Fractions in Order of Value

OBJECTIVE

24. List a given group of fractions in order of value from smallest to largest.

APPLICATION

A container of a chemical was weighed by two people. Mary recorded the weight as $3\frac{1}{8}$ lb. George read the weight as $3\frac{3}{16}$ lb. Whose measurement was heavier?

VOCABULARY

When two or more fractions have the same denominator, we say that they have a *common denominator.*

HOW AND WHY

It is clear from the illustrations in Figure 2.5 that $\frac{4}{5}$ is larger than $\frac{2}{5}$.

Figure 2.5

Whenever two fractions have the same denominator, the one with the larger numerator has the larger value.

If fractions to be compared do not have a common denominator, then one or more must be "built" up so that all have a common denominator and can be compared easily. The preferred common denominator of a group of fractions is the least common multiple of all the denominators.

List $\frac{5}{8}, \frac{7}{16}, \frac{1}{2}$, and $\frac{9}{16}$ from smallest to largest. The least common multiple (LCM) of 8, 16, and 2 is 16.

$$\frac{5}{8} = \frac{10}{16} \quad \text{fourth (largest)}$$

$$\frac{7}{16} = \frac{7}{16} \quad \text{first (smallest)}$$ *Build the fractions so that all have 16 for a denominator*

$$\frac{1}{2} = \frac{8}{16} \quad \text{second}$$

$$\frac{9}{16} = \frac{9}{16} \quad \text{third}$$

The list is $\frac{7}{16}, \frac{1}{2}, \frac{9}{16}, \frac{5}{8}$.

To list fractions from smallest to largest:

1. **Build the fractions so that they have a common denominator. Use the LCM of the denominators.**
2. **List the fractions (with common denominators) with numerators from smallest to largest.**
3. **Reduce each fraction in the list to lowest terms.**

EXAMPLES

a. Which has larger value: $\dfrac{6}{11}$ or $\dfrac{8}{11}$? $\dfrac{8}{11}$ is larger because 8 is larger than 6.

b. Which has larger value: $\dfrac{6}{11}$ or $\dfrac{1}{2}$? It is easier to determine the answer if we build both fractions to a denominator of 22. (The LCM of 2 and 11 is 22.) $\dfrac{6}{11} = \dfrac{12}{22}$ and $\dfrac{1}{2} = \dfrac{11}{22}$, so $\dfrac{6}{11}$ is larger.

c. List the fractions $\dfrac{2}{3}, \dfrac{3}{8}$, and $\dfrac{3}{4}$ in order from smallest to largest. (24 is the LCM of 3, 8, and 4.) $\dfrac{2}{3} = \dfrac{16}{24}, \dfrac{3}{8} = \dfrac{9}{24}$, and $\dfrac{3}{4} = \dfrac{18}{24}$, so the list is $\dfrac{3}{8}, \dfrac{2}{3}, \dfrac{3}{4}$.

APPLICATION SOLUTION

To find out whose measurement was heavier, we must determine which fraction is larger, $\dfrac{1}{8}$ or $\dfrac{3}{16}$.

The LCM of 8 and 16 is 16.

$\dfrac{1}{8} = \dfrac{2}{16}, \quad \dfrac{3}{16} = \dfrac{3}{16}$

So George's measurement ($3\dfrac{3}{16}$ lb) was heavier.

WARM UPS

Which fraction is larger?

1. $\dfrac{4}{11}, \dfrac{6}{11}$ 2. $\dfrac{5}{13}, \dfrac{4}{13}$ 3. $\dfrac{1}{4}, \dfrac{1}{5}$

4. $\dfrac{1}{6}, \dfrac{1}{7}$ 5. $\dfrac{1}{2}, \dfrac{4}{6}$ 6. $\dfrac{3}{10}, \dfrac{1}{5}$

List these fractions from smallest to largest:

7. $\dfrac{5}{8}, \dfrac{3}{8}, \dfrac{6}{8}$ 8. $\dfrac{5}{9}, \dfrac{2}{9}, \dfrac{3}{9}, \dfrac{7}{9}$ 9. $\dfrac{1}{6}, \dfrac{1}{3}, \dfrac{1}{2}$

10. $\dfrac{1}{2}, \dfrac{3}{8}, \dfrac{1}{4}$ 11. $\dfrac{3}{5}, \dfrac{7}{10}, \dfrac{1}{2}$ 12. $\dfrac{1}{2}, \dfrac{5}{6}, \dfrac{2}{3}$

Exercises 2.6

ANSWERS

1. _____

2. _____

3. _____

4. _____

5. _____

6. _____

7. _____

8. _____

9. _____

10. _____

11. _____

12. _____

13. _____

14. _____

15. _____

16. _____

17. _____

18. _____

19. _____

20. _____

21. _____

22. _____

23. _____

24. _____

A

Which fraction is larger?

1. $\dfrac{3}{4}, \dfrac{1}{4}$ 　　　　**2.** $\dfrac{13}{17}, \dfrac{18}{17}$ 　　　　**3.** $\dfrac{1}{2}, \dfrac{5}{6}$

List these fractions from smallest to largest:

4. $\dfrac{3}{4}, \dfrac{5}{8}, \dfrac{7}{16}$ 　　　　**5.** $\dfrac{1}{3}, \dfrac{2}{9}, \dfrac{4}{9}$ 　　　　**6.** $\dfrac{2}{3}, \dfrac{3}{4}, \dfrac{7}{12}$

B

Which fraction is larger?

7. $\dfrac{5}{6}, \dfrac{2}{3}$ 　　**8.** $\dfrac{3}{5}, \dfrac{1}{2}$ 　　**9.** $\dfrac{2}{7}, \dfrac{7}{2}$ 　　**10.** $\dfrac{2}{3}, \dfrac{2}{5}$

List these fractions from smallest to largest:

11. $\dfrac{2}{5}, \dfrac{2}{3}, \dfrac{4}{7}$ 　　　　**12.** $\dfrac{4}{5}, \dfrac{2}{3}, \dfrac{3}{4}$ 　　　　**13.** $\dfrac{5}{8}, \dfrac{7}{10}, \dfrac{3}{4}$

14. $\dfrac{5}{7}, \dfrac{7}{9}, \dfrac{9}{11}$

C

Which fraction is larger?

15. $\dfrac{8}{9}, \dfrac{9}{10}$ 　　　　**16.** $\dfrac{6}{7}, \dfrac{6}{11}$ 　　　　**17.** $1\dfrac{3}{4}, 1\dfrac{7}{10}$

18. $3\dfrac{5}{12}, 3\dfrac{3}{7}$

List these fractions from smallest to largest:

19. $\dfrac{11}{24}, \dfrac{17}{36}, \dfrac{35}{72}$ 　　　　**20.** $\dfrac{3}{5}, \dfrac{8}{25}, \dfrac{31}{50}, \dfrac{59}{100}$

21. $\dfrac{13}{28}, \dfrac{17}{35}, \dfrac{6}{14}$ 　　　　**22.** $\dfrac{11}{15}, \dfrac{17}{20}, \dfrac{9}{12}$

23. $\dfrac{7}{18}, \dfrac{2}{5}, \dfrac{11}{30}, \dfrac{17}{45}$ 　　　　**24.** $\dfrac{47}{80}, \dfrac{9}{16}, \dfrac{13}{20}, \dfrac{5}{8}$

25. _____

26. _____

27. _____

28. _____

29. _____

30. _____

31. _____

32. _____

33. _____

34. _____

35. _____

D

25. A container of garden fertilizer was weighed by two apprentice horticulturists. Betty read the weight as $11\frac{11}{32}$ lb and Pedro read the weight as $11\frac{3}{8}$ lb. Whose reading was heavier?

26. Two rulers are marked in inches, but on one the spaces are divided into tenths and on the other they are divided into sixteenths. Both are used to measure a line on a scale drawing. The nearest mark on one ruler is $5\frac{7}{10}$ inches, and the nearest mark on the other is $5\frac{11}{16}$ inches. Which is the larger (longer) measurement?

27. The night nurse at Malcolm X Community Hospital found bottles containing codeine tablets out of the usual order. The bottles contained tablets having the following strengths of codeine: $\frac{1}{8}, \frac{3}{32}, \frac{5}{16}, \frac{3}{8}, \frac{9}{16}, \frac{1}{2}$, and $\frac{1}{4}$ grain, respectively. Arrange the bottles in order of the strength of codeine from the smallest to the largest.

28. Joe, an apprentice helper, was given the task of sorting a bin of bolts according to their diameters. The bolts had the following diameters: $\frac{11}{16}, \frac{7}{8}, 1\frac{1}{16}, \frac{3}{4}, 1\frac{1}{8}$, and $1\frac{3}{32}$ inches. How should he list the diameters from smallest to largest?

29. Four partly filled cans of milk were delivered to the local dairy. They contained $\frac{3}{8}$ gallon, $\frac{7}{16}$ gallon, $\frac{1}{4}$ gallon, and $\frac{1}{2}$ gallon. Which is the largest amount and which is the smallest amount of milk?

30. Four pick-up trucks were advertised in the local car ads. The load capacities listed were $\frac{3}{4}$ ton, $\frac{5}{8}$ ton, $\frac{7}{16}$ ton, and $\frac{1}{2}$ ton. Which capacity is the smallest and which is the largest?

E   *Maintain Your Skills*

Identify each of the following as prime or composite:

31. 91   **32.** 101   **33.** 201   **34.** 301

35. 401

ANSWERS TO WARM UPS (2.6) **1.** $\dfrac{6}{11}$ **2.** $\dfrac{5}{13}$ **3.** $\dfrac{1}{4}$ **4.** $\dfrac{1}{6}$ **5.** $\dfrac{4}{6}$

6. $\dfrac{3}{10}$ **7.** $\dfrac{3}{8}, \dfrac{5}{8}, \dfrac{6}{8}$ **8.** $\dfrac{2}{9}, \dfrac{3}{9}, \dfrac{5}{9}, \dfrac{7}{9}$ **9.** $\dfrac{1}{6}, \dfrac{1}{3}, \dfrac{1}{2}$ **10.** $\dfrac{1}{4}, \dfrac{3}{8}, \dfrac{1}{2}$

11. $\dfrac{1}{2}, \dfrac{3}{5}, \dfrac{7}{10}$ **12.** $\dfrac{1}{2}, \dfrac{2}{3}, \dfrac{5}{6}$

2.7 Addition of Fractions

OBJECTIVES

25. Add two or more like fractions.
26. Add two or more unlike fractions.

APPLICATION

What is the distance around (perimeter of) this triangle?

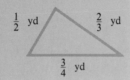

$\frac{1}{2}$ yd $\frac{2}{3}$ yd

$\frac{3}{4}$ yd

VOCABULARY

Like fractions are fractions with the same (or common) denominators. For instance, $\dfrac{1}{4}$ and $\dfrac{3}{4}$ are like fractions.

HOW AND WHY

The sum $\dfrac{1}{5} + \dfrac{2}{5}$ is ? Remember that the denominators tell the number of equal parts in the rectangle. The numerator tells us how many of these parts are shaded. By adding the numerators we find the total number of shaded parts, and the common denominator keeps track of the correct size of the parts. See Figure 2.6.

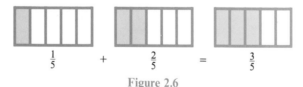

$\frac{1}{5}$ + $\frac{2}{5}$ = $\frac{3}{5}$

Figure 2.6

To add two or more like fractions,

1. Add the numerators and write the sum over the common denominator.
2. Reduce, if possible.

$$\frac{a}{c} + \frac{b}{c} = \frac{a+b}{c}$$

To add unlike fractions,

1. Build each fraction using a common denominator.
2. Add.
3. Reduce, if possible.

EXAMPLES

a. $\dfrac{3}{8} + \dfrac{4}{8} = \dfrac{7}{8}$

b. $\dfrac{1}{3} + \dfrac{2}{3} + \dfrac{1}{3} = \dfrac{4}{3}$

c. $\dfrac{3}{8} + \dfrac{1}{4} = ?$ The LCM of 8 and 4 is 8. Now we have $\dfrac{3}{8} = \dfrac{3}{8}$ and $\dfrac{1}{4} = \dfrac{1}{4} \cdot \dfrac{2}{2} = \dfrac{2}{8}$, so $\dfrac{3}{8} + \dfrac{1}{4} = \dfrac{3}{8} + \dfrac{2}{8} = \dfrac{5}{8}$.

d. $\dfrac{5}{12} + \dfrac{2}{9} = ?$ The LCM of 12 and 9 is 36. Now we have $\dfrac{5}{12} = \dfrac{5}{12} \cdot \dfrac{3}{3} = \dfrac{15}{36}$ and $\dfrac{2}{9} = \dfrac{2}{9} \cdot \dfrac{4}{4} = \dfrac{8}{36}$, so $\dfrac{5}{12} + \dfrac{2}{9} = \dfrac{15}{36} + \dfrac{8}{36} = \dfrac{23}{36}$.

e. $\dfrac{3}{4} + \dfrac{7}{10} = ?$ The LCM of 4 and 10 is 20. Now we have $\dfrac{3}{4} + \dfrac{7}{10} = \dfrac{15}{20} + \dfrac{14}{20} = \dfrac{29}{20}$.

f. $\dfrac{1}{32} + \dfrac{3}{4} + \dfrac{3}{16} + \dfrac{5}{32} = ?$ The LCM of 32, 4, and 16 is 32. $\dfrac{1}{32} + \dfrac{3}{4} + \dfrac{3}{16} + \dfrac{5}{32} = \dfrac{1}{32} + \dfrac{24}{32} + \dfrac{6}{32} + \dfrac{5}{32} = \dfrac{36}{32} = \dfrac{9}{8}$.

APPLICATION SOLUTION

To find the perimeter of the triangle, add the lengths of the sides.

Formula: $\quad P = a + b + c, a = \dfrac{1}{2}, b = \dfrac{2}{3}, c = \dfrac{3}{4}$

Substitute: $\quad P = \dfrac{1}{2} + \dfrac{2}{3} + \dfrac{3}{4}$

Solve: $\quad = \dfrac{6}{12} + \dfrac{8}{12} + \dfrac{9}{12}$

$\quad = \dfrac{23}{12}$

Answer: $\quad$ So the perimeter of the triangle is $\dfrac{23}{12}$ yards or $1\dfrac{11}{12}$ yards.

WARM UPS

Add. Write answers as proper or improper fractions reduced to lowest terms.

1. $\dfrac{1}{2} + \dfrac{1}{2}$

2. $\dfrac{1}{4} + \dfrac{2}{4}$

3. $\dfrac{3}{10} + \dfrac{4}{10} + \dfrac{1}{10}$

4. $\dfrac{5}{12} + \dfrac{4}{12}$

5. $\dfrac{8}{11} + \dfrac{2}{11}$

6. $\dfrac{2}{15} + \dfrac{4}{15} + \dfrac{3}{15}$

7. $\dfrac{1}{2} + \dfrac{1}{4}$

8. $\dfrac{2}{3} + \dfrac{1}{6}$

9. $\dfrac{1}{2} + \dfrac{1}{3} + \dfrac{1}{6}$

10. $\dfrac{3}{5} + \dfrac{1}{4}$

11. $\dfrac{2}{7} + \dfrac{3}{14}$

12. $\dfrac{1}{2} + \dfrac{3}{4} + \dfrac{3}{16}$

Exercises 2.7

A

Add. Reduce to lowest terms.

1. $\dfrac{1}{9} + \dfrac{4}{9} + \dfrac{1}{9}$ **2.** $\dfrac{3}{8} + \dfrac{1}{8} + \dfrac{2}{8}$ **3.** $\dfrac{5}{24} + \dfrac{2}{24} + \dfrac{1}{24}$

4. $\dfrac{1}{3} + \dfrac{1}{4} + \dfrac{1}{12}$ **5.** $\dfrac{1}{6} + \dfrac{3}{8} + \dfrac{1}{2}$ **6.** $\dfrac{3}{10} + \dfrac{2}{5} + \dfrac{1}{2}$

B

7. $\dfrac{8}{30} + \dfrac{9}{30} + \dfrac{1}{30}$ **8.** $\dfrac{13}{50} + \dfrac{7}{50} + \dfrac{2}{50}$ **9.** $\dfrac{3}{15} + \dfrac{7}{12}$

10. $\dfrac{5}{9} + \dfrac{5}{12}$ **11.** $\dfrac{3}{35} + \dfrac{8}{21}$ **12.** $\dfrac{7}{16} + \dfrac{3}{20} + \dfrac{1}{5}$

13. $\dfrac{5}{12} + \dfrac{9}{16} + \dfrac{7}{24}$ **14.** $\dfrac{3}{10} + \dfrac{7}{20} + \dfrac{11}{30}$

C

15. $\dfrac{9}{14} + \dfrac{5}{21}$ **16.** $\dfrac{1}{10} + \dfrac{2}{5} + \dfrac{5}{6} + \dfrac{1}{15}$ **17.** $\dfrac{5}{12} + \dfrac{3}{16} + \dfrac{7}{20}$

18. $\dfrac{5}{18} + \dfrac{4}{21}$ **19.** $\dfrac{8}{65} + \dfrac{7}{39}$ **20.** $\dfrac{19}{54} + \dfrac{7}{48}$

21. $\dfrac{21}{25} + \dfrac{11}{50} + \dfrac{7}{75}$ **22.** $\dfrac{1}{12} + \dfrac{1}{36} + \dfrac{1}{54}$ **23.** $\dfrac{23}{96} + \dfrac{14}{64} + \dfrac{7}{80}$

24. $\dfrac{117}{240} + \dfrac{41}{270}$

D

25. Find the perimeter of the triangle.

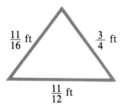

26. Find the distance around (perimeter of) the given figure.

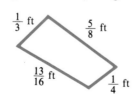

1. _____
2. _____
3. _____
4. _____
5. _____
6. _____
7. _____
8. _____
9. _____
10. _____
11. _____
12. _____
13. _____
14. _____
15. _____
16. _____
17. _____
18. _____
19. _____
20. _____
21. _____
22. _____
23. _____
24. _____
25. _____
26. _____

ANSWERS

27. _____

28. _____

29. _____

30. _____

31. _____

32. _____

33. _____

34. _____

35. _____

27. On the New York Stock Exchange, Vern's stock rose $\frac{1}{8}$ of a point in the morning and an additional $\frac{3}{16}$ of a point during the afternoon. What was the total rise for the day?

28. A Boise Cascade lumber mill produces dry lumber that measures $\frac{11}{16}$ inch thick. What was the original thickness if the shrinkage from green size is $\frac{3}{32}$ inch?

29. Three lamps are connected in series. The total resistance is the sum of the individual resistances. What is the total resistance in ohms if the individual resistances are $\frac{1}{4}$ ohm, $\frac{5}{8}$ ohm, and $\frac{5}{6}$ ohm?

30. Jonnie Lee Simms is assembling a rocking horse for his granddaughter. He needs a bolt to reach through a $\frac{7}{8}$ inch piece of steel tubing, a $\frac{1}{16}$ inch bushing, a $\frac{1}{2}$ inch piece of tubing, a $\frac{1}{8}$ inch thick washer and a $\frac{1}{4}$ inch thick nut. How long a bolt does he need?

E   *Maintain Your Skills*

31. *Add:*
345
360
45
707
9
432

32. *Subtract:*
457923
376874

33. *Multiply:*
5092
308

34. *Divide:*
36)345694

35. Do the indicated operations:
$23 - 18 + 4(4) - 20$

ANSWERS TO WARM UPS (2.7) **1.** 1 **2.** $\frac{3}{4}$ **3.** $\frac{4}{5}$ **4.** $\frac{3}{4}$ **5.** $\frac{10}{11}$ **6.** $\frac{3}{5}$

7. $\frac{3}{4}$ **8.** $\frac{5}{6}$ **9.** 1 **10.** $\frac{17}{20}$ **11.** $\frac{1}{2}$ **12.** $\frac{23}{16}$

2.8 Addition of Mixed Numbers

OBJECTIVE

27. Add two or more mixed numbers.

APPLICATION

Michelle worked $4\frac{1}{2}$ hours on both Monday and Wednesday, $6\frac{3}{4}$ hours on Tuesday, and 8 hours on Saturday. How many hours did she work during the week?

VOCABULARY

Recall that a *mixed number* is the sum of a whole number and a fraction.

HOW AND WHY

$3\frac{1}{4} + 5\frac{1}{6} = ?$ This can also be written $\left(3 + \frac{1}{4}\right) + \left(5 + \frac{1}{6}\right)$. Grouping the whole numbers and grouping the fractions, we get:

$$\left(3 + \frac{1}{4}\right) + \left(5 + \frac{1}{6}\right) = (3 + 5) + \left(\frac{1}{4} + \frac{1}{6}\right) = 8 + \left(\frac{3}{12} + \frac{2}{12}\right) = 8\frac{5}{12}$$

If we write the problem in vertical form, this grouping takes place naturally:

$3\frac{1}{4} = 3\frac{3}{12}$ *Add the whole numbers and add the fractions as above.*

$\underline{5\frac{1}{6} = 5\frac{2}{12}}$

$\qquad 8\frac{5}{12}$

When we are adding mixed numbers, it is possible that the fractions have a sum that is greater than 1. In this case, the resulting improper fraction can be changed to a mixed number and added to the whole number part. For example:

$11\frac{7}{10} = 11\frac{21}{30}$

$\underline{23\frac{8}{15} = 23\frac{16}{30}}$

$\qquad 34\frac{37}{30} = 34 + 1\frac{7}{30} = 35\frac{7}{30}$

> **To add mixed numbers;**
>
> **1. Add the whole numbers.**
> **2. Add the fractions. If the sum of the fractions is more than 1, change the fraction to a mixed number and add again.**

EXAMPLES

a. $5\dfrac{5}{12} = 5\dfrac{5}{12}$

$1\dfrac{3}{4} = 1\dfrac{9}{12}$

$6\dfrac{14}{12} = 6 + 1\dfrac{2}{12} = 7\dfrac{2}{12} = 7\dfrac{1}{6}$

b. $25\dfrac{7}{8} = 25\dfrac{63}{72}$

$13\dfrac{5}{9} = 13\dfrac{40}{72}$

$7\dfrac{1}{6} = 7\dfrac{12}{72}$

$45\dfrac{115}{72} = 45 + 1\dfrac{43}{72} = 46\dfrac{43}{72}$

c. $8 = 8\dfrac{0}{3}$

$5\dfrac{2}{3} = 5\dfrac{2}{3}$

$13\dfrac{2}{3}$

8 can be written as a mixed number by adding on a fraction whose numerator is zero and whose denominator is any natural number. The value of such a fraction is zero. 3 is chosen so the denominators will be the same.

APPLICATION SOLUTION

To find the number of hours Michelle worked during the week, total the hours she worked each day.

Monday $4\dfrac{1}{2} = 4\dfrac{2}{4}$

Tuesday $6\dfrac{3}{4} = 6\dfrac{3}{4}$

Wednesday $4\dfrac{1}{2} = 4\dfrac{2}{4}$

Saturday $8 \ = 8\dfrac{0}{4}$

$22\dfrac{7}{4} = 22 + 1\dfrac{3}{4} = 23\dfrac{3}{4}$

So Michelle worked $23\dfrac{3}{4}$ hours during the week.

WARM UPS

Add. State all answers as mixed numbers where possible.

1. $7\dfrac{3}{8}$

   $4\dfrac{2}{8}$

2. $15\dfrac{3}{5}$

   $17\dfrac{1}{5}$

3. $8\dfrac{5}{12}$

   $6\dfrac{1}{12}$

4. $5\dfrac{1}{4}$

$2\dfrac{3}{4}$

$7\dfrac{1}{4}$

5. $6\dfrac{2}{7}$

$4\dfrac{3}{7}$

6. $12\dfrac{2}{3}$

$1\dfrac{1}{3}$

7. $3\dfrac{4}{9}$

$2\dfrac{1}{9}$

$1\dfrac{1}{9}$

8. $8\dfrac{2}{3}$

$1\dfrac{2}{3}$

9. $5\dfrac{1}{2}$

$2\dfrac{1}{4}$

10. $7\dfrac{1}{6}$

$2\dfrac{1}{3}$

11. $14\dfrac{7}{10}$

$11\dfrac{3}{5}$

12. $9\dfrac{5}{24}$

$101\dfrac{7}{12}$

Exercises 2.8

A

Add. Write answers as mixed numbers where possible.

1. $3\frac{2}{3}$ **2.** $6\frac{1}{6}$ **3.** $5\frac{5}{9}$ **4.** $6\frac{5}{6}$

$2\frac{1}{4}$ $8\frac{2}{3}$ $\frac{2}{3}$ $8\frac{2}{3}$

5. $3\frac{2}{3}$ **6.** $7\frac{3}{10}$

$2\frac{1}{2}$ $8\frac{2}{5}$

$4\frac{5}{6}$ $6\frac{1}{2}$

B

7. $14\frac{7}{10}$ **8.** $9\frac{5}{24}$ **9.** $213\frac{5}{8}$ **10.** $213\frac{5}{9}$

$11\frac{3}{5}$ $52\frac{3}{8}$ $506\frac{7}{12}$ $312\frac{5}{36}$

11. $47\frac{1}{5} + 23\frac{2}{3} + 15\frac{1}{2}$ **12.** $47\frac{3}{8} + 23 + 42\frac{5}{12}$

13. $213\frac{5}{18}$ **14.** $213\frac{5}{6}$

$506\frac{7}{12}$ $347\frac{3}{10}$

C

15. $142\frac{7}{11} + 93$ **16.** $82\frac{41}{45}$

                                                  $97\frac{25}{27}$

17. $47\frac{3}{8}$ **18.** $7\frac{7}{20}$

$23\frac{1}{3}$ $\frac{3}{4}$

$42\frac{5}{12}$ $8\frac{9}{40}$

19. $1\frac{1}{5} + 3\frac{7}{10} + \frac{1}{2} + 7\frac{16}{25}$ **20.** $2\frac{2}{5} + 7\frac{1}{6} + 1\frac{4}{15} + 3\frac{1}{10}$

21. $6\frac{4}{11} + 7\frac{5}{6}$ **22.** $24\frac{8}{9} + 423\frac{5}{6}$

1. _____
2. _____
3. _____
4. _____
5. _____
6. _____
7. _____
8. _____
9. _____
10. _____
11. _____
12. _____
13. _____
14. _____
15. _____
16. _____
17. _____
18. _____
19. _____
20. _____
21. _____
22. _____

ANSWERS

23. _____

24. _____

25. _____

26. _____

27. _____

28. _____

29. _____

30. _____

31. _____

32. _____

33. _____

34. _____

35. _____

23. $16 \frac{7}{16}$

$72 \frac{8}{9}$

$52 \frac{5}{144}$

24. $17 \frac{3}{35}$

$8 \frac{1}{5}$

$9 \frac{5}{7}$

$12 \frac{1}{2}$

D

25. Maria worked the following schedule in her part time job. How many hours did she work during the month?

Oct 1 – Oct 7 $25\frac{1}{2}$ hours

Oct 8 – Oct 14 $16\frac{2}{3}$ hours

Oct 15 – Oct 21   10   hours

Oct 22 – Oct 28 $19\frac{5}{6}$ hours

Oct 29 – Oct 31 $4\frac{3}{4}$ hours

26. Lea worked $9\frac{3}{10}$ hours on Sunday and Thursday, $6\frac{5}{6}$ hours on Monday and Friday, $4\frac{1}{4}$ hours on Tuesday and Wednesday. Lea had Saturday off. How many hours did she work that week at the Friendly Family Taco House?

27. On successive days, Wendy picked $21\frac{3}{4}$, $31\frac{5}{8}$, and $27\frac{3}{16}$ pounds of beans. How many pounds of beans did she pick during the three days?

28. John rode his bicycle $2\frac{3}{8}$ miles Tuesday, $7\frac{3}{4}$ miles Wednesday, $10\frac{2}{3}$ miles Thursday, and $4\frac{5}{6}$ miles Friday. What was his total mileage during the four days?

29. A brass rod was cut into lengths of $3\frac{1}{2}$, $6\frac{1}{4}$, $4\frac{5}{8}$, and $6\frac{3}{4}$ inches. How long was the original piece of stock if $\frac{1}{16}$ inch is allowed for each cut?

30. In a recent road test, three cars used the following amounts of gasoline: $9\frac{3}{10}$ gal, $10\frac{1}{2}$ gal, and $9\frac{3}{4}$ gal. How much gas was used in the road test?

E   *Maintain Your Skills*

Write the prime factorization of each of the following whole numbers.

31. 345   **32.** 364   **33.** 675   **34.** 256

35. 456

ANSWERS TO WARM UPS (2.8)

1. $11\frac{5}{8}$ **2.** $32\frac{4}{5}$ **3.** $14\frac{1}{2}$ **4.** $15\frac{1}{4}$ **5.** $10\frac{5}{7}$

6. 14 **7.** $6\frac{2}{3}$ **8.** $10\frac{1}{3}$ **9.** $7\frac{3}{4}$ **10.** $9\frac{1}{2}$

11. $26\frac{3}{10}$ **12.** $110\frac{19}{24}$

2.9 Subtraction of Fractions

OBJECTIVE

28. Subtract two fractions.

APPLICATION

Lumber mill operators must plan for shrinkage of "green" (wet) boards when they cut logs. If the shrinkage for a $\frac{5}{8}$-inch-thick board is expected to be $\frac{1}{16}$ inch, what will be the thickness of the dried board?

VOCABULARY

No new words.

HOW AND WHY

$\frac{2}{3} - \frac{1}{3} = ?$ Looking at the rectangles in Figure 2.7, we can see that all that is necessary is to subtract the numerators and retain the common denominator.

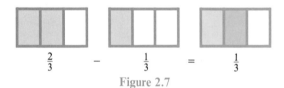

$$\frac{2}{3} \quad - \quad \frac{1}{3} \quad = \quad \frac{1}{3}$$

Figure 2.7

To subtract two fractions,

1. **Build the fractions so that they have a common denominator.**
2. **Subtract the numerators and write the difference over the common denominator.**
3. **Reduce if possible.**

So,

$$\frac{3}{4} - \frac{1}{3}$$

$$\frac{9}{12} - \frac{4}{12} \qquad \textit{Build fractions to a common denominator.}$$

$$\frac{5}{12} \qquad \textit{Subtract the numerators.}$$

EXAMPLES

a. $\dfrac{11}{20} - \dfrac{5}{20} = \dfrac{6}{20} = \dfrac{3}{10}$

b. $\dfrac{7}{8} - \dfrac{2}{3} = \dfrac{21}{24} - \dfrac{16}{24} = \dfrac{5}{24}$

c. $\dfrac{7}{15} - \dfrac{1}{4} = \dfrac{28}{60} - \dfrac{15}{60} = \dfrac{13}{60}$

d. $\dfrac{11}{15} - \dfrac{1}{12} = \dfrac{44}{60} - \dfrac{5}{60} = \dfrac{39}{60} = \dfrac{13}{20}$

APPLICATION SOLUTION

If the "green" board is originally $\dfrac{5}{8}$ inch thick and shrinks $\dfrac{1}{16}$ inch while drying, the dried board will be $\dfrac{5}{8} - \dfrac{1}{16}$ inch thick.

$\dfrac{5}{8} = \dfrac{10}{16}$

$\dfrac{1}{16} = \dfrac{1}{16}$

$\dfrac{9}{16}$

The dried board will be $\dfrac{9}{16}$ inch thick.

WARM UPS

Subtract. Reduce answers to lowest terms.

1. $\dfrac{5}{7} - \dfrac{3}{7}$ **2.** $\dfrac{12}{15} - \dfrac{4}{15}$ **3.** $\dfrac{7}{2} - \dfrac{1}{2}$

4. $\dfrac{15}{16} - \dfrac{7}{16}$ **5.** $\dfrac{3}{14} - \dfrac{1}{14}$ **6.** $\dfrac{9}{10} - \dfrac{7}{10}$

7. $\dfrac{7}{4} - \dfrac{3}{4}$ **8.** $\dfrac{5}{16} - \dfrac{1}{8}$ **9.** $\dfrac{13}{20} - \dfrac{3}{10}$

10. $\dfrac{3}{4} - \dfrac{5}{12}$ **11.** $\dfrac{5}{6} - \dfrac{3}{5}$ **12.** $\dfrac{17}{30} - \dfrac{7}{30}$

Exercises 2.9

ANSWERS

1. _____

2. _____

3. _____

4. _____

5. _____

6. _____

7. _____

8. _____

9. _____

10. _____

11. _____

12. _____

13. _____

14. _____

15. _____

16. _____

17. _____

18. _____

19. _____

20. _____

21. _____

22. _____

23. _____

24. _____

A

Subtract. Reduce to lowest terms.

1. $\dfrac{8}{9} - \dfrac{3}{9}$

2. $\dfrac{17}{30} - \dfrac{7}{30}$

3. $\dfrac{7}{10} - \dfrac{1}{10}$

4. $\dfrac{5}{12} - \dfrac{1}{12}$

5. $\dfrac{3}{15} - \dfrac{2}{45}$

6. $\dfrac{8}{9} - \dfrac{5}{18}$

B

7. $\dfrac{11}{15} - \dfrac{2}{5}$

8. $\dfrac{2}{3} - \dfrac{3}{8}$

9. $\dfrac{3}{4} - \dfrac{5}{16}$

10. $\dfrac{11}{12} - \dfrac{11}{15}$

11. $\dfrac{12}{21} - \dfrac{5}{14}$

12. $\dfrac{7}{10} - \dfrac{7}{15}$

13. $\dfrac{5}{6} - \dfrac{7}{15}$

14. $\dfrac{6}{14} - \dfrac{5}{21}$

C

15. $\dfrac{9}{50} - \dfrac{3}{40}$

16. $\dfrac{13}{12} - \dfrac{17}{18}$

17. $\dfrac{33}{35} - \dfrac{17}{28}$

18. $\dfrac{7}{10} - \dfrac{9}{20}$

19. $\dfrac{3}{10} - \dfrac{1}{15}$

20. $\dfrac{7}{16} - \dfrac{1}{6}$

21. $\dfrac{7}{24} - \dfrac{5}{18}$

22. $\dfrac{23}{48} - \dfrac{21}{80}$

23. $\dfrac{5}{9} - \dfrac{2}{75}$

24. $\dfrac{22}{33} - \dfrac{9}{44}$

ANSWERS

25. _____

26. _____

27. _____

28. _____

29. _____

30. _____

31. _____

32. _____

33. _____

34. _____

35. _____

D

25. If the shrinkage of a $\frac{5}{4}$-inch-thick "green" board is $\frac{1}{8}$ inch, what will be the thickness of the dried board?

26. If the shrinkage of an $\frac{11}{4}$-inch-thick "green" board is $\frac{3}{16}$ inch, what will be the thickness of the dried board?

27. A cake recipe calls for $\frac{3}{4}$ cup of milk. Carol has $\frac{1}{3}$ cup of milk. How much milk will she need to borrow from her neighbor in order to make the cake?

28. Mary is given $\frac{7}{8}$ lb of dried fruit. She uses $\frac{2}{3}$ lb to make a fruit cake. How much dried fruit does she have left?

29. A loan is to be paid as follows: $\frac{1}{5}$ the first year, $\frac{1}{4}$ the second year, $\frac{1}{3}$ the third year, and the remainder the fourth year. How much of the loan is paid the fourth year?

30. Dan has a bolt that is $\frac{3}{8}$ of an inch in length. He found that it is $\frac{3}{16}$ of an inch too long. What is the length of the bolt Dan needs?

E   *Maintain Your Skills*

Find the least common multiple (LCM) of each of the following groups of numbers.

31. 6, 12, 24   **32.** 18, 24, 32   **33.** 45, 50, 60

34. 25, 50, 75   **35.** 57, 76, 95, 114

2.10 Subtraction of Mixed Numbers

OBJECTIVE

29. Subtract mixed numbers.

APPLICATION

Frank poured $8\dfrac{3}{10}$ yards of cement for a fountain. Another fountain took $5\dfrac{3}{8}$ yards. How much more cement was needed for the larger fountain than for the smaller one?

VOCABULARY

No new words.

HOW AND WHY

In general, the pattern for subtraction of mixed numbers is similar to the pattern for addition of mixed numbers.

> **To subtract mixed numbers,**
>
> 1. **Build the fraction parts so they have a common denominator.**
> 2. **If the second fraction part is larger than the first fraction part, rename the first mixed number by "borrowing" 1 from the whole number part and adding it to the fraction part.**
> 3. **Subtract the fraction parts and then subtract the whole number parts.**

To rename $32\dfrac{1}{3}$ with a larger fractional part, borrow one (1) from 32 and add it to the fraction.

$$32\dfrac{1}{3} = 31 + 1\dfrac{1}{3} = 31 + \dfrac{4}{3} = 31\dfrac{4}{3}$$

Any mixed number can be given a new name in this way.

EXAMPLES

a. $\begin{aligned} 47\dfrac{5}{8} &= 47\dfrac{35}{56} \\ \underline{36\dfrac{3}{7}} &= \underline{36\dfrac{24}{56}} \\ & 11\dfrac{11}{56} \end{aligned}$

b. $\begin{aligned} 28\dfrac{9}{16} &= 28\dfrac{27}{48} \\ \underline{13\dfrac{1}{3}} &= \underline{13\dfrac{16}{48}} \\ & 15\dfrac{11}{48} \end{aligned}$

c. $15\dfrac{3}{10} = 14 + 1\dfrac{3}{10} = 14\dfrac{13}{10}$

$\dfrac{7\dfrac{7}{10} =}{}\dfrac{7\dfrac{7}{10} = 7\dfrac{7}{10}}{}$

$7\dfrac{6}{10} = 7\dfrac{3}{5}$

d. $32\dfrac{1}{3} = 32\dfrac{4}{12} = 31 + 1\dfrac{4}{12} = 31\dfrac{16}{12}$

$\dfrac{27\dfrac{3}{4} = 27\dfrac{9}{12} =}{}\dfrac{27\dfrac{9}{12} = 27\dfrac{9}{12}}{}$

$4\dfrac{7}{12}$

In the next two examples, one of the numbers is a whole number. Any whole number can be written as a mixed number by adding on a fraction whose numerator is zero and whose denominator is a natural number (remember that the value of such a fraction is zero). In these examples, the natural number is chosen so the denominators will be the same.

e. $4\dfrac{2}{3} = 4\dfrac{2}{3}$

$\dfrac{3 = 3\dfrac{0}{3}}{}$

$1\dfrac{2}{3}$

f. $11 = 11\dfrac{0}{9} = 10 + 1\dfrac{0}{9} = 10\dfrac{9}{9}$

$\dfrac{2\dfrac{2}{9} = 2\dfrac{2}{9} =}{}\dfrac{2\dfrac{2}{9} = 2\dfrac{2}{9}}{}$

$8\dfrac{7}{9}$

APPLICATION SOLUTION

To find the additional cement used in the larger fountain, subtract the amount used in the smaller fountain from the amount used in the larger one.

$8\dfrac{3}{10} = 8\dfrac{12}{40} = 7\dfrac{52}{40}$

$\dfrac{5\dfrac{3}{8} = 5\dfrac{15}{40} = 5\dfrac{15}{40}}{}$

$2\dfrac{37}{40}$

So the larger fountain needed $2\dfrac{37}{40}$ yards of cement more than the smaller fountain.

WARM UPS

Subtract. Reduce answers to lowest terms.

1. $3\frac{3}{5}$
   $\underline{1\frac{2}{5}}$

2. $7\frac{8}{9}$
   $\underline{2\frac{4}{9}}$

3. $17\frac{5}{6}$
   $\underline{6\frac{1}{6}}$

4. $1\frac{7}{12}$
   $\underline{\frac{3}{12}}$

5. $6\frac{28}{29}$
   $\underline{5\frac{19}{29}}$

6. $9\frac{3}{5}$
   $\underline{2\frac{3}{5}}$

7. $23\frac{5}{8}$
   $\underline{15}$

8. 13
   $\underline{2\frac{1}{2}}$

9. $12\frac{5}{8}$
   $\underline{3\frac{1}{4}}$

10. $32\frac{5}{9}$
    $\underline{25\frac{1}{3}}$

11. $13\frac{6}{7}$
    $\underline{8\frac{4}{7}}$

12. $27\frac{5}{9}$
    $\underline{23\frac{2}{9}}$

Exercises 2.10

A

Subtract. Reduce to lowest terms.

1. $3\frac{2}{3}$
$\underline{1\frac{1}{3}}$

2. $5\frac{3}{4}$
$\underline{3\frac{1}{4}}$

3. $11\frac{4}{5}$
$\underline{7\frac{3}{10}}$

4. $2\frac{7}{12}$
$\underline{2\frac{1}{3}}$

5. $5\frac{1}{4}$
$\underline{3\frac{3}{4}}$

6. $10\frac{1}{3}$
$\underline{1\frac{2}{3}}$

B

7. 6
$\underline{2\frac{3}{8}}$

8. 12
$\underline{11\frac{4}{7}}$

9. $2\frac{1}{3}$
$\underline{1\frac{7}{12}}$

10. $10\frac{1}{10}$
$\underline{3\frac{2}{5}}$

11. $145\frac{2}{3}$
$\underline{27\frac{1}{2}}$

12. $6\frac{5}{6}$
$\underline{3\frac{3}{10}}$

13. $4\frac{7}{9}$
$\underline{1\frac{5}{12}}$

14. $28\frac{3}{10}$
$\underline{14\frac{7}{10}}$

C

15. $212\frac{1}{9}$
$\underline{57}$

16. $36\frac{7}{16}$
$\underline{19\frac{11}{16}}$

17. 45
$\underline{16\frac{2}{3}}$

18. $81\frac{3}{14}$
$\underline{16\frac{5}{7}}$

19. $15\frac{5}{12}$
$\underline{13\frac{17}{18}}$

20. $9\frac{2}{11}$
$\underline{6\frac{23}{44}}$

21. $14\frac{3}{14} - 5$

22. $14 - 5\frac{3}{14}$

23. $113\frac{29}{36} - 85\frac{13}{24}$

24. $18\frac{5}{24} - 11\frac{3}{40}$

1. _____
2. _____
3. _____
4. _____
5. _____
6. _____
7. _____
8. _____
9. _____
10. _____
11. _____
12. _____
13. _____
14. _____
15. _____
16. _____
17. _____
18. _____
19. _____
20. _____
21. _____
22. _____
23. _____
24. _____

ANSWERS

25. _____

26. _____

27. _____

28. _____

29. _____

30. _____

31. _____

32. _____

33. _____

34. _____

35. _____

36. _____

37. _____

D

25. A compact car used $9\frac{3}{4}$ gallons of gas to travel 340 miles. A large car used $18\frac{8}{9}$ gallons to travel the same distance. How much less gas did the compact car use?

26. Rene Archibald used $3\frac{5}{8}$ yards of cement to pour her patio. Robin White used $4\frac{3}{16}$ yards to pour his patio. How much more cement did Robin use?

27. Cecil has a piece of lumber that measures $10\frac{7}{12}$ ft, and it is to be used in a spot that calls for a length of $8\frac{3}{4}$ ft. How much must be cut off of the board?

28. Dick harvested $30\frac{3}{4}$ tons of wheat. He sold $12\frac{3}{10}$ tons to the Cartwright Flour Mill. How many tons of wheat does he have left?

29. Larry starts with a bar of steel of length $21\frac{3}{4}$ inches and cuts off pieces of length $3\frac{1}{8}$ inches, $2\frac{1}{4}$ inches, and $5\frac{1}{2}$ inches. How much of the bar is left, allowing for a waste of $\frac{3}{32}$ inch for each cut?

30. The actual weight of a box of Whammy O's cereal must be within $\frac{1}{4}$ oz of the weight shown on the box. What are the maximum and minimum weights allowable if the weight shown on the box is $10\frac{1}{2}$ oz?

31. The outlet voltage on a 110 volt line that has a voltage drop of $6\frac{7}{8}$ volts from the panel bus to the outlet is given by $V = 110 - 6\frac{7}{8}$. What is the voltage at the outlet?

32. A metal bar weighed $6\frac{1}{8}$ lb. It was machined down to $5\frac{1}{2}$ lb. How many pounds were removed?

E   *Maintain Your Skills*

Change each improper fraction to a mixed number.

33. $\frac{25}{12}$ **34.** $\frac{34}{7}$ **35.** $\frac{49}{22}$

Change each mixed number to an improper fraction.

36. $12\frac{3}{7}$ **37.** $34\frac{5}{12}$

ANSWERS TO WARM UPS (2.10)

1. $2\frac{1}{5}$ 2. $5\frac{4}{9}$ 3. $11\frac{2}{3}$ 4. $1\frac{1}{3}$ 5. $1\frac{9}{29}$ 6. 7

7. $8\frac{5}{8}$ 8. $10\frac{1}{2}$ 9. $9\frac{3}{8}$ 10. $7\frac{2}{9}$ 11. $5\frac{2}{7}$ 12. $4\frac{1}{3}$

2.11 Order of Operations

OBJECTIVE

30. Do any combination of operations with fractions.

APPLICATION

The inside height of a china hutch is to be $66\frac{1}{4}$ inches. Five shelves are to be inserted, each one $\frac{3}{4}$ inch in thickness and spaced equally. The distance between the shelves is given by

$$D = \left(66\frac{1}{4} - 5 \cdot \frac{3}{4}\right) \div (5 + 1)$$

Calculate the distance.

VOCABULARY

No new vocabulary.

HOW AND WHY

The conventional order of operations for fractions is the same as for whole numbers. (See Section 1.5.)

EXAMPLES

a. $\dfrac{5}{6} - \dfrac{1}{2} \cdot \dfrac{2}{3} = \dfrac{5}{6} - \dfrac{1}{3}$ *Multiplication performed first*

$\phantom{\dfrac{5}{6} - \dfrac{1}{2} \cdot \dfrac{2}{3}} = \dfrac{3}{6}$ *Subtraction performed next*

$\phantom{\dfrac{5}{6} - \dfrac{1}{2} \cdot \dfrac{2}{3}} = \dfrac{1}{2}$ *Reduce to lowest terms*

b. $\dfrac{1}{2} \div \dfrac{2}{3} \cdot \dfrac{1}{4} = \dfrac{3}{4} \cdot \dfrac{1}{4}$ *Division performed first*

$\phantom{\dfrac{1}{2} \div \dfrac{2}{3} \cdot \dfrac{1}{4}} = \dfrac{3}{16}$ *Multiplication performed next*

APPLICATION SOLUTION

To find the distance between the shelves, calculate D.

$$D = \left(66\frac{1}{4} - 5 \cdot \frac{3}{4}\right) \div (5 + 1)$$

$$= \left(66\frac{1}{4} - \frac{15}{4}\right) \div 6$$

$$= \left(65\frac{5}{4} - 3\frac{3}{4}\right) \div 6$$

$$= 62\frac{1}{2} \div 6$$

$$= 10\frac{5}{12}$$

So the distance between the shelves is $10\frac{5}{12}$ inches.

WARM UPS

Do the indicated operations:

1. $\dfrac{4}{5} - \dfrac{2}{5} - \dfrac{1}{5}$

2. $\dfrac{4}{5} + \dfrac{2}{5} - \dfrac{1}{5}$

3. $\dfrac{4}{5} \cdot \dfrac{2}{5} - \dfrac{1}{25}$

4. $\dfrac{19}{25} - \dfrac{2}{5} \cdot \dfrac{1}{5}$

5. $\dfrac{4}{5} \div \dfrac{2}{5} \cdot \dfrac{1}{5}$

6. $\dfrac{4}{5} \div \dfrac{2}{5} + \dfrac{1}{5}$

7. $\dfrac{4}{5} + \dfrac{2}{5} \div \dfrac{1}{5}$

8. $\dfrac{4}{5} \div \dfrac{2}{5} \div \dfrac{1}{5}$

9. $\dfrac{4}{5} \cdot \left(\dfrac{2}{5} - \dfrac{1}{5}\right)$

10. $\left(\dfrac{4}{5} + \dfrac{2}{5}\right) \div \dfrac{1}{5}$

11. $\dfrac{11}{6} + \dfrac{1}{2} \div \dfrac{3}{2}$

12. $\dfrac{1}{4} \div \dfrac{3}{8} + \dfrac{1}{2}$

Exercises 2.11

ANSWERS

1. _____

2. _____

3. _____

4. _____

5. _____

6. _____

7. _____

8. _____

9. _____

10. _____

11. _____

12. _____

13. _____

14. _____

15. _____

16. _____

17. _____

18. _____

19. _____

20. _____

21. _____

22. _____

23. _____

24. _____

25. _____

A

Do the indicated operations.

1. $\dfrac{6}{7} - \dfrac{3}{7} - \dfrac{1}{7}$

2. $\dfrac{6}{7} + \dfrac{3}{7} - \dfrac{1}{7}$

3. $\dfrac{1}{2} \cdot \left(\dfrac{3}{7} - \dfrac{1}{7}\right)$

4. $\dfrac{1}{3} \div \dfrac{1}{2} \cdot \dfrac{1}{3}$

5. $\dfrac{2}{6} - \dfrac{1}{2} \cdot \dfrac{1}{3}$

6. $\dfrac{1}{4} + \dfrac{3}{8} \div \dfrac{1}{2}$

B

7. $\dfrac{1}{3} \cdot \dfrac{4}{5} + \dfrac{4}{15}$

8. $\dfrac{3}{10} + \dfrac{4}{5} \cdot \dfrac{1}{2}$

9. $\dfrac{7}{8} - \dfrac{1}{2} \cdot \dfrac{3}{4}$

10. $\dfrac{3}{4} \div 2 + \dfrac{1}{8}$

11. $\dfrac{1}{3} \cdot \dfrac{3}{7} \div \dfrac{1}{4}$

12. $\dfrac{1}{3} \div \dfrac{2}{3} \cdot \dfrac{4}{7}$

13. $\dfrac{5}{3} \cdot \dfrac{2}{3} - \dfrac{7}{9}$

14. $\dfrac{3}{4} \div \dfrac{1}{3} \cdot \dfrac{1}{6}$

C

15. $\dfrac{3}{4} \div \left(\dfrac{1}{3} \cdot \dfrac{1}{6}\right)$

16. $\dfrac{1}{3} \cdot \dfrac{1}{6} \div \dfrac{3}{4}$

17. $\dfrac{4}{5} - \dfrac{2}{5} \cdot \dfrac{1}{2}$

18. $\dfrac{4}{5} - \dfrac{2}{5} + \dfrac{1}{2}$

19. $\dfrac{7}{9} - \dfrac{1}{3} + \dfrac{1}{2} \div \dfrac{2}{5} \cdot \dfrac{4}{5}$

20. $\dfrac{1}{2} \cdot \dfrac{2}{3} \div 3 \cdot \dfrac{3}{4} \div \dfrac{1}{2}$

21. $\left(\dfrac{7}{9} - \dfrac{1}{3} + \dfrac{1}{2}\right) \div \dfrac{2}{5} \cdot \dfrac{4}{5}$

22. $\dfrac{17}{9} - \left(\dfrac{1}{3} + \dfrac{1}{2} \div \dfrac{5}{2}\right) \cdot \dfrac{5}{4}$

23. $\left(\dfrac{5}{8} - \dfrac{1}{2} \cdot \dfrac{3}{4}\right) \div \dfrac{1}{2} + \dfrac{1}{2}$

24. $\left(\dfrac{3}{4} + \dfrac{5}{8} \div \dfrac{3}{4}\right) \cdot \dfrac{6}{5} - \dfrac{4}{5}$

D

25. The inside height of a bookcase is to be $81\dfrac{5}{8}$ inches. Six shelves are to be inserted, each $\dfrac{5}{8}$ inch thick and equally spaced. The distance between the shelves is given by D. Find D.

$$D = \left(81\dfrac{5}{8} - 6 \cdot \dfrac{5}{8}\right) \div (6 + 1)$$

26. The inside height of a china hutch is to be $32\frac{3}{4}$ inches. Three shelves, each $\frac{3}{4}$ inch thick, are to be inserted and spaced equally. What will be the distance between the shelves?

27. The resistance of a series-parallel electric circuit is given by

$$\Omega = 2\frac{1}{5} + 4\frac{3}{10} + 1 \div \left(\frac{1}{4} + \frac{1}{2} + \frac{2}{5}\right) \text{ ohms}$$

Find the resistance of the circuit.

28. The resistance of a series-parallel electric circuit is given by

$$\Omega = 4\frac{1}{3} + 3\frac{1}{6} + 1 \div \left(\frac{1}{2} + \frac{1}{4} + \frac{2}{3}\right) \text{ ohms}$$

Find the resistance of the circuit.

29. Gwen, Sam, Carlos, and Sari have equal shares in a florist shop. Sam decides to sell out his share. He sells $\frac{3}{8}$ of his share to Gwen, $\frac{1}{2}$ to Carlos, and the rest to Sari. What is Gwen's share of the florist shop now?

30. In exercise 29, what fraction of the shop does Carlos now own?

31. To check that $x = \frac{2}{3}$ is a solution to the equation $6 - 5x + 3(x + 1) = 7\frac{2}{3}$, we substitute $\frac{2}{3}$ for x and find the value of

$$6 - 5\left(\frac{2}{3}\right) + 3\left(\frac{2}{3} + 1\right)$$

Verify that the above expression equals $7\frac{2}{3}$.

32. Substitute $x = 1\frac{1}{6}$ in the following equation and check that it is a solution (makes the equation true).

$$3x - 2\frac{1}{4} = 1\frac{1}{4}$$

E   *Maintain Your Skills*

Multiply:

33. $\frac{5}{9} \cdot \frac{7}{3} \cdot \frac{2}{1}$

34. $\frac{8}{5} \cdot \frac{3}{7} \cdot \frac{4}{5}$

35. $\frac{11}{13} \cdot \frac{5}{9} \cdot \frac{8}{3}$

36. An article is priced to sell for $8. If the price is reduced by $\frac{1}{4}$, how many dollars was the price reduced?

37. In a certain school $\frac{2}{3}$ of the students take math. One-fifth of those who take math are enrolled in Basic Math. What part of the student body takes Basic Math?

Chapter 2
POST-TEST

1. (Obj. 27) *Add:* $4\dfrac{3}{5}$

$1\dfrac{2}{3}$

$7\dfrac{1}{2}$

$4\dfrac{5}{6}$

2. (Obj. 30) Do the indicated operations: $\dfrac{2}{3} \cdot \dfrac{9}{10} - \dfrac{4}{5} \div \dfrac{8}{3}$

3. (Obj. 22) *Divide:* $\dfrac{2}{3} \div \dfrac{5}{8}$

4. (Obj. 20) *Multiply:* $3\dfrac{1}{2} \cdot 4\dfrac{2}{3} \cdot 1\dfrac{5}{7}$

5. (Obj. 26) *Add:* $\dfrac{1}{2} + \dfrac{1}{4} + \dfrac{1}{3}$

6. (Obj. 22) *Divide:* $3\dfrac{1}{2} \div 3\dfrac{1}{4}$

7. (Obj. 29) *Subtract:* $15\dfrac{5}{7}$

$5\dfrac{5}{9}$

8. (Obj. 23) Find the missing numerator: $\dfrac{2}{9} = \dfrac{?}{108}$

9. (Obj. 29) *Subtract:* $17 - 4\dfrac{5}{6}$

10. (Obj. 21) What is the reciprocal of $3\dfrac{2}{5}$?

11. (Obj. 28) *Subtract:* $\dfrac{7}{8} - \dfrac{2}{3}$

12. (Obj. 20) *Multiply:* $\dfrac{3}{4} \cdot \dfrac{2}{7} \cdot \dfrac{14}{9}$

13. (Obj. 30) Do the indicated operations: $\dfrac{1}{2} + \dfrac{1}{3} \cdot \dfrac{5}{6}$

14. (Obj. 20) *Multiply:* $\dfrac{4}{5} \cdot \dfrac{9}{7}$

15. (Obj. 24) List the following fractions in order from smallest to largest.

$\dfrac{3}{10}, \dfrac{3}{8}, \dfrac{2}{5}$

16. (Obj. 29) *Subtract:* $11\dfrac{2}{3}$

$5\dfrac{7}{8}$

17. (Obj. 27) *Add:* $3\frac{5}{8} + 4\frac{1}{3} + \frac{5}{6}$

18. (Obj. 22) *Divide:* $\frac{7}{8} \div 21$

19. (Obj. 25) *Add:* $\frac{27}{96} + \frac{13}{96}$

20. (Obj. 24) List the following fractions in order from smallest to largest.

$$\frac{5}{9}, \frac{4}{7}, \frac{2}{3}, \frac{10}{21}$$

21. (Obj. 19) Reduce to lowest terms: $\frac{42}{63}$

22. (Obj. 27) Four poured brass castings have weights of $38\frac{3}{4}$ lb, $37\frac{1}{2}$ lb, $42\frac{7}{8}$ lb, and

$27\frac{5}{16}$ lb. What is the total weight of the metal in the four castings?

23. (Obj. 29) A metal bar weighing $48\frac{5}{8}$ lb is machined and then reweighed. If the bar

then weighs $42\frac{3}{4}$ lb, how much metal was removed?

24. (Obj. 20) The distance around a circle is about $3\frac{1}{7}$ times its diameter. What is the

approximate distance around a circle whose diameter is $4\frac{1}{2}$ inches?

25. (Obj. 22) How many pieces may be cut from a 2×4 that is 16 ft long if each piece

is to be $1\frac{1}{3}$ ft in length? Disregard any waste from the saw cut.

ANSWERS TO WARM UPS (2.11)   **1.** $\frac{1}{5}$ **2.** 1   **3.** $\frac{7}{25}$ **4.** $\frac{17}{25}$ **5.** $\frac{2}{5}$

6. $2\frac{1}{5}$ **7.** $2\frac{4}{5}$ **8.** 10   **9.** $\frac{4}{25}$ **10.** 6

11. $\frac{13}{6}$ or $2\frac{1}{6}$ **12.** $\frac{7}{6}$ or $1\frac{1}{6}$

3

Decimals

1. _____

2. _____

3. _____

4. _____

5. _____

6. _____

7. _____

8. _____

9. _____

10. _____

11. _____

12. _____

13. _____

14. _____

15. _____

16. _____

17. _____

18. _____

19. _____

20. _____

21. _____

22. _____

23. _____

24. _____

25. _____

Chapter 3
PRE-TEST

The problems in the following pre-test are a sample of the material in the chapter. You may already know how to work some of these. If so, this will allow you to spend less time on those parts. As a result, you will have more time to give to the sections that you find more difficult or that are new to you. The answers are in the back of the text.

1. **(Obj. 32)** Write the word name of 81.0081.

2. **(Obj. 33)** List the following in order of value from the smallest to largest:

   .0678, .068, .07, .067

3. **(Obj. 34)** Round 36.498 to the nearest hundredth.

4. **(Obj. 35)** *Add:* $16.72 + 8.8 + 0.06 + 7.33$

5. **(Obj. 35)** *Add:* $17 + 0.6 + 3.9 + 5.6228 + 1$

6. **(Obj. 36)** *Subtract:* $1.067 - .762$

7. **(Obj. 36)** *Subtract:* $27 - 8.301$

8. **(Obj. 36)** *Subtract:* $13.028 - 12.82$

9. **(Obj. 37)** *Multiply:* $8.92(6.3)$

10. **(Obj. 37)** *Multiply:* $.0055(2.34)$ 11. **(Obj. 37)** *Multiply:* $(1.8)(2.34)(7.1)$

12. **(Obj. 38)** *Divide:* $2185.4 \div 35$ 13. **(Obj. 38)** *Divide:* $6.17684 \div 6.14$

14. **(Obj. 38)** *Divide.* Round answer to the nearest tenth. $87.6 \div .78$

15. **(Obj. 38)** *Divide.* Round answer to the nearest hundredth. $1.669 \div .88$

16. **(Obj. 39)** Write $\dfrac{71}{80}$ as a decimal.

17. **(Obj. 39)** Write $\dfrac{19}{17}$ as an approximate decimal rounded to the nearest thousandth.

18. **(Obj. 40)** Do the indicated operations. $8.14 - .3(7.6 - 5.9)$

19. **(Obj. 40)** Do the indicated operations. $.001 + 1.8(5.3) - .01 \div 5$

20. **(Obj. 40)** Do the indicated operations. $1.1^3 - .6^2$

21. **(Obj. 40)** Yanell bought 12 motto T-shirts on sale. The advertised price was four for $13.87. How much did she pay for them?

22. **(Obj. 38)** If Rana can average 46.5 miles per hour, how long will it take her to drive 217 miles?

23. **(Obj. 38)** An airplane carries 215 gallons of fuel. It uses approximately 21.7 gallons per hour at cruising speed. At this rate, how many hours can the plane fly? (Round to the nearest tenth of an hour.)

24. **(Obj. 36)** The normal barometric pressure for Carville is 30.11 inches of mercury for the month of August. If a reading of 31.68 is recorded, how much above normal is this?

25. **(Obj. 40)** A foundry worker is paid $13.75 per hour and time and one-half for any time over 40 hours in any one week. What is his gross pay if he works the following hours during one week: 9, 8, 12.5, 10.5, and 7.5?

114

3.1 Decimal Numbers: Numeral Form and Word Names

OBJECTIVES

31. Write numeral forms from word names.
32. Write word names from numeral forms.

APPLICATION

An accountant must write a company check for $73.07. How should she write the amount in words on the check?

VOCABULARY

Recall that the symbols 0, 1, 2, 3, 4, 5, 6, 7, 8, and 9 are called digits. These digits, and a period called a *decimal point,* are used to write other names for fractions and mixed numbers $\left(\frac{1}{4} = .25\right)$. These other names are called *decimal numerals,* or more commonly *decimal numbers* or *decimals.* This is an extension of the place value system for whole numbers, since place value in decimals is used in a similar manner.

The number of digits to the right of the decimal point is sometimes referred to as the number of *decimal places.* For example, 3.47 is said to have two decimal places.

HOW AND WHY

Decimals are written by using a standard place value in the same manner as the whole numbers. The place value in the case of decimals is:

1. the same as whole numbers for digits to the left of the decimal point, and
2. a fraction whose denominator is a power of 10 for digits to the right of the decimal point.

The digits to the right of the decimal point have place values of $\frac{1}{10^1}$, $\frac{1}{10^2}$, $\frac{1}{10^3}$, . . . in that order from left to right. When the denominator of a fraction is a power of ten, it can be written as a decimal. Any fraction equivalent to such a fraction can also be written as a decimal. This includes all whole numbers, since any whole number can be written with a denominator of 1 (or 10^0).

Using the ones place as the central position (the place value 10^0), the place value of a decimal looks like the following:

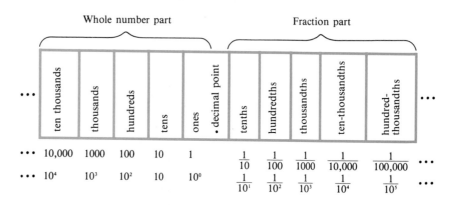

Note that the decimal point separates the whole number part from the fraction part.

Consider 226.35 and 0.127:

	Number to Left of Decimal Point	Decimal Point	Number to Right of Decimal Point	Place Value of Last Digit
Numeral Form	226	.	35	$\frac{1}{100}$
Word Name of Each	two hundred twenty-six	and	thirty-five	hundredths
Word Name of Decimal	Two hundred twenty-six and thirty-five hundredths			

	Number to Left of Decimal Point	Decimal Point	Number to Right of Decimal Point	Place Value of Last Digit
Numeral Form	0	.	127	$\frac{1}{1000}$
Word Name of Each	omit	omit	one hundred twenty-seven	thousandths
Word Name of Decimal	One hundred twenty-seven thousandths			

To write the word name for a decimal:

1. Write the name for the whole number to the left of the decimal point.
2. Write the word "and" for the decimal point.
3. Write the whole number name for the number to the right of the decimal point.
4. Write the place value of the digit farthest to the right.

If the decimal has only zero or no number to the left of the decimal point, omit steps 1 and 2.

Work backwards to find the numeral form.

EXAMPLES

a. The word name for .58 is "fifty-eight hundredths." This can also be written $\frac{58}{100}$.

b. The word name for .003 is "three thousandths." This can also be written $\frac{3}{1000}$.

c. The word name for 2.24 is "two and twenty-four hundredths." This can also be written $2\frac{24}{100}$.

d. The numeral form for "fifteen ten-thousandths" is .0015.

e. The numeral form for "four hundred five and four hundred five thousandths" is 405.405.

f. A single wire strand in an insulated cable is fifty ten-thousandths inch in diameter. In numeral form this is written .0050 inch.

APPLICATION SOLUTION

The word name the accountant will write for $73.07 is

Seventy-three and seven hundredths

(Note: Many people use a combination of word and numeral names on checks such as

Seventy-three and $\dfrac{07}{100}$)

WARM UPS

Write (or read) the word name of the following.

1. 2.5   **2.** .016   **3.** 36.05

4. 16.16   **5.** .0018   **6.** 3.003

Write (or read) the numeral form of each of the following.

7. Four tenths

8. Two and two hundredths

9. Thirteen and fifteen thousandths

10. Six hundredths

11. One hundred and one hundredth

12. Sixteen and sixteen hundredths

ANSWERS

1. _____

2. _____

3. _____

4. _____

5. _____

6. _____

7. _____

8. _____

9. _____

10. _____

11. _____

12. _____

13. _____

14. _____

15. _____

16. _____

17. _____

18. _____

19. _____

20. _____

21. _____

22. _____

23. _____

24. _____

Exercises 3.1

A

Write the word name of each of the following.

1. .4   **2.** 0.219   **3.** 3.26

Write the numeral form for each of the following.

4. Forty-five thousandths   **5.** Eleven hundredths

6. Four and five hundredths

B

Write the word name of each of the following.

7. 5.04   **8.** 50.4   **9.** 500.4   **10.** 50.04

Write the numeral form for each of the following.

11. Fifteen hundredths

12. Five hundred and five thousandths

13. Five thousand and five hundredths

14. Five hundred-thousandths

C

Write the word name for each of the following.

15. 18.0205   **16.** 48.0004   **17.** 300.045

18. 110.0345   **19.** 10.1011

Write the numeral form of each of the following.

20. Five thousand

21. One thousand and five thousandths

22. Two hundred and thirty-one thousandths

23. Two hundred thirty-one thousandths

24. One hundred five and one hundred five thousandths

25. _____

26. _____

27. _____

28. _____

29. _____

30. _____

31. _____

32. _____

33. _____

D

25. Dan Ngo bought a toaster-oven that had a marked price of $73.85. What word name will he write on the check?

26. Fari Alhadet bought a pick-up truck load of organic fertilizer for her garden. The price of the load was $93.99. What word name will she write on the check?

27. John Benhry orders plastic inserts for his storm windows. What does he write for the numeral form if the windows are three and six tenths feet by five and one hundred twenty-five thousandths feet?

28. Pete Rios is typing up the bid specification for a new press from an audio tape. How does he record the following specifications in numeral form? The minimum sheet size is to be five and twenty-five hundredths by eight and thirty hundredths. The maximum sheet size is to be ten and six hundredths by twenty-one and seventy-five hundredths.

E   *Maintain Your Skills*

Find the missing numerators.

29. $\dfrac{27}{21} = \dfrac{?}{147}$ 30. $\dfrac{13}{11} = \dfrac{?}{55}$ 31. $\dfrac{8}{9} = \dfrac{?}{72}$

32. What fraction with denominator 48 is equivalent to $\dfrac{5}{6}$?

33. Write a fraction that is equivalent to $\dfrac{3}{4}$ and that has a denominator of 128.

ANSWERS TO WARM UPS (3.1)   **1.** Two and five tenths   **2.** Sixteen thousandths

3. Thirty-six and five hundredths   **4.** Sixteen and sixteen hundredths

5. Eighteen ten-thousandths   **6.** Three and three thousandths

7. .4   **8.** 2.02   **9.** 13.015   **10.** .06   **11.** 100.01   **12.** 16.16

3.2 Listing Decimals in Order of Value

OBJECTIVE

33. List a given group of decimals in order of value from smallest to largest.

APPLICATION

An Insapik camera has a fixed exposure time of .042 second. A Trupix camera has an exposure time of .0325 second. Which camera has the shortest exposure time?

VOCABULARY

A group of numbers is said to be *ordered* when they are arranged in order of value from smallest to largest. The decimals .2, .35, and .471 are in order from smallest to largest.

HOW AND WHY

Fractions can be ordered when they have a common denominator by ordering the numerators. This idea can be extended to decimals when they have the same number of places to the right of the decimal point. This assures us of common denominators when we write them as fractions. Thus, $.26 = \dfrac{26}{100}$ and $.37 = \dfrac{37}{100}$ have a common denominator when written in fraction form; therefore we know that .26 is less than .37. If the decimals do not contain the same number of places to the right of the decimal point, zeros can be inserted on the right-hand end. So, .3 and .15 have a common denominator when written in fraction form as $.3 = .30 = \dfrac{30}{100}$ and $.15 = \dfrac{15}{100}$. Therefore, .15 is less than .3.

Decimals can be listed in order by the following procedure:

1. **Make sure that all numbers have the same number of decimal places by inserting zeros on the right when necessary.**
2. **Write the numbers in order from smallest to largest, ignoring the decimal point.**
3. **Replace the decimal points and remove the extra zeros.**

EXAMPLES

List the decimals in each group from smallest to largest.

a. .62, .637, .6159, .621. First write all numbers with the same number of decimal places to the right of the decimal point.

.6200, .6370, .6159, .6210

Now order the whole numbers formed by ignoring the decimal points.

6159, 6200, 6210, 6370

Remove the extra zeros. The numbers are now ordered.

.6159, .62, .621, .637

b. 1.357, 1.361, 1.3534, 1.358

1.3570, 1.3610, 1.3534, 1.3580

13534, 13570, 13580, 13610

1.3534, 1.357, 1.358, 1.361

APPLICATION SOLUTION

To find out which camera has the shorter exposure time, list the times in order of value.

Insapik: .042 sec   Trupix: .0325 sec

.042, .0325
.0420, .0325
.0325, .0420
.0325, .042

Answer: Trupix has the shorter exposure time.

WARM UPS

List the decimals in each group from smallest to largest.

1. .8, .81, .801

2. .06, .05, .07

3. 1.3, 1.29, 1.31

4. 17.0, 16.9, 17.5

5. .0051, .0049, .00491

6. .11, .09, .13

7. .113, .115, .112, .119

8. 100.6, 99.9, 99.08

9. 4.35, 3.96, 4.87, 2.49

10. 1.67, 1.65, 1.6, 1.7

11. .565, .556, .566, .555

12. 3.2, 3.19, 3.21, 3.27

Exercises 3.2

A

List the decimals in each group from smallest to largest.

1. .12, .14, .11

2. .3, .31. .301

3. .5, .503, .510

4. .7, .72, .713

5. .02, .022, .019, .021

6. 0.2, 0.25, 0.21, 0.23

B

7. 0.75, 0.692, 0.748

8. .014, .104, .041

9. 6.14, 6.141, 6.139

10. 5.18, 5.183, 5.179

11. 7.3, 7.19, 7.21, 7.27

12. .785, .587, .875, .857

13. .0907, .0097, .0709, .079

14. .0103, .0013, .0301, .031

C

15. .9, .899, .86, .91, .903

16. .0031, .00305, .003, .00312

17. .1163, .116, .1159, .117

18. 17.05, 17.16, 17.0506, 17.057

19. .0729, .073001, .072, .073, .073015

20. .30009, .301, .3008, .30101

21. .888, .88799, .8881, .88759

22. 1.999, 2., 1., 1.5, 2.006

23. 8.36, 8.2975, 8.3599, 8.3401

24. 10.43, 10.34, 10.429, 10.339

1. _____
2. _____
3. _____
4. _____
5. _____
6. _____
7. _____
8. _____
9. _____
10. _____
11. _____
12. _____
13. _____
14. _____
15. _____
16. _____
17. _____
18. _____
19. _____
20. _____
21. _____
22. _____
23. _____
24. _____

25. _____

26. _____

27. _____

28. _____

29. _____

30. _____

31. _____

32. _____

33. _____

D

25. A Quickpik camera has an exposure time of .0375 sec and a Trepid camera has an exposure time of .0358 sec. Which camera has the shorter exposure time?

26. Greg and Larry each measured the diameter of a coin and got the following measurements respectively: .916 centimeter and .921 centimeter. Which measurement is the smallest?

27. The Davis Meat Co. bid 98.375 cents per pound to provide meat to the Ajax Grocery. Circle K Meats put in a bid of 98.35 cents, and J & K Meats made a bid of 98.3801 cents. Which is the best bid for the Ajax Grocery?

28. A vial contains 2.3059 grams of charcoal. Dan weighs the charcoal and determines that it weighs 2.31 grams. Is Dan's weighing too light or too heavy?

E *Maintain Your Skills*

Which fraction has the larger value?

29. $\dfrac{5}{8}, \dfrac{16}{25}$

30. $\dfrac{7}{9}, \dfrac{39}{50}$

Arrange these fractions in order of increasing value.

31. $\dfrac{7}{8}, \dfrac{19}{25}, \dfrac{3}{4}$

32. $\dfrac{5}{9}, \dfrac{11}{20}, \dfrac{27}{50}$

33. $\dfrac{4}{5}, \dfrac{7}{8}, \dfrac{2}{3}$

124

ANSWERS TO WARM UPS (3.2)

1. .8, .801, .81	**2.** .05, .06, .07	**3.** 1.29, 1.3, 1.31
4. 16.9, 17.0, 17.5	**5.** .0049, .00491, .0051	**6.** .09, .11, .13
7. .112, .113, .115, .119	**8.** 99.08, 99.9, 100.6	**9.** 2.49, 3.96, 4.35, 4.87
10. 1.6, 1.65, 1.67, 1.7	**11.** .555, .556, .565, .566	**12.** 3.19, 3.2, 3.21, 3.27

3.3 Approximation by Rounding

OBJECTIVE

34. Round off a given decimal to a specified place value.

APPLICATION

The displacement of a V-8 diesel is about 4941.3829 cc. What is the displacement to the nearest hundredth?

VOCABULARY

Decimals are exact or approximate. *Exact decimals* name the numbers under discussion. *Approximate decimals* are approximations of numbers to a specified place value.

The exact decimal 15.3 is said to be *rounded off* to the nearest whole number when expressed as 15. In this case, 15 is an approximate decimal. $\approx$ is a symbol for "approximately equal to," so 15.3 $\approx$ 15.

HOW AND WHY

Look at this ruler:

To the nearest tenth the arrow shows 2.6 because the arrow is more than half way between 2.5 and 2.6. To the nearest hundredth the arrow shows 2.56, because the arrow is less than half way between 2.56 and 2.57.

To round off, all we need to do is see if the number we have is less than half way between the two approximate measurements. If it is less than half way, choose the smaller approximation. If it is not less than half way, choose the larger approximation.

To round 6.3265 to the nearest hundredth, draw an arrow under the hundredths place to identify the round off position.

6.3265
  ↑

We must choose between 6.32 and 6.33. Since the first digit to the right of the round off position is 6, the number is more than half way to 6.33. So,

6.3265 $\approx$ 6.33

To round off a decimal number to a given place value:

1. Draw an arrow under the given place value.
2. Change all digits following the given place value to zeros.
3. a. If the arrow is to the *left* of the decimal point, omit all zeros following the decimal point.
   b. If the arrow is to the *right* of the decimal point, omit all zeros following the arrow.
4. Select two approximations. The first is the number found in step 3. The second is found by adding one (1) to the digit in the place value indicated by the arrow in the first approximation.
5. Choose the approximation by using

the larger approximation $\begin{cases} \text{if the digit in the place value} \\ \text{after the arrow in step 1 is} \\ \text{5, 6, 7, 8, or 9} \end{cases}$

the smaller approximation: $\begin{cases} \text{if the digit in the place value} \\ \text{after the arrow in step 1 is} \\ \text{0, 1, 2, 3, or 4} \end{cases}$

EXAMPLES

a. .3582 rounded to the nearest hundredth is .36.

  .3582   *Draw an arrow under the given place value*
    ↑

  .3500   *Change all digits to the right of the arrow to zero*
    ↑

  .35   *Drop zeros*

  .35 or .36   *Add one to the hundredths place*

  .36   *Chosen because the digit after the arrow is 8*

b. 3582 rounded to the nearest thousand is 4000, because the digit to the right
    ↑
  of the round off position is 5.

c. 16.349 rounded to the nearest tenth is 16.3, because the digit to the right of
      ↑
  the round off position is 4.

d. 249.7 rounded to the nearest unit is 250. The choice is between 249 and 250;
      ↑
  the 7 tells us to choose the larger approximation.

e. 3.996 rounded to the nearest hundredth is 4.00. The choice is between 3.99
      ↑
  and 3.99 + .01 = 4.00. The 6 tells us to choose the larger approximation.

APPLICATION SOLUTION

To find the displacement to the nearest hundredth, round off.

4941.3829
⎯↑⎯

4941.3800
⎯↑⎯

4941.38

4941.38 or 4941.39

4941.38

So the displacement of the V-8 diesel is 4941.38 cc to the nearest hundredth.

WARM UPS

Round each of the following to the nearest unit, tenth, and hundredth:

	Unit	Tenth	Hundredth
23.461	23	23.5	23.46
1. 321.222			
2. 46.777			
3. 529.655			
4. 8.496			
5. 0.6493			
6. 11.399			

Exercises 3.3

ANSWERS

1. _____
2. _____
3. _____
4. _____
5. _____
6. _____
7. _____
8. _____
9. _____
10. _____
11. _____
12. _____
13. _____
14. _____
15. _____
16. _____
17. _____
18. _____
19. _____
20. _____
21. _____
22. _____
23. _____
24. _____
25. _____
26. _____

A

Round each of the following decimals to the nearest tenth, hundredth, and thousandth.

1. 2.6532   **2.** .8359   **3.** 12.3015   **4.** 1.3347

5. 9.9892   **6.** 10.0752

B

7. 53.3125   **8.** 9.7765   **9.** 2.1789

10. 3.0007   **11.** 0.7934   **12.** 14.5553

13. 0.7891   **14.** .9999

C

Round each of the following to the nearest thousand, hundred, and ten.

15. 6,921   **16.** 75,671   **17.** 3,971

18. 26,775   **19.** 965.0348   **20.** 999,501

Round each of the following to the nearest whole number, tenth, and hundredth.

21. 36.58   **22.** 100.495   **23.** 5.399

24. 78.905

D

25. Instructions for listing deductions on the 1040 IRS form call for rounding each deduction to the nearest dollar. What amount should Sam enter for charitable expenses if his donations total $483.72?

26. A chemist made an analysis of the ingredients in a box of Wheaties. He found 10.063 grams of non-nutritive fiber. Round this weight to the nearest tenth of a gram.

ANSWERS

27. _____

28. _____

29. _____

30. _____

31. _____

32. _____

33. _____

34. _____

35. _____

36. _____

27. A micrometer is used to measure the thickness of a metal bar. The operator reads the measurement as .4978 inch. What is the thickness of the bar to the nearest hundredth of an inch?

28. A vial contains 2.3059 grams of charcoal. What is its weight to the nearest hundredth of a gram?

29. The cost of material needed to install storm windows on Jane's house was estimated at $1374.82. What was the estimated cost to the nearest ten dollars?

30. A sprinter, in preparation for the Olympics, was timed at 9.871 seconds for the 100 meter dash. What was his time to the nearest tenth of a second?

31. Jose used 13 gallons of gasoline to drive 487 miles. This averages to 37.3846 miles per gallon. What was Jose's mpg to the nearest mile?

E   *Maintain Your Skills*

Reduce to lowest terms.

32. $\dfrac{250}{300}$ **33.** $\dfrac{128}{256}$ **34.** $\dfrac{48}{360}$ **35.** $\dfrac{75}{225}$

36. $\dfrac{100}{480}$

ANSWERS TO WARM UPS (3.3)   **1.** 321, 321.2, 321.22   **2.** 47, 46.8, 46.78   **3.** 530, 529.7, 529.66

4. 8, 8.5, 8.50   **5.** 1, .6, .65   **6.** 11, 11.4, 11.40

3.4 Addition and Subtraction of Decimals

OBJECTIVES

35. Add two or more decimals.
36. Subtract two decimals.

APPLICATION

A train has, in part, 2 carloads of wheat which weigh 41.93 tons each. There are 3 carloads of corn which weigh 38.09 tons each. How much weight do the five carloads add to the train's load? How much more does a carload of wheat weigh than a carload of corn?

VOCABULARY

Recall that the number of digits after the decimal point is called the number of decimal places.

HOW AND WHY

The addition facts commonly listed in the "addition table" are used in the addition of decimals in much the same way they are used in the addition of whole numbers.

$6.3 + 2.5 = ?$

We make use of the expanded notation of our place value system to explain the process.

$6.3 = 6$ ones $+ 3$ tenths
$\underline{2.5 = 2$ ones $+ 5$ tenths$}$
         8 ones $+ 8$ tenths $= 8.8$

The vertical form gives us a natural grouping of the ones and tenths.

Decimals can be added or subtracted as if they were whole numbers by writing them in vertical columns. Align the decimal points so the columns have the same place value.

Thus, $2.8 + 13.4 + 6.22$ can be written $2.80 + 13.40 + 6.22$ (with the same number of decimal places as you use when ordering decimals) and then,

2.80
13.40
 6.22   (see Example a)

Consider $6.59 - 2.34$. Written vertically in expanded form, we have:

$6.59 = 6$ ones $+ 5$ tenths $+ 9$ hundredths
$\underline{2.34 = 2$ ones $+ 3$ tenths $+ 4$ hundredths$}$
         4 ones $+ 2$ tenths $+ 5$ hundredths

So, $6.59 - 2.34$ is 4.25. Note that we subtracted ones from ones, tenths from tenths, and hundredths from hundredths.

When necessary, we can regroup or "borrow" as with whole numbers. $6.271 - 3.845 = ?$ In expanded form we have:

$6.271 = 6$ ones $+ 2$ tenths $+ 7$ hundredths $+ 1$ thousandth
$\underline{3.845 = 3$ ones $+ 8$ tenths $+ 4$ hundredths $+ 5$ thousandths$}$

Since we cannot subtract 5 thousandths from 1 thousandth, or 8 tenths from 2 tenths, we regroup. "Borrow" one of the hundredths from the seven hundredths and combine with the 1 thousandth (.01 = .010, and .010 + .001 = .011 or 11 thousandths). Also "borrow" one of the ones from the 6 ones and combine with the 2 tenths (1 = 1.0, and 1.0 + .2 = 1.2 or 12 tenths). Then we have:

$$6.271 = 5 \text{ ones} + 12 \text{ tenths} + 6 \text{ hundredths} + 11 \text{ thousandths}$$
$$\underline{3.845 = 3 \text{ ones} + 8 \text{ tenths} + 4 \text{ hundredths} + 5 \text{ thousandths}}$$
$$ 2 \text{ ones} + 4 \text{ tenths} + 2 \text{ hundredths} + 6 \text{ thousandths}$$

So, $6.271 - 3.845 = 2.426$. The examples show the common shortcuts.

EXAMPLES

a.
```
  11
  2.80
 13.40
  6.22
 22.42
```
Think:

hundredths column: $0 + 0 + 2 = 2$, *no carry*

tenths column: $8 + 4 + 2 = 14$, *carry the "1" to the ones column and write the "4" in the tenths column* (14 tenths = 1.4)

ones column: $1 + 2 + 3 + 6 = 12$, *carry the "1" to the tens column and write the "2" in the ones column*

tens column: $1 + 1 = 2$, *no carry*

b. $1.05 + .723 + 72.6 + 8 = ?$ Write each decimal with the same number (three) of decimal places: $1.050 + .723 + 72.600 + 8.000$ or

```
  1.050
   .723
 72.600
  8.000
 82.373
```

c. *Subtract:*

```
 5.831
  .287
```

Think:

We must regroup since we cannot subtract 7 thousandths from 1 thousandth. Therefore, we borrow 1 hundredth from the hundredths to regroup with the thousandths.

```
    211
 5.8 3̸ 1̸
  .2 8 7
```

```
     12
   7 2̸ 11
 5.8̸ 3̸ 1̸
  .2 8 7
 5.5 4 4
```

Now we cannot subtract 8 hundredths from 2 hundredths, so we borrow one tenth from the tenths to regroup with the hundredths.

Check:
```
  .287
 5.544
 5.831
```

d. *Subtract:*

```
 6
 2.94
```

5 10
$\cancel{6}.\cancel{0}0$ *Think:*
2.94   *The "6" can be rewritten as 6.00 so both numbers will have the same*
    number of decimal places. We borrow 1 one and add it to the tenths
9
5 $\cancel{10}$10   *place (10 tenths) and then borrow 1 tenth and add it to the hun-*
$\cancel{6}.\cancel{0}\cancel{0}$ *dredths place. We can now subtract in each column.*
2.94
3.06

Check: 2.94
        3.06
        6.00

e. Calculator example

$6.3975 + 0.0116 + 3.41 + 18.624 = ?$

ENTER	DISPLAY
6.3975	6.3975
+	6.3975
0.0116	0.0116
+	6.4091
3.41	3.41
+	9.8191
18.624	18.624
=	28.4431

The sum is 28.4431.

f. Calculator example

$127.9635 - 96.9358 = ?$

ENTER	DISPLAY
127.9635	127.9635
−	127.9635
96.9358	96.9358
=	31.0277

The difference is 31.0277.

APPLICATION SOLUTION

To find the weight added to the train add the weight of the five cars.

41.93   *Two carloads of wheat*
41.93
38.09   *Three carloads of corn*
38.09
38.09
198.13

So the five cars increase the weight of the train by 198.13 tons.
    To find how much more the carload of wheat weighs, subtract.

41.93
38.09
3.84

So a carload of wheat weighs 3.84 tons more than a carload of corn.

WARM UPS

Add:

1. .6
   .5
	2. .03
	    .27
		3. 6.11
		     .98

4. 7.23
    2.54
	5. 5.42
	    3.5
		6. 5.9
		    4.31

Subtract:

7. .8
   .5
	8. .75
	    .04
		9. 16.31
		     3.20

10. 4.63
    1.22
	11. 7.7
	    5.23
		12. 6.35
		    3.3

Exercises 3.4

ANSWERS

1. _____

2. _____

3. _____

4. _____

5. _____

6. _____

7. _____

8. _____

9. _____

10. _____

11. _____

12. _____

13. _____

14. _____

15. _____

16. _____

17. _____

18. _____

19. _____

20. _____

21. _____

22. _____

23. _____

24. _____

A

Add:

1. .38
.91

2. .135
.026

3. 5.678
3.157

Subtract:

4. .75
.33

5. 4.93
2.86

6. 12.5
8.79

B

Add:

7. 5.302
9.01
.9479
14.85

8. 5.19
9.957
4.3

9. 87.5521
572.63
98.007
113.98

10. 213.6
48.089
117.35
92.91

Subtract:

11. 3.457
2.509

12. 9.006
3.257

13. 12.1
9.34

14. 10
3.582

C

Add:

15. 12.385 + 2.03 + 9.0015 + 14

16. 110.607 + 5.31 + 18.115 + 12.012

17. 39.75 + .8 + 0.95 + 12.1 + 5

18. 15.715 + 33.33 + 12.012 + 6.9 (Round sum to nearest hundredth.)

19. 101.101 + 49.49 + 6.6 + 10.01 + 3.295 (Round sum to nearest tenth.)

Subtract:

20. 38.06 − 14.5

21. 114.95 − 89.365

22. .9371 − .38

23. 162.375 − 87.496. Round difference to nearest hundredth.

24. 3 − 2.651. Round difference to nearest tenth.

25. _____

26. _____

27. _____

28. _____

29. _____

30. _____

31. _____

32. _____

33. _____

34. _____

35. _____

36. _____

37. _____

D

25. The train in the application also has three coal cars each carrying 63.48 tons of coal and three cars each carrying 28.75 tons of fertilizer. How much weight do the six cars add to the train? How much more does a car of coal weigh than a car of fertilizer?

26. Three farmers brought in the following amounts of beans to the processing plant: 34.63 tons, 47.18 tons, and 23.6 tons. How many tons of beans were brought to the processing plant by the three farmers?

27. During a chemistry experiment, Ross had 10.3 cc of a solution. He used 7.6 cc. How much of the solution did he have left?

28. Jack went shopping with $25.50 in cash. He bought a record for $4.89 and a sweater for $11.88. On the way home he stopped for $4.75 worth of gas. How much cash did he have left?

29. Russ was hitchhiking across the city. The first car to offer him a ride took him 3.8 miles. The second took him 8.5 miles, and the third car went 7.4 miles before he got out. How many miles did he ride altogether?

30. On a short vacation trip, Paul stopped for gas four times. The first time he got 8.6 gallons. At the second station he got 14.9 gallons, and at the third he got 15.4 gallons. At the last stop he got 13.5 gallons. How much gas did he buy on the trip?

31. Heather wrote five checks in the amounts of $34.85, $17.37, $67.48, $21.10, and $8.10. She has $152.29 in her checking account. Does she have enough money deposited to cover the five checks?

32. The local Boy Scout Troop has $536.72 in its treasury. For a campout they spent $83.15 for food, $30 for the campground, $17.95 for transportation, and $25 for projects. How much was left in the treasury after these bills were paid?

E   *Maintain Your Skills*

Multiply and reduce each answer to lowest terms.

33. $\left(\dfrac{5}{8}\right)\left(\dfrac{4}{9}\right)\left(\dfrac{3}{10}\right)$

34. $(5)\left(\dfrac{1}{12}\right)\left(\dfrac{3}{5}\right)$

35. $\left(3\dfrac{3}{4}\right)\left(5\dfrac{4}{5}\right)$

36. $\left(2\dfrac{2}{3}\right)(12)\left(1\dfrac{1}{2}\right)$

37. $\left(\dfrac{3}{4}\right)\left(\dfrac{8}{9}\right)\left(\dfrac{5}{6}\right)\left(\dfrac{7}{8}\right)$

3.5 Multiplication of Decimals

OBJECTIVE

37. Multiply two or more decimals.

APPLICATION

Gasoline costs 140.9 cents a gallon. How much will 56.3 gallons cost?

VOCABULARY

No new vocabulary. (See multiplication of whole numbers.)

HOW AND WHY

The "multiplication table" for whole numbers is used in the multiplication of decimals in the same way it is used to multiply whole numbers. In fact, the multiplication procedure for decimals is, with one exception, the same as for whole numbers. The one exception is in the placement of the decimal point in the product. To locate the decimal point in $(.3)(.8)$, we write the decimals as fractions:

$$\frac{3}{10} \times \frac{8}{10} = \frac{24}{100}$$

In decimal form this becomes: $.3 \times .8 = .24$

To multiply two decimals:

1. **Multiply the numbers as if they were whole numbers.**
2. **Locate the decimal point by counting the number of decimal places (to the right of the decimal point) in both factors. The total of these two counts is the number of decimal places the product must have.**
3. **If necessary, zeros are inserted at the *left of the numeral* so there are enough decimal places (see Example g.)**

This is true because in fraction form the denominators are multiples of ten, and the product of these is the same or a larger multiple of ten. Thus, the product $7.2 \times .13$ requires three decimal places. Since $72 \times 13 = 936$, the product is

$$7.2 \times .13 = .936 \quad \text{or} \quad \begin{array}{r} 7.2 \\ .13 \\ \hline 216 \\ 72 \\ \hline .936 \end{array}$$

It may happen that the product of two decimals does not contain enough digits for the necessary number of decimal places. For instance, for $.2 \times .3$ the product of 2×3, which is 6, is a *one* digit numeral and we need *two* decimal places in the product. In this case, we insert a zero before the six:

$.2 \times .3 = .06$

This is true because

$$\frac{2}{10} \times \frac{3}{10} = \frac{6}{100} = .06$$

EXAMPLES

a. $.7 \times 6 = 4.2$ **b.** $7 \times .6 = 4.2$

c. $.7 \times .6 = .42$

d. $2.5 \times .2 = .50 = .5$ *Note that the zero was counted as one of the decimal places.*

e.   .53   **f.**   2.31
            15   3.4
           ───   ───
           265   924
           53   693
           ───   ─────
          7.95   7.854

g.   .21
           .14
           ───
           84
           21
          ─────
          .0294   *Note that we must insert a zero before the "2" because the product must have four decimal places.*

h. Calculator example

$(62.75)(136.492) = ?$

ENTER	DISPLAY
62.75	62.75
✕	62.75
136.492	136.492
=	8564.873

The product is 8564.873

APPLICATION SOLUTION

To find the cost of the gasoline, multiply the price per gallon times the number of gallons.

Formula: $C = pn$, $p = 140.9$ cents, $n = 56.3$ gallons

Substitute: $C = (140.9)(56.3)$

Solve: 14 0.9
                5 6.3
               ──────
               42 2 7
               845 4
               7045
              ────────
              7932.6 7

Answer: The gasoline cost 7932.67 cents or $79.33 to the nearest hundredth of a dollar (nearest cent).

WARM UPS

Multiply:

1. .3
        .6

2. 3(3.24)

3. 1.05
           7

4. 5(2.16)

5. 2.6
        .003

6. 7.51
        .03

7. 8.65
       .001

8. .33
       .06

9. .5(2.3)

10. .6(1.09)

11. 1.4
        .03

12. 8.97
        .04

Exercises 3.5

A

Multiply:

1. .7
   .2

2. 9(.03)

3. 8(.7)

4. (9.1)(6)

5. 3.6
   .6

6. 1.73
   .4

B

7. 34.6
   6.5

8. .272
   3.1

9. 9.04
   .47

10. 7.41
   .206

11. 8.2
   .68

12. 7.213
   88

13. 3.32
   .04

14. 6.005
   75

C

15. 98
   .079

16. .345
   4.8

17. (5.31)(5.9)

18. (.011)(.032)

19. (2.52)(1.37)

20. (6.02)(3.59)

21. (9.08)(.078)

22. (.0649)(2.5)

23. (18)(5.09)(9.3)

24. (3.99)(9.4)(2.1)

ANSWERS

1. _____
2. _____
3. _____
4. _____
5. _____
6. _____
7. _____
8. _____
9. _____
10. _____
11. _____
12. _____
13. _____
14. _____
15. _____
16. _____
17. _____
18. _____
19. _____
20. _____
21. _____
22. _____
23. _____
24. _____

D

25. Mary fills her car with gas costing 135.9 cents per gallon. If it took 12.3 gallons, how much did she pay for the gas?

26. If gasoline costs 145.9 cents per gallon, how much must Dale pay for 12.7 gallons?

27. Joe earns $4.47 an hour. How much did he earn during the week if he worked 30.25 hours? (Round to the nearest hundredth of a dollar.)

28. Amanda earns $14.57 per hour on her job as a welder. How much does she make in a 35 hour week?

29. Melanie bought six records on sale. The advertised sale price was two records for $3.87. How much did she pay for them?

30. Jeff buys a dozen golf balls. The advertised price of a package of three balls was $4.97. How much did Jeff pay for the dozen balls?

31. Chia has $700 in a savings account that pays 8% interest compounded annually. The value of her savings in two years is given by

$$A = 700 \times (1.08)^2$$

Find the value of her savings at the end of the two years.

32. Fred and Freda deposit $300 in a savings account that pays 6% interest compounded annually. The formula for finding the value of their savings at the end of three years is

$$A = 300 \times (1.06)^3$$

Find the value of their savings to the nearest cent.

E   *Maintain Your Skills*

Divide. Reduce answers to lowest terms and express as a mixed number where possible.

33. $\left(\dfrac{7}{9}\right) \div \left(\dfrac{14}{45}\right)$ 34. $36 \div \left(5\dfrac{1}{4}\right)$ 35. $\left(3\dfrac{5}{11}\right) \div \left(2\dfrac{10}{33}\right)$

36. $\left(7\dfrac{1}{8}\right) \div \left(1\dfrac{3}{16}\right)$ 37. $\left(15\dfrac{1}{3}\right) \div \left(3\dfrac{5}{6}\right)$

ANSWERS TO WARM UPS (3.5)   **1.** .18   **2.** 9.72   **3.** 7.35   **4.** 10.8   **5.** .0078   **6.** .2253

7. .00865   **8.** .0198   **9.** 1.15   **10.** .654   **11.** .042   **12.** .3588

3.6 Division of Decimals

OBJECTIVE

38. Divide two decimals.

APPLICATION

A compact car averages 39.4 miles per gallon for a long trip. How many gallons would be needed for a trip of 1,480 miles? State the number of gallons to the nearest tenth of a gallon.

VOCABULARY

No new vocabulary.

HOW AND WHY

Division of a decimal by a counting number is the same as division of whole numbers except for the placement of the decimal point. The decimal point is written above the decimal point of the dividend.

Division is the inverse of multiplication; therefore, the quotient of a decimal divided by a counting number whose prime factors are 2 and/or 5 must have the same number of decimal places as the dividend. The decimal point can be placed correctly by writing it directly above the one in the dividend. After that, the division can be done as if the numbers were whole numbers. In the dividend, zeros may be inserted after the last digit following the decimal point, if necessary, to continue the division process. It should be noted that the quotients where the divisors are counting numbers with only prime factors of 2 and/or 5 are always exact decimals.

Consider:

$$
\begin{array}{r}
.0019 \\
20\overline{)\,.0380} \\
\underline{20} \\
180 \\
\underline{180} \\
0
\end{array}
$$

Check: $\begin{array}{r} .0019 \\ \underline{20} \\ .0380 \end{array}$

Think: Write the decimal point for the quotient above the decimal point in the dividend. Zero divided by twenty is zero. Write "0" above the "0". There are no groups of twenty in 3, so write "0" above the "3". Next, divide thirty-eight by twenty. There is one group of twenty in thirty-eight, so write "1" above the "8." Multiply and subtract as with whole numbers. The difference is 180. 180 divided by 20 is nine. Write "9" above the "0." Multiply and subtract. Since the remainder is zero, the division is complete. Note that the quotient and the dividend both have four decimal places.

When a decimal is divided by 7, or any other whole number that has a prime factor other than 2 or 5, the division process may not have a remainder of zero at any step.

$$
\begin{array}{r}
.33 \\
7\overline{)\,2.34} \\
\underline{2\,1} \\
24 \\
\underline{21} \\
3
\end{array}
$$

At this step we can insert zeros after 2.34 (for example, 2.34000) as before.

```
      .33428
  7)2.34000
    2 1
      24
      21
       30
       28
        20
        14
         60
         56
          4
```

It appears that we might go on inserting zeros and continue endlessly. This, indeed, is what would happen. In practical applications, we stop the division process one place value beyond the accuracy required by the situation, and then round off. Therefore,

$2.34 \div 7 \approx .33$ to the nearest hundredth and

$2.34 \div 7 \approx .3343$ to the nearest ten-thousandth.

Division problems in which the divisor also contains a decimal point are easiest to handle by changing to an equivalent problem in which the divisor is a counting number. This can be done by multiplying both the divisor and the dividend by a power of 10. Since every indicated division can be treated as a fraction, this is the same as multiplying by some form of 1 which results in an equivalent fraction. For instance,

$$2.338 \div .7 = \frac{2.338}{.7} \cdot \frac{10}{10} = \frac{23.38}{7}$$

$$7.8 \div 0.14 = \frac{7.8}{.014} \cdot \frac{10^3}{10^3} = \frac{7800}{14}$$

$$4.865 \div .23 = 486.5 \div 23$$

To divide two numbers,

1. **If the divisor is not a whole number, move both decimal points to the right the same number of places until the divisor is a whole number.**
2. **Place the decimal point in the quotient above the decimal point in the dividend.**
3. **Divide as if both numbers were whole numbers.**
4. **Round off to the given place value. (If no round-off place is given, divide until the remainder is 0 or round as appropriate in the problem. For instance, in problems with money round to the nearest cent.)**

EXAMPLES

a.
```
     .23
 8)1.88
   1 6
     28
     24
      4
```
Here the remainder is not zero, so the division is not complete. We may insert a zero (1.880) without changing the value of the dividend, and continue dividing.

$$\begin{array}{r} .235 \\ 8\overline{)1.880} \\ \underline{1\ 6} \\ 28 \\ \underline{24} \\ 40 \\ \underline{40} \\ 0 \end{array}$$

Note that the quotient (.235) and the rewritten dividend (1.880) both have three decimal places.

Check:
$$\begin{array}{r} .235 \\ \underline{\times\ 8} \\ 1.880 \end{array}$$

b. $486.5 \div 23$ *Round to the nearest hundredth.*

$$\begin{array}{r} 21.152 \\ 23\overline{)486.500} \\ \underline{46} \\ 26 \\ \underline{23} \\ 35 \\ \underline{23} \\ 120 \\ \underline{115} \\ 50 \\ \underline{46} \\ 4 \end{array}$$

It is necessary to insert two zeros in order to round off to hundredths.

Hence, $486.5 \div 23 \approx 21.15.$

c. Find the quotient correct to the nearest thousandth.

$.072\overline{)\,.47891}$

$$\begin{array}{r} 6.6515 \\ 72\overline{)478.9100} \\ \underline{432} \\ 46\ 9 \\ \underline{43\ 2} \\ 3\ 71 \\ \underline{3\ 60} \\ 110 \\ \underline{72} \\ 380 \\ \underline{360} \\ 20 \end{array}$$

Multiply dividend and divisor by 1000. (As the arrows show, move the decimal point in each number three places to the right.)

Hence, $.47891 \div .072 \approx 6.652.$

d. Calculator example

$78.1936 \div 8.705 = ?$ (Round to the nearest thousandth.)

ENTER	DISPLAY
78.1936	78.1936
$\div$	78.1936
8.705	8.705
$=$	8.9826077

The quotient is 8.983 to the nearest thousandth.

APPLICATION SOLUTION

To find the number of gallons needed for the trip, divide.

```
              37.56
    39.4)1480.0 00
         1182
          298 0
          275 8
           22 2 0
           19 7 0
            2 5 00
            2 3 64
              1 36
```

So 37.6 gallons of gas are needed for the trip.

WARM UPS

Divide:

1. 6).24	**2.** 6)2.4	**3.** .6)24
4. .6).24	**5.** .06)2.4	**6.** .3)6.96
7. .5)1.2	**8.** .032 ÷ 4	**9.** 25)10
10. .25)1.5	**11.** 25)12.3	**12.** 4)2.95

Exercises 3.6

ANSWERS
1. _____
2. _____
3. _____
4. _____
5. _____
6. _____
7. _____
8. _____
9. _____
10. _____
11. _____
12. _____
13. _____
14. _____
15. _____
16. _____
17. _____
18. _____
19. _____
20. _____
21. _____
22. _____
23. _____
24. _____

A

Divide:

1. $5 \overline{)5.5}$ **2.** $7 \overline{)8.4}$ **3.** $4 \overline{).0016}$

4. $5 \overline{)0.0035}$ **5.** $.3 \overline{).339}$ **6.** $.7 \overline{)16.38}$

B

7. $80 \overline{)104.8}$ **8.** $20 \overline{)7}$ **9.** $200 \overline{)3.6}$

10. $16.64 \div 32$ **11.** $6.5 \div 125$ **12.** $64 \overline{)6211.84}$

13. $32 \overline{)201.824}$ **14.** $.41 \overline{).7093}$

C

Divide. Round answers to the nearest tenth.

15. $9 \overline{)2.202}$ **16.** $7.4 \overline{)17.7}$

Divide. Round answers to the nearest thousandth.

17. $1.09 \overline{)6.48}$ **18.** $2.2 \overline{)34.22}$ **19.** $24.9 \overline{)60.363}$

20. $131 \overline{)29}$

Divide. Round answer to nearest hundredth.

21. $.029 \overline{).226}$ **22.** $3.46 \overline{)56.99}$ **23.** $518 \overline{)3582}$

24. $72.9 \overline{)486.5}$

ANSWERS

25. _____

26. _____

27. _____

28. _____

29. _____

30. _____

31. _____

32. _____

33. _____

34. _____

35. _____

36. _____

37. _____

D

25. A mid-size car averages 26.3 miles per gallon for a long trip. How many gallons, to the nearest tenth, would be needed for a trip of 1480 miles?

26. A luxury car averages 19.4 miles per gallon for a long trip. How many gallons, to the nearest tenth, would be needed for a trip of 1480 miles?

27. Vern bought a pair of red socks. The socks were on sale at 3 pairs for $2.60. How much did he pay for the socks?

28. Ross drove 214.6 miles on 11.3 gallons of gas. What was his mileage (miles per gallon)? Round to the nearest mile.

29. Pat rides his motor bike 237.8 miles in 5 hours. What is his average speed? (Round to the nearest mile per hour.)

30. Eighty alumni of Tech U. donated $3560.72 to the University. What was the average donation?

31. A partly filled spool contains 1348 pounds of cable. If the cable weighs 2.3 pounds per foot, how many feet of cable remain on the spool? (Round to the nearest foot.)

32. If 19 dry-cell batteries connected in series (so that their voltages add) yield 28.5 volts, what is the yield of each battery?

E *Maintain Your Skills*

Add. Write answers as proper or improper fractions reduced to lowest terms.

33. $\dfrac{5}{8} + \dfrac{3}{4} + \dfrac{5}{6}$

34. $\dfrac{1}{5} + \dfrac{2}{3} + \dfrac{5}{6}$

35. $\dfrac{11}{15} + \dfrac{3}{4} + \dfrac{5}{8}$

36. $\dfrac{8}{39} + \dfrac{7}{72}$

37. $\dfrac{113}{120} + \dfrac{13}{18}$

3.7 Fractions to Decimals

OBJECTIVE

39. Change a fraction or mixed number either to an exact decimal or to an approximate decimal to the nearest tenth, hundredth, or thousandth.

APPLICATION

Chef Carl is in charge of food preparation at a banquet for several clubs. He has already decided to bake 25 loaves of bread when he is told that one club, about $\frac{1}{8}$ of the total, cannot come. What decimal $\left(\text{equal to } \frac{7}{8}\right)$ should he multiply by 25 to find how many loaves of bread will be needed now?

VOCABULARY

No new vocabulary.

HOW AND WHY

Recall that every fraction can be interpreted as the indicated division of two whole numbers, $\frac{a}{b} = a \div b$.

> To change a fraction to a decimal, divide the numerator by the denominator.
>   To change a mixed number to a decimal, change the fraction part to a decimal and add to the whole number part.
>   To find an approximate decimal name for a fraction, divide the numerator by the denominator and round off the result to the desired decimal place.
>   To change a mixed number to an approximate decimal, find the decimal approximation of the fraction part and add it to the whole number part.

EXAMPLES

a. $\frac{7}{20}$

$$
\begin{array}{r}
.35 \\
20\overline{)7.00} \\
6\ 0 \\
\hline
1\ 00 \\
1\ 00 \\
\hline
0
\end{array}
$$

$\frac{7}{20} = .35$

b. $\frac{9}{16}$

$$
\begin{array}{r}
.5625 \\
16\overline{)9.0000} \\
8\ 0 \\
\hline
1\ 00 \\
96 \\
\hline
40 \\
32 \\
\hline
80 \\
80 \\
\hline
0
\end{array}
$$

$\frac{9}{16} = .5625$

c. $\dfrac{7}{12}$ *If we divide 7 by 12 we can find the desired approximate decimal.*

$$\begin{array}{r} .583 \\ 12\overline{)7.000} \\ \underline{6\,0} \\ 1\,00 \\ \underline{96} \\ 40 \\ \underline{36} \\ 4 \end{array}$$

$\dfrac{7}{12}$ *to the nearest hundredth is .58*

d. $21\dfrac{3}{7} = 21 + \dfrac{3}{7}$

$$\begin{array}{r} .428 \\ 7\overline{)3.000} \\ \underline{2\,8} \\ 20 \\ \underline{14} \\ 60 \\ \underline{56} \\ 4 \end{array}$$

Therefore $\dfrac{3}{7} \approx .43$ *to the nearest hundredth.*

$21\dfrac{3}{7} = 21 + \dfrac{3}{7} \approx 21 + .43 = 21.43$

e. Calculator Example

$32\dfrac{49}{72}$ (To the nearest thousandth.)

ENTER	DISPLAY
49	49.
÷	49.
72	72.
=	0.6805556

So $32\dfrac{49}{72} \approx 32.681$ to the nearest thousandth.

APPLICATION SOLUTION

Chef Carl needs to change $\dfrac{7}{8}$ to a decimal. To do this divide.

$$\begin{array}{r} .875 \\ 8\overline{)7.000} \\ \underline{6\,4} \\ 60 \\ \underline{56} \\ 40 \\ \underline{40} \end{array}$$

So Chef Carl needs to multiply by .875.

WARM UPS

Write each of the following as a decimal.

1. $\dfrac{3}{5}$ **2.** $\dfrac{4}{5}$ **3.** $\dfrac{3}{4}$ **4.** $\dfrac{1}{4}$

5. $\dfrac{1}{2}$ **6.** $\dfrac{1}{8}$ **7.** $\dfrac{11}{20}$ **8.** $\dfrac{1}{20}$

9. $\dfrac{3}{8}$ **10.** $\dfrac{14}{20}$ **11.** $\dfrac{11}{40}$ **12.** $\dfrac{3}{16}$

Exercises 3.7

A

Write each of the following as a decimal.

1. $\dfrac{13}{20}$ 2. $\dfrac{17}{20}$ 3. $\dfrac{99}{125}$ 4. $\dfrac{7}{8}$

5. $3\dfrac{5}{8}$ 6. $\dfrac{77}{200}$

B

7. $\dfrac{59}{800}$ 8. $\dfrac{7}{625}$ 9. $\dfrac{837}{250}$ 10. $\dfrac{393}{400}$

Find the approximate decimal to the nearest tenth for each of the following.

11. $\dfrac{1}{3}$ 12. $\dfrac{5}{6}$ 13. $\dfrac{78}{81}$ 14. $\dfrac{28}{33}$

C

Find the approximate decimal to the nearest hundredth for each of the following.

15. $\dfrac{2}{7}$ 16. $\dfrac{18}{23}$ 17. $5\dfrac{8}{27}$ 18. $\dfrac{45}{67}$

19. $\dfrac{28}{33}$

Find the approximate decimal to the nearest thousandth for each of the following.

20. $\dfrac{15}{43}$ 21. $\dfrac{7}{9}$ 22. $\dfrac{7}{12}$ 23. $\dfrac{9}{11}$

24. $16\dfrac{63}{87}$

ANSWERS
1.
2.
3.
4.
5.
6.
7.
8.
9.
10.
11.
12.
13.
14.
15.
16.
17.
18.
19.
20.
21.
22.
23.
24.

151

25. _____

26. _____

27. _____

28. _____

29. _____

30. _____

31. _____

32. _____

33. _____

34. _____

35. _____

36. _____

37. _____

D

25. Chef Carl was told that $\frac{1}{32}$ of the guests at a banquet were vegetarians. What decimal number represents the part who were vegetarians?

26. The diameter of a wrist pin is $\frac{15}{16}$ inch. A micrometer measures in decimal units. What is the diameter of the pin read on a micrometer?

27. What reading, to the nearest thousandth, will show on a micrometer for a measure of $\frac{25}{32}$ inch?

28. Stock prices are listed in dollars and fractions of dollars. If a stock is selling for $\$6\frac{7}{8}$, write the price in decimal form.

29. In playing the game of craps, the mathematical probability that the person throwing the dice will win is $\frac{244}{495}$, and the probability he will lose is $\frac{251}{495}$. Express these fractions as approximate decimals to the nearest hundredth.

30. The chances of filling an inside straight in a single poker hand are 4 out of 47 $\left(\text{that is, } \frac{4}{47}\right)$. Express the chances as a decimal to the nearest hundredth.

31. The probability of throwing a six 3 times in a row with the toss of a die is $\frac{1}{216}$. Express this probability as a decimal to the nearest thousandth.

32. The probability of drawing three hearts from a deck of 52 playing cards is $\frac{11}{850}$. Express this probability as a decimal to the nearest thousandth.

E   *Maintain Your Skills*

Add. State all answers as mixed numbers where possible.

33. $5\frac{5}{8}$
    $4\frac{1}{2}$

34. $6\frac{8}{9}$
    $3\frac{2}{3}$
    $2\frac{1}{6}$

35. $8\frac{5}{12}$
    $1\frac{3}{4}$
    $3\frac{1}{3}$

36. $3\frac{3}{7}$
    $3\frac{4}{21}$
    $5\frac{2}{3}$
    $2\frac{5}{14}$

37. $2\frac{4}{9}$
    $7\frac{5}{6}$
    $1\frac{1}{3}$
    $4\frac{1}{2}$

ANSWERS TO WARM UPS (3.7)   **1.** .6   **2.** .8   **3.** .75   **4.** .25   **5.** .5   **6.** .125

7. .55   **8.** .05   **9.** .375   **10.** .7   **11.** .275   **12.** .1875

3.8 Review: Order of Operations

OBJECTIVE

40. Do any combination of operations (addition, subtraction, multiplication, and/or division) with decimals.

APPLICATION

A mechanic is assigned a service call that requires him to be out of town for four days. He is paid $.25 per mile and $75.00 a day for food and lodging while on service calls. At the beginning of his trip the odometer reading of his car is 24872.3 miles, and when he returns it is 25264.8 miles. How much expense money was he paid?

VOCABULARY

No new vocabulary.

HOW AND WHY

The conventional order of operations for decimals is the same as for whole numbers.

EXAMPLES

a. $.45 - (.32)(.75) = .45 - .24$ *Multiplication performed first*
$= .21$ *Subtraction performed next*

b. $.75 \div (.03)(1.6) = (25)(1.6)$ *Division performed first*
$= 40$ *Multiplication performed next*

c. Calculator Example

$2.55 \div 17 + (2.3)(4.5) + 1.37 = ?$

If your calculator has algebraic logic:

ENTER	DISPLAY
2.55	2.55
÷	2.55
17	17.
+	0.15
2.3	2.3
×	2.3
4.5	4.5
+	10.5
1.37	1.37
=	11.87

If your calculator is a four-function (basic), non-algebraic model:

	ENTER	DISPLAY	WRITE DOWN FOR LATER USE
	2.55	2.55	
	÷	2.55	
	17	17.	
	=	0.15	0.15
	AC	0.	
	2.3	2.3	
	×	2.3	
	4.5	4.5	
	=	10.35	10.35
	AC	0.	
	0.15	0.15	
	+	0.15	
	10.35	10.35	
	+	10.5	
	1.37	1.37	
	=	11.87	

The answer is 11.87.

APPLICATION SOLUTION

To find the expense money paid, we first set up a simpler word form of the problem.

Simpler Word Form: $\text{Expense money} = \left(\begin{array}{c}\text{No. of}\\\text{days}\end{array}\right)\left(\begin{array}{c}\text{Allowance}\\\text{per day}\end{array}\right)$

$+ \left(\begin{array}{c}\text{No. of}\\\text{miles}\end{array}\right)\left(\begin{array}{c}\text{Allowance}\\\text{per mile}\end{array}\right)$

Substitute: $E = (4)(75) + (25264.8 - 24872.3)(.25)$

Solve: $= (4)(75) + (392.5)(.25)$

$= 300 + 98.125$

$= 398.125$

Answer: The mechanic was paid $398.13.

WARM UPS

Do the indicated operations.

1. $(.3)(2) + .4$

2. $6.2 - 3(.2)$

3. $.6 \div .2 + 4(.2)$

4. $(.2)(.5) + .3(.7)$

5. $.30 \div .3 - .2$

6. $(.6)(.2) \div 2$

7. $.2(.4 - .2)$

8. $.1(.65 - .34)$

9. $.5 + .3(.2 - .1)$

10. $10 - 3(.8)$

11. $(.6)(.4) \div (.5)$

12. $(.18) \div (6)(.5)$

Exercises 3.8

A

Do the indicated operations.

1. $.45 \div 5 - .08$ **2.** $.35 \div .7 + 1.1$

3. $4.8 + 2.6 \div .13$ **4.** $6.9 - .56 \div .8$

5. $(1.05)(.75) + .18$ **6.** $.08 + 2.37 - 1.6 + .98$

B

7. $(6.75)(1.3) - 2.61 + (3.2)(.5)$ **8.** $4.97 \div (.07)(3.1)$

9. $1.56 - .216 \div .18$ **10.** $.0033 \div .88 + 1.075 - .0976$

11. $(.2)^2 - (.3)^3$ **12.** $(5.3 - 1.2)^2 - (2.1)^2$

13. $4.067 - (3.7)(.33) + 1.108$ **14.** $(7.86)(5.06) - 3.4 + 7.09$

C

15. $(2.1)(4.3)^2 - (6.8 + 5.03) + (3.4)^2$

16. $[5.1(6.3 - 3.5)] + (8.3)^2 \div .05$

17. $(2.06)(.13) - (1.17)(.08)$

18. $18.5 - [3.4 + (2.1)^2 \div 3]$

19. $4.33 - 2.75 \div 5$

20. $3.62 \div .02 + (8.6)(.51) - 82.6$

21. $4.8 \div 1.6 + .8(5.5) - 3.5$

22. $(1.26)(.3) \div .06 - 3.9 + 2.4 \div 16$

23. $3.6 \div .09(.3) - (9.6)(.16) \div 24$

24. $110 \div 5.5 - 1.5(3) \div 1.8 + .05$

1.
2.
3.
4.
5.
6.
7.
8.
9.
10.
11.
12.
13.
14.
15.
16.
17.
18.
19.
20.
21.
22.
23.
24.

ANSWERS

25. _____

26. _____

27. _____

28. _____

29. _____

30. _____

31. _____

32. _____

33. _____

34. _____

35. _____

D

25. An electronics firm pays its sales people a fee of $65 a day for lodging, $24.50 a day for meals, and $.30 a mile for transportation. How much should Jose report as expenses for his recent trip of 6 days in which his odometer went from 73586.6 to 75443.8 miles?

26. On another trip the salesman in exercise 25 was gone for eleven days and his odometer went from 81004.6 to 83751.9 miles. What expenses should he report?

27. An electrician is paid $9.68 per hour and time and a half for any hours over 40 hours in a given week. What was his gross pay if he worked the following hours in one week: 6.5, 10.75, 8.5, 12, 9.25?

28. A welder is paid $12.32 per hour and time and a half for any hours over 40 in any one week. What is his gross pay if he worked the following hours in one week: 8, 9.5, 8.75, 10, 9.25, and 4.5?

29. A house is valued at $48,500. The annual rate for fire insurance is $3.42 per thousand. What is the cost of a five year policy if the charge is 4.2 times the annual rate?

30. Ten years ago a land developer purchased a piece of land that contained 57.04 acres for $85,560. He now plans to divide it into lots of .62 acre each. If he sells each lot for $15,000, what is the difference between the original purchase price and the total selling price?

E   *Maintain Your Skills*

Subtract. Reduce answers to lowest terms.

31. $15\frac{5}{8} - 3\frac{1}{3}$

32. $25\frac{9}{10} - 12\frac{14}{15}$

33. $37\frac{1}{2}$

$28\frac{2}{3}$

34. $65\frac{5}{6}$

$45\frac{7}{8}$

35. 13

$8\frac{3}{4}$

1. _____

2. _____

3. _____

4. _____

5. _____

6. _____

7. _____

8. _____

9. _____

10. _____

11. _____

12. _____

13. _____

14. _____

15. _____

16. _____

17. _____

18. _____

19. _____

20. _____

21. _____

22. _____

23. _____

24. _____

25. _____

Chapter 3
POST-TEST

1. (Obj. 37) *Multiply:* 2.3
4.6

2. (Obj. 36) *Subtract:* 7.3862
7.2986

3. (Obj. 36) *Subtract:* .1572 − .0213

4. (Obj. 37) *Multiply:* (4.2)(8.3)(.614)

5. (Obj. 3) List the following in order of value from smallest to largest.

.762, .0763, .7062, .0702

6. (Obj. 36) *Subtract:* 15 − 2.14

7. (Obj. 38) *Divide:* Round answer to nearest tenth. $2.3\overline{)16.219}$

8. (Obj. 35) *Add:* 2.34 + .582 + 15.21 + 31 + 2.6

9. (Obj. 39) Write $\frac{3}{7}$ as an approximate decimal rounded to the nearest thousandth.

10. (Obj. 38) *Divide:* $7\overline{).861}$

11. (Obj. 40) Do the indicated operations. 15.36 ÷ .04 − (.27)(1.8)

12. (Obj. 38) *Divide:* Round the answer to the nearest hundredth. $.25\overline{).2731}$

13. (Obj. 33) List the following in order of value from smallest to largest.

.0043, .00425, .004, .00431

14. (Obj. 38) *Divide:* $.21\overline{).6342}$

15. (Obj. 40) Do the indicated operations. 7.82 − (1.8)(2.05) + 6.214

16. (Obj. 35) *Add:* 13.264
.8132
7.1431
.021

17. (Obj. 39) Write $\frac{5}{8}$ as a decimal.

18. (Obj. 35) *Add:* .612
2.145
3
4.1

19. (Obj. 37) *Multiply:* 6.234
.025

20. (Obj. 40) Do the indicated operations. (2.3)(1.4) − .62

21. (Obj. 35) Find the perimeter (distance around) of a five-sided building lot that has the following measurements: 150.6 m, 81.34 m, 79.6 m, 72.93 m, and 74.9 m.

22. (Obj. 38) A copper wire has a diameter of .125 inch. How many turns of the wire can be wound on a metal core that is 4.5 inches long?

23. (Obj. 37) Norm's pay rate is $4.24 per hour. If he works 7.5 hours per day, what is his salary per day? How much would he make in a five-day week?

24. (Obj. 36) A piece of metal rod 3.375 inches long is cut from a rod that is 29 inches long. Allowing .0625 inch for waste, what is the length of the remaining piece?

25. (Obj. 40) What is the total height of a pile that has 25 sheets of metal that are each .045 inch thick and 42 sheets that are each .025 inch thick?

ANSWERS TO WARM UPS (3.8)   **1.** 1.0   **2.** 5.6   **3.** 3.8   **4.** .31   **5.** .8   **6.** .06

7. .04   **8.** .031   **9.** .53   **10.** 7.6   **11.** .48   **12.** .015

4

Measurement

1. _____

2. _____

3. _____

4. _____

5. _____

6. _____

7. _____

8. _____

9. _____

10. _____

11. _____

12. _____

13. _____

14. _____

15. _____

16. _____

Chapter 4
PRE-TEST

The problems in the following pre-test are a sample of the material in the chapter. You may already know how to work some of these. If so, this will allow you to spend less time on those parts. As a result, you will have more time to give to the sections that you find more difficult or that are new to you. The answers are in the back of the text.

1. **(Obj. 41)** 9.5 yd = ? feet

2. **(Obj. 41)** 2.5 gal = ? pints

3. **(Obj. 41)** 4 lb 3 oz = ? ounces

4. **(Obj. 42)** *Add:* 6 gal 3 qt and 2 gal 3 qt

5. **(Obj. 42)** *Subtract:* 2 yd 2 ft 5 in and 2 ft 11 in

6. **(Obj. 43)** 7844 g = ? kg

7. **(Obj. 43)** 32.77 km = ? cm

8. **(Obj. 43)** 3475 mℓ = ? ℓ

9. **(Obj. 44)** *Add:* 75 g 2 dg 7 cg and 54 g 7 dg 6 cg

10. **(Obj. 44)** *Subtract:* 32 kℓ 730 ℓ and 17 kℓ 820 ℓ

11. **(Obj. 45)** Convert 17.5 cents per minute to dollars per hour.

12. **(Obj. 45)** Convert 85 kilometres per hour to metres per second (to the nearest tenth of a metre).

13. **(Obj. 45)** A passenger jet flies at a speed of 500 miles per hour. How many feet per second is this, to the nearest foot per second?

14. **(Obj. 46)** Evaluate the following formula if $r = 4$, $h = 9$, and $\pi \approx 3.14$:

$$V = \frac{1}{3} \pi r^2 h$$

15. **(Obj. 46)** Evaluate the following formula if $n = 50$, $a = 40$, and $\ell = 250$:

$$S = \frac{1}{2} n(a + \ell)$$

16. **(Obj. 47)** Find the circumference of a circle whose diameter is 24 metres. (Let $\pi \approx 3.14$.)

17. _____

18. _____

19. _____

20. _____

21. _____

22. _____

23. _____

24. _____

25. _____

17. (Obj. 47) Find the perimeter of the following figure. (Let $\pi \approx 3.14$.)

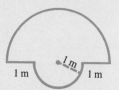

3.6 m

8.4 m

18. (Obj. 48) Find the area of a triangle whose base is 13 feet and whose height is 7 feet.

19. (Obj. 48) Find the area of a circle whose diameter is 18 yards. (Let $\pi \approx 3.14$.)

20. (Obj. 49) Find the area of the following figure.

1 m   _1 m_   1 m

21. (Obj. 49) Find the area of the shaded portion of the following figure. (Let $\pi \approx 3.14$.)

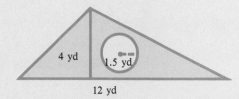

4 yd   1.5 yd

12 yd

22. (Obj. 50) Find the volume of a sphere whose radius is 3.2 metres, to the nearest tenth. (Let $\pi \approx 3.14$.)

23. (Obj. 50) Find the volume of a pyramid whose base is a rectangle with sides 125 metres and 75 metres and whose height is 25 metres.

24. (Obj. 51) What is the total surface area of a right circular cone whose radius is 10 feet and whose slant height is 7 feet?

25. (Obj. 51) A tent, which will sleep six, is in the form of a rectangular solid with a pyramid top. How many square feet of canvas are needed to make the tent (with floor), given the following dimensions? The floor is 20 feet by 18 feet, the tent walls are 5.5 feet high, and the slant height from the top of the wall to the peak is 10.3 feet.

4.1 English Measurements

41. Express an English measurement in an equivalent English measure when the relationship between the two units of measure is known.

42. Express in specified units the sum or difference of English measurements.

APPLICATION

The Corner Grocery sold 20 lb 6 oz of hamburger on Wednesday, 13 lb 8 oz on Thursday, and 17 lb 10 oz on Friday. How much hamburger was sold during the three days?

VOCABULARY

A *measurement* is a symbol formed by using a number along with a unit of measure to indicate "how many" or "how much of a measurable quantity." 5 miles is a measurement. *English measurements* are measurements whose unit of measure has its origin in the English-speaking nations. Examples of English measures are: weight, 1 pound; length, 1 foot, volume, 1 quart.

Equivalent measurements are different measures of the same quantity or object using different units of measure (for example, 12 inches = 1 foot).

The *measure* of a measurement is the numeral without its label. "5" is the measure of "5 miles."

HOW AND WHY

English measures and their equivalents that you should know are listed in Table 4–1. Using the equivalent measures in this table allows us to express equivalent measurements. Consider converting 4 feet to inches:

$$4 \text{ feet} = 4 \cdot (1 \text{ foot})$$
$$= 4 \cdot (12 \text{ inches})$$
$$= 48 \text{ inches}$$

TABLE 4–1

LENGTH	TIME
12 inches (in) = 1 foot (ft)	60 seconds (sec) = 1 minute (min)
3 feet (ft) = 1 yard (yd)	60 minutes (min) = 1 hour (hr)
5280 feet (ft) = 1 mile (mi)	24 hours (hr) = 1 day
	7 days = 1 week

LIQUID	WEIGHT
3 teaspoons (tsp) = 1 tablespoon (tbs)	16 ounces (oz) = 1 pound (lb)
2 cups (c) = 1 pint (pt)	2000 pounds (lb) = 1 ton
2 pints (pt) = 1 quart (qt)	
4 quarts (qt) = 1 gallon (gal)	

To express a measurement in a new measure, replace the unit of measure by an equivalent measure and find the indicated product.

When 15 cents is added to 2 dollars, the sum is neither 17 cents nor 17 dollars. First we convert 2 dollars to cents:

2 dollars = 2 (100 cents) = 200 cents

Then we can add 15 cents and 200 cents and find the total measure to be 215 cents.

163

> The sum (or difference) of two measurements with common units of measure is the sum (difference) of their *measures* followed by the unit of measure.

EXAMPLES

a. $13\frac{3}{4}$ minutes expressed in seconds is

$$13\frac{3}{4} \text{ (1 minute)} = 13\frac{3}{4} \text{ (60 seconds)} = 825 \text{ seconds.}$$

b. 6 feet + 7 feet + 11 feet = 24 feet.

c. *Add:*   3 gal 2 qt
   5 gal 3 qt
   8 gal 5 qt = 8 gal + 4 qt + 1 qt
      = 8 gal + 1 gal + 1 qt
      = 9 gal 1 qt

d. If a carpenter cuts off a board of length 2 ft 5 in from a board of length 8 ft 3 in, how much board is left?

8 ft   3 in   *Since 5 cannot be subtracted from 3, borrow 1 ft from the 8 ft*
2 ft   5 in

7 ft 15 in   *(1 ft = 12 in)*
2 ft   5 in
5 ft 10 in

APPLICATION SOLUTION

To find out how much hamburger was sold in the three days, add.

20 lb   6 oz
13 lb   8 oz
17 lb 10 oz
50 lb 24 oz = 50 lb + 1 lb + 8 oz
      = 51 lb 8 oz

So the Corner Grocery sold 51 lb 8 oz of hamburger in the three days.

WARM UPS

Find the missing number.

1. 2 ft = ? in

2. 2 yd = ? ft

3. 2 cups = ? pt

4. 2 pt = ? cups

5. 4 qt = ? gal

6. 2 qt = ? pt

7. $1\frac{1}{2}$ days = ? hours

8. $1\frac{1}{2}$ hours = ? minutes

9. $2\frac{1}{4}$ lb = ? oz

10. $3\frac{1}{2}$ tons = ? lb

Add:

11. 3 days 20 hr
   2 days 14 hr

12. 1 qt 1 pt
   1 qt 1 pt

Exercises 4.1

A

Find the missing number.

1. 5 ft = ? in   **2.** 7 yd = ? ft

3. 32 pt = ? qt   **4.** 2 hours = ? minutes

Add:

5. 4 hours 40 minutes   **6.** 1 ft 10 in
      1 hour 25 minutes   8 in

B

Find the missing number.

7. 3.5 gal = ? qt   **8.** 2 miles = ? yd

9. $4\frac{3}{16}$ lb = ? ounces   **10.** 6.5 yd = ? inches

11. 2.13 tons = ? pounds   **12.** 2.5 hr = ? seconds

Add:

13. 2 hr 20 min 35 sec   **14.** 3 yd 2 ft 8 in
       3 hr 52 min 41 sec   1 yd 1 ft 9 in

C

Find the missing number.

15. $89\frac{2}{3}$ yd = ? ft   **16.** 4 ft 9 in = ? inches

17. 7 yd 2 ft = ? inches   **18.** 40 minutes = ? seconds

Add:

19. 2 ft 5 in   **20.**   6 ft 5 in
       6 ft 4 in   2 ft 7 in
                                      10 ft 7 in

21. 9 gal 3 qt 1 pt   **22.**   5 lb 11 oz
       4 gal 2 qt 1 pt   11 lb 9 oz
       3 gal 1 qt 1 pt

ANSWERS

1. _____
2. _____
3. _____
4. _____
5. _____
6. _____
7. _____
8. _____
9. _____
10. _____
11. _____
12. _____
13. _____
14. _____
15. _____
16. _____
17. _____
18. _____
19. _____
20. _____
21. _____
22. _____

23. _____

24. _____

25. _____

26. _____

27. _____

28. _____

29. _____

30. _____

31. _____

32. _____

33. _____

34. _____

35. _____

Subtract:

23. 6 yd 2 ft
    3 yd 1 ft

24. 21 min 39 sec
    14 min 47 sec

D

25. The Generous Beef Distributors delivered 240 lb 12 oz of meat to one fast-food chain, 433 lb 10 oz to a second chain, and 353 lb 2 oz to a third chain. How many pounds of meat were delivered to the three food chains?

26. At a recent Old Timers relay race, the four-person team turned in times of 4 min 12 sec, 4 min 35 sec, 5 min 10 sec, and 5 min 11 sec. What was the total time of the relay team?

27. If a car travels 55 miles in one hour, how far will it travel in 7 hours at the same speed?

28. A buyer purchased 52 quarts of a rare wine for four clients. If each client is to share equally, how many quarts of wine will they each get?

29. Paul pledged 7 dollars a month to the United Fund. What was his annual donation?

30. Carol bought one gallon of milk at the local store. When she got home, Dan drank 3 cups of milk, Larry drank $1\frac{1}{2}$ cups, Greg drank 2 cups, and Carol used $\frac{1}{2}$ cup in cooking. How much milk remained at the end of the day?

E   *Maintain Your Skills*

Write the word name for each of the following.

31. 2.0005   **32.** 234.87952   **33.** .0003876

Write the numeral form of each of the following.

34. Thirteen and four hundred thirty-two ten-thousandths.

35. Four thousand three hundred thirty-five millionths

ANSWERS TO WARM UPS (4.1) **1.** 24 **2.** 6 **3.** 1 **4.** 4 **5.** 1 **6.** 4

7. 36 **8.** 90 **9.** 36 **10.** 7000 **11.** 6 days 10 hr **12.** 3 qt

4.2 Metric Measurements

OBJECTIVES

43. Express a metric measurement in an equivalent metric measure when the relationship between the two units of measure is known.
44. Express in specified units the sum or difference of metric measurements.

APPLICATION

Paul used a metre stick to measure the distance from his receiver to the place where he wants to put his stereo speakers. He found that each speaker will be 1 metre and 15 centimetres from the receiver. How many metres of wire does he need to connect the two speakers?

VOCABULARY

Metric measurements are expressed in basic units developed in France almost 200 years ago. Most nations today use the metric system. In the United States, scientists and an increasing number of industries are using the metric system.

The standard units in the metric system are

length: 1 metre or 1 m

weight: 1 gram or 1 g

volume: 1 litre or 1 ℓ

HOW AND WHY

Every measuring system has one or more standard units of length, weight, volume, and other quantities. The metric system was invented to take advantage of our base ten place value system in the same way that our monetary system does. (See Table 4–2.)

TABLE 4–2

	1 mil	= .001 dollar
10 mils	= 1 cent	= .01 dollar
10 cents	= 1 dime	= .1 dollar
10 dimes	= 1 dollar	= 1 dollar

For most purposes, the units of metre, gram, and litre along with their multiples are sufficient. Some of the metric measures are shown in Table 4–3, which should be memorized.

TABLE 4-3

LENGTH (basic unit is 1 m)

1 millimetre (mm) =		.001 m
1 centimetre (cm)	= 10 millimetres =	.01 m
1 decimetre (dm)	= 10 centimetres =	.1 m
1 METRE (m)	= 10 decimetres =	1 m
1 dekametre (dam)	= 10 metres =	10 m
1 hectometre (hm)	= 10 dekametres =	100 m
1 kilometre (km)	= 10 hectometres =	1000 m

WEIGHT (basic unit is 1 g)

1 milligram (mg) =		.001 g
1 centigram (cg)	= 10 milligrams =	.01 g
1 decigram (dg)	= 10 centigrams =	.1 g
1 GRAM (g)	= 10 decigrams =	1 g
1 dekagram (dag)	= 10 grams =	10 g
1 hectogram (hg)	= 10 dekagrams =	100 g
1 kilogram (kg)	= 10 hectograms =	1000 g
1000 kg = 1 metric ton		

LIQUID AND DRY MEASURE (basic unit is 1 ℓ)

1 millilitre (mℓ) =		.001 ℓ
1 centilitre (cℓ)	= 10 millilitres =	.01 ℓ
1 decilitre (dℓ)	= 10 centilitres =	.1 ℓ
1 LITRE (ℓ)	= 10 decilitres =	1 ℓ
1 dekalitre (daℓ)	= 10 litres =	10 ℓ
1 hectolitre (hℓ)	= 10 dekalitres =	100 ℓ
1 kilolitre (kℓ)	= 10 hectolitres =	1000 ℓ

Since the metric system takes advantage of the base ten place value system, the conversion of units can take place by moving the decimal point. The following listing of the prefixes and the base unit will help.

```
                b
                a
                s
                e
k   h   da      d   c   m
                u
                n
                i
                t
```

To convert to a new metric measure, move the decimal point the same number of places and in the same direction as the number of places and direction it takes you to go from the original prefix to the new one on the above chart.

For instance,

23 hg = ? dg

The d prefix is three places to the right of the h prefix, so move the decimal point three places to the right.

```
                base
k   h   da   unit   d   c   m
2   3    0    0    0.
```

23 hg = 23000 dg

Also,

56 mℓ = ? kℓ

56 mℓ = .000056 kℓ

The k prefix is six places to the left of the m prefix, so move the decimal point six places to the left.

EXAMPLES

a. 50 dm = ? cm   *Since c is one space to the right of d.*
    50 dm = 500 cm

b. How many metres are in .4 kilometres?

    .4 km = 400 m   *Since m is three spaces to the right of k.*

c. What is the measure of 8 metres + 45 cm in metres?

8 metres + 45 cm = 8 m + .45 m   *Since m is two spaces to the left of c.*
            = 8.45 m

d. What is the measure of 1.3 litres − 135 millilitres in millilitres?

1.3 litres − 135 millilitres = 1300 mℓ − 135 mℓ *Since m is three spaces to the right of the base unit.*
            = 1165 mℓ

APPLICATION SOLUTION

To find out how many metres of wire Paul will need to connect the speakers, add the two lengths and then convert to metres.

1 m 15 cm
1 m 15 cm
2 m 30 cm = 2 m + .3 m = 2.3 m

So Paul will need 2.3 m of wire.

WARM UPS

Find the missing number.

1. 100 mils = ? cents   **2.** 100 mm = ? cm

3. 100 mg = ? g   **4.** 100 mℓ = ? cℓ

5. .5 km = ? m   **6.** .5 kg = ? g

7. 2 km = ? m   **8.** 3 kg = ? g

Add:

9. 3 cm 2 mm + 4 cm 9 mm

10. 5 cℓ 7 mℓ + 3 cℓ 8 mℓ

11. 2 m 95 cm + 3 m 10 cm

12. 7 kg 80 g + 2 kg 60 g

Exercises 4.2

A

Find the missing number.

1. 5 kℓ = ? ℓ   **2.** 50 g = ? mg

3. 1.3 km = ? m   **4.** 1.3 kg = ? g

Add:

5. 3 g 6 dg + 3 g 5 dg   **6.** 3 m 7 dm + 1 m 6 dm

B

Find the missing number.

7. 1.3 kℓ = ? ℓ   **8.** 244 mm = ? m

9. 245 mℓ = ? ℓ   **10.** 246 mg = ? g

11. 8 g 6 dg = ? dg   **12.** 5 ℓ 3 mℓ = ? mℓ

Add:

13. 4 cℓ 8 mℓ + 5 cℓ 3 mℓ = ? mℓ   **14.** 5 g 3 dg + 3 g 4 dg = ? dg

C

Find the missing number.

15. .4 km + 22 m = ? m   **16.** 2 kg + 45 hg = ? g

17. 2.6 km − 1900 m = ? m   **18.** 17 ℓ − 444 mℓ = ? mℓ

Add:

19. 3 km 250 m
   3 km 900 m
        = ? m

20. 2 g 8 dg
   14 g 5 dg
        = ? dg

21. 4 kℓ 8 ℓ
   3 kℓ 5 ℓ
        = ? ℓ

ANSWERS	
1.	
2.	
3.	
4.	
5.	
6.	
7.	
8.	
9.	
10.	
11.	
12.	
13.	
14.	
15.	
16.	
17.	
18.	
19.	
20.	
21.	

ANSWERS

22. _____

23. _____

24. _____

25. _____

26. _____

27. _____

28. _____

29. _____

30. _____

31. _____

32. _____

33. _____

34. _____

35. _____

Subtract:

22. 3 kℓ 5 hℓ
  1 kℓ 7hℓ
    = ? hℓ

23. 15 km 700 m
   3 km 850 m
    = ? m

24. 9 g 7 dg
  5 g 8 dg
    = ? dg

___

D

25. Cary wants to connect four speakers to her stereo. Two speakers are 3 m 6 cm from the stereo and the other two are 4 m 5 cm away. How many metres of wire should she buy?

26. Steve is going to string lights for his annual barbecue. One light is to be placed 11 m 7 dm from the outlet, a second light will be placed 15 m 8 dm 6 cm away, and the third one will be 21 m 8 dm away. How many metres of wire should he buy?

27. The internist at St. Vincent's Hospital ordered three 0.25 mg tablets of reserpine for a patient. How many milligrams of reserpine did the patient receive?

28. If a can contains 298 grams of soup, how many grams of soup are contained in 7 cans?

29. Tim, who is a lab assistant, has 282 mℓ of acid that is to be divided among 24 students. How many mℓ will each student receive?

30. Gayle purchased a package of ground beef that cost $1.43 and weighed .5 kg. What was the price per kilogram?

___

E   *Maintain Your Skills*

List the following decimals from smallest to largest.

31. 2.345, 0.485, 0.0572, 0.0099

32. 0.053, 4.9, 0.09, 0.11

33. 2.04, 2.4, 2.004, 0.24

34. 0.34, 1.03, 1.2, .099

35. 1.2234, 1.2233, 1.2244, 1.2345

ANSWERS TO WARM UPS (4.2)

1. 10	**2.** 10	**3.** .1	**4.** 10	**5.** 500
6. 500	**7.** 2000	**8.** 3000	**9.** 8 cm 1 mm	**10.** 9 cℓ 5 mℓ
11. 6 m 5 cm	**12.** 9 kg 140 g or 9 kg 1 hg 4 dag			

4.3 Conversion of Units

OBJECTIVE

45. **Convert a given measurement to an equivalent measurement with specified units.**

APPLICATION

The Concorde can fly at a speed of 3240 ft/sec, which is approximately three times the speed of sound. How many miles per hour does the Concorde fly? (To the nearest mile per hour.)

VOCABULARY

Recall that equivalent measurements measure the same quantity but involve different basic units of measure. For instance, 6 feet and 2 yards are equivalent measurements.

HOW AND WHY

Since 12 inches = 1 foot, these two are equivalent measurements. If we consider the indicated division, (12 inches) ÷ (1 foot), and think of these as numbers, the division asks "how many" units of measure 1 foot it will take to make 12 inches. Since they measure the same length, the answer is 1.

$$\frac{12 \text{ inches}}{1 \text{ foot}} = 1$$

The indicated division of two equivalent measurements can always be thought of as a name for 1. This concept, along with the multiplication property of one, will be used to convert from one measure to its equivalent in another measure. Consider converting 48 inches to feet.

48 inches = (48 inches)(1)

$$= 48 \text{ inches} \cdot \frac{1 \text{ foot}}{12 \text{ inches}}$$

$$= \frac{48 \cdot 1 \text{ inch} \cdot 1 \text{ foot}}{12 \cdot 1 \text{ inch}}$$

$$= \frac{48 \cdot 1 \text{ foot}}{12} \cdot \frac{1 \text{ inch}}{1 \text{ inch}}$$

$$= 4 \cdot 1 \text{ foot} \cdot 1$$

$$= 4 \text{ feet}$$

Recall that, in working with fractions, $\frac{ab}{ac} = \frac{b}{c}$. Using this concept, the details of the problem can be simplified.

$$48 \text{ inches} = \frac{48 \text{ inches}}{1} \cdot \frac{1 \text{ foot}}{12 \text{ inches}}$$

$$= \frac{48}{12} \text{ feet}$$

$$= 4 \text{ feet}$$

In some cases it may be necessary to multiply by several different names of one. This is how we convert 7200 seconds to hours:

$$7200 \text{ seconds} = \frac{7200 \text{ seconds}}{1} \cdot \frac{1 \text{ minute}}{60 \text{ seconds}} \cdot \frac{1 \text{ hour}}{60 \text{ minutes}}$$

$$= \frac{7200}{3600} \text{ hours}$$

$$= 2 \text{ hours}$$

To convert units of measurement,

1. **Multiply the measurement by fractions formed by equivalent measurements (names for one) to get the required units.**
2. **Multiply and reduce.**

EXAMPLES

a. Convert 4 gallons to pints.

$$4 \text{ gallons} = \frac{4 \text{ gallons}}{1} \cdot \frac{4 \text{ quarts}}{1 \text{ gallon}} \cdot \frac{2 \text{ pints}}{1 \text{ quart}}$$

$$= 4 \cdot 4 \cdot 2 \text{ pints}$$

$$= 32 \text{ pints}$$

b. Convert 3 ft² to in². (See page 193 for an explanation of the units in this example.) Since 1 ft² = (1 foot) (1foot) and 1 in² = (1 inch) (1 inch), we proceed as follows:

$$3 \text{ ft}^2 = 3 \cdot 1 \text{ foot} \cdot 1 \text{ foot} \cdot \frac{12 \text{ inches}}{1 \text{ foot}} \cdot \frac{12 \text{ inches}}{1 \text{ foot}}$$

$$= 3 \cdot 12 \cdot 12 \text{ in}^2$$

$$= 432 \text{ in}^2$$

c. Convert 60 miles per hour to feet per second.

$$60 \text{ miles per hour} = \frac{60 \text{ miles}}{1 \text{ hour}} \cdot \frac{1 \text{ hour}}{60 \text{ min}} \cdot \frac{1 \text{ min}}{60 \text{ sec}} \cdot \frac{5280 \text{ ft}}{1 \text{ mile}}$$

$$= \frac{60 \cdot 5280}{60 \cdot 60} \frac{\text{ft}}{\text{sec}}$$

$$= 88 \text{ feet per second}$$

60 miles per hour is equivalent to 88 feet per second.

d. Convert 48 g per square metre to g per square centimetre.

$$\frac{48 \text{ g}}{\text{m}^2} = \frac{48 \text{ g}}{\text{m}^2} \cdot \frac{.1 \text{ m}}{\text{cm}} \cdot \frac{.1 \text{ m}}{\text{cm}}$$

$$= (48)(.1)(.1) \frac{\text{g}}{\text{cm}^2}$$

$$= .48 \frac{\text{g}}{\text{cm}^2}$$

APPLICATION SOLUTION

To convert the speed of the Concorde to miles per hour, multiply by the names of one as shown.

$$\frac{3240 \text{ ft}}{\text{sec}} = \frac{3240 \text{ ft}}{\text{sec}} \cdot \frac{1 \text{ mile}}{5280 \text{ ft}} \cdot \frac{60 \text{ sec}}{\text{min}} \cdot \frac{60 \text{ min}}{\text{hr}}$$

$$\approx 2209.09 \frac{\text{mi}}{\text{hr}}$$

So the Concorde flies at approximately 2209 mph.

WARM UPS

Convert each of the following to the indicated unit of measure.

1. 30 inches = ? feet

2. $1\frac{1}{4}$ ft = ? in

3. 15 ft = ? yd

4. 270 minutes = ? hours

5. 240 sec = ? minutes

6. 3600 sec = ? hr

7. 11 cups = ? quarts

8. 15 pints = ? quarts

9. 3 ft 9 in = ? in

10. 4 min 15 sec = ? sec

11. $\frac{60 \text{ mi}}{\text{hr}} = ? \frac{\text{mi}}{\text{min}}$

12. $\frac{120 \text{ ft}}{\text{sec}} = ? \frac{\text{yd}}{\text{sec}}$

Exercises 4.3

A

Convert each of the following to the indicated unit of measure.

1. 42 inches = ? feet

2. $1\frac{1}{3}$ feet = ? inches

3. 13 feet = ? yards

4. 300 min = ? hr

5. 6000 lb = ? tons

6. $2\frac{1}{2}$ gal = ? qt

B

7. 5 min 36 sec = ? sec

8. 3 ft 9 in = ? in

9. 7 km 528 m = ? km

10. 7 miles 528 feet = ? miles

11. 1 mile to inches

12. 10,080 minutes to days

13. $\dfrac{44 \text{ feet}}{\text{second}}$ to $\dfrac{\text{miles}}{\text{hour}}$

14. $\dfrac{84 \text{ m}}{\text{min}}$ to $\dfrac{\text{m}}{\text{sec}}$

C

15. 5 yd 2 ft 7 in to inches

16. 4 hr 56 min 10 sec to sec

17. 75 min to hours

18. 12 ft 9 inches to yards

19. $\dfrac{9 \text{ tons}}{\text{ft}}$ to $\dfrac{\text{pounds}}{\text{in}}$

20. $\dfrac{10 \text{ ounces}}{\text{cup}}$ to $\dfrac{\text{pounds}}{\text{gallon}}$

21. $\dfrac{144 \text{ kilometres}}{\text{hour}}$ to $\dfrac{\text{metres}}{\text{second}}$

22. $\dfrac{36 \text{ km}}{\text{hour}}$ to $\dfrac{\text{cm}}{\text{sec}}$

23. 3540 cm² to m²

24. 153 ft² to yd²

177

ANSWERS

1. _____

2. _____

3. _____

4. _____

5. _____

6. _____

7. _____

8. _____

9. _____

10. _____

11. _____

12. _____

13. _____

14. _____

15. _____

16. _____

17. _____

18. _____

19. _____

20. _____

21. _____

22. _____

23. _____

24. _____

ANSWERS

25. _____

26. _____

27. _____

28. _____

29. _____

30. _____

31. _____

32. _____

33. _____

34. _____

35. _____

36. _____

37. _____

D

25. Shirley is going on a diet that will cause her to lose 4 ounces every day. At this rate, how many pounds will she lose in 6 weeks?

26. The local candy manufacturer packs 30 pieces of candy in a box. To ship these to market he puts 40 boxes in a case. How many pieces of candy are in a shipment of 27 cases?

27. Carol has a recipe that calls for 12 cubic centimetres of milk. She has only a litre measure. How much of a litre should she add? (1 cubic centimetre = 1 millilitre)

28. If Dan averages 50 miles per hour during an eight hour day of driving, how many days will it take him to drive 2000 miles?

29. Greg's Sport Shop decides to donate to charity 1 cent for every yard Joanne jogs during one week. Joanne jogs 2 miles every day. How much does Greg donate to charity?

30. If water weighs 64 pounds per ft^3, how many ounces (to the nearest tenth) does one in^3 weigh? Hint: 1 ft^3 = (1 foot)(1 foot)(1 foot) and 1 in^3 = (1 inch)(1 inch)(1 inch).

31. Mr. Smith's car averages 12.5 miles per gallon of gasoline. How many gallons of gas are needed to make a trip of 412 miles?

32. If a secretary can type 60 words per minute and the average page contains 500 words, how many pages can she type in 4 hours?

E   *Maintain Your Skills*

Round each of the following to the nearest thousandth.

33. 0.234723   **34.** 2345.871173   **35.** 12.89787777

Round each of the following to the nearest hundred.

36. 2345   **37.** 1234557

ANSWERS TO WARM UPS (4.3)

1. 2.5 ft	**2.** 15 in	**3.** 5 yd	**4.** 4.5 hr	**5.** 4 min
6. 1 hr	**7.** 2.75 qt	**8.** 7.5 qt	**9.** 45 in	**10.** 255 sec
11. 1 mph	**12.** $40\dfrac{\text{yd}}{\text{sec}}$			

4.4 Evaluating Expressions and Formulas

OBJECTIVE

46. Evaluate an algebraic expression or formula.

APPLICATION

An outdoor stage will be constructed of bricks with a concrete base. If it will be circular with a 30 ft diameter, how many square feet of stage area will be available? ($A = \pi r^2$, $\pi \approx 3.14$.)

VOCABULARY

An *algebraic expression* consists of numerals, operation signs (for addition, subtraction, multiplication, and division) and letters of the alphabet used as placeholders for numbers. Algebraic expressions do not contain equal signs (they are not equations.) The placeholders (letters) are called *variables* or *unknowns*. When the variables have been replaced by numbers (substitution), the expression can be *evaluated* or *simplified*. *Formulas* consist of two algebraic expressions separated by an equal sign (=). (Since formulas contain equal signs, they are also equations.)

HOW AND WHY

The order in which the operations should be carried out is the same as in Chapter 1. Multiplication signs between variables or between numerals and variables are usually omitted. For instance: $2t$ means $2 \cdot t$ or $2(t)$.

$3 + 2t$ has the value 15 when t is replaced by 6 since

$$3 + 2 \cdot 6 = 3 + 12 = 15$$

The formula for the perimeter of a rectangle is

$$p = 2\ell + 2w$$

Therefore, a rectangle whose length is 13 cm and whose width is 7 cm has a perimeter of

$$p = 2\ell + 2w = 2 \cdot 13 + 2 \cdot 7$$
$$= 26 + 14$$
$$= 40 \text{ cm}$$

To evaluate an expression or formula, replace each variable by its given value and perform the indicated operation(s).

EXAMPLES

Evaluate the following expressions given $a = 2$, $b = 3$, $x = 5$, and $z = 0$.

a. $x + ab = 5 + 2(3) = 11$

b. $az + \dfrac{x}{b} = 2 \cdot 0 + \dfrac{5}{3} = 0 + \dfrac{5}{3} = \dfrac{5}{3}$

c. $ab - a + z = 6 - 2 + 0 = 4$

d. $ax + bx + xz = (10) + (15) + 0 = 25$

Evaluate the following formulas.

e. $A = \dfrac{1}{2} bc$ (area of a triangle); $b = 7$ and $c = 5$

$$A = \dfrac{1}{2}(7)(5) = 17.5$$

f. $S = \dfrac{1}{2}n(a + \ell)$ (sum of an arithmetic sequence); $n = 100$, $a = 1$, and $\ell = 100$.

$$S = \dfrac{1}{2} \cdot 100(1 + 100) = 50 \cdot 101 = 5050$$

(The sum $1 + 2 + 3 + \ldots + 98 + 99 + 100 = 5050$)

g. $R = \dfrac{1}{\dfrac{1}{R_1} + \dfrac{1}{R_2}}$; $R_1 = 2$, $R_2 = 1$

$$R = \dfrac{1}{\dfrac{1}{2} + \dfrac{1}{1}} = \dfrac{1}{\dfrac{3}{2}} = 1 \div \dfrac{3}{2} = \dfrac{2}{3}$$

h. Calculator example

$$V = \dfrac{1}{2} \pi r^2 h, \pi \approx 3.14, r = 16, h = 20$$

$$V \approx \left(\dfrac{1}{2}\right)(3.14)(16)^2(20)$$

ENTER	DISPLAY
.5	0.5
×	0.5
3.14	3.14
×	1.57
16	16.
x^2 or × 16	256. or 16.
×	401.92
20	20.
=	8038.4

So $V \approx 8038.4$.

APPLICATION SOLUTION

To find the number of square feet in the stage, replace r by 15 and evaluate. (r is one-half of the diameter.)

Formula: $A = \pi r^2$, $r = 15$

Substitute: $A \approx (3.14)(15)^2$

Solve: ≈ 706.5

Answer: The stage will contain approximately 706.5 sq ft.

WARM UPS

Evaluate the following expressions given $d = 10$, $f = 5$, $h = 2$, and $t = 1000$.

1. h^2

2. $6.3t$

3. $\dfrac{100d}{t} + h$

4. $\dfrac{2f}{d} \div h$

5. $2t - 10d$

6. $df + h$

7. $13.5t$

8. $2d + 3f + 4h$

9. $\dfrac{t}{2df} - h$

10. $t \div h \div f \div d$

11. $d^2 - f^2$

12. $\dfrac{t}{h(d^2 + f^2)}$

Exercises 4.4

A

Evaluate the following expressions given $a = 7$, $b = 10$, $c = 8$, $x = 1$, $y = 6$, and $z = 4$.

1. a^2x^2 **2.** $a^2 + x^2$ **3.** $ab + yz$

4. $cy - cx$ **5.** $cbx - ay$ **6.** $ay + bz$

B

7. $cx + cz$ **8.** $abz - abx$ **9.** $c(x + z)$

10. $ab(z - x)$ **11.** $(a + x)(c + y)$ **12.** $b + y(a + z)$

13. $2c + 4y$ **14.** $(3a + 5)z$

C

15. $\dfrac{6 - z}{b}$ **16.** $\dfrac{a + b + c}{x + y}$ **17.** $\dfrac{(x + y + z)^2}{c - x}$

18. $\dfrac{ax + by + cz}{3c - a}$ **19.** $\dfrac{b^2 - a^2 + c^2}{x^2 + y^2}$ **20.** $\dfrac{(a + b)^2 - c^2}{y^2 - x^2}$

Evaluate the following formulas.

21. $A = \dfrac{1}{2} bc$ if $b = 32$ and $c = 29$ (area of triangle)

22. $S = \dfrac{1}{2} n(a + \ell)$ if $n = 100$, $a = 2$, and $\ell = 200$ (sum of first 100 even numbers)

23. $R = \dfrac{1}{\dfrac{1}{R_1} + \dfrac{1}{R_2}}$ if $R_1 = .3$ and $R_2 = .6$ (resistance in parallel)

24. $V = \pi r^2 h$ if $\pi \approx 3.14$, $r = 2$, and $h = 3$ (volume of a cylinder)

1. _____
2. _____
3. _____
4. _____
5. _____
6. _____
7. _____
8. _____
9. _____
10. _____
11. _____
12. _____
13. _____
14. _____
15. _____
16. _____
17. _____
18. _____
19. _____
20. _____
21. _____
22. _____
23. _____
24. _____

ANSWERS

25. _____

26. _____

27. _____

28. _____

29. _____

30. _____

31. _____

32. _____

33. _____

34. _____

35. _____

D

25. The stage at the outdoor amphitheatre is in the shape of a semicircle of radius 100 ft. Find the area (A) of the stage. $\left(A = \dfrac{1}{2}\, \pi r^2,\ \pi \approx 3.14. \right)$

26. The plaza at the new mall in downtown Seattle is in the shape of a square with a semicircle at one end. If the side of the square is 60 feet, how many square feet are in its area (A)? $\left(A = s^2 + \dfrac{1}{2}\, \pi r^2,\ \pi \approx 3.14,\ s = 60,\ r = 30. \right)$

27. How much simple interest is earned on a savings account of $150 ($p = 150$) at the rate of $6\dfrac{1}{2}\%$ ($r = .065$) for 2 years ($t = 2$)? [Formula: $I = prt$]

28. How far does a plane travel at the rate of 425 mph ($r = 425$) in 9 hours ($t = 9$)? [Formula: $D = rt$]

29. The first ounce of mail costs 20¢ and each additional ounce costs 17¢. How much does an envelope weighing 9 ounces cost ($w = 8$)? [Formula: $C = 20 + 17w$]

30. A real estate salesperson receives a commission of 5% on the first $40,000 of a sale and 3% for all above $40,000. How much commission is earned from the sale of an $85,000 home? [$C = 2000 + .03(h - 40000)$, $h = 85000$]

E   *Maintain Your Skills*

Add:

31.	2.345	32.	.0034	33.	23.4
	3		4.12		4.765
	33.45		.005		23.5
	14.126		1.34		.0043
			45.7632		100.0034

34. $.0024 + 1.45 + 9.869 + .07345 + .567 =$

35. $23.5 + 87 + .986 + 5.75 + 3.5 =$

184

ANSWERS TO WARM UPS (4.4) **1.** 4   **2.** 6300   **3.** 3   **4.** $\frac{1}{2}$ **5.** 1900   **6.** 52

7. 13500   **8.** 43   **9.** 8   **10.** 10   **11.** 75   **12.** 4

4.5 Perimeter and Circumference

OBJECTIVE

47. Find the perimeters of geometric figures.

APPLICATION

A kitchen floor is in the shape of a rectangle with dimensions 9 feet by 11 feet. A mop board costing 38¢ per foot is to be installed around the perimeter of the room. Allowing 5 feet for door space, what will be the cost of the mop board?

VOCABULARY

Circles, rectangles, triangles, and *squares* are four examples of *geometric figures.* An example of each is shown in Figure 4.1. Different parts of the figures are also shown.

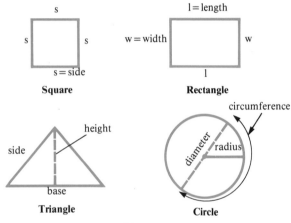

Figure 4.1

The *perimeter* (*P*) of a geometric figure is the total distance around the figure. The *circumference* (*C*) of a circle is the total distance around the circle. The *radius* (*r*) of a circle is a line segment from the center of a circle to any point on the circle. The *diameter* (*d*) of a circle is a line segment passing through the center of a circle joining any two points on the circle. The diameter is twice the radius.

HOW AND WHY

The perimeter of any figure in which all of the sides are straight is the sum of the lengths of the sides. Recall that to add measurements the unit of measure must be the same. For example, find the perimeter of Figure 4.2.

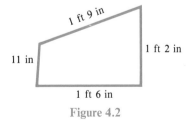

Figure 4.2

P (the perimeter) is the sum of the lengths of the four sides, that is,

```
1 ft   9 in
1 ft   2 in
1 ft   6 in
       11 in
3 ft 28 in   (28 in = 2 ft 4 in)
```

So $P = 5$ ft 4 in.

There are special formulas for the perimeters of squares, rectangles, and circles and for the length of a semicircle.

Square	If "P" is the perimeter and "s" is the length of one side, then the formula is

$$P = 4s$$

Rectangle	If "P" is the perimeter, "ℓ" is the length, and "w" is the width, then the formula is

$$P = 2\ell + 2w$$

Circle	IF "C" is the circumference and "d" is the diameter, then the formula is

$$C = \pi d = 2\pi r$$

(π is read "pi.")

Semicircle	If "L" is the length and "r" is the radius, then the formula is

$$L = \pi r = \frac{1}{2}\pi d$$

To find the perimeter of a geometric figure that is not one of the above, add the lengths of the sides.

There is no fraction or exact decimal that names the number π, so we will use 3.14 as an approximate value.

EXAMPLES

Find the perimeter of each of the following.

a.

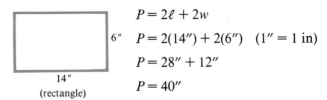

14"
(rectangle)

$P = 2\ell + 2w$
$P = 2(14'') + 2(6'')$ (1" = 1 in)
$P = 28'' + 12''$
$P = 40''$

b.

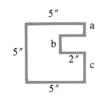

$P = 3(5'') + 2(2'') + a + b + c$

Note $a + b + c = 5''$ (same length as opposite side)

$P = 4(5'') + 2(2'')$
$P = 20'' + 4''$
$P = 24''$

c.

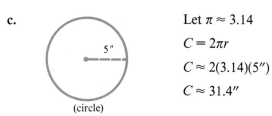

Let $\pi \approx 3.14$

$C = 2\pi r$

$C \approx 2(3.14)(5'')$

$C \approx 31.4''$

d.

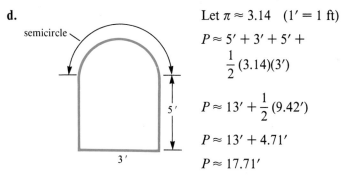

Let $\pi \approx 3.14$ $(1' = 1 \text{ ft})$

$P \approx 5' + 3' + 5' +$
$\quad \dfrac{1}{2}(3.14)(3')$

$P \approx 13' + \dfrac{1}{2}(9.42')$

$P \approx 13' + 4.71'$

$P \approx 17.71'$

e. Calculator example

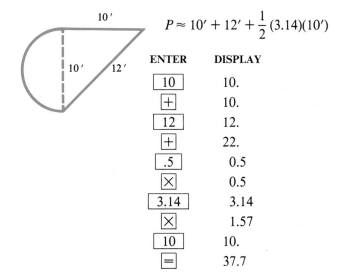

$P \approx 10' + 12' + \dfrac{1}{2}(3.14)(10')$

ENTER	DISPLAY
10	10.
+	10.
12	12.
+	22.
.5	0.5
×	0.5
3.14	3.14
×	1.57
10	10.
=	37.7

The perimeter is approximately 37.7′.

APPLICATION SOLUTION

To find the cost of the mop board, first find the perimeter of the room and subtract 5 feet for the door space.

Simpler Word Form:	$\dfrac{\text{Length of}}{\text{mop board}} = \dfrac{\text{Perimeter}}{\text{of room}} - \text{Door space}$
Formula:	$L = 2w + 2\ell - 5,\ w = 9,\ \ell = 11$
Substitute:	$L = 2(9) + 2(11) - 5$
Solve:	$= 18 + 22 - 5$
	$= 35$

So 35 feet of mop board are needed.

Continued on next page.

To find the cost, multiply the length of the mop board by 38¢ per foot.

Simpler Word Form:	$\text{Cost} = \left(\dfrac{\text{Cost per}}{\text{foot}}\right)\left(\dfrac{\text{Length}}{\text{in feet}}\right)$
Formula:	$C = cL,\ c = 38,\ L = 35$
Substitute:	$C = (38)(35)$
Solve:	$= 1330$
Answer:	The cost of the mop board is 1330¢ or $13.30.

WARM UPS

Find the perimeter or circumference of each of the following. Let $\pi \approx 3.14$.

1. A square with side 3 cm.

2. A square with side 1.2 yd.

3. A rectangle that is 1 ft by 3 ft.

4. A rectangle that is 3 m by 2 m.

5. A triangle with sides 2 in, 4 in, and 5 in.

6. A triangle with sides 2.3 cm, 2.2 cm, and 3.1 cm.

7. A circle with radius 1 dm.

8. A circle with radius $\dfrac{1}{2}$ yd.

9.

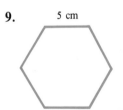

5 cm

10.

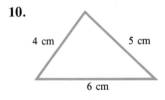

4 cm   5 cm

6 cm

11.

8′

4′   5′

6′

12.

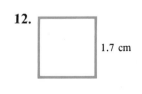

1.7 cm

Exercises 4.5

A

Find the perimeter or circumference of each of the following. Let $\pi \approx 3.14$.

1. A square with side 13 cm.

2. A rectangle that is 3 ft by 7 ft.

3. A circle with diameter 200 in.

4. A triangle that is 9 m on each side.

5. A four-sided figure (quadrilateral) with sides 12 in, 13 in, 15 in, and 17 in.

6. A five-sided figure (pentagon) with sides of 3 ft, 3 ft, 5 ft, 6 ft, and 8 ft.

B

Find the perimeter or circumference of each of the following figures. Let $\pi \approx 3.14$.

7.

16 ft

8.

9.5 cm
15 cm

9.
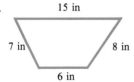
15 in
7 in
8 in
6 in

10.
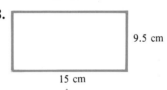
30 ft

11.
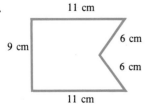
11 cm
9 cm
6 cm
6 cm
11 cm

12.
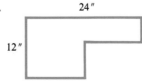
24″
12″

13.

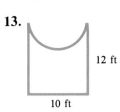

12 ft

10 ft

14.

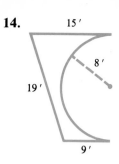

15 ′

8 ′

19 ′

9 ′

C

15.

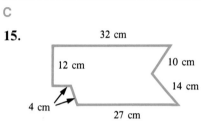

32 cm

12 cm

10 cm

14 cm

4 cm

27 cm

16.

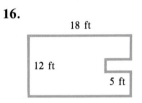

18 ft

12 ft

5 ft

17.

5 in

2 in

18.

2.7 cm

1 cm

6.2 cm

1.6 cm

9.4 cm

19.

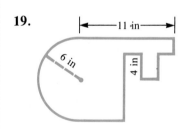

11 in

6 in

4 in

20.

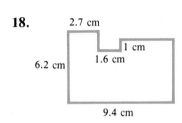

25 m

27 m

30 m

16 m

21.

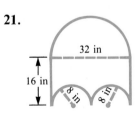

32 in

16 in

8 in   8 in

22.

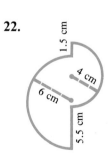

1.5 cm

4 cm

6 cm

5.5 cm

23.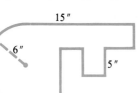

24.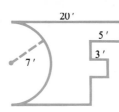

D

25. A family room is in the shape of a rectangle with dimensions 24 ft by 15 ft. A mop board costing 52¢ per foot is to be installed around the perimeter of the room. Allowing 11 ft for door space, what will be the cost of the mop board?

26. It costs 28¢ per foot to have the perimeter of a house sprayed for ants. What would it cost to spray the house whose outline is given below?

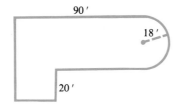

27. What is the distance around a baseball infield if it is 90 ft between the bases?

28. How many feet of picture framing does Bonnie need to frame 4 pictures, each measuring 8 inches by 10 inches?

29. How far will the tip of the minute hand of a clock travel in 15 minutes if the hand is one inch in length?

ANSWERS

30. _____

31. _____

32. _____

33. _____

34. _____

35. _____

36. _____

37. _____

30. If fencing costs $5 per yard, what will be the cost of fencing a rectangular lot that is 407 ft long and 82 ft wide?

31. If Atha needs 2 minutes to put one foot of binding on a rug, how long will it take her to put the binding on a rug that is 15 ft by 12 ft?

32. A wheel on Mary's automobile is 16 inches in diameter. To the nearest whole number, how many revolutions will the wheel turn if the automobile travels one mile? ($\pi \approx 3.14$)

E   *Maintain Your Skills*

Subtract:

33. 15.2
    4.234

34. 345.8009
    39.874

35. 18
    5.87

36. 398.078 − 123.7654

37. 4.0078 − 1.876

4.6 Area of Common Geometric Figures

OBJECTIVE

48. Find the area of common geometric figures.

APPLICATION

A concrete retaining wall is 6 feet high and 596 feet long. Find the cost of paint for covering the wall with two coats of paint if one gallon of paint costs $13.25 and covers 75 square feet of concrete.

VOCABULARY

Area is a measure of a surface, and surface is measured in square units. Two examples of surface measure are shown in Figure 4.3.

The unit of measure on the right is called a *"square inch,"* since the square is one inch on each side. The "square inch" measures the surface that is contained within the square. "Square inch" is written "in^2."

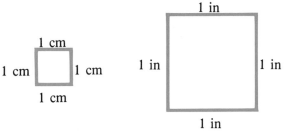

Figure 4.3

The unit of measure on the left is called a *square centimetre,* since the square is one centimetre on each side. The "square centimetre" measures the surface that is contained within the square. "Square centimetre" is written "cm^2."

There are other units of surface measure, such as *"square foot," "square mile," "square metre,"* and *"square kilometre."* In each case the unit measures the amount of surface within a square having that length on each side.

In Figure 4.4, two different geometric figures are shown: the parallelogram and the trapezoid. Note that the trapezoid has two bases and one altitude, while the parallelogram has one base and one altitude.

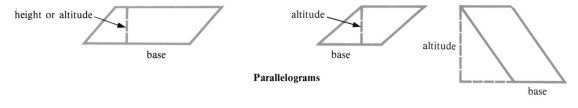

Parallelograms

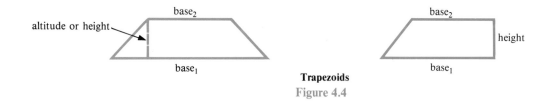

Trapezoids

Figure 4.4

HOW AND WHY

To find the area of a geometric figure means to determine how many square units of measure (surface units) are contained within that figure. Consider a square that is 2 inches on each side.

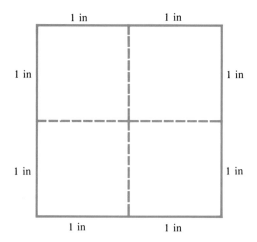

We can see from the drawing that this square can be divided into four squares, each of which is one inch on a side. This indicates that the area of a square that is two inches on each side is four square inches. The area (4 in²) is the square of the length of a side. That is, $A = (2\text{ in})^2 = (2\text{ in})(2\text{ in}) = 4\text{ in}^2$. (Notice that "in" times "in" gives "in².")

> **The area of a square can be found by squaring the length of one of its equal sides. That is,**
>
> $A = s^2$

Consider a rectangle that has a length of 3 centimetres and a width of 2 centimetres. To find the area of this rectangle, we consider the following diagram.

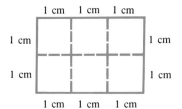

The diagram indicates that the rectangle can be divided into six squares, each of which is one centimetre on a side. This tells us that the area of the rectangle is 6 cm². Notice that if we multiply the length (3 cm) times the width (2 cm), we have (3 cm) × (2 cm), which is 6 cm².

> **The area of a rectangle can be found by multiplying the length times the width. Stated as a formula,**
>
> $A = \ell \cdot w$

There are formulas to find the area of a parallelogram, a triangle, a trapezoid, and a circle. We could consider each in a manner similar to the square and the rectangle, but instead the formulas will be stated.

The area of a parallelogram	$A = b \cdot h$
The area of a triangle	$A = \dfrac{1}{2} b \cdot h$
The area of a trapezoid	$A = \dfrac{1}{2} (b_1 + b_2) \cdot h$
The area of a circle	$A = \pi r^2$ or $A = \dfrac{\pi d^2}{4}$

Here, b indicates the base (or bases, in the case of the trapezoid), h indicates the height or altitude, r is the radius, and d is the diameter of the circle.

EXAMPLES

Find the area of each of the following.

a. A square that is 4 in on each side.

$A = s^2$

$A = (4 \text{ in})^2$

$A = 16 \text{ in}^2$

b.

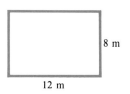

The figure is a rectangle with $\ell = 12$ m and $w = 8$ m.

$A = \ell \cdot w$

$A = (12 \text{ m})(8 \text{ m})$

$A = 96 \text{ m}^2$

c.

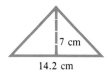

The figure is a triangle with $b = 14.2$ cm and $h = 7$ cm.

$A = \dfrac{1}{2} \cdot b \cdot h$

$A = \dfrac{1}{2} \cdot (14.2 \text{ cm})(7 \text{ cm})$

$A = 49.7 \text{ cm}^2$

d. A parallelogram whose base is 1 ft and whose height is 4 in.

$A = b \cdot h$

$A = (1 \text{ ft})(4 \text{ in})$ or $A = (1 \text{ ft})(4 \text{ in})$

$A = (12 \text{ in})(4 \text{ in})$ $A = (1 \text{ ft})\left(\dfrac{1}{3} \text{ ft}\right)$

$A = 48 \text{ in}^2$ $A = \dfrac{1}{3} \text{ ft}^2$

e. Calculator example

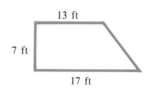

13 ft

7 ft

17 ft

The figure is a trapezoid with $b_1 = 17$ ft, $b_2 = 13$ ft, and $h = 7$ ft.

$$A = \frac{(b_1 + b_2) \cdot h}{2}$$

$$A = \frac{(17 \text{ ft} + 13 \text{ ft}) \cdot 7 \text{ ft}}{2}$$

ENTER	DISPLAY	or	ENTER	DISPLAY
(	0.		17	17.
17	17.		+	17.
+	17.		13	13.
13	13.		=	30.
)	30.		×	30.
×	30.		7	7.
7	7.		÷	210.
÷	210.		2	2.
2	2.		=	105.
=	105.			

The area of the trapezoid is 105 ft².

f. The figure is a circle with $r = 8$ cm.

8 cm

$A = \pi r^2$, where $\pi \approx 3.14$

$A \approx (3.14)(8 \text{ cm})^2$

$A \approx (3.14)(64 \text{ cm}^2)$

$A \approx 200.96 \text{ cm}^2$

APPLICATION SOLUTION

To find the cost of painting the retaining wall, first find the area of its exposed wall.

Formula: $A = \ell w$, $\ell = 596$ ft, $w = 6$ ft

Substitute: $A = (596 \text{ ft})(6 \text{ ft})$

Solve: $A = 3576 \text{ ft}^2$

Continued on next page.

Now find the number of gallons of paint needed by dividing the area by the coverage of one gallon. The area is doubled because the wall is to be painted with two coats.

Simpler Word Form: Number of gallons of paint = $\left(\begin{array}{c}\text{Area to}\\\text{be}\\\text{painted}\end{array}\right)(2) \div \left(\begin{array}{c}\text{Coverage}\\\text{per}\\\text{gallon}\end{array}\right)$

Substitute: $N = (3576 \text{ ft}^2)(2) \div \dfrac{75 \text{ ft}^2}{\text{gallon}}$

Solve: $= 95.36$ gallons

To find the cost, multiply the cost per gallon times the number of gallons.

Formula: $C = cN, \; c = \dfrac{\$13.25}{\text{gallon}}, \; N = 95.36$ gallons

Substitute: $C = \left(\dfrac{\$13.25}{\text{gallon}}\right)(95.36 \text{ gallons})$

Solve: $= \$1263.52$

Answer: The cost of the paint to paint the retaining wall is $1263.52.

WARM UPS

Find the area of each of the following. Let $\pi \approx 3.14$.

1. A square with side 8 cm.

2. A square with side $1\frac{1}{2}$ in.

3. A rectangle that is 4 ft by 11 ft.

4. A rectangle that is 1 ft by 6 in.

5. A parallelogram with base 16 cm and height 4 cm.

6. A triangle with base 16 cm and height 4 cm.

7. A circle with radius 1 dm.

8. A square with side 2.2 cm.

9.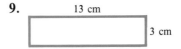

10.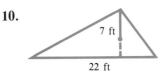

11.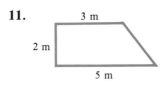

12.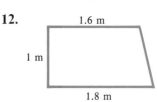

Exercises 4.6

A

Find the area of each of the following. Let $\pi \approx 3.14$.

1. A circle whose diameter is 20 cm.

2. A square that is 4 metres on a side.

3. A parallelogram with base 4 in and height $1\frac{1}{2}$ in.

4. A triangle with base 4 in and height $1\frac{1}{4}$ in.

5.

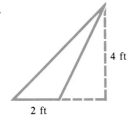

2 ft, 4 ft

6.

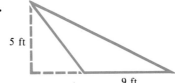

5 ft, 9 ft

B

Find the area of each of the following figures. Let $\pi \approx 3.14$.

7.

6 in, 16 in

8.

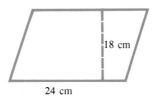

18 cm, 24 cm

9.

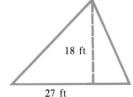

18 ft, 27 ft

10.

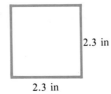

2.3 in, 2.3 in

11.
2.4 m
1.2 m
3.2 m

12.
12 in
18 in

13.
12 in

14.
19 yd
10 yd
26 yd

C

15.
24.8′
36.5′

16.
8.7 in
11.3 in

17.
9.7 ft

18.
36.8 cm

19.
21.6 ft

20.
15.7 m

21.
58.9 in
32.1 in

22.
22.5 cm
17.8 cm

23.
15 cm
12.3 cm
18.7 cm

24.
4.6 ft
5.7 ft
11.3 ft

200

D

25. A stone retaining wall is 4 feet high and 120 feet long. Find the cost of water sealer for covering the wall with two coats of sealer if one gallon of the sealer costs $11.75 and covers 45 square feet of stone.

26. The top of a rectangular vanity table is to be covered with ceramic tile. If the table top measures 40 inches by 32 inches, and each tile is 6 inches on a side and costs $2.35, what is the cost of the tile? Assume that a whole number of tiles must be purchased.

27. Joan is making a circular mirror of diameter 36 inches. If the glass costs $.11 per square inch, what is the cost of the glass in the mirror?

28. Peggy is recarpeting a rectangular floor that measures 21 ft by 30 ft. What is the total cost of the carpet if it sells for $22.50 per square yard?

29. One ounce of weed killer treats one square meter of lawn and costs $.87 an ounce. What will it cost Debbie to treat a rectangular lawn that measures 30 m by 8 m?

30. What will it cost to buy a rectangular plot of ground if the length is 1320 ft and the width is 528 ft, and one acre of ground sells for $5250? (43560 ft^2 = 1 acre.)

31. At $.37 a tile, each 12 in on a side, what will it cost to cover a floor that is 8 ft wide and 16 ft long?

E   *Maintain Your Skills*

Multiply:

32. 46.87
    1.7

33. 6.098
    .076

34. .134
    3.56

35. (2.4506)(12.3)

36. (123.8)(.097)

ANSWERS TO WARM UPS (4.6)

1. 64 cm²	**2.** 2.25 in²	**3.** 44 ft²	**4.** 72 in² or .5 ft²	**5.** 64 cm²
6. 32 cm²	**7.** 3.14 dm²	**8.** 4.84 cm²	**9.** 39 cm²	**10.** 77 ft²
11. 8 m²	**12.** 1.7 in²			

4.7 A Second Look at Areas

OBJECTIVE

49. **Find the area of a geometric figure that can be considered as a combination of two or more common geometric figures.**

APPLICATION

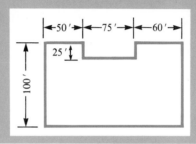

The floor of a shop (floor plan shown below) is to be poured concrete. If concrete costs $3.35 a square yard, what will the floor cost?

VOCABULARY

No new vocabulary.

HOW AND WHY

It is possible to find the areas of geometric figures that are not squares, rectangles, circles, or any of the other more common figures. Some of these figures can be divided into two or more of the common shapes. The sum of the areas of each of these common figures is the area of the entire region.

Consider the following shapes:

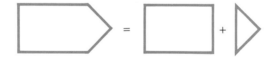

The figure on the left can be divided into a rectangle with a triangle attached, as shown on the right. If we can find the area of the rectangle and the area of the triangle, then the sum of those areas will be the area of the entire region.

Consider the following shapes:

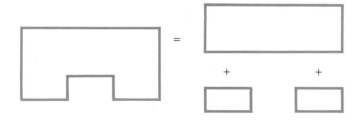

The figure on the left can be divided into three rectangles, as shown on the right. If we can find the area of each rectangle, the sum of those areas will be the area of the entire region.

In some figures it is helpful to add a region to the original figure so that it may be divided into the common figures. For example:

The figure on the left can be considered as a semicircle, a rectangle, and a triangle, as shown on the right. First we find the area of the semicircle, the rectangle, and the triangle. Then we find the sum of the area of the semicircle and the rectangle, and subtract the area of the triangle to find the area of the entire region.

> **To find the area of a geometric figure that is composed of the sum or the difference of two or more common geometric figures:**
>
> 1. **Divide the figure into common geometric figures; or add a region or regions, and then divide into common geometric figures.**
> 2. **Find the area of each of the common figures.**
> 3. **Find the sum or difference of those areas.**

EXAMPLES

Find the area of each of the following.

a.

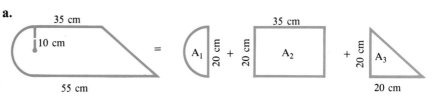

The area of the entire region is

A = Area of semicircle + Area of rectangle + Area of triangle

$A = A_1 + A_2 + A_3$

$$A_1 = \frac{\pi r^2}{2} \approx \frac{(3.14)(10 \text{ cm})^2}{2} = 157 \text{ cm}^2$$

$$A_2 = \ell \cdot w = (35 \text{ cm})(20 \text{ cm}) = 700 \text{ cm}^2$$

$$A_3 = \frac{1}{2} \cdot b \cdot h = \frac{1}{2}(20 \text{ cm})(20 \text{ cm}) = 200 \text{ cm}^2$$

$A \approx 157 \text{ cm}^2 + 700 \text{ cm}^2 + 200 \text{ cm}^2 = 1057 \text{ cm}^2$

b.

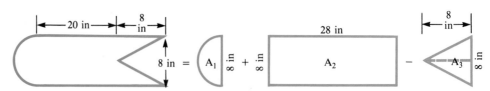

A = Area of semicircle + Area of rectangle − Area of triangle

$A = A_1 + A_2 - A_3$

$\approx 25.12 \text{ in}^2 + 224 \text{ in}^2 - 32 \text{ in}^2$

$\approx 217.12 \text{ in}^2$

APPLICATION SOLUTION

To find the surface area to be poured, break up the diagram as shown below.

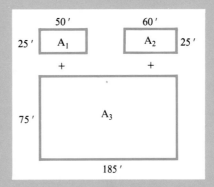

Formula: $A = A_1 + A_2 + A_3$

Substitute: $A = (25')(50') + (60')(25') + (75')(185')$

Solve: $A = 16625 \text{ ft}^2$

Now calculate the number of square yards in the surface.

$$A = 16625 \text{ ft}^2 \cdot \frac{1 \text{ yd}}{3 \text{ ft}} \cdot \frac{1 \text{ yd}}{3 \text{ ft}}$$

$$A = 1847\frac{2}{9} \text{ yd}^2$$

The cost is now found by multiplying the number of square yards by the price per square yard.

Formula: $C = pn$

Substitute: $C = \frac{\$3.35}{\text{yd}^2} \cdot 1847\frac{2}{9} \text{ yd}^2$

Solve: $C = \$6188.19$

Answer: The cost of pouring the cement is $6188.19.

WARM UPS

Find the area of each of the following figures. Let $\pi \approx 3.14$.

1.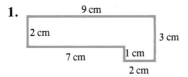

2.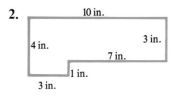

3.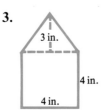

4.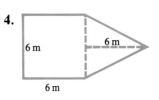

5.

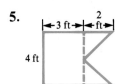
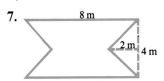

6.

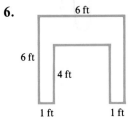

7.

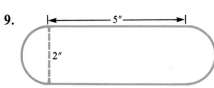

8.

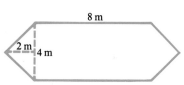
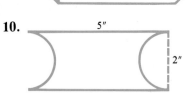

9.

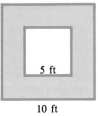

10.

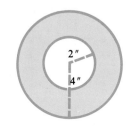

11. The shaded area.

12.

19. Round to the nearest hundredth.   **20.** Round to the nearest tenth.

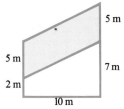

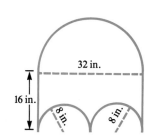

Find the area of the shaded portion.

21.

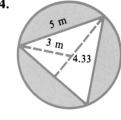

22.

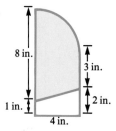

23.

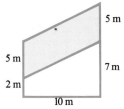

24.

ANSWERS

19. _____

20. _____

21. _____

22. _____

23. _____

24. _____

25. _____

D

25. A sign board at the local baseball park is a 24′ × 12′ rectangle that is topped with three equal semi-circles as shown. The local painter charges $.65 a square foot to paint a sign. How much will it cost to paint a new sign on the board?

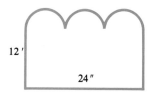

12 ′

24 ″

209

ANSWERS

26. _____

27. _____

28. _____

29. _____

30. _____

31. _____

32. _____

33. _____

34. _____

35. _____

26. It costs $2.10 a square foot to install an instant lawn from Sod Brothers. How much will it cost to install lawn on the given area?

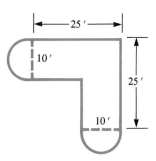

27. The following diagram shows the Jones' yard with respect to their house. How much grass seed is needed to sow the lawn if one pound of seed will sow 1000 ft²? Find the answer to the nearest pound.

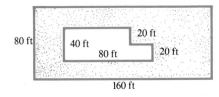

28. What is the area of the infield of a race track in which the straight-aways are 100 yards long and the ends are semicircles with 30-yard diameters?

29. The floor of a shop (floor plan shown below) is to be poured concrete. If concrete costs $3.35 a square yard, what will the floor cost?

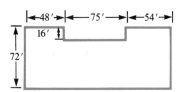

30. How many squares of aluminum siding are needed for the shed shown below? Assume that there are no windows and that the door will be made of the siding. A different material will be used for the roof. (100 ft² = 1 square of siding)

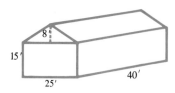

E *Maintain Your Skills*

Divide:

31. 2.5)‾47.3‾   **32.** 3.5)‾1.05‾   **33.** 13.35)‾27.3675‾

Divide, round to the nearest tenth.

34. 3.05)‾50.4‾   **35.** 2.01)‾5.34‾

ANSWERS TO WARM UPS (4.7)   **1.** 20 cm²   **2.** 33 in²   **3.** 22 m²   **4.** 54 m²   **5.** 16 ft²

6. 20 ft²   **7.** 24 m²   **8.** 40 m²   **9.** 13.14 in²   **10.** 6.86 in²

11. 75 ft²   **12.** 37.68 in²

4.8 Volume of Common Geometric Solids

OBJECTIVE

50. Find the volumes of common geometric solids.

APPLICATION

How many tons of coal will a boxcar hold if the inside dimensions of the car are 8 feet wide and 35 feet long? The car is to be filled to a depth of 5 feet and coal weighs 50 pounds per cubic foot.

VOCABULARY

Volume is the name given to the amount of space that is contained inside a three-dimensional figure. A box, for example, has volume. The question "How much does the box hold?" is referring to its volume. Volume is measured in terms of a cubic unit, that is, a *cube* (a box) that is one unit on each edge. Volume can be thought of as the number of cubic units needed to form the figure. An example of a unit of volume measure is a *cubic inch* (in³), which is shown in Figure 4.5.

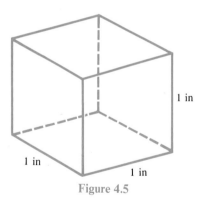

1 in

1 in

1 in

Figure 4.5

A cube that is one inch on each edge is a cubic inch. A cube that is 1 centimetre on each edge is a cubic centimetre (cm³). A cube that is one foot on each edge is a cubic foot (ft³). Each of the cubes is a measure of volume. In fact, any cube may be thought of as a unit measure of volume.

To find the volume of a geometric solid means to determine how many cubes are contained within the solid. Each cube is the same size and is called a cubic unit.

HOW AND WHY

To find the volume of a cube that is of length e on each edge, we use the formula $V = e^3$, that is: $V = e \cdot e \cdot e$.

The volume of a rectangular solid (a box is a rectangular solid) is $V = \ell \cdot w \cdot h$, where V is the volume, ℓ is the length, h is the height, and w is the width. (See Figure 4.6.)

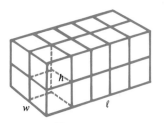

Figure 4.6

The length, width, and height must all be measured with the same unit of measure or converted to the same unit.

A cube is a rectangular solid in which the length, width, and height are all equal.

Pictures of a cylinder, a sphere, a cone, and a pyramid are shown in Figure 4.7. The formula for the volume of each is as follows:

Cylinder	$V = \pi r^2 h$	
Sphere	$V = \dfrac{4}{3}\pi r^3$	
Cone	$V = \dfrac{1}{3}\pi r^2 h$	
Pyramid	$V = \dfrac{1}{3}Bh$	**(where B is the area of the base)**

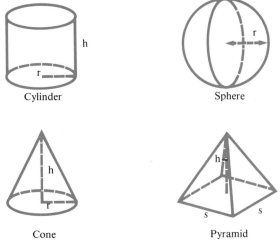

Figure 4.7

EXAMPLES

a. The volume of a cube that is 6 inches on each edge is:

$V = e^3$ or $V = e \cdot e \cdot e$

$V = (6\text{ in})^3$

$V = (6\text{ in})(6\text{ in})(6\text{ in})$

$V = 216\text{ in}^3$

b.

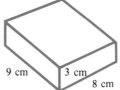

9 cm   3 cm

8 cm

The volume of the rectangular solid (box) is:

$V = \ell \cdot w \cdot h$

$V = (9 \text{ cm})(8 \text{ cm})(3 \text{ cm})$

$V = 216 \text{ cm}^3$

c.   2 ft

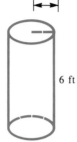

6 ft

The volume of the cylinder is:

$V = \pi r^2 h$ (Let $\pi \approx 3.14$.)

$V \approx (3.14)(2 \text{ ft})^2(6 \text{ ft})$

$V \approx 75.36 \text{ ft}^3$

d. The volume of a sphere whose diameter is 18 centimetres is:

$V = \dfrac{4}{3} \pi r^3$ (Let $\pi \approx 3.14$.)

$V \approx \dfrac{4}{3}(3.14)(9 \text{ cm})^3$ ($r = 9$ cm since $d = 18$ cm)

$V \approx \dfrac{4}{3}(3.14)(729 \text{ cm}^3)$

$V \approx 3052.08 \text{ cm}^3$

e. Calculator example

5 m

3 m

The volume of the cone is:

$V = \dfrac{1}{3} \pi r^2 h$ (Let $\pi \approx 3.14$.)

$V \approx \dfrac{1}{3}(3.14)(3 \text{ m})^2(5 \text{ m})$

ENTER	DISPLAY
1	1.
$\div$	1.
3	3.
$\times$	0.3333333
3.14	3.14
$\times$	1.0466667
3	3.
x^2	9.
$\times$	9.4199999
5	5.
$=$	47.1

The volume is approximately 47.1 m³.

f. Let us find the volume of a pyramid whose base is a rectangle with length 6 in and width 4 in, and which has a height of 5 in. Since the base is a rectangle, the area of the base is $\ell \cdot w$, and we have

$$V = \frac{1}{3}(\ell \cdot w)h$$

$$V = \frac{1}{3}(6 \text{ in})(4 \text{ in})(5 \text{ in})$$

$$V = 40 \text{ in}^3$$

APPLICATION SOLUTION

The coal in the boxcar will fill a rectangular solid as shown.

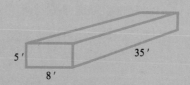

First find the volume of the coal.

Formula: $V = \ell \cdot w \cdot h$

Substitute: $V = (35')(8')(5')$

Solve: $V = 1400 \text{ ft}^3$

Now find the weight of the coal in tons.

Formula: $W = \text{ft}^3 \cdot \dfrac{\text{lb}}{\text{ft}^3} \cdot \dfrac{\text{ton}}{\text{lb}}$

Substitute: $W = 1400 \text{ ft}^3 \cdot \dfrac{50 \text{ lb}}{\text{ft}^3} \cdot \dfrac{1 \text{ ton}}{2000 \text{ lb}}$

Solve: $W = 35 \text{ tons}$

Answer: The box car will be filled with 35 tons of coal.

WARM UPS

Find the volume.

1.

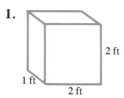

2 ft

1 ft

2 ft

2.

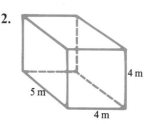

4 m

5 m

4 m

3. Find the volume of a rectangular solid (box) with length 5 yd, width 2 yd, and height 3 yd.

4. Find the volume of a rectangular solid (box) with length 25 cm, width 4 cm, and height 6 cm.

5. Find the volume of a cube that is 3 ft on each edge.

6. Find the volume of a cube that is 10 cm on each edge.

Find the volume. Let $\pi \approx 3.14$.

7.

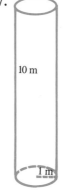

10 m

1 m

8.

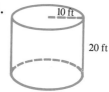

10 ft

20 ft

9.

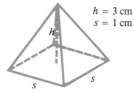

$h = 3$ cm
$s = 1$ cm

h

s s

10.

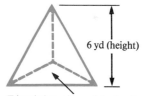

6 yd (height)

Triangle has a base of 4 yd and a height of 2 yd

11. Find the volume, in cubic yards, of a cube that measures 6 feet on an edge.

12. Find the volume, in cubic feet, of a rectangular solid that has dimensions 6 in $\times$ 1 ft $\times$ 8 in.

Exercises 4.8

A

Find the volume. Let $\pi \approx 3.14$.

1.

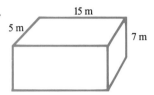

2.

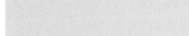

3.

4.

5. Find the volume of a cube that is 5 inches on each edge.

6. Find the volume of a cube that is 7.3 cm on each edge.

B

7. Find the volume.

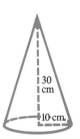

8. Find the volume of a cone with radius 6 ft and height 8 ft.

1. _____

2. _____

3. _____

4. _____

5. _____

6. _____

7. _____

8. _____

9. _____

10. _____

11. _____

12. _____

13. _____

14. _____

15. _____

16. _____

17. _____

9. Find the volume. Round to the nearest tenth.

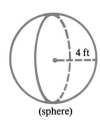

(sphere)

10. Find the volume of a sphere with diameter 30 metres.

11. Find the volume.

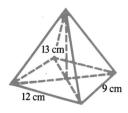

12. Find the volume of a pyramid whose base is a square with side 150 feet and whose height is 30 feet.

13. Find the volume, in cubic metres, of a rectangular solid with $\ell = 131.6$ cm, $w = 119.5$ cm, and $h = 12.7$ cm. (To the nearest tenth of a cubic metre.)

14. Find the volume, in cubic yards, of a cube that is 11.75 ft on a side. (To the nearest tenth of a cubic yard.)

C

15. Find the volume, in cubic inches, of a cylinder that has radius 9 inches and height 2 feet.

16. Find the volume of a cylinder that has a 13″ diameter and is 4′ tall. (To the nearest cubic inch.)

17. Find the volume of a pyramid with a height of 37 ft and the triangular base shown below.

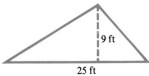

18. Find the volume of a pyramid with height 37 metres and a square base that is 25 metres on each side.

18. _____

19. Find the volume. Round to the nearest tenth.

19. _____

20. Find the volume of a sphere with a diameter of 2 ft 4 in. (To the nearest cubic inch.)

20. _____

21. Find the volume of a geometric solid that is a cube 9.6 ft on a side with a pyramid on top of the cube. The pyramid has a height of 10 ft and uses one face of the cube as its base.

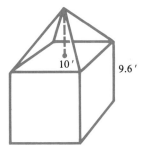

21. _____

22. _____

22. Find the volume of a geometric solid that is a rectangular solid with dimensions of 6′ × 8′ × 4′, with a pyramid on top. The pyramid is 9′ tall and uses a 6′ × 8′ face of the rectangular solid as its base.

23. _____

23. Find the volume of a geometric solid that has a cylindrical base of radius 3.4 m and a height of 8 m, with a cone on top. The cone is 6 m high, and uses the end of the cylinder as its base. (Let $\pi \approx 3.14$)

24. Find the volume of a geometric solid in the shape of an ice cream cone with conical base of radius .75 in and height 3 in. The top is a hemisphere of the same radius. (Let $\pi \approx 3.14$, and round to the nearest hundredth.)

24. _____

ANSWERS

25. _____

26. _____

27. _____

28. _____

29. _____

30. _____

31. _____

32. _____

33. _____

34. _____

35. _____

D

25. How many tons of iron ore will a boxcar hold if the inside dimensions of the car are 10 ft × 36 ft × 5.6 ft? Iron ore weighs 75 lb per cubic foot.

26. How many cubic yards of concrete are needed to pour a sidewalk that is 3 ft wide, 4 in thick, and 54 ft long?

27. A water tank is a cylinder that is 22 inches in diameter and $3\frac{1}{2}$ feet high. If there are 231 in³ in a gallon, how many gallons of water will the tank hold?

28. How many boxes that are 12 in wide, 8 in high, and 18 in long can be loaded into a truck bed that is 8 ft wide, 6 ft high, and 21 ft long?

29. An excavation is being made for a basement. The hole is 24 ft wide, 35 ft long, and 9 ft deep. If the bed of a truck holds 14 yd³, how many truckloads of dirt will need to be hauled away?

30. A swimming pool that is 30 ft long and 10 ft wide is filled to a depth of 5 ft.
   a. How many cubic ft of water are in the pool?
   b. If one cubic foot of water is approximately 7.5 gallons, how many gallons of water are in the pool?

E   *Maintain Your Skills*

Write each of the following as a decimal.

31. $\dfrac{48}{50}$ **32.** $\dfrac{19}{32}$ **33.** $\dfrac{28}{140}$

Find the approximate decimal to the nearest thousandth for the following.

34. $\dfrac{27}{47}$ **35.** $\dfrac{45}{97}$

ANSWERS TO WARM UPS (4.8) **1.** 4 ft³   **2.** 80 m³   **3.** 30 yd³   **4.** 600 cm³   **5.** 27 ft³

6. 1000 cm³   **7.** 31.4 cm³   **8.** 6380 ft³   **9.** 1 cm³   **10.** 8 yd³

11. 8 yd³   **12.** $\frac{1}{3}$ ft³

4.9 Surface Area

OBJECTIVE

51. Compute the surface areas of given geometric solids.

APPLICATION

How many square feet of sheet metal are needed to make a cylindrical heat duct that is 6″ in diameter and 15′ in length?

VOCABULARY

Lateral surface area is the area of the sides (lateral faces) of geometric solids such as cylinders, rectangular solids, cones, and pyramids. *Total surface area* is the lateral surface area plus the area of the top and/or bottom.

HOW AND WHY

The sides of many geometric solids are actually simple geometric figures. The rectangular solid has six faces, each of which is a square or a rectangle. The cylinder has circles for top and bottom and a rectangle or square for the curved surface. The solids that we will consider, along with their formulas for surface area, follow.

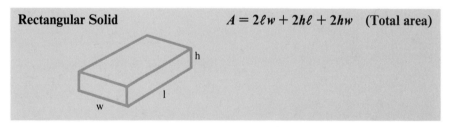

Rectangular Solid $A = 2\ell w + 2h\ell + 2hw$ **(Total area)**

Cube $A = 6e^2$ **(Total area)**

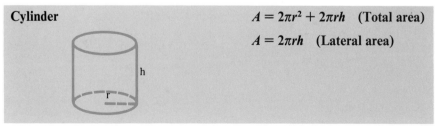

Cylinder $A = 2\pi r^2 + 2\pi rh$ **(Total area)**

$A = 2\pi rh$ **(Lateral area)**

r is the radius of the base and *h* is the height of the cylinder.

Right Circular Cone

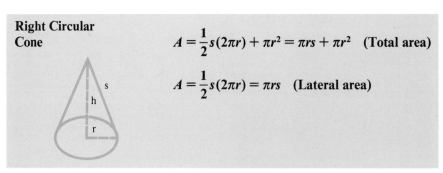

$$A = \frac{1}{2}s(2\pi r) + \pi r^2 = \pi rs + \pi r^2 \quad \text{(Total area)}$$

$$A = \frac{1}{2}s(2\pi r) = \pi rs \quad \text{(Lateral area)}$$

s is the slant height and *r* is the radius of the base.

Regular Pyramid

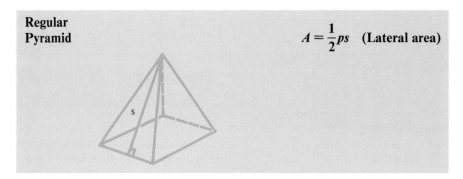

$$A = \frac{1}{2}ps \quad \text{(Lateral area)}$$

p is the perimeter of the base and *s* is the slant height.

Sphere

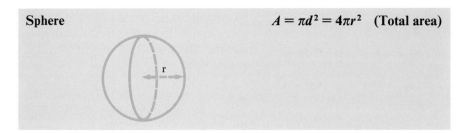

$$A = \pi d^2 = 4\pi r^2 \quad \text{(Total area)}$$

d is the diameter of the sphere and *r* is the radius of the sphere.

EXAMPLES

a. Find the lateral surface area of a cylinder that is 14 cm high and has a radius of 4 cm.

$A = 2\pi rh$

$A \approx 2(3.14)(4)(14)$

$A \approx 351.68 \text{ cm}^2$

b. Find the total surface area of a cube that is 18″ on each edge.

$A = 6e^2$

$A = 6 \cdot 18^2$

$A = 6 \cdot 324$

$A = 1944 \text{ in}^2$

c. What is the lateral surface area of a pyramid whose base is a square that is 3 metres on each side and whose slant height is 2 metres?

$$A = \frac{1}{2}ps$$

$$A = \frac{1}{2}(4 \cdot 3) \cdot 2$$

$$A = \frac{1}{2}(12) \cdot 2$$

$$A = 12 \text{ m}^2$$

APPLICATION SOLUTION

To find the number of square feet of sheet metal needed to make the heat duct, find the lateral area of the cylindrical duct.

Formula: $\quad A = 2\pi rh, \quad r = 3'' \text{ or } .25', \ h = 15', \ \pi \approx 3.14$

Substitute: $\quad A \approx (2)(3.14)(.25')(15')$

Solve: $\quad A \approx 23.55 \text{ ft}^2$

Answer: $\quad$ Approximately 23.55 ft² of sheet metal is needed to make the duct.

Exercises 4.9

A

1. Find the total surface area of a cube whose edge is 50 cm.

2. Find the lateral surface area of a cone whose radius is 2 feet and whose slant height is 3 feet.

3. Find the total surface area of a cylinder whose radius is 8 inches and whose height is 48 inches.

4. Find the lateral surface area of a pyramid whose base is a square 16 cm on each side and whose slant height is 20 cm.

5. What is the total surface area of a rectangular solid whose length, width, and height are 8 inches, 6 inches, and 10 inches respectively?

6. What is the total surface area of a sphere whose radius is 2 m?

B

7. Find the total surface area of a cube that is 17.6 in on a side.

8. Find the outside surface area of a rectangular box (with an open top) that is 1.4 ft wide, 2 ft long, and 1 ft high.

9. Find the outside lateral surface area of a circular tube of radius 1.3 m and length 17 m.

10. Find the outside lateral surface area of a cylindrical container that has a radius of 8 in and a height of 10 in.

11. Find the outside surface area of a bowl that is in the shape of one-half a sphere of radius 4 in.

12. Find the surface area of a sphere that has a radius of 12.3 cm.

13. Find the lateral surface area of a pyramid with slant height of 8 in and whose base is a square 6 in on a side.

14. Find the lateral surface area of a pyramid with slant height of 12 in and whose base is a rectangle with dimensions 9 in by 15 in.

C

15. Find the total surface area, in square feet, of a cube that is 9.6 in on a side.

16. Find the total surface area, in square metres, of a rectangular solid with dimensions 124 cm × 96 cm × 32 cm. (To the nearest tenth.)

17. Find the total surface area of a cylinder with a diameter of 16 in and a height of 3.5 ft.

ANSWERS

1. _____

2. _____

3. _____

4. _____

5. _____

6. _____

7. _____

8. _____

9. _____

10. _____

11. _____

12. _____

13. _____

14. _____

15. _____

16. _____

17. _____

18. Find the total surface area of a right circular cone whose base has a diameter of 43 cm and whose slant height is 130 mm.

19. Find the total surface area of a regular pyramid whose base is a square 6 ft on a side and whose slant height is 3 yd.

20. Find the surface area, in square yards, of a sphere whose diameter is 10.6 ft. (To the nearest tenth.)

21. Find the total surface area of a geometric solid whose base is a cube with an edge of 9 in, topped with a regular pyramid. The pyramid has one face of the cube as its base and has a slant height of 11 in.

22. Find the total surface area of a geometric solid whose base is a cylinder of radius 10 in and height of 14 in, topped by a right circular cone. The cone uses the end of the cylinder as its base and has a slant height of 9 in.

23. Find the total surface area of the geometric solid (ice cream cone) that is a right circular cone with a slant height of 5 in and a diameter of 8 in, topped with a hemisphere. The hemisphere has the same diameter as the cone.

24. Find the total surface area of a geometric solid that has a rectangular base with dimensions $5' \times 10' \times 4'$ topped with a cylinder. The cylinder sits on a $10' \times 5'$ face of the rectangle and has a radius of $3'$ and a height of $2'$.

D

25. How many square feet of sheet metal are needed to make a cylindrical duct that is $8'$ in diameter and $35'$ in length?

26. How many square feet of sheet metal will it take to make a cylindrical tank that is to be 20 inches in diameter and 4 feet in height? (Total surface area.)

27. A T.P. tent is in the shape of a cone. How many square feet of canvas are needed to make a T.P. tent (with floor) whose base has a radius of 8 feet and whose slant height is 15 feet? (Total surface area.)

28. The interior of a storage tank which is a rectangular solid is to be painted. One gallon of paint will cover 300 square feet of surface. How many gallons are needed if the tank's length, width, and height are 20 feet, 18 feet, and 12 feet respectively? (Total surface area.)

29. A three-sided shed is to be painted with redwood stain. Two sides have dimensions $10' \times 15'$ and the back has dimensions $10' \times 25'$. If only the outside is to be painted and a gallon of stain covers 60 sq ft, what is the cost of the stain if it sells for $15.75 a gallon?

30. A circular swimming pool that is uniformly 6 ft deep is to be painted. The pool is 15 ft in diameter, and one gallon of paint will cover 42 sq ft. What will be the cost of the paint if it sells for $21.98 a gallon?

E   *Maintain Your Skills*

Do the indicated operations.

31. $(2.3)(3.2) - (.34)(.1)$

32. $.15 \div (.3 + .2) - .01$

33. $2.45 + (1.2)(2.1) - 8(.2)$

34. $32.1 + (9.8)(.02)$

35. $(14.82) \div .3 + 2.3 - .8$

226

Chapter 4
POST-TEST

1. **(Obj. 47)** Find the circumference of a circle whose radius is 6 inches. (Let $\pi \approx 3.14$.)

2. **(Obj. 46)** Evaluate the following formula if $a = 32$ and $t = 3$.

$$d = \frac{1}{2}\,at^2 - \frac{1}{2}\,a(t-1)^2$$

3. **(Obj. 49)** Find the area of the following figure.

4. **(Obj. 45)** Convert 480 minutes to hours.

5. **(Obj. 41)** 4 yd 1 ft 3 in = ? in

6. **(Obj. 43)** 4682 mℓ = ? ℓ

7. **(Obj. 42)** *Subtract:*   5 yd 1 ft
                2 yd 2 ft

8. **(Obj. 50)** The bed of a truck is 9 feet long, 6 feet wide, and 4 feet deep. How many scoops of gravel, each containing 1 cubic yard (27 cubic feet) will it take to fill the truck bed?

9. **(Obj. 50)** What is the weight of a metal ingot that is 2 ft long, 18 in wide, and 6 in high, if the metal weighs 58.75 lb per cubic foot?

10. **(Obj. 42)** *Add:*   6 ft 4 in
           2 ft 5 in
           3 ft 8 in

11. **(Obj. 48)** Find the area of a circle whose radius is 6 inches. (Let $\pi \approx 3.14$.)

12. **(Obj. 44)** *Subtract:*   28 km 450 m
                  6 km 580 m

13. **(Obj. 41)** 2.75 gal = ? quarts

14. **(Obj. 43)** 2.3 km = ? m

ANSWERS

15. _____

16. _____

17. _____

18. _____

19. _____

20. _____

21. _____

22. _____

23. _____

24. _____

25. _____

15. **(Obj. 44)** *Add:* 4 km 350 m
　　　　　　　　　　 9 km 840 m

16. **(Obj. 45)** Convert 50 miles per hour to feet per second. (To the nearest tenth.)

17. **(Obj. 45)** Norm's automobile averages 25 miles per gallon as he travels to and from work. If he travels 52 miles each day, round trip, how many gallons of gasoline does he use in one week? He works five days each week.

18. **(Obj. 47)** What is the perimeter of the following figure?

19. **(Obj. 51)** How many gallons of paint are needed to paint the walls and ceiling of a room that is 18 feet long, 15 feet wide, and 7.5 feet high if one gallon will cover 450 square feet? (Assume that an allowance has already been made for windows and doorways.)

20. **(Obj. 49)** What is the area of the shaded portion of the following figure?

21. **(Obj. 41)** 6 ft = ? inches

22. **(Obj. 51)** What is the lateral surface area of a cylinder that has a radius of 16 in and a height of 48 in? (Let $\pi \approx 3.14$.)

23. **(Obj. 46)** Evaluate the following formula if $r = 2$ and $\pi \approx 3.14$. (Round to nearest tenth.)

$$V = \frac{4}{3} \pi r^3$$

24. **(Obj. 43)** 462 cm = ? m

25. **(Obj. 48)** Find the area of a rectangle whose length is 18 inches and whose width is 12 inches.

5

Signed
Numbers

b. _____

2. _____

3. _____

4. _____

5. _____

6. _____

7. _____

8. _____

9. _____

10. _____

11. _____

12. _____

13. _____

14. _____

15. _____

16. _____

17. _____

18. _____

19. _____

20. _____

21. _____

22. _____

23. _____

24. _____

25. _____

Chapter 5
PRE-TEST

The problems in the following pre-test are a sample of the material in the chapter. You may already know how to work some of these. If so, this will allow you to spend less time on those parts. As a result, you will have more time to give to the sections that you find more difficult or that are new to you. The answers are in the back of the text.

Do the indicated operations.

1. **(Obj. 52, 53)** a. $-(+32) = ?$
   b. $|+32| = ?$

2. **(Obj. 54)** $(-17) + (-33) + (45)$

3. **(Obj. 54)** $(-7.5) + (8.33)$

4. **(Obj. 54)** $\left(\dfrac{7}{8}\right) + \left(-\dfrac{3}{4}\right) + \left(-\dfrac{1}{8}\right)$

5. **(Obj. 54)** $(-123) + (65) + (-77) + (8)$

6. **(Obj. 55)** $(-38) - (44)$

7. **(Obj. 55)** $(134) - (-97)$

8. **(Obj. 55)** $\left(\dfrac{7}{8}\right) - \left(-\dfrac{1}{4}\right)$

9. **(Obj. 55)** $(-35.87) - (-21.04)$

10. **(Obj. 56)** $(-8)(5)(-16)$

11. **(Obj. 56)** $(-2)^3(-12)$

12. **(Obj. 56)** $\left(\dfrac{5}{9}\right)\left(-\dfrac{2}{5}\right)$

13. **(Obj. 57)** $(-98) \div (7)$

14. **(Obj. 57)** $(-455) \div (-13)$

15. **(Obj. 57)** $(-6.45) \div (.43)$

16. **(Obj. 57)** $\left(\dfrac{5}{8}\right) \div \left(-\dfrac{2}{5}\right)$

17. **(Obj. 58)** $(8 - 21)(21 - 8)$

18. **(Obj. 58)** $(8^2 - 33)(17 - 3^2)$

19. **(Obj. 58)** $(-81 + 18) \div (-7) + (5)(-15)$

20. **(Obj. 58)** $(-7)^2 - (-4^3)$

21. **(Obj. 58)** $\left(\dfrac{5}{3} - 4\right) + (-51) \div 17$

22. **(Obj. 58)** $\left(\dfrac{9}{14}\right) \div \left(-\dfrac{3}{7}\right) + (4)\left(-\dfrac{2}{3}\right)$

23. **(Obj. 55)** The temperature in Cheyenne ranged from a high of 26° to a low of $-5°$ within a three-day period. What was the drop in temperature expressed as a signed number?

24. **(Obj. 57)** The Downtown Stompers lost 40.8 yards (-40.8) in 12 plays during an inter-league football game. What was their average loss per play, expressed as a signed number?

25. **(Obj. 58)** What Celsius temperature is equal to a reading of $-4°F$? Use the formula

$$C = \frac{5}{9}(F - 32)$$

5.1 Signed Numbers, Opposites, and Absolute Value

OBJECTIVES

52. Determine the opposite of a signed number.
53. Determine the absolute value of a signed number.

APPLICATION

At the New York Stock Exchange, positive and negative numbers are used to record changes in stock prices on the board. What is the opposite of a gain of three-eighths $\left(+\dfrac{3}{8}\right)$?

VOCABULARY

Positive numbers are numbers greater than zero. Negative numbers are numbers less than zero. Zero is neither positive nor negative. Positive numbers, negative numbers, and zero together are called *signed numbers.*

The *opposite* of a signed number is that number which, on the number line, is the same number of units from zero but on the opposite side of zero. Zero is its own opposite. The symbol -5 can be read as the opposite of 5.

Now that we have the negative numbers (the opposites of the numbers of arithmetic), we have two more useful classifications of numbers. The *integers* are the whole numbers and their opposites ($\ldots -3, -2, -1, 0, 1, 2, 3, 4, \ldots$). The *rational numbers* are the numbers of arithmetic and their opposites. The rational numbers are described as all numbers that can be written in the form $\dfrac{a}{b}$, where a and b are integers and $b \neq 0$. The following chart shows the classification of these numbers:

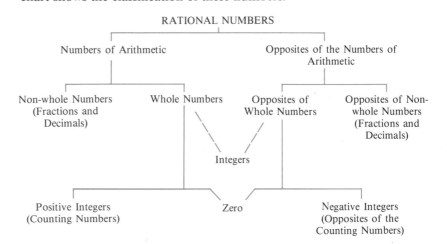

The *absolute value* of a signed number is the number of units (on the number line) between the number and zero. The symbol $|7|$ is read as the absolute value of 7.

HOW AND WHY

The application in this section involves numbers other than whole numbers, fractions, or decimals. These numbers are sometimes called signed numbers since there is another symbol (sign) that is used to identify them. Consider the following number line:

Notice that those symbols to the right of zero are those that were used in arithmetic, while those to the left of zero have a dash, or minus sign, in front of them. The dash is read as "negative" and indicates that the number is to the left of zero on the number line. Those numbers to the right of zero are called positive (and can be written with or without a plus sign). Zero is neither positive nor negative.

The following symbols name the numbers indicated in the column on the right.

7   Seven or positive seven

$+\dfrac{1}{2}$ One-half or positive one-half

-3 Negative three

$-.12$ Negative twelve hundredths

0   Zero (neither positive nor negative)

> **The opposite of a signed number is located on the opposite side of zero and the same number of units from zero on the number line.**
> **If a is a positive number, then**
>
> $-(a) = -a$ **(The opposite of a positive number is negative.)**
>
> $-(-a) = a$ **(The opposite of a negative number is positive.)**

For instance: $-(-3) = 3$

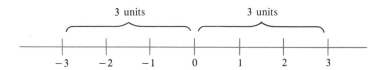

> **The absolute value of a signed number is determined by the number of units between the number and zero on the number line.**
> **If a is a positive number (or zero), then**
>
> $|a| = a$
>
> $|-a| = a$

For instance, $|-5| = 5$

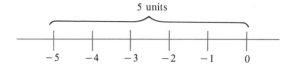

EXAMPLES

a. The opposite of 8 is written -8, which is negative eight (on the opposite side and 8 units from 0).

b. The opposite of negative 9 is written $-(-9)$, and this is 9 (on the opposite side and 9 units from 0).

c. -12 is read negative twelve or the opposite of twelve.

d. The absolute value of six is written $|6|$ and is 6.

e. The absolute value of negative sixteen is written $|-16|$ and is 16.

f. $|0| = 0$; the absolute value of zero is zero.

APPLICATION SOLUTION

The opposite of a positive number $\left(+\dfrac{3}{8}\right)$ is a negative number, so the opposite of the stock gain is a loss of three-eighths or $-\dfrac{3}{8}$.

WARM UPS

Find the opposite of each of the following numbers.

1. -6	**2.** 4	**3.** 7
4. $-\dfrac{1}{2}$	**5.** -1.5	**6.** 3.2

Find the absolute value of each of the following numbers.

7. $|-2|$	**8.** $|3|$	**9.** $\left|-\dfrac{7}{8}\right|$
10. $|-5.3|$	**11.** $\left|\dfrac{1}{9}\right|$	**12.** $|-92|$

Exercises 5.1

A

Find the opposite of each of the following numbers.

1. -3　　　　　**2.** 5　　　　　**3.** -2.1

Find the absolute value of each of the following numbers.

4. $|-1|$　　　**5.** $\left|\dfrac{1}{6}\right|$　　　　**6.** $|-1.2|$

B

Find the opposite of each of the following numbers.

7. -31　　　　**8.** 13　　　　**9.** $-\dfrac{2}{3}$

10. -2.35

Find the absolute value of each of the following numbers.

11. $|-7|$　　　**12.** $\left|-\dfrac{4}{5}\right|$　　　**13.** $|.035|$

14. $\left|-\dfrac{25}{8}\right|$

C

Find the opposite of each of the following numbers.

15. $3\dfrac{1}{8}$　　　**16.** $-22\dfrac{2}{3}$　　　**17.** 0.23

18. -103.6　　　**19.** $-\left(4\dfrac{15}{16}\right)$

Find the absolute value of each of the following numbers.

20. $\left|-\dfrac{7}{18}\right|$　　　**21.** $|-21.75|$　　　**22.** $|2.03|$

23. $|6-1.5|$　　　**24.** $|0|$

ANSWERS

1. _____
2. _____
3. _____
4. _____
5. _____
6. _____
7. _____
8. _____
9. _____
10. _____
11. _____
12. _____
13. _____
14. _____
15. _____
16. _____
17. _____
18. _____
19. _____
20. _____
21. _____
22. _____
23. _____
24. _____

25. _____

26. _____

27. _____

28. _____

29. _____

30. _____

31. _____

32. _____

33. _____

34. _____

35. _____

D

25. At the New York Stock Exchange a stock is shown to have made a gain of two and seven-eighths points $\left(+ 2\frac{7}{8} \right)$. What is the opposite of this gain?

26. At the American Stock Exchange a stock is shown to have taken a loss of five-eighths points $\left(- \frac{5}{8} \right)$. What is the opposite of this loss?

27. On a thermometer, temperatures above zero are listed as positive and those below zero as negative. What is the opposite of a reading of $-12°C$?

28. On a thermometer like that in problem 27, what is the opposite of a reading of $23°C$?

29. The modern calendar counts the years after the birth of Christ as positive numbers (A.D. 1976 or $+1976$). Years before Christ are listed using negative numbers (2045 B.C. or -2045). What is the opposite of 1875 B.C. or -1875?

30. The empty weight center of gravity of an airplane is determined. A generator is installed at a moment of -300. At what moment could a weight be placed so that the center of gravity remains the same? (Moment is the product of a quantity, such as weight, and its distance to a fixed point. In this application the moments must be opposites to keep the same center of gravity.)

E *Maintain Your Skills*

Find the missing number.

31. 5 ft = ? in   **32.** 3.5 gal = ? pt   **33.** 8 yd 1 ft = ? in

34. *Add:*

  5 gal 2 qt 1 pt
  3 gal 1 qt 1 pt
  2 gal 3 qt 1 pt

35. *Subtract:*

  48 min 37 sec
  18 min 49 sec

5.2 Addition of Signed Numbers

OBJECTIVE	**54.** Find the sum of two signed numbers if one or both are negative.

APPLICATION

> While an airplane is being reloaded, 577 pounds of baggage and mail are removed (-577 pounds) and 482 pounds of baggage and mail are loaded on ($+482$ pounds). What net change in weight should the cargo master report?

VOCABULARY

No new vocabulary.

HOW AND WHY

Positive and negative numbers are used to show quantities having opposite characteristics.

$+482$ lb may show 482 pounds loaded.
-577 lb may show 577 pounds unloaded.
$+27$ dollars may show 27 dollars earned.
-19 dollars may show 19 dollars spent.

Using these characteristics we can find the sum of signed numbers. The various combinations follow.

$(27) + (-19) = ?$

If you think of this as 27 dollars earned (positive) and 19 dollars spent (negative), the end result is 8 dollars left in your pocket (positive). So $(27) + (-19) = 8$.

$(-23) + (15) = ?$

If you think of this as 23 dollars spent (negative) and 15 dollars earned (positive), the end result is that you still owe 8 dollars (negative). So $(-23) + (15) = -8$.

$(-5) + (-2) = ?$

If you think of this as 5 dollars spent (negative) and 2 dollars spent (negative), the end result is 7 dollars spent (negative). So $(-5) + (-2) = -7$. Shortcut: $-(5 + 2) = -7$.

The sum of two signed numbers can also be illustrated on the number line.

$(-2) + (-5) = -7$ *Start at zero*

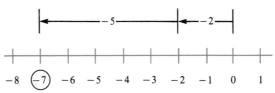

$$(-8) + (3) = -5$$

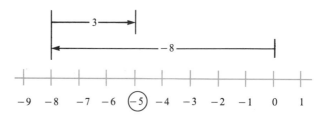

The sum of two positive numbers is positive.	$(+6) + (+9) = +15$
The sum of two negative numbers is negative.	$(-10) + (-11) = -21$

The sum of a positive and a negative number is found by subtracting and

a. choosing a positive answer if the larger number (in absolute value) is positive; $(-10) + (+15) = \ +5$

b. choosing a negative answer if the larger number (in absolute value) is negative; $(-20) + \ (+7) = -13$

c. choosing zero if the numbers are opposites. $(-11) + (+11) = \quad 0$

EXAMPLES

a. $(35) + (-15) = 20$

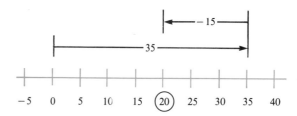

b. $(-48) + (29) = -19$

c. $(-.33) + (-.17) = -.50$

d. $(-8.3) + (25) = 16.7$

e. $\dfrac{4}{5} + -\dfrac{3}{2} = -\dfrac{7}{10}$

f. $(-13) + (72) + (-20) + (-23) = 16$

g. $(-.1) + (.5) + (-3.4) + (.8) = -2.2$

h. $-\dfrac{3}{4} + \dfrac{7}{8} + \left(-\dfrac{1}{2}\right) + \left(-\dfrac{7}{8}\right) = -\dfrac{5}{4}$

i. Calculator example

If your calculator has a $\boxed{+/-}$ key, you can use it to add positive and negative numbers.

$$(-63) + (48) + (-61) + (-14) = \,?$$

ENTER	DISPLAY
$\boxed{63}$	63.
$\boxed{+/-}$	$-63.$
$\boxed{+}$	$-63.$
$\boxed{48}$	48.
$\boxed{+}$	$-15.$

61	61.
+/−	−61.
+	−76.
14	14.
+/−	−14.
=	−90.

APPLICATION SOLUTION

The net weight change of the airplane is the sum of the pounds of baggage removed and loaded so

net change = (−577) + (+482) = −95

The cargo master should report a net change of −95 pounds or 95 pounds lighter.

WARM UPS

Add:

1. $(-3) + (-2)$ **2.** $(-4) + (5)$

3. $(6) + (-2)$ **4.** $(-1) + (-1)$

5. $(-8) + (10)$ **6.** $(7) + (-3)$

7. $(-6) + (0)$ **8.** $(-8) + (8)$

9. $(-7) + (-8)$ **10.** $(-7) + (4) + (-1)$

11. $(11) + (-9) + (-3)$ **12.** $(-5) + (-7) + (5)$

Exercises 5.2

ANSWERS

1. _____

2. _____

3. _____

4. _____

5. _____

6. _____

7. _____

8. _____

9. _____

10. _____

11. _____

12. _____

13. _____

14. _____

15. _____

16. _____

17. _____

18. _____

19. _____

20. _____

21. _____

22. _____

23. _____

24. _____

25. _____

26. _____

A

Add:

1. $(-10) + (-3)$ **2.** $(7) + (-14)$ **3.** $(11) + (-3)$

4. $(-15) + (-12)$ **5.** $(-17) + (9)$ **6.** $(-39) + (17)$

B

Add:

7. $(48) + (-39)$ **8.** $(-28) + (-47)$

9. $(-432) + (439)$ **10.** $(-239) + (225)$

11. $(-41) + (-144)$ **12.** $(-197) + (200)$

13. $(-8.2) + (-3.2)$ **14.** $(6.3) + (-3.7)$

C

Add:

15. $\left(\frac{5}{12}\right) + \left(-\frac{7}{12}\right)$ **16.** $\left(-\frac{4}{15}\right) + \left(-\frac{6}{15}\right)$

17. $(-1.4) + (4.1)$ **18.** $(-31) + (18) + (-63) + (22)$

19. $(-19) + (54) + (-68) + (6)$ **20.** $(-17.3) + (14.6) + (6.9)$

21. $(-11.23) + (15.36) + (-27.22)$ **22.** $\left(1\frac{2}{3}\right) + \left(-4\frac{5}{6}\right) + \left(-2\frac{1}{4}\right)$

23. $\left(13\frac{2}{9}\right) + \left(-8\frac{2}{3}\right) + \left(-2\frac{2}{3}\right)$ **24.** $(-.035) + (.751) + (-.111)$

D

25. While an airplane is being reloaded, 1234 pounds of baggage and mail are removed (-1234 pounds) and 1184 pounds of baggage and mail are loaded on ($+1184$ pounds). What net change in weight should the cargo master report?

26. At another stop the plane in problem 25 unloads 977 pounds of baggage and mail and takes on 1055 pounds. What net change should the cargo master at this airport report?

27. During the current fiscal year the LeBaroque Coffee House recorded the following quarterly earnings (positive numbers represent profit, negative numbers represent loss): $3456, −$507, −$498, $4007. What was the total profit (or loss) for the year?

28. The Central West Book Depository handles most textbooks for local schools. On September 1, the inventory was 18,340 volumes. During the month the depository had the following transactions (positive numbers represent volumes received, negative numbers represent shipments): 1800, −356, −843, −500, 250, −650. What is the inventory at the end of the month?

29. The Central Southwest Book Depository had 8,066 volumes on September 1. During the month the depository had the following transactions: −1044, 213, −555, −178, −840, 33, −229. What is the inventory at the end of the month?

30. The change in altitude of a plane in flight was measured every ten minutes. The figures betwen 3 PM and 4 PM were:

3:00 PM 30,000 ft initially	(+30,000)
3:10 PM increase of 220 ft	(+220)
3:20 PM decrease of 200 ft	(−200)
3:30 PM increase of 55 ft	(+55)
3:40 PM decrease of 110 ft	(−110)
3:50 PM decrease of 25 ft	(−25)
4:00 PM increase of 40 ft	(+40)

What was the altitude of the plane at 4 PM? (Hint: Find the sum of the initial altitude and the six measured changes between 3 and 4 PM).

E   *Maintain Your Skills*

Find the missing number.

31. 350 mℓ = ? ℓ   32. 3 kg + 35 hg = ? g

33. 3.5 km − 1600 m = ? km

34. *Add:*   35. *Subtract:*

5 km 520 m   18 kℓ 3 hℓ
7 km 370 m   13 kℓ 5 hℓ

242

ANSWERS TO WARM UPS (5.2) **1.** -5 **2.** 1   **3.** 4   **4.** -2 **5.** 2   **6.** 4

7. -6 **8.** 0   **9.** -15 **10.** -4 **11.** -1 **12.** -7

5.3 Subtraction of Signed Numbers

OBJECTIVE

55. Find the difference of two signed numbers.

APPLICATION

Viking II recorded high and low temperatures of $-22°$ and $-107°$ for one day on the surface of Mars. What was the change in temperature for that day?

VOCABULARY

Recall that subtraction is the inverse of addition, that is, $10 - 4 = 6$ because $4 + 6 = 10$.

HOW AND WHY

$(11) - (8) = ?$ asks what must be added to 8 to equal 11.
$(11) - (8) = 3$, because $8 + 3 = 11$.
    $(-3) - (+5) = ?$ asks what number must be added to $+5$ to equal -3.
$(-3) - (+5) = -8$, because $5 + (-8) = -3$.
    $(-4) - (-7) = ?$ asks what number must be added to -7 to equal -4.
$(-4) - (-7) = 3$, because $(-7) + (3) = -4$.
Compare:

$11 - 8$ and $11 + (-8)$ [both equal 3]

$(-3) - (5)$ and $(-3) + (-5)$ [both equal -8]

$(-4) - (-7)$ and $(-4) + (7)$ [both equal 3]

Every subtraction problem can be turned into an addition problem by adding the opposite of the second number. This is the common procedure for subtracting signed numbers.

    The difference of two signed numbers can also be shown on a number line. (Draw the arrow for the second number in the opposite direction as you would for addition.)

$(-4) - (-7) = 3$

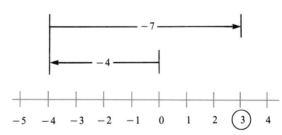

To find the difference between two numbers, add the opposite of the number that is to be subtracted.
    If a and b are any two numbers,

$a - b = a + (-b)$

EXAMPLES

a. $(45) - (-33) = (45) + (33) = 78$

b. $(-47) - (29) = (-47) + (-29) = -76$

c. $(-38) - (-56) = (-38) + (56) = 18$

d. $(88) - (114) = -26$

e. $\dfrac{4}{5} - \left(-\dfrac{1}{2}\right) = \dfrac{4}{5} + \left(\dfrac{1}{2}\right) = 1\dfrac{3}{10}$

f. $(16) - (-4) - (7) = 13$

g. $(-.21) - (-3.4) - (-.15) - (.42) = 2.92$

h. $-\dfrac{3}{4} - \dfrac{7}{8} - \left(-\dfrac{1}{2}\right) - \left(-\dfrac{1}{8}\right) = -1$

i. Calculator example

$(-34.8) - (-49.3)$

ENTER	DISPLAY
34.8	34.8
+/−	−34.8
−	−34.8
49.3	49.3
+/−	−49.3
=	14.5

APPLICATION SOLUTION

The change in temperature for one day can be found by subtracting the lowest temperature from the highest one.

net change $= (-22) - (-107) = 85$

The change in temperature recorded for that day was 85°.

WARM UPS

Subtract:

1. $(6) - (4)$ **2.** $(-6) - (4)$

3. $6 - (-4)$ **4.** $(-6) - (-4)$

5. $(-8) - (-5)$ **6.** $(10) - (7)$

7. $(-6) - (-1)$ **8.** $(-8) - (8)$

9. $(-7) - (-8)$ **10.** $(-7) - (4) - (-1)$

11. $(11) - (-9) - (-3)$ **12.** $(-5) - (-7) - (5)$

Exercises 5.3

A

Subtract:

1. $(-10) - (-3)$ **2.** $(10) - (-3)$ **3.** $(-10) - (3)$

4. $(-15) - (-12)$ **5.** $(-15) - (12)$ **6.** $(15) - (-12)$

B

Subtract:

7. $(16) - (-25)$ **8.** $(56) - (77)$

9. $(-42) - (-33)$ **10.** $(-49) - (43)$

11. $(-45) - (45)$ **12.** $(-83) - (-83)$

13. $(-8.2) - (-3.2)$ **14.** $(6.3) - (-3.7)$

C

Subtract:

15. $\left(\dfrac{5}{12}\right) - \left(-\dfrac{7}{12}\right)$ **16.** $\left(-\dfrac{1}{4}\right) - \left(\dfrac{1}{2}\right)$

17. $\left(-\dfrac{5}{8}\right) - \left(-\dfrac{2}{5}\right)$ **18.** $(3.45) - (4.22)$

19. $(7.33) - (-2.48)$ **20.** $(-43.8) - (-62.9) - (1)$

21. $(-25) - (-18.7) - (-2)$ **22.** $\left(-\dfrac{2}{3}\right) - \left(\dfrac{1}{2}\right) - \left(\dfrac{1}{4}\right)$

23. $\left(\dfrac{6}{7}\right) - \left(\dfrac{9}{10}\right) - (-1)$ **24.** $(-.035) - (.751) - (-.111)$

1. ——————————
2. ——————————
3. ——————————
4. ——————————
5. ——————————
6. ——————————
7. ——————————
8. ——————————
9. ——————————
10. ——————————
11. ——————————
12. ——————————
13. ——————————
14. ——————————
15. ——————————
16. ——————————
17. ——————————
18. ——————————
19. ——————————
20. ——————————
21. ——————————
22. ——————————
23. ——————————
24. ——————————

ANSWERS

25. _____

26. _____

27. _____

28. _____

29. _____

30. _____

31. _____

32. _____

33. _____

34. _____

35. _____

D

25. The surface of one of Jupiter's satellites was measured for one week. The highest temperature recorded was $-75°$ and the lowest was $-138°$. What was the difference in the extreme temperatures for that week?

26. A satellite of Saturn showed a high temperature of $-31°$ and a low of $-122°$. What was the change in temperature for this period?

27. At the beginning of the month, Joe's bank account had a balance of $315.65. At the end of the month, the account was overdrawn by $63.34 ($-63.34). If there were no deposits during the month, what was the total amount of the checks Joe wrote? (Hint: subtract the ending balance from the original balance.)

28. At the beginning of the month, Jana's bank account had a balance of $237.65. At the end of the month, the account was overdrawn by $14.22 ($-14.22). If there were no deposits during the month, what was the total amount of the checks Jana wrote? (Hint: subtract the ending balance from the original balance.)

29. What is the difference in altitude between the highest point in the United States and the lowest point?
    Highest point: Mt. McKinley is 20,320 ft above sea level ($+20,320$)
    Lowest point: Death Valley is 282 ft below sea level (-282)

30. What is the difference in altitude between the highest point in the world and the lowest point?
    Highest point: Mt. Everest is 29,028 ft above sea level ($+29,028$)
    Lowest point: Part of the Dead Sea shore is 1291 ft below sea level (-1291)

E   *Maintain Your Skills*

Convert each of the following to the indicated unit of measure.

31. 75 in $= ?$ ft   32. 5 m 9 cm $= ?$ cm

33. 8 mi 675 ft $= ?$ ft   34. $\dfrac{154 \text{ feet}}{\text{second}}$ to $\dfrac{\text{miles}}{\text{hour}}$

35. $\dfrac{6 \text{ tons}}{\text{ft}}$ to $\dfrac{\text{pounds}}{\text{in}}$

5.4 Multiplication of Signed Numbers

OBJECTIVE

56. Find the product of two signed numbers.

APPLICATION

The formula for converting a temperature measurement from Fahrenheit to Celsius is $C = \frac{5}{9}(F - 32)$. What Celsius measure is equal to $18°F$?

VOCABULARY

When the sum of two numbers is zero, the numbers are opposites. So if $a + b = 0$, then $a = -b$ (a is the opposite of b). The distributive law of multiplication over addition is $a(b + c) = ab + ac$. So $2(3 + 4) = 2 \cdot 3 + 2 \cdot 4$.

HOW AND WHY

Consider the following problem:

$3 + (-3) = 0$ *Sum of opposites*

$4[3 + (-3)] = 4 \cdot 0$ *Multiply by 4*

$4[3 + (-3)] = 0$ *Multiplication property of zero*

$4(3) + 4(-3) = 0$ *Distributive law*

$12 + 4(-3) = 0$ *Multiply*

So $4(-3) = -12$ *$4(-3)$ must be the opposite of 12 because their sum is 0*

This example can be repeated for any sum equal to zero, and it leads to the conclusion that the product of a positive number and a negative number is a negative number. The fact that $4(-3) = -12$ can also be shown by interpreting multiplication as repeated addition: $4(-3) = (-3) + (-3) + (-3) + (-3) = -12$.

The product of two negatives can be found in a similar fashion. Consider:

$(2) + (-2) = 0$ *Sum of opposites*

$(-3)[(2) + (-2)] = (-3)(0)$ *Multiply by -3*

$(-3)[(2) + (-2)] = 0$ *Multiplication property of 0*

$(-3)(2) + (-3)(-2) = 0$ *Distributive law*

$-6 + (-3)(-2) = 0$ *Product of a positive and a negative*

So $(-3)(-2) = -(-6) = 6$ *$(-3)(-2)$ is added to -6 for a sum of 0*

Therefore, we conclude that the product of two negative numbers is a positive number.

In summary:

The product of two positive numbers is positive. $(+3)(+11) = +33$
The product of two negative numbers is positive. $(-8)(-12) = +96$
The product of a positive and a negative number
is negative. $(+7)(-18) = (-18)(+7)$
                                                    $= -126$

EXAMPLES

a. $(-6)(5) = -30$

b. $(2)(-3.3) = -6.6$

c. $\left(-\dfrac{1}{3}\right)\left(\dfrac{1}{5}\right) = -\dfrac{1}{15}$

d. $(5)(-.7) = -3.5$

e. $(-6)(-3)(-4) = (18)(-4) = -72$

f. $(-6)(-7)^2 = (-6)(49) = -294$

g. $\left(-\dfrac{2}{7}\right)\left(-\dfrac{3}{8}\right) = \dfrac{6}{56} = \dfrac{3}{28}$

h. $(-121)(-5) = 605$

i. $(-2)^2(-3)^3 = (-2)(-2)(-3)(-3)(-3)$

$\qquad\qquad = (4)(-27)$

$\qquad\qquad = -108$

j. $(-6)(-3)^2(-4) = (-6)(9)(-4) = 216$

k. **Calculator example**

$(-28)(-4.6)(-2.9)$

ENTER	DISPLAY
28	28.
+/−	−28.
×	−28.
4.6	4.6
+/−	−4.6
×	128.8
2.9	2.9
+/−	−2.9
=	−373.52

APPLICATION SOLUTION

Formula: $\quad C = \dfrac{5}{9}(F - 32),\ F = 18°$

Substitute: $\quad C = \dfrac{5}{9}(18 - 32)$

Simplify: $\quad C = \dfrac{5}{9}(-14) = -7\dfrac{7}{9} \approx -7.8$

Answer: $\quad$ 18° Fahrenheit is approximately −7.8° Celsius.

WARM UPS

Multiply:

1. $(7)(-9)$

2. $(-7)(-9)$

3. $(-7)(9)$

4. $5(-12)$

5. $(-5)(12)$

6. $(-5)(-12)$

7. $(2)(-17)$

8. $(-2)(-17)$

9. $(-2)(17)$

10. $(6)(-9)(-1)$

11. $(-6)(-9)(-1)$

12. $(6)(9)(-1)$

Exercises 5.4

ANSWERS

1. _____

2. _____

3. _____

4. _____

5. _____

6. _____

7. _____

8. _____

9. _____

10. _____

11. _____

12. _____

13. _____

14. _____

15. _____

16. _____

17. _____

18. _____

19. _____

20. _____

21. _____

22. _____

23. _____

24. _____

A

Multiply:

1. $(-5)(-2)$ **2.** $(-1)(18)$ **3.** $(-17)(0)$

4. $(8)(-6)$ **5.** $(-5)(19)$ **6.** $(-1)(-1)(-1)$

B

Multiply:

7. $(-12)(-3)^2$ **8.** $(-7)(-2)^2(-3)$ **9.** $(2)(5)(-6)$

10. $(-3.02)(6)$ **11.** $(-.6)(-.4)(.2)$ **12.** $(-6)^3$

13. $\left(-\frac{3}{4}\right)\left(-\frac{1}{2}\right)$ **14.** $\left(-\frac{2}{3}\right)\left(\frac{3}{8}\right)$

C

Multiply:

15. $(-20.16)(-1.5)$ **16.** $(3.8)(-15.2)$

17. $(5-3)(6-8)$ **18.** $(3)(-.7)(-5)$

19. $\left(\frac{1}{2}+\frac{1}{3}\right)\left(5-6\frac{1}{3}\right)$ **20.** $(-2)^3(-4)(-5)$

21. $(5)(-1)^7(2)(-11)$ **22.** $\left(-\frac{2}{3}\right)\left(\frac{1}{2}\right)\left(\frac{1}{4}\right)$

23. $\left(\frac{6}{7}\right)\left(\frac{9}{10}\right)(-1)$ **24.** $(-1)^5(2.99)^2$

249

25. _____

25. Use the formula in the application to find the Celsius measure that is equal to 14°F.

26. _____

26. Use the formula in the application to find the Celsius measure that is equal to −13°F.

27. _____

27. For five consecutive weeks Ms. Obese recorded a loss of 2.5 lb. If each loss is represented by −2.5 lb, what was the total weight loss for the five weeks, expressed as a signed number?

28. _____

28. For six consecutive weeks Mr. Obese recorded a loss of 1.9 lb. If each loss is represented by −1.9 lb, what was the total weight loss for the six weeks, expressed as a signed number?

29. _____

29. The Industrial Stock Average sustained twelve straight days of a 2.83 point decline (−2.83) each day. What was the total decline during the twelve-day period, expressed as a signed number?

30. _____

31. _____

30. Safeway Stores offered, as a loss leader, 10 lb of sugar at a loss of twelve cents per bag (−12¢). If 560 bags were sold during the sale, what was the total loss, expressed as a signed number?

32. _____

33. _____

E   *Maintain Your Skills*

Evaluate the following expressions, given that $a = 5$, $b = 7$, $c = 12$, and $d = 800$.

34. _____

31. $2a + 3b + 4c$ **32.** $ab + bc$ **33.** a^2b^2

35. _____

34. $2d - 10c$ **35.** $(b - a)(d - c)$

ANSWERS TO WARM UPS (5.4) **1.** -63 **2.** 63 **3.** -63 **4.** -60 **5.** -60 **6.** 60

7. -34 **8.** 34 **9.** -34 **10.** 54 **11.** -54 **12.** -54

5.5 Division of Signed Numbers

OBJECTIVE

57. Find the quotient of two signed numbers.

APPLICATION

The coldest temperature in Eycee Northland for each of five days was $-15°$, $-5°$, $-4°$, $2°$, and $-3°$. What was the average of these temperatures?

VOCABULARY

Recall that division is the inverse of multiplication, that is, $12 \div 3 = 4$ because $3 \cdot 4 = 12$.

HOW AND WHY

To divide two signed numbers we must find the number that when multiplied by the divisor equals the dividend.

$(-8) \div (4) = ?$ asks what number when multiplied by 4 equals -8. So, $(-8) \div (4) = -2$, because $(4)(-2) = -8$.

$(-21) \div (-3) = ?$ asks what number when multiplied by -3 equals -21. So, $(-21) \div (-3) = 7$, because $(-3)(7) = -21$.

$15 \div (-5) = ?$ asks what number when multiplied by -5 equals 15. So, $15 \div (-5) = -3$, because $(-5)(-3) = 15$.

These examples lead to rules similar to those for multiplication.

The quotient of two positive numbers is positive. $(+18) \div (+6) = +3$
The quotient of two negative numbers is positive. $(-32) \div (-8) = +4$
The quotient of a positive and a negative number is negative. $(+68) \div (-4) = -17$

EXAMPLES

a. $(-6) \div (3) = -2$

b. $(-8.6) \div (4.3) = -2$

c. $\left(-\frac{3}{4}\right) \div \left(-\frac{1}{2}\right) = \frac{3}{2}$

d. $(-7) \div \left(-\frac{4}{3}\right) = \frac{21}{4}$

e. $(1.44) \div (-.3) = -4.8$

f. $(68) \div (-4) = -17$

g. Calculator example

$(-4428) \div (-36)$

ENTER	DISPLAY
4428	4428.
+/−	−4428.
÷	−4428.
36	36.
+/−	−36.
=	123.

APPLICATION SOLUTION

The average temperature is the sum of the five temperatures divided by 5:

$(-15) + (-5) + (-4) + (2) + (-3) = -25$

$(-25) \div 5 = -5$

So the average low temperature was $-5°$.

WARM UPS

Divide:

1. $(-56) \div (8)$ **2.** $(-56) \div (-8)$

3. $(56) \div (-8)$ **4.** $(-10) \div (5)$

5. $(10) \div (-2)$ **6.** $(22) \div (-2)$

7. $(-44) \div (2)$ **8.** $(-44) \div (-4)$

9. $(18) \div (-1)$ **10.** $(-33) \div (11)$

11. $(-56) \div (2)$ **12.** $(-51) \div (-3)$

Exercises 5.5

A

Divide:

1. $(-56) \div (4)$ **2.** $(57) \div (-3)$ **3.** $(-58) \div (-2)$

4. $(60) \div (-4)$ **5.** $(-63) \div (-3)$ **6.** $(-68) \div (4)$

B

Divide:

7. $(6.06) \div (-6)$ **8.** $(-4.95) \div (-.9)$

9. $(-15.5) \div (.05)$ **10.** $(.65) \div (-.13)$

11. $(-6.5) \div (.13)$ **12.** $(-65) \div (-1.3)$

13. $(.065) \div (-8)$ **14.** $(-650) \div (16)$

C

Divide:

15. $[(-5)(-2)] \div (-10)$ **16.** $36 \div [(-3)(-6)]$

17. $[28(-1)] \div (-7)$ **18.** $(-1.2) \div (.06)$

19. $(100) \div (-5)^2$ **20.** $(6 - 15) \div 3$

21. $(3 - 8 - 5) \div (-2)$ **22.** $\left(-\frac{3}{4}\right) \div \left(-\frac{3}{5}\right)$

23. $\left(-\frac{2}{3}\right) \div \left(5\frac{1}{6}\right)$ **24.** $[(-8.2)(3.6)] \div (-1.8)$

1.
2.
3.
4.
5.
6.
7.
8.
9.
10.
11.
12.
13.
14.
15.
16.
17.
18.
19.
20.
21.
22.
23.
24.

ANSWERS

25.

26.

27.

28.

29.

30.

31.

32.

33.

34.

35.

D

25. In a remote part of northern Canada the daily high temperature was recorded for one week. The temperatures were $-10°$, $-4°$, $-1°$, $-1°$, $3°$, $-6°$, and $-9°$. What was the average of these temperatures?

26. In northern Norway the daily low temperature was recorded for six days. The temperatures were $-22°$, $-18°$, $-19°$, $-14°$, $-18°$, and $-11°$. What was the average of these temperatures?

27. Mr. Gambit lost \$836 (-836) in twelve straight hands of poker at a casino in Reno. What was his average loss per hand, expressed as a signed number?

28. Ms. Gambit lost \$122 (-122) in seven straight hands of blackjack at a casino in Las Vegas. What was her average loss per hand, expressed as a signed number?

29. The membership of the Burlap Baggers Investment Club took a loss of \$183.66 ($-$183.66$) on the sale of stock. If there are six equal members in the club, what is each member's share of the loss, expressed as a signed number?

30. The membership of the Back-to-the-Wall Investment Club took a loss of \$98.34 ($-$98.34$) on the sale of stock. If there are eleven equal members in the club, what is each member's share of the loss, expressed as a signed number.

E   *Maintain Your Skills*

Find the perimeter or circumference of each of the following. Let $\pi \approx 3.14$.

31. A rectangle that is 2 ft by 5 ft.

32. A triangle with sides 5 m, 4 m, and 8 m.

33. A circle with radius 10 in.

34. A square with side 8 cm.

35. A five-sided figure (pentagon) with sides 5 ft, 8 ft, 6 ft, 4 ft, and 7 ft.

5.6 Review: Order of Operations

OBJECTIVE

58. Do any combination of operations (addition, subtraction, multiplication, or division) on signed numbers in the conventional order.

APPLICATION

Does $-40°$ F equal $-40°$ C? To check, substitute -40 for F and also for C in the formula

$$F = \frac{9}{5}C + 32$$

and tell whether the statement is true.

VOCABULARY

No new vocabulary.

HOW AND WHY

The conventional order of operations for signed numbers is the same as that for whole numbers. (See Section 1.5.)

EXAMPLES

a. $(-64) + (-22) \div (2) = (-64) + (-11)$ *Division performed first*
$= -75$ *Addition performed next*

b. $(-12)(5) - (54) \div (-3) = (-60) - (-18)$ *Multiplication and division performed first*
$= -42$ *Subtraction performed next*

c. $(3) \div (-.8) + 2(-1.7) = (-3.75) + (-3.4)$
$= -7.15$

d. $6 - \left(\frac{1}{2}\right)(-5) = 6 - \left(-2\frac{1}{2}\right)$
$= 6 + 2\frac{1}{2}$
$= 8\frac{1}{2}$

e. $(-2)^3 + 4^2 - 51 \div 3 = -8 + 16 - 51 \div 3$ *Exponents first*
$= -8 + 16 - 17$ *Division next*
$= -9$ *Addition and subtraction last*

f. $75 - (5^2 - 6^2) = 75 - (25 - 36)$ *Operations inside parentheses first (exponents)*
$= 75 - (-11)$ *Subtraction next (still inside parentheses)*
$= 86$

g. Calculator example

$(13)(-12) - (-42) \div (-7) = ?$

If your calculator has algebraic logic:

ENTER	DISPLAY
13	13.
×	13.
12	12.
+/−	−12.
−	−156.
42	42.
+/−	−42.
÷	−42.
7	7.
+/−	−7.
=	−162.

If your calculator has parentheses:

ENTER	DISPLAY
(	0.
13	13.
×	13.
12	12.
+/−	−12.
)	−156.
−	−156.
(	0.
42	42.
+/−	−42.
÷	−42.
7	7.
+/−	−7.
)	6.
=	−162.

APPLICATION SOLUTION

Formula: $F = \dfrac{9}{5}C + 32,\ F = -40°$ and $C = -40°$

Substitute: $(-40) = \dfrac{9}{5}(-40) + 32$

Simplify: $-40 = -72 + 32$ ***Multiply first***

$-40 = -40$ ***Add***

Answer: It is true that $-40°$ is the same temperature on both Fahrenheit and Celsius scales.

WARM UPS

Do the indicated operations.

1. $(-3)(-1) + 4$
2. $(-4) \div (-2) + 1$
3. $(-4) + (-6)(-2)$
4. $(-4) - (-6)(-2)$
5. $(3) + (6)(-1)$
6. $(8) - (5) \div (-5)$
7. $(-3)(-4) - (2)$
8. $(-9)(-2) + (-3)$
9. $(10) - (-3) + (3)$
10. $(-5) + (7) - (2)$
11. $(3)(-2)^2 + (-12)$
12. $(-2)^3 - (6)^2$

Exercises 5.6

A

Do the indicated operations.

1. $(2)(-8) + 9$

2. $19 - (4)(-2)$

3. $(-2)(3) + (5)(-2)$

4. $(-3)(-7) - 10$

5. $2(-6 + 5) - 3$

6. $(-7)(6) \div (-3)$

B

Do the indicated operations.

7. $(-35) \div (-5) + (-2)(-3)$

8. $6(-10 + 4) - 33 \div (-11)$

9. $(-20) \div (-8)(-0.75)$

10. $28 \div (-4) - 7 + (-1)(-8) - 6$

11. $5^2 - 8^2$

12. $(-4)^3 \div (-2)^4$

13. $(7 - 8)^5 + (-5 + 3)^5$

14. $(3^2 - 2^3)(-2) - (-3)^3$

C

Do the indicated operations.

15. $[(-8)9 + 20](-3) - 25$

16. $[6 + (-8)(-5)] - (-8)$

17. $(-126) - [(-6)(6) - 6]$

18. $[(-5)(-6) - 3(8)] \div \{6(-3) + 6(6)\}$

19. $(-8)9 + [20(-3) - 25]$

20. $[(-8)(-6) + 4(11)] \div [7(-8) - 6(-8)]$

21. $-1(-2)^2 + (-11)^2$

22. $(-1)^2(2) - (11)^2$

23. $[14^2 - 20^2] - [75 - 12^2]$

24. $84 - 9^2 + 6(-2) - (-5)^3$

1. _____
2. _____
3. _____
4. _____
5. _____
6. _____
7. _____
8. _____
9. _____
10. _____
11. _____
12. _____
13. _____
14. _____
15. _____
16. _____
17. _____
18. _____
19. _____
20. _____
21. _____
22. _____
23. _____
24. _____

25. _____

26. _____

27. _____

28. _____

29. _____

30. _____

31. _____

32. _____

33. _____

34. _____

35. _____

D

25. Does $20°F$ equal $-20°C$? To check, substitute 20 for F and -20 for C in the formula $F = \frac{9}{5}C + 32$ and tell whether the statement is true.

26. Does $14°F$ equal $-10°C$? To check, substitute 14 for F and -10 for C in the formula $F = \frac{9}{5}C + 32$ and tell whether the statement is true.

27. The temperature at 5 PM was $18°C$. If the temperature dropped $.6°$ every hour until midnight, we can find the midnight temperature by calculating the value of

$T = 18 + (7)(-0.6)$

What was the midnight temperature?

28. If the temperature at 6 AM was $-10°F$ and rose $3.4°F$ every hour until 11 AM, what was the temperature at 11 AM?

29. To check whether $x = -6$ is a solution to the equation

$x^2 - 4x - 7 = 53$

we substitute -6 for x and find the value of

$(-6)(-6) - 4(-6) - 7$

Check whether this expression equals 53.

30. Substitute $x = -1\frac{1}{2}$ in the following equation and check whether it is a solution.

$\frac{3}{4}\left(x + \frac{2}{3}\right) - \frac{3}{8} = -1$

E   *Maintain Your Skills*

Find the area of each of the following. Let $\pi \approx 3.14$.

31. A square with sides 12 cm.

32. A rectangle that is 12 ft by 18 ft.

33. A triangle with base 18 cm and height 12 cm.

34. A parallelogram with base 16 ft and height 16 ft.

35. A circle with radius 25 cm.

1. _____

2. _____

3. _____

4. _____

5. _____

6. _____

7. a. _____

b. _____

8. _____

9. _____

10. _____

11. _____

12. _____

13. _____

14. _____

15. _____

16. _____

17. _____

18. _____

19. _____

20. _____

21. _____

22. _____

23. _____

24. _____

25. _____

Chapter 5
POST-TEST

Do the indicated operations.

1. **(Obj. 54)** $(-21) + (-17) + (42) + (-18)$

2. **(Obj. 57)** $\left(-\dfrac{3}{8}\right) \div \left(\dfrac{3}{10}\right)$ 3. **(Obj. 55)** $\left(-\dfrac{7}{15}\right) - \left(-\dfrac{3}{5}\right)$

4. **(Obj. 58)** $(36 - 42)(-18 + 6)$

5. **(Obj. 58)** $(-3 + 18) - (21 - 7) + 3$

6. **(Obj. 54)** $(-3.65) + (4.72)$

7. **(Obj. 52, 53)** a. $-(-21) = ?$
                     b. $|-21| = ?$

8. **(Obj. 55)** $(-37) - (-41)$

9. **(Obj. 58)** $(-16 + 4) \div 3 + 2(-6)(-3)$

10. **(Obj. 57)** $(-88) \div (-22)$

11. **(Obj. 56)** $(-7)(-9)(2)$

12. **(Obj. 56)** $(|-6|)(-4)(-1)(-1)$

13. **(Obj. 55)** $(-57.9) - (32.5)$

14. **(Obj. 58)** $(-21)(3) - (-6)(-7)$

15. **(Obj. 57)** $(30.66) \div (-0.6)$

16. **(Obj. 54)** $\left(-\dfrac{1}{3}\right) + \left(\dfrac{5}{6}\right) + \left(-\dfrac{1}{2}\right) + \left(-\dfrac{1}{6}\right)$

17. **(Obj. 57)** $(-56) \div (-7)$

18. **(Obj. 55)** $(18) - (-25)$

19. **(Obj. 58)** $-7(3 - 11)(-2) - 4(-3 - 5) + 18 \div (-6)$

20. **(Obj. 56)** $\left(-\dfrac{1}{3}\right)\left(\dfrac{6}{7}\right)$

21. **(Obj. 54)** $(-17) + (-36)$ 22. **(Obj. 58)** $3(-7) + 25$

23. **(Obj. 55)** The temperature in Chicago ranged from a high of 12°F to a low of -9°F within a twenty-four hour period. What was the drop in temperature expressed as a signed number?

24. **(Obj. 54)** A stock on the West Coast Exchange opened at $6\dfrac{5}{8}$ on Monday. It recorded the following changes during the week: Monday, $+\dfrac{1}{8}$; Tuesday, $-\dfrac{3}{8}$; Wednesday, $+1\dfrac{1}{4}$; Thursday, $-\dfrac{7}{8}$; Friday, $+\dfrac{1}{4}$. What was its closing price on Friday?

25. **(Obj. 58)** What Fahrenheit temperature is equal to a reading of -10°C? use the formula

$$F = \dfrac{9}{5}C + 32$$

ANSWERS TO WARM UPS (5.6)   **1.** 7   **2.** 3   **3.** 8   **4.** -16 **5.** -3 **6.** 9

7. 10   **8.** 15   **9.** 16   **10.** 0   **11.** 0   **12.** -44

6

Equations
and
Polynomials

1. _____

2. . _____

3. _____

4. _____

5. _____

6. _____

7. _____

8. _____

9. _____

10. _____

11. _____

12. _____

13. _____

14. _____

15. _____

16. _____

17. _____

18. _____

19. _____

20. _____

21. _____

22. _____

23. _____

24. _____

25. _____

Chapter 6
PRE-TEST

The problems in the following pre-test are a sample of the material in the chapter. You may already know how to work some of these. If so, this will allow you to spend less time on those parts. As a result, you will have more time to give to the sections that you find more difficult or that are new to you. The answers are in the back of the text.

1. **(Obj. 59)** Solve for x: $6x - 12 = 54$

2. **(Obj. 59)** Solve for x: $x + 3y = 5$

3. **(Obj. 60)** Express as a power of y: $(y^5)(y^2)(y^8)$

4. **(Obj. 60)** Express as a power of z: $\dfrac{(z^3)^3}{z^3}$

5. **(Obj. 61)** *Evaluate:* $(3^3)^3$

6. **(Obj. 62)** *Add:* $8y + 6y + 7y$

7. **(Obj. 62)** Do the indicated operations:
$$(2abc)(3ac) - \frac{12a^2bc^3}{2c} + 4a^2bc^2$$

8. **(Obj. 63)** *Simplify:* $(6a + 2b - c) - (4a - b + c)$

9. **(Obj. 63)** *Add:* $4x + 4y + 5$, $8x - 2y + 7$, and $3x + 4y + 16$

10. **(Obj. 63)** Solve for x: $8x + 12 = 5x - 24$

11. **(Obj. 64)** Solve for a: $(4a - 12) - (5a + 4) = 6$

12. **(Obj. 66)** *Multiply:* $(x - 7)(x + 4)$

13. **(Obj. 66)** *Multiply:* $(3x + 5)(2x + 3)$

14. **(Obj. 66)** Solve for x: $4(x - 2) - 3(x + 2) = 3$

15. **(Obj. 66)** Solve for x: $8(2x - 3) + 2(x - 4) = 4$

16. **(Obj. 67)** *Divide:* $(-5a^2bc + 15abc^2 - 10ab^2c) \div 5abc$

17. **(Obj. 67)** *Divide:* $(2x^2 + 11x + 12) \div (x + 4)$

18. **(Obj. 68)** *Factor:* $12ac - 6abc + 18abc^2$

19. **(Obj. 69)** *Factor:* $x^2 - x - 56$

20. **(Obj. 69)** *Factor:* $12x^2 + 5x - 2$

21. **(Obj. 70)** *Factor:* $25x^2 - 64$

22. **(Obj. 71)** Solve for x: $x^2 - 26x + 25 = 0$

23. **(Obj. 71)** Solve for x: $x^2 - 15x + 56 = 0$

24. **(Obj. 65)** The sum of three times a number and 5 is 50. What is the number?

25. **(Obj. 65)** The perimeter of a rectangle is 94 m. The length is two more than two times the width. What are the length and the width of the rectangle?

6.1 Linear Equations

59. Solve a linear equation that can be written in the form $ax + b = c$ when a, b, and c are given.

APPLICATION

A Fahrenheit thermometer scale shows a temperature of 95 degrees. Use the formula $F = \dfrac{9}{5}C + 32$ to find the corresponding temperature on the Celsius scale. (That is, solve for C in $95 = \dfrac{9}{5}C + 32$.)

VOCABULARY

An *equation* is a statement that two expressions are equal. Every equation has a left side (or member) and a right side (or member) with an "=" between them.

$$2 + 87 \quad = \quad 98 - 9$$
LEFT SIDE   RIGHT SIDE

An equation may be true. For example, $8 + 17 = 25$ is always true, while $3x = 21$ is true when $x = 7$.

An equation may be false. For example, $7 - 11 = 4$ is always false, while $3x = 21$ is false when $x = -2$.

A *literal equation* such as $a + b = c$ contains more than one letter.

Recall that a *variable* is a placeholder for a number.

In a *linear equation* the largest exponent of the variable is the number 1 ($3x^1 - 2 = 13$ or $3y - 7 = 2y$, but not $x^2 - 3x = 4$).

An *identity* is an equation that is true for all replacements of the variable ($2x = 2x$ or $3y + 4y = 7y$).

A *conditional equation* is true for some replacements, but not for all. For example, $2x = 7$ is true if $x = 3.5$ or $3\dfrac{1}{2}$ but not if $x = 2$.

The *solution* or root of an equation is the set of all number replacements that, when substituted for the variable, make the equation true. In the equation $2x = 6$, $x = 3$ is the solution.

HOW AND WHY

To find the solution of a linear equation, we write a series of equations (called equivalent equations) that all have the same solution, until we arrive at an equation of the form $x = $ (some number) or (some number) $= x$. For instance, $x = 6$ or $-\dfrac{13}{4} = x$.

Fundamental Laws of Equations

1. **Adding the same number to (or subtracting the same number from) both sides of an equation does not change the solution. (That is, it yields an equivalent equation.)**
2. **Multiplying or dividing both sides of an equation by the same number (except zero) does not change the solution. (This also yields an equivalent equation.)**

To solve the equation $3x - 2 = 13$, we can write this series of equations:

$3x - 2 = 13$	*Original equation*
$3x - 2 + 2 = 13 + 2$	*Add 2 to both sides*
$3x = 15$	*Simplify*
$\dfrac{3x}{3} = \dfrac{15}{3}$	*Divide both sides by 3*
$x = 5$	*Simplify (5 is the solution or root)*
$3(5) - 2 = 13$	*To check, replace x with 5 in the original equation*
$15 - 2 = 13$	
$13 = 13$	

The shortcuts for this process involve doing some of the steps mentally (see Example d).

To find the solution of an equation that can be written in the form $ax + b = c$, apply the fundamental laws of equations until the variable is isolated on one side of the equation.

Literal equations are solved using the same rules. For example:

Solve $y + z = k$ for y.

$y + z = k$	*Original Equation*
$y + z - z = k - z$	*Subtract z from both sides*
$y = k - z$	*Simplify*

The solution is $k - z$.

EXAMPLES

Solve.

a. $6x = 372$

$$\frac{6x}{6} = \frac{372}{6} \qquad \text{\textit{Divide both sides by 6}}$$

$$x = 62 \qquad \text{\textit{Simplify}}$$

The solution is 62.

b. $\qquad 2x + \dfrac{5}{6} = \dfrac{2}{3}$

$$2x + \frac{5}{6} - \frac{5}{6} = \frac{2}{3} - \frac{5}{6} \qquad \text{\textit{Subtract} } \frac{5}{6} \text{ \textit{from both sides}}$$

$$2x = -\frac{1}{6} \qquad \text{\textit{Simplify}}$$

$$\frac{2x}{2} = -\frac{1}{6} \div 2 \qquad \text{\textit{Divide both sides by 2}}$$

$$x = -\frac{1}{12} \qquad \text{\textit{Simplify}}$$

The solution is $-\dfrac{1}{12}$.

c.
$$7.4 = 1.8x - 16$$
$$7.4 + 16 = 1.8x - 16 + 16$$
$$23.4 = 1.8x$$
$$\frac{23.4}{1.8} = \frac{1.8x}{1.8}$$
$$13 = x$$

d. $13 - 6x = 5$
$$-6x = -8$$
$$x = \frac{-8}{-6}$$
$$x = \frac{4}{3}$$

e. Solve for x.

$$2x + 3y = 5$$

$2x + 3y - 3y = 5 - 3y$ *Subtract 3y from both sides*

$2x = 5 - 3y$ *Simplify*

$\dfrac{2x}{2} = \dfrac{5 - 3y}{2}$ *Divide both sides by 2*

$x = \dfrac{5 - 3y}{2}$ *Simplify*

APPLICATION SOLUTION

Formula:	$F = \dfrac{9}{5}C + 32$	*Original formula*
Substitute:	$95 = \dfrac{9}{5}C + 32$	*Replace F by 95*
Solve:	$95 - 32 = \dfrac{9}{5}C + 32 - 32$	*Subtract 32 from both sides*
	$63 = \dfrac{9}{5}C$	*Simplify*
	$\dfrac{5}{9} \cdot 63 = \dfrac{5}{9} \cdot \dfrac{9}{5}C$	*Multiply both sides by $\dfrac{5}{9}$*
	$35 = C$	*Simplify*
Check:	$95 = \dfrac{9}{5}(35) + 32$	*To check, replace C by 35 and F by 95 in the original formula*
	$95 = 9(7) + 32$	*Simplify (reduce or divide 35 by 5)*
	$95 = 63 + 32$	
	$95 = 95$	
Answer:	So $95°$F measures the same temperature as $35°$C.	

WARM UPS

Solve:

1. $x + 7 = 19$ **2.** $y + 10 = 14$ **3.** $w - 4 = 11$

4. $z - 5 = 12$ **5.** $5x = 55$ **6.** $8y = -48$

7. $x + 15 = 7$ **8.** $y - 5 = -3$ **9.** $-3w = 39$

10. $\dfrac{x}{3} = 5$

Solve for x:

11. $x + y = 12$ **12.** $x + 2y = 15$

Exercises 6.1

A

Solve:

1. $5x = 30$

2. $2x + 6 = 30$

3. $x - 9 = 21$

4. $3x - 11 = -2$

5. $3 = 7 + x$

6. Solve for a: $a + b = c$

B

Solve:

7. $7x - 1 = 8$

8. $4x + 5 = 21$

9. $3x - 8 = 7$

10. $5a - 12 = -3$

11. $6b + 16 = -2$

12. $3y + 4 = 13$

Solve for x:

13. $2x + y = 4$

14. $2x + 5y = 7$

C

Solve:

15. $\frac{1}{4}y + \frac{1}{4} = \frac{1}{2}$

16. $\frac{5}{3}z - \frac{1}{2} = \frac{7}{6}$

17. $(3)(4) = 2x + 12$

18. $4a + 1 = 3$

19. $4 - 3b = 1$

20. $15 = 6 - 3z$

21. $.2x + .8 = 1.4$

22. $\frac{1}{4} - \frac{1}{2}a = \frac{2}{3}$

23. Solve for ℓ:

$2\ell + 2w = P$

24. Solve for b:

$A = \frac{1}{2}bh$

1.
2.
3.
4.
5.
6.
7.
8.
9.
10.
11.
12.
13.
14.
15.
16.
17.
18.
19.
20.
21.
22.
23.
24.

25. _____

D

25. A Fahrenheit thermometer scale shows a temperature of 140°. Use the formula $F = \frac{9}{5}C + 32$ to find the corresponding temperature on the Celsius scale.

26. _____

26. A Fahrenheit thermometer scale shows a temperature of 41 degrees. Use the formula $F = \frac{9}{5}C + 32$ to find the corresponding temperature on the Celsius scale.

27. _____

27. The formula for the perimeter of a rectangle is $P = 2\ell + 2w$. If the perimeter of a certain rectangle is 40 cm and its length is 12 cm, what is the measure of its width?

28. _____

28. The formula for the perimeter of a rectangle is $P = 2\ell + 2w$. If the perimeter of a certain rectangle is 52 ft and its width is 12 ft, what is the measure of its length?

29. _____

29. The velocity of a falling object at any time t with initial velocity v_0 is given by $v_t = v_0 + 32.2t$. What is the initial velocity if the velocity after 5 seconds is 188 feet per second?

30. _____

30. The velocity of a falling object at any time t with initial velocity v_0 is given by $v_t = v_0 + 32.2t$. What is the initial velocity if the velocity after 8 seconds is 300 feet per second?

31. _____

E   *Maintain Your Skills*

31. What is the area of the infield of a race track in which the straightaways are 200 yards long and the ends are semicircles? The radius of each of the semicircles is 30 yards. Let $\pi \approx 3.14$.

32. _____

32. A piece of metal is 12 inches square. A circle whose diameter is 10 inches is cut from the square. What is the area of the piece remaining? Let $\pi \approx 3.14$.

33. _____

33. Find the volume of a sphere whose diameter is 10 m. Let $\pi \approx 3.14$. (To the nearest tenth.)

34. _____

34. Find the volume of a cylinder that has a radius of 22 inches and is $3\frac{1}{2}$ feet high. Let $\pi \approx 3.14$.

35. _____

35. Find the volume of a pyramid whose base is a square with side 60 feet and whose height is 40 feet.

ANSWERS TO WARM UPS (6.1)

ANSWERS TO WARM UPS (6.1)	**1.** $x = 12$	**2.** $y = 4$	**3.** $w = 15$	**4.** $z = 17$	**5.** $x = 11$
	6. $y = -6$	**7.** $x = -8$	**8.** $y = 2$	**9.** $w = -13$	**10.** $x = 15$
	11. $x = 12 - y$	**12.** $x = 15 - 2y$			

6.2 The Laws of Exponents

OBJECTIVES

60. Multiply or divide powers with the same base.
61. Raise a power to a power.

APPLICATION

The volume of a cylinder is given by the formula

$V = \pi r^2 h$

where r represents the radius and h represents the height. If a certain cylinder has a height that is the cube of the radius, express the volume in terms of the radius.

VOCABULARY

See Chapter 1 for the meanings of *exponent, base,* and *power.*

HOW AND WHY

The following laws of exponents are shortcuts for algebraic multiplication and division. Each law is followed by an example that is done first the long way and second by using the law.

Law 1: $x^a \cdot x^b = x^{a+b}$

Example:

$x^3 \cdot x^5 = (xxx)(xxxxx)$

$\qquad = xxxxxxxx$

$\qquad = x^8$

or more quickly, $x^3 \cdot x^5 = x^{3+5} = x^8$

Law 2: $\dfrac{x^a}{x^b} = x^{a-b}$

Example:

$\dfrac{x^9}{x^3} = \dfrac{xxxxxxxxx}{xxx}$

$\qquad = xxxxxx \qquad \text{(by reducing)}$

$\qquad = x^6$

or more quickly, $\dfrac{x^9}{x^3} = x^{9-3} = x^6$

Law 3: $(x^a)^b = x^{ab}$

Example:

$(x^3)^5 = x^3 \cdot x^3 \cdot x^3 \cdot x^3 \cdot x^3$

$\qquad = x^{3+3+3+3+3}$

$\qquad = x^{15}$

or more quickly, $(x^3)^5 = x^{3 \cdot 5} = x^{15}$

Law 4: $(xy)^a = x^a y^a$

Example:

$(wz)^5 = wz \cdot wz \cdot wz \cdot wz \cdot wz$

$\qquad = wwwww \cdot zzzzz$

$\qquad = w^5 z^5$

or more quickly, $(wz)^5 = w^5 z^5$

Law 5: $\left(\dfrac{x}{y}\right)^a = \dfrac{x^a}{y^a}$

Example:

$\left(\dfrac{x}{4}\right)^3 = \dfrac{x}{4} \cdot \dfrac{x}{4} \cdot \dfrac{x}{4}$

$\qquad = \dfrac{x^3}{4^3}$ or $\dfrac{x^3}{64}$

or more quickly, $\left(\dfrac{x}{4}\right)^3 = \dfrac{x^3}{4^3}$

EXAMPLES

a. $y^5 \cdot y^6 \cdot y = y^{5+6+1} = y^{12}$ *Law 1*

b. $w^a \cdot w^2 = w^{a+2}$

c. $\dfrac{t^{10}}{t^2} = t^{10-2} = t^8$ *Law 2*

d. $\dfrac{a^7}{a^5} = a^{7-5} = a^2$

e. $(3^2)^{10} = 3^{2 \cdot 10} = 3^{20}$ *Law 3*

f. $(4w^3)^2 = 4^2 \cdot (w^3)^2 = 16\,w^6$ *Law 4*

g. $(x^2 y^5)^3 = (x^2)^3 (y^5)^3 = x^6 y^{15}$

h. $\left(\dfrac{2}{r^2}\right)^3 = \dfrac{2^3}{(r^2)^3} = \dfrac{8}{r^6}$ *Law 5*

i. $\left(\dfrac{a^3}{b^4}\right)^5 = \dfrac{(a^3)^5}{(b^4)^5} = \dfrac{a^{15}}{b^{20}}$

APPLICATION SOLUTION

To express the volume of the cylinder in terms of the radius r, replace h by r^3 and multiply.

Formula: $V = \pi r^2 h,\ h = r^3$

Substitute: $V = \pi r^2 (r^3)$

                 $V = \pi r^5$

Answer: The volume of the cylinder in terms of the radius is πr^5.

WARM UPS

Evaluate:

1. $(3^2)^2$

2. $(2^3)^2$

3. $\left(\dfrac{1}{3}\right)^2$

4. $\dfrac{3^7}{3^4}$

5. $[(2^2)(3^2)]^2$

6. $\left(\dfrac{2}{3}\right)^2$

7. $\left(\dfrac{1}{4}\right)^3$

8. $(5^3)^2$

Express each of the following as a power or a product of powers of x, y, or z.

9. $(x^3)^3$

10. $(z^4)^2$

11. $(x^3)(x^4)$

12. $\dfrac{y^5}{y^2}$

Exercises 6.2

A

Evaluate:

1. $(2^2)^3$

2. $(2^3)(2^2)$

3. $\dfrac{2^5}{2^3}$

4. $(x^3)(x^2)$

5. $(y^2)^5$

6. $(x^2y^3)^6$

B

Evaluate:

7. $[(2^3)(3^2)]^2$

8. $(6^3)^2$

9. $\left(\dfrac{2}{3}\right)^4$

Express each of the following as a power or a product of powers of a, b, x, or y.

10. $\dfrac{(x^2)^2}{y^3}$

11. $(x^4)(x^5)(x^2)$

12. $(x^7)(x^2)$

13. $(y^3)^4$

14. $(a^3b^3)^2$

C

15. $\dfrac{x^6}{x^3}$

16. $\left(\dfrac{a^2}{b}\right)^3$

17. $(2ab^2)^3$

18. $\dfrac{a^4b^7}{a^2b^6}$

19. $(3a^4b^2x^3)^3$

20. $\dfrac{(x^3y^2)^2}{x^2y^3}$

21. $(3ax)^4(b^2y^2)^2$

22. $\dfrac{(a^3b^4)^2}{(ab^2)^3}$

23. $(x^2)^3(x)^4(x^3)^5$

24. $(y^3)^2(y^5)(y^4)^3$

D

25. The volume of a sphere is given by $V = \dfrac{1}{6}\pi d^3$ (d is the diameter). Express the volume in terms of the radius ($d = 2r$).

26. The volume of a box is given by $V = \ell wh$ (ℓ is the length; w is the width; and h is the height). If the length is the square of the width ($\ell = w^2$) and the height is the cube of the width ($h = w^3$), express the volume in terms of w, using a single exponent, and then find the volume if $w = 2$ ft.

ANSWERS
1.
2.
3.
4.
5.
6.
7.
8.
9.
10.
11.
12.
13.
14.
15.
16.
17.
18.
19.
20.
21.
22.
23.
24.
25.
26.

273

27. Express the volume of a box in terms of the width (w) if the length is equal to the width ($\ell = w$) and the height is the square of the width ($h = w^2$).

28. The area of a rectangle is given by the formula $A = \ell w$. Express the area in terms of the width (w) when the length (ℓ) is the cube of the width.

29. The volume of a right circular cylinder is given by the formula $V = \pi r^2 h$. Express the volume in terms of the radius (r) when the height (h) is the cube of the radius.

30. Express the volume of a box in terms of the width if the length is the cube of the width and the height is the square of the width.

E   *Maintain Your Skills*

31. Find the total surface area of a cube whose edge is 100 cm.

32. Find the total surface area of a cylinder whose radius is 8 m and whose height is 10 m. Let $\pi \approx 3.14$.

33. Find the lateral surface area of a pyramid whose base is a square that is 6 m on each side and whose slant height is 4 m.

34. Find the total surface area of a sphere whose diameter is 18 inches. Let $\pi \approx 3.14$.

35. Find the lateral surface area of a cone whose radius is 6 feet and whose slant height is 9 feet. Let $\pi \approx 3.14$.

ANSWERS TO WARM UPS (6.2)   **1.** 81   **2.** 64   **3.** $\dfrac{1}{9}$ **4.** 27   **5.** 1296   **6.** $\dfrac{4}{9}$

7. $\dfrac{1}{64}$ **8.** 15625   **9.** x^9 **10.** z^8 **11.** x^7 **12.** y^3

6.3 Operations on Monomials

OBJECTIVE

62. Add, subtract, multiply, and divide monomials.

APPLICATION

To help pay for Ron's education, when he was born Ron's father invested enough money to amount to $15,000 when Ron is 20 years of age. If the money was invested at a 10% (.10 or .1) interest rate, how much did Ron's father invest? (If A is the total amount of money, P is the amount invested originally, r is the rate of interest, and t is the time in years, we can use the formula $A = P + Prt$ and solve for P).

VOCABULARY

The *terms* of an algebraic expression are connected by the symbols $+$ and $-$. (The expression $3x - by + 2m^2w$ has three terms: the first is $3x$, the second is by, and the third is $2m^2w$.)

A single term is called a *monomial* and contains numerals (numerical coefficients), variable factors, and symbols for multiplication, but no indicated division by a variable. For instance, in $-\dfrac{8}{3}x^2y$, the variables are x and y; $-\dfrac{8}{3}$ is the numerical coefficient, usually called the *coefficient*.

Like terms are terms that have common variable factors. ($3xy$ and $4xy$ are like terms, but $5xy$ and $5x^2y$ are not like terms because x^2 is a factor of only $5x^2y$.)

HOW AND WHY

To add (or subtract) two or more monomials, we use a law of algebra:

DISTRIBUTIVE LAW

$ab + ac = a(b + c)$ or

$ab - ac = a(b - c)$

This law states that in each of these pairs of expressions:

$3 \cdot 5 + 3 \cdot 7$ and $3(5 + 7)$

$2\dfrac{1}{2} \cdot 4 + 3\dfrac{1}{2} \cdot 4$ and $\left(2\dfrac{1}{2} + 3\dfrac{1}{2}\right) \cdot 4$

$6(8.7) - 6(3.8)$ and $6(8.7 - 3.8)$

both expressions have the same value. Thus both sides of

$3 \cdot 5 + 3 \cdot 7 = 3(5 + 7)$

$15 + 21 = 3 \cdot 12$

$36 = 36$

have the value 36. Both sides of

$$2\frac{1}{2} \cdot 4 + 3\frac{1}{2} \cdot 4 = \left(2\frac{1}{2} + 3\frac{1}{2}\right) \cdot 4$$

$$10 + 14 = (6) \cdot 4$$

$$24 = 24$$

have the value 24, and both sides of the third example have the value 29.4.

To simplify (add) $6x + 3x$, we can use the distributive law to rewrite:

$6x + 3x = (6 + 3)x$ *Distributive law*

$ = 9x$ *Substitution*

> **To add (or subtract) like terms, add (or subtract) the coefficients and attach the common variable factor (or factors).**

To multiply two monomials we use two laws of algebra. The associative law of multiplication permits the rearrangement of parentheses.

$$(3x^3)(7xy) = (3x^3 \cdot 7)(xy) = (3x^3)(7x)(y)$$

The commutative law of multiplication permits the interchanging of factors.

$$(3x^3 \cdot 7)(x7) = (3 \cdot 7 \cdot x^3)(x7)$$

By using these laws repeatedly, the product of two monomials can be written so that the numerical coefficient precedes the variable factors.

$$(3x^3)(7xy) = (3 \cdot 7)(x^3 \cdot x)(y)$$
$$= 21x^4 y$$

> **To multiply monomials, multiply the numerical coefficients and multiply the variable factors.**

To divide two monomials we use the definition of division ($a \div b = c$, provided $bc = a$) and the law for dividing powers of the same base $\left(\dfrac{x^a}{x^b} = x^{a-b}\right)$.

For instance, $(-51a^3bc^2) \div (3abc)$ can be written

$$\frac{-51a^3bc^2}{3abc} = \frac{-51}{3} \cdot \frac{a^3}{a} \cdot \frac{b}{b} \cdot \frac{c^2}{c}$$
$$= -17a^{3-1} \cdot 1 \cdot c^{2-1}$$
$$= -17a^2c$$

> **To divide monomials, divide the numerical coefficients and divide the variable factors.**

When an equation is not in the form $ax + b = c$, we must use the rules for combining like terms along with the Fundamental Laws of Equations to solve the equation. The word "simplify" is often used to describe such operations on monomials. For instance,

$6x - 14x + 3 = 51$

$-8x + 3 = 51$ *Combine the monomials containing x or simplify*

$-8x = 48$

$x = -6$

If terms containing the unknown appear on both sides of the equation, use the Fundamental Laws of Equations to isolate the unknown on one side. The equation can then be solved as before.

$$8x - 12 = 4x + 8$$

$$8x + (-4x) - 12 = 4x + (-4x) + 8$$ *Add $-4x$ to both sides to isolate the unknown on the left side*

$$4x - 12 = 8$$ *Simplify*

$$4x = 20$$

$$x = 5$$

EXAMPLES

Do the indicated operations.

a. $4d + 5d = (4 + 5)d = 9d$

b. $7bx - 19bx = (7 - 19)bx = -12bx$

c. $13x^2 + 4x^2 = (13 + 4)x^2 = 17x^2$

d. $8x^2 - 3x$ *Cannot be simplified (or subtracted) because they are not like terms*

e. $8.3m^2n + 5.2m^2n - 12m^2n = (8.3 + 5.2 - 12)m^2n$
$$= 1.5m^2n$$

f. $(-3by)(by^2) = -3(bb)(yy^2) = -3b^2y^3$

g. $3x(-4xy)(5xy^2) = (3 \cdot -4 \cdot 5)(xxx)(yy^2) = -60x^3y^3$

h. $(-6cw^2)^3 = (-6 \cdot -6 \cdot -6)(ccc)(w^2w^2w^2) = -216c^3w^6$
$$\text{or} = (-6)^3(c)^3(w^2)^3 = -216 \ c^3w^6$$

i. $\dfrac{7x^6}{3x^2} = 2\dfrac{1}{3}x^4 \text{ or } \dfrac{7}{3}x^4 \text{ or } \dfrac{7x^4}{3}$

j. $\dfrac{-5m^3n^2}{-m} = \dfrac{-5m^3n^2}{-1 \cdot m} = \dfrac{-5}{-1} \cdot \dfrac{m^3}{m} \cdot n^2 = 5m^2n^2$

k. $\dfrac{6w^2y}{-4w^2y} = -1\dfrac{1}{2} \text{ or } -\dfrac{3}{2} \text{ or } -1.5$

l. *Solve:* $5x + 14 - 3x + 9x - 4 = 43$

$$11x + 10 = 43$$

$$11x = 33$$

$$x = 3$$

m. *Solve:* $13x + 16 = 7x + 20$

$$13x + -7x + 16 = 7x + -7x + 20$$ *Add $-7x$ to each side to isolate the terms with x on one side*

$$6x + 16 = 20$$

$$6x = 4$$

$$x = \dfrac{4}{6} = \dfrac{2}{3}$$

APPLICATION SOLUTION

Formula:	$A = P + Prt$, $A = \$15,000$, $r = .1$, $t = 20$
Substitute:	$15,000 = P + P(.1)(20)$
Solve:	$15,000 = P + 2P$
	$15,000 = 3P$
	$5,000 = P$
Answer:	Ron's father invested $5,000 when Ron was born.

WARM UPS

Do the indicated operations.

1. $2x + 5x$ **2.** $5y - 2y$

3. $8a^2 + 4a^2$ **4.** $15a - 17a$

5. $-7z + 12z$ **6.** $(3ab)(-2ab)$

7. $(2x)(3xy)(-2y)$ **8.** $(-4ab)(-3b)(-5ab)$

9. $(-4x^2)^2$ **10.** $\dfrac{8x^4}{2x^2}$

11. $\dfrac{48x^3}{6}$ **12.** $\dfrac{-64x^3}{4x^4}$

Exercises 6.3

A

Do the indicated operations.

1. $6a + 3a$

2. $3x - 5x$

3. $5a - 7a$

4. $(3ab)(-2a)(-b)$

5. $\dfrac{-18a^2b}{9a}$

6. $(-2xy^2)^2$

B

7. $6y + 11y + (-y) + (-18y)$

8. $5a^2 + 13a^2 - 11a^2 + 4a^2$

9. $st + 7st - 15st$

10. $-3xy^2 + 7xy^2 - 17xy^2 + 10xy^2$

11. $3ab + 4a + 2ab$

12. $5xy^2 - 9xy^2 + 4x^2y^2$

13. $(6ab)(-2a)(4b)$

14. $\dfrac{2ab^2}{2ab}$

C

15. $\dfrac{-16x^2y^2}{6xy}$

16. $\dfrac{-15ax^3}{5x^2}$

17. $(3x)^3 - 4x(2x)^2 + 6x^3$

18. $(3abc)(2ab) - \dfrac{15a^2b^2c^2}{5c} + 12a^2bc^2$

19. $25z^3 - \dfrac{12xyz^3}{-6xy} + \dfrac{18xy^2z^4}{-9x^2yz}$

ANSWERS

1. _____

2. _____

3. _____

4. _____

5. _____

6. _____

7. _____

8. _____

9. _____

10. _____

11. _____

12. _____

13. _____

14. _____

15. _____

16. _____

17. _____

18. _____

19. _____

ANSWERS

20.

21.

22.

23.

24.

25.

26.

27.

28.

29.

30.

31.

32.

33.

Solve:

20. $10x + 5 = 2x + 29$

21. $13a - 4 = 7a + 8$

22. $-2x + 5 = -8x + 6$

23. $\frac{1}{5}x - \frac{1}{2}x + \frac{3}{4}x = 9$

24. $x + 5 - 3x = 2x + 8 - 5x$

D

25. How much money must be invested to amount to $27,000 in 10 years if the rate of interest is 8% (.08)? ($A = P + Prt$)

26. Millie will need $26,500 in 15 years to pay the remaining mortgage on her home. How much money should she invest today at 11% (.11) to have that amount in 15 years?

27. The formula for the total surface area of a rectangular solid is $A = 2\ell w + 2\ell h + 2hw$ where ℓ is the length, w the width, and h is the height. If the width of the solid is 8 m, the length is 12 m, and the total surface area is 416 m², what is the height of the solid?

28. Using the formula in problem 27, if the height is 36 ft, the width is 16 ft and the total surface area is 3232 ft², what is the length?

E *Maintain Your Skills*

Find the opposite of the following.

29. -5 **30.** $\frac{3}{4}$

Find the absolute value of each of the following numbers.

31. $\left| -\frac{3}{4} \right|$ **32.** $|-.003|$ **33.** $|8 - 12|$

6.4 Combining Polynomials and Removing Parentheses

OBJECTIVES

63. Add and subtract polynomials.

64. Remove parentheses and brackets in an algebraic expression.

APPLICATION

The cost of manufacturing q widgets at the Super Whammy Widget Factory, using people to assemble the widgets, is given by the formula

$$C = \frac{1}{2}q^2 - 25q + 4000$$

The cost is reduced when robots are used to assemble the widgets, and is given by the formula

$$C = \frac{1}{4}q^2 - 30\,q + 5000$$

The savings that result from using robots is the difference between the costs and is given by

$$S = \left(\frac{1}{2}q^2 - 25q + 4000\right) - \left(\frac{1}{4}q^2 - 30q + 5000\right)$$

Simplify the right side of the formula.

VOCABULARY

A *polynomial* is an algebraic expression containing one or more monomials.
A *binomial* is a two-term polynomial ($3x + 4y$).
A *trinomial* is a three-term polynomial ($3x^2 - 4x + 7$).

HOW AND WHY

To combine polynomials (add, subtract, or both) we use one or more of the following laws of algebra:

Distributive law
    Example: $-3(-4x + 5) = (-3)(-4x) + (-3)(5)$

Associative law of addition
    Example: $(6x + 3y) + (4x + 5y) = 6x + (3y + 4x) + 5y$

Commutative law of addition
    Example: $6x + (3y + 4x) + 5y = 6x + (4x + 3y) + 5y$

Opposite of a monomial
    Example: $-x = -1x$

Opposite of a polynomial
    Example: $-(2x - 7y) = -1(2x - 7y) = -2x + 7y$

The following examples show how these laws are used to add polynomials.

The examples do not show all the steps, but rather show how shortcuts may be used.

$$(6x + 3y) + (4x + 5y) = (6x + 4x) + (3y + 5y)$$ *Use both the commutative and associative laws of addition*

$$= 10x + 8y$$ *Add the monomials as before*

The example can also be arranged vertically so that the monomials with a common variable are lined up.

$$
\begin{array}{r}
6x + 3y \\
4x + 5y \\
\hline
10x + 8y
\end{array}
$$

It is possible to combine more polynomials by similar procedures.

$$(-.2a + .4c) + (.6a + .5b) + (-.7a + .5b)$$
$$= [(-.2a) + .6a + (-.7a)] + (.5b + .5b) + .4c$$ *Commutative and associative laws of addition*

$$= -.3a + b + .4c$$

A vertical arrangement may again be used.

$$
\begin{array}{r}
-.2a \qquad + .4c \\
.6a + .5b \\
-.7a + .5b \\
\hline
-.3a + \quad b + .4c
\end{array}
$$

> **To add two or more polynomials, add the like terms.**

To subtract polynomials we use the same method as in the section on subtracting signed numbers. Every subtraction problem can be changed to an addition problem; then the laws of addition can be used.

$$(6x + 3y) - (4x + 5y) = (6x + 3y) + -(4x + 5y)$$ *Change to addition problem, that is, add the "opposite"*

$$= (6x + 3y) + (-4x + -5y)$$ *Find the opposite or find $-1(4x + 5y)$*

$$= (6x + -4x) + (3y + -5y)$$ *Use the commutative and associative laws of addition*

$$= 2x + -2y$$
$$= 2x - 2y$$ *Change the addition of the opposite to a subtraction problem*

When only one subtraction problem is involved, the work can be arranged vertically.

SUBTRACT   or   ADD

$$
\begin{array}{r}
6x + 3y \\
4x + 5y \\
\hline
2x - 2y
\end{array}
\qquad
\begin{array}{r}
6x + 3y \\
-4x - 5y \\
\hline
2x - 2y
\end{array}
$$

> **To find the difference of two polynomials, add the opposite of the polynomial to be subtracted from the first.**

Exercises with more than one addition or subtraction, as indicated by braces { }, brackets [ ], or parentheses ( ), can be done by using the method for each operation, working from inside out.

$$10 + 3x - \{9 - [6x - (4x + 7)]\} = 10 + 3x + -\{9 + -[6x + -(4x + 7)]\}$$
$$= 10 + 3x + -\{9 + -[6x + -4x + -7]\}$$
$$= 10 + 3x + -\{9 + -6x + 4x + 7\}$$
$$= 10 + 3x + -9 + 6x + -4x + -7$$
$$= (10 + -9 + -7) + (3x + 6x + -4x)$$
$$= -6 + 5x \text{ or } 5x - 6$$

> **To remove braces, brackets, and parentheses, perform the indicated operations, working from inside out.**

When an equation contains polynomials within parentheses, the parentheses can be removed and the equation solved as before.

$$(x + 3) + (2x - 5) - (x - 4) = 16$$

$(x + 3) + (2x + -5) + -(x + -4) = 16$ *Change all subtractions to additions*

$(x + 3) + (2x + -5) + (-x + 4) = 16$ *Find the opposite or find $-1(x + -4)$*

$2x + 2 = 16$ *Remove parentheses and combine monomials*

$2x = 14$

$x = 7$

EXAMPLES

a. *Add:* $6x, 11x, -x, -12x$

$$6x + 11x + -1x + -12x = (6 + 11 + -1 + 12)x$$
$$= 4x$$

b. *Simplify:* $(4a - 5b) + (a - 2b) + (3 - 3a)$

$$(4a - 5b) + (a - 2b) + (3 - 3a) = (4a + -5b) + (a + -2b) + (3 + -3a)$$
$$= (4a + a + -3a) + (-5b + -2b) + 3$$
$$= 2a + -7b + 3 \text{ or } 2a - 7b + 3$$

c. *Add:* $2a - 3x + 7y, 4x - 2a + 8y,$ and $3y - 5x + 3a$

$$
\begin{array}{c}
2a - 3x + 7y \\
-2a + 4x + 8y \\
\underline{3a - 5x + 3y} \\
3a - 4x + 18y
\end{array}
\qquad \text{or} \qquad
\begin{array}{c}
2a + -3x + 7y \\
-2a + 4x + 8y \\
\underline{3a + -5x + 3y} \\
3a + -4x + 18y
\end{array}
$$

or

$$3a - 4x + 18y$$

d. *Subtract:* $-6x + y^2$ from $4y^2 - 3x$

$$(4y^2 - 3x) - (-6x + y^2) = (4y^2 + -3x) + -(-6x + y^2)$$
$$\text{or } (4y^2 + -3x) + (-1)(-6x + y^2)$$
$$= (4y^2 + -3x) + (6x + -y^2)$$
$$= 3y^2 + 3x$$

e. *Simplify:* $m + [5t - (2t + 4m)]$

$$
\begin{aligned}
m + [5t - (2t + 4m)] &= m + [5t + -(2t + 4m)] \\
&= m + [5t + -2t + -4m] \\
&= m + 5t + -2t + -4m \\
&= -3m + 3t
\end{aligned}
$$

f. *Solve:* $5x - (3x + 5) = x + 13$

$$5x + -(3x + 5) = x + 13$$

$$5x + (-3x + -5) = x + 13$$

$$2x + -5 = x + 13$$

$$2x + -x + -5 + 5 = x + -x + 13 + 5 \qquad \text{\textit{Add} } -x \text{ \textit{and} 5 \textit{to each side}}$$
to isolate the terms with x on one side.

$$x = 18$$

g. *Solve:* $(3x - 7) - (4x + 9) = 3x - (2 - 4x)$

$$(3x + -7) + -(4x + 9) = 3x + -(2 + -4x)$$

$$(3x + -7) + (-4x + -9) = 3x + (-2 + 4x)$$

$$-x + -16 = 7x + -2$$

$$-x + -7x + -16 + 16 = 7x + -7x + -2 + 16$$

$$-8x = 14$$

$$x = -\frac{14}{8} = -\frac{7}{4}$$

APPLICATION SOLUTION

To simplify the right side of the formula, subtract.

$$S = \left(\frac{1}{2} q^2 - 25q + 4000\right) - \left(\frac{1}{4} q^2 - 30q + 5000\right)$$

$$S = \left(\frac{1}{2} q^2 - 25q + 4000\right) + \left(-\frac{1}{4} q^2 + 30q - 5000\right)$$

$$S = \left(\frac{1}{2} q^2 - \frac{1}{4} q^2\right) + (-25q + 30q) + (4000 - 5000)$$

$$S = \frac{1}{4} q^2 + 5q - 1000$$

WARM UPS

Simplify:

1. $4x - (3x + 1)$

2. $(a + 2) + (a - 7)$

3. $(4x + 2) + (2x - 1)$

4. $(4a + 1) - (2a + 1)$

5. $(6x - 1) - (3x - 4)$

6. $4x - (3x - 4) - 1$

7. $5 + [6x + (3x + 2)]$

8. $5a - [3a + (2a + 1)]$

Add:

9. $4x + 3y$
   $\underline{2x + 6y}$

10. $3a - 2b$
   $\underline{4a + b}$

11. $3y + 2z$
   $\underline{-2y - z}$

12. $-8x - 7y$
   $\underline{-2x - 3y}$

Exercises 6.4

A

Simplify:

1. $3x - (2x + 6)$

2. $(a - 4) + (a - 5)$

3. $(2x + 6) + (3x + 9) + (x - 1)$

4. $(2a - 4) - (a + 2)$

5. $(5a - 2) - 3a$

6. $4 + [2 - (3 - x)]$

B

Add:

7. $\begin{array}{r} 7a + 3w \\ -5a + w \end{array}$

8. $\begin{array}{r} x + y - z \\ 3x - 4y + 5z \end{array}$

9. $\begin{array}{r} 6x^2 - 7x - 8 \\ -7x^2 + 9x - 2 \end{array}$

10. $\begin{array}{r} 10r + 9s - 8t \\ -r + 5s + 2t \\ -12r - 7s - 3t \end{array}$

11. $\begin{array}{r} -5a^2 - 3b^2 + 2ac \\ -4a^2 + b^2 - 5ac \\ -a^2 + b^2 - 5ac \end{array}$

12. $\begin{array}{r} 3p + 5r \\ -6p + 7q \\ -12q - 11r \end{array}$

13. $\begin{array}{r} .8bc - 3.5ad \\ -1.1bc - .8ad \end{array}$

14. $\begin{array}{r} 6.22w - 8.91 \\ -9.13w + 5.83 \end{array}$

C

Simplify:

15. $(6x - 2y) + (7x - 8y) + 3y$

16. $(9a - b + c) - (3a + 2b - 2c) + (8b + 2c)$

17. $\left(\dfrac{1}{4}x + 2y - 1\dfrac{1}{2}z \right) + \left(1\dfrac{1}{4}x - \dfrac{3}{4}y + z \right) - \left(-\dfrac{1}{2}x - \dfrac{1}{2}y - \dfrac{1}{2}z \right)$

18. $(.34r - .55s - .17t) + (.28r + .32s - .83t) - (-.12r + s + .19t)$

19. $16 - [3x - (4x - 1)]$

20. $[8at + 3 - (-2a - 4b)] - [(a + 2b) - 4a]$

21. $2 - \{[4x - (2x + 2y)] + 6y\} + (3x - 4y)$

Solve:

22. $(2b + 8) - 3 = (4b - 8) + 1$

23. $(x - 1) - (2x - 4) - (3x + 5) = 9$

24. $6x - (2x - 6) + (x - 5) = 30 - (2x + 4)$

287

ANSWERS

1. _____
2. _____
3. _____
4. _____
5. _____
6. _____
7. _____
8. _____
9. _____
10. _____
11. _____
12. _____
13. _____
14. _____
15. _____
16. _____
17. _____
18. _____
19. _____
20. _____
21. _____
22. _____
23. _____
24. _____

25. _____

26. _____

27. _____

28. _____

29. _____

30. _____

31. _____

32. _____

33. _____

34. _____

35. _____

36. _____

D

25. The cost of manufacturing n brushes is given by $C = \frac{1}{5}n^2 - 14n + 20$. The cost of marketing the same items is given by $C = 5n^2 - 30n + 50$. Find the formula for the total cost of manufacturing and marketing the brushes.

26. The cost of manufacturing n carburetors before automation is given by $C = 15000 - 1000n + n^2$. The cost after automation is given by $C = 1000 - 750n + \frac{1}{2}n^2$. Write the formula for the savings due to automation.

27. Barbara's house has an odd-shaped yard, as shown in the diagram.

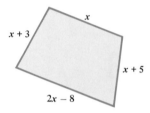

Find the perimeter of the yard in terms of x, and simplify.

28. Express the perimeter of the trapezoid shown in the diagram in terms of w, and simplify.

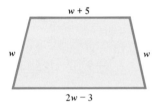

29. In a Gobblebug game, Cindy's score was four more than five times Minh's score. If the difference in their scores was 4900, what did each score?

30. Cynthia and Max are comparing their present fortunes. Cynthia has $3.82 less than twice as much as Max has. If the difference in the amount of money each has is $2.33, how much does each have?

31. Two rockets are fired at the same time from ground level. The height of the first rocket at any time t is $H_1 = 420t - 16t^2$. The height of the second and slower rocket is $H_2 = 212t - 16t^2$. Write the formula for the difference (D) in their heights at any given time.

E   *Maintain Your Skills*

Add:

32. $(-25) + (-18) + (-63) + (25)$

33. $(48) + (-26) + (-82) + (-68)$

34. $(-18.5) + (22.7) + (18.3) + (-29.7)$

35. $\left(3\frac{2}{3}\right) + \left(-7\frac{1}{2}\right) + \left(-8\frac{1}{3}\right)$

36. $(-11.23) + (18.36) + (-27.22)$

ANSWERS TO WARM UPS (6.4)   **1.** $x - 1$ **2.** $2a - 5$ **3.** $6x + 1$ **4.** $2a$ **5.** $3x + 3$

6. $x + 3$ **7.** $9x + 7$ **8.** -1 **9.** $6x + 9y$ **10.** $7a - b$

11. $y + z$ **12.** $-10x - 10y$

6.5 Solving Word Problems Using Equations

OBJECTIVE	**65. Solve word problems using equations.**

APPLICATION	Three lamps are connected in series. The first has a resistance of 150 ohms more than the second, and the third has a resistance of 100 ohms less than the second. Their total resistance (when connected in series, it is the sum of the individual resistances) is 800 ohms. What is the resistance of each of the lamps?

VOCABULARY	A *word phrase,* such as "one number is six larger than another number," can be translated into a *mathematical expression* (x is the smaller number and $x + 6$ is the larger number) by letting a letter (x) represent one of the unknowns.

| HOW AND WHY | Applications of mathematics are often written in word form. The word form usually contains an implied equation that can be solved. In order to set up the equation it is necessary to translate the word phrases into mathematical expressions.

Consider the following phrases and the mathematical translations: |
|---|---|

One number is five less than another	x is one number $x - 5$ is the other
One number is three more than twice another number	x is one number $2x + 3$ is the other
The cost of a TV set is twice the cost of a hi-fi set	x is the cost of the hi-fi set $2x$ is the cost of the TV set

Once the unknowns have been established and represented by mathematical expressions, a simpler word form of the equation can be found. If two different ways of expressing the same relationship can be found, the desired equation can be written.

Consider the following problems and the simpler word forms with the corresponding equations:

The sum of two numbers is 35. One number is five less than the other. Find the two numbers.

Simpler word form: (first number) + (second number) is 35

Equation: (x) + $(x - 5)$ $= 35$

One number is three more than twice a second number. If the smaller number is subtracted from the larger, the result is 23. Find the two numbers.

Simpler word form: (larger number) − (smaller number) is 23

Equation: $(2x + 3)$ − (x) $= 23$

The cost of a new TV set is $678.50. The cost of the TV is twice the cost of a hi-fi set. What is the cost of the hi-fi set?

Simpler word form: 2(cost of hi-fi) is (cost of TV)

Equation: $2(x)$ $= 678.50$

After solving the equation, a check should be made into the original wording of the problem rather than into the equation. The solution is then written in terms of the problem. For example, solving the second problem above we have:

$$2x + 3 - x = 23$$

$$x + 3 = 23$$

$$x = 20 \quad \text{(one number)}$$

$$2x + 3 = 43 \quad \text{(the other number)}$$

Check: The difference is 23; $43 - 20 = 23$.
The two numbers are 20 and 43.

Some formulas that are often used to solve applications are:

[speed (r)][time (t)] = distance (d)

$$rt = d$$

[rate $(R\%)$][base (B)] = amount (A)

$$(R\%)(B) = A$$

[number of items (n)][cost per item (c)] = total cost (t)

$$nc = t$$

[percent of mixture $(R\%)$][total amount of mixure (M)] =
$$\text{amount of substance in mixture } (S)$$

$$(R\%)(M) = S$$

An example of using one of the above formulas is: If an alcohol solution is 35% alcohol, how much alcohol is in 5 gallons of the solution?

$$\begin{pmatrix} \% \text{ of solution} \\ \text{that is} \\ \text{alcohol} \end{pmatrix} \cdot \begin{pmatrix} \text{total amount} \\ \text{of solution} \end{pmatrix} = \begin{pmatrix} \text{amount of} \\ \text{alcohol in} \\ \text{solution} \end{pmatrix}$$

If S represents the amount of alcohol in the solution, then

$$(35\%)(5) = S$$

$$(.35)(5) = S$$

$$1.65 = S$$

So the solution contains 1.65 gallons of alcohol

Seven basic steps for solving word problems are:

1. **Read the problem carefully.**
2. **Write a simpler word form of the problem (or use the appropriate formula).**
3. **Represent one of the unknowns with a letter (x) and all the others in terms of the same letter. (In some problems a sketch, diagram, or table might be helpful.)**

Continued on next page.

4. **Write the equation of the problem by replacing the words in the simpler word form by the mathematical expressions.**
5. **Solve the equation.**
6. **Check the solution(s) in the original wording of the problem.**
7. **Write the answer.**

EXAMPLES

a. The sum of two numbers is 35. One number is five less than the other. Find the two numbers.

Simpler word form: (first number) + (second number) is 35

Select variable: Let x represent one number and $x - 5$ represent the other.

Write equation: $x + (x - 5) = 35$

Solve:
$$2x - 5 = 35$$
$$2x = 40$$
$$x = 20 \quad \text{(one number)}$$
$$x - 5 = 15 \quad \text{(the other number)}$$

Check: The sum is 35; $20 + 15 = 35$.

Answer: The two numbers are 20 and 15.

b. The cost of a new TV set is $678.50. The cost of the TV is twice the cost of a hi-fi set. What is the cost of the hi-fi set?

Simpler word form: 2(cost of hi-fi) is (cost of TV)

Select variable: Let x represent the cost of the hi-fi

Write equation: $2x = 678.50$

Solve: $x = 339.25$ (cost of the hi-fi)

Check: Cost of TV ($678.50) is twice the cost of the hi-fi; ($339.25)(2) = $678.50.

Answer: The hi-fi set costs $339.25.

c. If the sum of three consecutive odd numbers is 129, what are the numbers? (Hint: consecutive numbers are in order of value and can be represented by $x, x + 1, x + 2, \ldots$; consecutive odd or even numbers differ by 2, and can be represented by $x, x + 2, x + 4, \ldots$)

Simpler word form: (first number) + (second number) + (third number) is 129

Select variable: Let x represent the first number; then $x + 2$ represents the second number and $x + 4$ represents the third number

Write equation: $x + (x + 2) + (x + 4) = 129$

Solve:
$$3x + 6 = 129$$
$$3x = 123$$
$$x = 41 \quad \text{(first number)}$$
$$x + 2 = 43 \quad \text{(second number)}$$
$$x + 4 = 45 \quad \text{(third number)}$$

Check: The sum is 129: $41 + 43 + 45 = 129$.

Answer: The three consecutive odd numbers are 41, 43, and 45.

d. Car A and car B start together from the same place and drive in the same direction. If car A travels at 40 mph and car B travels at 52 mph, in how many hours will they be 60 miles apart?

Simpler word form:

$$\left(\begin{array}{c}\text{Distance traveled}\\\text{by car B}\end{array}\right) - \left(\begin{array}{c}\text{Distance traveled}\\\text{by car A}\end{array}\right) = \left(\begin{array}{c}\text{Distance}\\\text{apart}\end{array}\right)$$

Select variable: Let t represent the time the cars are driven; then $52t$ is the distance of car B and $40t$ is the distance of car A.

Write equation: $52t - 40t = 60$

Solve: $12t = 60$

$t = 5$

Check: In 5 hours car B will travel $(52)(5)$ or 260 miles and car A will travel $(40)(5)$ or 200 miles; $260 - 200 = 60$.

Answer: In 5 hours car B will be 60 miles ahead of car A.

APPLICATION SOLUTION

To find the resistance of each lamp, we use the simpler word form.

Simpler word form:

$$\left(\begin{array}{c}\text{Resistance of}\\\text{first lamp}\end{array}\right) + \left(\begin{array}{c}\text{Resistance of}\\\text{second lamp}\end{array}\right) + \left(\begin{array}{c}\text{Resistance of}\\\text{third lamp}\end{array}\right) = \left(\begin{array}{c}\text{Total}\\\text{resistance}\end{array}\right)$$

Select variable: If r represents the resistance of the second lamp, then $r + 150$ is the resistance of the first lamp and $r - 100$ is the resistance of the third lamp.

Write equation: $r + 150 + r + r - 100 = 800$

Solve: $3r + 50 = 800$

$3r = 750$

$r = 250$ (the second lamp)

$r + 150 = 400$ (the first lamp)

$r - 100 = 150$ (the third lamp)

Check: The sum of the resistances is $400 + 250 + 150 = 800$ ohms.

Answer: The lamps have 400, 250, and 150 ohms of resistance respectively.

Exercises 6.5

1. Four resistors in series have a total resistance of 1940 ohms. If each one after the first is 150 ohms more than the previous one, find the resistance of each resistor.

2. Four lamps connected in series have a total resistance of 1050 ohms. One lamp has a resistance of 250 ohms, and a second has a resistance of 200 ohms. The third has a resistance that is twice the fourth. Find the resistance of the third and fourth lamps.

3. A student bought two books; one cost $3.85 more than the other. If the total cost was $14.75, what was the cost of each book?

4. A roofer laid shingles for three days. The second day he was able to lay two bundles more than on the first day. The third day was cut short by rain and he laid five bundles less than on the first day. If he laid a total of 27 bundles, how many bundles did he lay each day?

5. The difference between two numbers is 123. If the larger number is five less than three times the smaller, find the two numbers.

6. The perimeter of a triangle is 48 ft. If the first side is three feet longer than the second, and the second is the same length as the third, what is the length of each of the sides? (See the diagram below.)

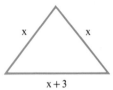

7. A farmer has 890 ft of fencing. He wants to use it to fence in a rectangular lot where the length of the lot is 175 ft more than the width. What are the dimensions of the lot he can fence in?

8. What are the length and width of a rectangle with perimeter of 48 in and width four inches less than its length?

9. At an agricultural test station three tests were conducted on germination of corn seed. The second test had twenty-five fewer germinations than the third test and five less than the first. If a total of 3000 seeds germinated, how many germinations were there in each test?

10. The current through one branch of an electrical circuit is 2.45 amperes more than through another circuit. When the circuits are joined, the total amperage is 8.75. What is the current in each of the circuits?

11. A strip of plate metal is to be cut into six pieces so that each successive piece is $\frac{1}{2}$ in longer than the preceding one. What is the length of each piece, if the original length of the plate was 36 in?

12. A metal bar $9\frac{5}{16}$ inches long is to be cut into two lengths so that one is $\frac{3}{16}$ inch shorter than the other. What is the length of each piece if $\frac{3}{16}$ inch allowance is made for the cut?

13. An alloy called "antifriction metal" contains twice as much antimony as copper, and six times as much tin as antimony. If a sample weighs 12 lb, how much of each metal does it contain?

ANSWERS

14. _____

15. _____

16. _____

17. _____

18. _____

19. _____

20. _____

21. _____

22. _____

23. _____

24. _____

25. _____

26. _____

27. _____

28. _____

29. _____

30. _____

14. Two cyclists start in opposite directions from the same place, one at 6 mph and the other at 9 mph. How long will it take for them to be 24 miles apart?

15. Two airplanes take off from Portland International Airport flying in opposite directions. The first plane flies at 340 mph and the second at 450 mph. How long will it take them to be 1185 miles apart?

16. Two cars leave from the same town at the same time traveling in the same direction. If the first car travels at a speed of 60 mph and the second car at 45 mph, how long will it take for the cars to be 80 miles apart?

17. Two hikers start out on the Pacific Crest Trail at the same location and walk in the same direction. If one hiker averages 13.4 miles per day and the other averages 9.6 miles per day, in how many days will they be 49.4 miles apart?

18. At the end of the Little League candy sale it was noted that the receipts contained the same number of quarters and half dollars. If the value of these coins was $101.25, how many of each were there?

19. The cashier at the National Women's Bank found that a certain deposit contained the same number of pennies, nickels, and dimes. If the value of these coins was $12, how many of each were in the deposit?

20. The local Grandmother's Market ordered the same number of cases of green beans and pork'n'beans. The green beans cost $9.60 per case, and the pork'n'beans cost $15.12 per case. How many cases of each were ordered if the total cost to the store was $543.84?

21. A shipment of rare crystal costing $1410 was received by the Exquisite Gift Shoppe. The shipment contained equal numbers of water, wine, and sherbet goblets. If the water goblets cost $21 each, the wine goblets cost $19.50 each, and the sherbets cost $18.25 each, how many of each were in the shipment?

22. The First National Bank offers 6.75% simple interest on its savings accounts. The Second National Bank offers 7.15% interest on its savings accounts. John Q Public invests the same amount in each bank. If the difference between the interest earned at the two banks is $10.24, how much does John Q Public have invested in each bank?

23. In Average City, USA, it is found that 8.6% of the residents are black, 6.7% are Hispanic, 5.7% are Asian, and .15% are American Indian. If the total number of these minorities is 13,536, how many people live in Average City?

24. Mr. Bar Fly had two drinks after work. Each drink contained the same number of ounces of liquid. One was 30% alcohol and the other was 25% alcohol. How many ounces of liquid were in each drink if he consumed 4.4 ounces of alcohol?

25. Mr. Peanut packages two varieties of mixed nuts. Each package contains the same number of ounces of nuts. One package contains 1 ounce more cashews than the other. If the packages contain 20% and 12% cashews, how many ounces of nuts are in each package?

E *Maintain Your Skills*

Subtract:

26. $(4.82) - (9.76)$ **27.** $(-454) - (-281)$

28. $(482) - (-624)$ **29.** $(-18.65) - (24.23)$

30. $\left(-\dfrac{7}{8}\right) - \left(-\dfrac{2}{3}\right)$

6.6 Multiplication of Polynomials

OBJECTIVE	**66. Find the product of polynomials.**

APPLICATION

A motorist drives from town A to town B at a rate of 45 mph. On the return trip he drives at a rate of 55 mph. If the round trip took five hours to complete, what is the distance between the two towns?

VOCABULARY

No new vocabulary.

HOW AND WHY

Products of polynomials can be found by repeated use of the distributive law and the other laws stated in previous sections. Study these three examples:

monomial times binomial

$$-6x(2x + 4y) = (-6x)(2x) + (-6x)(4y)$$
$$= -12x^2 - 24xy$$

binomial times binomial

$$(3x + 4)(2x - 5) = (3x + 4)(2x + -5)$$
$$= 3x(2x + -5) + 4(2x + -5)$$
$$= 3x(2x) + 3x(-5) + 4(2x) + 4(-5)$$
$$= 6x^2 - 15x + 8x - 20$$
$$= 6x^2 - 7x - 20$$

binomial times trinomial

$$(2x - 5y)(3x^2 - 4xy + 1) = (2x + -5y)(3x^2 + -4xy + 1)$$
$$= (2x + -5y)(3x^2) + (2x + -5y)(-4xy)$$
$$+ (2x + -5y)(1)$$
$$= (2x)(3x^2) + (-5y)(3x^2) + (2x)(-4xy)$$
$$+ (-5y)(-4xy) + (2x)(1) + (-5y)(1)$$
$$= 6x^3 - 23x^2y + 2x + 20xy^2 - 5y$$

In each example we first changed subtraction problems to addition of opposites (see section on combining polynomials) and then repeatedly applied the distributive law. In each case the last step shown is the sum of monomial products.

The steps can be cut short by realizing that every term of one polynomial is multiplied by each term of the second polynomial. A particular example of this is the use of the FOIL shortcut for two binomials.

FOIL means the product of the First terms plus the product of the Outside terms plus the product of the Inside terms plus the product of the Last terms.

The binomial product is labeled as follows:

F   L   F   L
$$(2m + 3)(3m - 1)$$
O   I   I   O

F refers to the first term of each binomial
O refers to the outside terms of the binomials
I refers to the inside terms of the binomials
L refers to the last term of each binomial

To find the product of $(2m + 3)(3m - 1)$ we do the following:

$$(2m + 3)(3m - 1) = (2m + 3)(3m + -1)$$

$$\begin{aligned}
&= \overset{\text{F}}{(2m)(3m)} + \overset{\text{O}}{(2m)(-1)} + \overset{\text{I}}{(3)(3m)} + \overset{\text{L}}{(3)(-1)} \\
&= 6m^2 + (-2m) + 9m + (-3) \\
&= 6m^2 + 7m - 3
\end{aligned}$$

The examples can also be written vertically.

$$\begin{array}{r} 2x + 4y \\ -6x \\ \hline \end{array}$$
$(-6x)(2x) + (-6x)(4y) = -12x^2 - 24xy$

$$\begin{array}{r} 2x - 5 \\ 3x + 4 \\ \hline \end{array}$$
$(3x)(2x) + (3x)(-5)$
$\qquad (4)(2x) + (4)(-5)$

or

$$\begin{array}{r} 6x^2 - 15x \\ +\ 8x + -20 \\ \hline 6x^2 - 7x\ -\ 20 \end{array}$$

$3x^2 + -4xy + 1$

$$\begin{array}{r} 2x + -5y \\ \hline 6x^3 + -\ 8x^2y + 2x \\ -15x^2y \qquad + 20xy^2 + -5y \\ \hline 6x^3 - \quad 23x^2y + 2x + 20xy^2 - 5y \end{array}$$

In general, to find the product of two polynomials,

1. **Write the product of each term of one polynomial and the terms of the other, or**
1a. **If both are binomials, multiply using the FOIL shortcut.**
2. **Simplify (combine terms) if possible.**

When an equation contains the product of polynomials, the product can be found and the equation solved as before if it is a linear equation. For example:

$$3(x + 2) - 2(x + 6) = 18$$
$$(3x + 6) - (2x + 12) = 18$$
$$(3x + 6) + (-2x + -12) = 18$$
$$x + -6 = 18$$
$$x = 24$$

To determine if the equation is linear it might be necessary to isolate the variable and combine monomials. (See Examples d and e.)

EXAMPLES

Multiply:

a. $3x(3a^2 + 4ab - 2b^2) = 3x(3a^2 + 4ab + -2b^2)$
$$= (3x)(3a^2) + (3x)(4ab) + (3x)(-2b^2)$$
$$= 9a^2x + 12abx - 6b^2x$$

b. $(3x + 2)(5x - 4) = (3x)(5x) + (3x)(-4) + (2)(5x) + (2)(-4)$
$$= 15x^2 + -12x + 10x + -8$$
$$= 15x^2 + -2x + -8$$
$$= 15x^2 - 2x - 8$$

c. $(.5y^2 + .3y - .5)(.2y - .75)$

$$
\begin{array}{r}
.5y^2 + \quad .3y + -.5 \\
.2y + -.75 \\
\hline
.1y^3 + \quad .06y^2 + -.1y \\
+ -.375y^2 + -.225y + .375 \\
\hline
.1y^3 + -.315y^2 + -.325y + .375
\end{array}
$$
or
$$.1y^3 - .315y^2 - .325y + .375$$

d. *Solve:*
$$x(x - 4) + 6 - x(x + 7) = 12$$
$$(x^2 - 4x) + 6 - (x^2 + 7x) = 12$$
$$(x^2 - 4x) + 6 + (-x^2 + -7x) = 12$$
$$-11x + 6 = 12$$
$$-11x = 6$$
$$x = -\frac{6}{11}$$

e. *Solve:*
$$(x + 2)(x - 3) = (x + 7)(x - 10)$$
$$x^2 + 2x - 3x - 6 = x^2 + 7x - 10x - 70$$
$$x^2 - x - 6 = x^2 - 3x - 70$$
$$2x = -64$$
$$x = -32$$

APPLICATION SOLUTION

In this instance we do not know the total distance, but we do know the total time. To find time we use the distance formula and solve for *t*.

$$d = rt \quad \text{so} \quad t = \frac{d}{r}$$

Simpler word form: $\left(\begin{array}{c}\text{The time to go}\\ \text{from A to B}\end{array}\right) + \left(\begin{array}{c}\text{The time to go}\\ \text{from B to A}\end{array}\right) = \left(\begin{array}{c}\text{Total}\\ \text{time}\end{array}\right)$

Select Variable: If *x* represents the distance going, then it will also represent the distance returning.

Write Equation: $\dfrac{x}{45} + \dfrac{x}{55} = 5$

Continued on next page.

$$\textit{Solve:} \quad 495\left(\frac{x}{45} + \frac{x}{55}\right) = 495(5)$$

$$11x + 9x = 2475$$

$$20x = 2475$$

$$x = 123.75$$

Answer: Therefore it is 123.75 or $123\frac{3}{4}$ miles between town A and town B.

WARM UPS

Multiply:

1. $4(3x + 1)$ **2.** $5(2x - 3)$

3. $x(3x + 4)$ **4.** $y(2y - 1)$

5. $4x + 2$ **6.** $5a + 3$
  $\underline{\quad -3}$ $\underline{\quad\quad a}$

7. $2x + 3y$ **8.** $4a - 3b$
  $\underline{\quad -2x}$ $\underline{\quad\quad 3c}$

9. $(x + 1)(x + 2)$ **10.** $(x + 1)(x - 1)$

11. $(x - 2)(x + 2)$ **12.** $(x + 3)(x + 4)$

Exercises 6.6

A

Multiply:

1. $3(2a - 1)$ **2.** $4(x + 6)$ **3.** $-2(3y - 1)$

4. $7x + 3$ **5.** $-5a + 2b$ **6.** $4x + 2xy$
   $\underline{-5}$ $\underline{3a}$ $\underline{-7x}$

B

7. $3x - 7$ **8.** $8y + 2w$
   $\underline{2x + 8}$ $\underline{-3y + 4}$

9. $-r + 4s$ **10.** $(x - 3)(x + 7)$
   $\underline{2r - 3s}$

11. $(x + 5)(x + 5)$ **12.** $(x + 2)(x - 1)$

13. $(3x + 5)(4x - 3)$ **14.** $(5x - 6)(5x + 6)$

C

15. $(3y - 4)(3y + 4)$ **16.** $(3x - 9y)(5x - 6y)$

17. $3w - 4x + 2y$ **18.** $(x + 4)(2x^2 - 7x - 7)$
   $\underline{3x - 5y}$

19. $(2a - 3b)(4a - 8ab + 5b)$

Solve:

20. $x(2x - 6) - x(x + 3) = x(x + 7) + 20$

21. $(x + 4)(x + 3) = (x - 6)(x + 2)$

Simplify:

22. $16 - 2[3x - 5(4x - 1)]$

23. $2 - 3\{-2[4x - (2x + 2y)] + 6y\} + 2(3y - 4y)$

24. $y - \{2 - 3[2y - 4(y + 4)]\}$

299

1.
2.
3.
4.
5.
6.
7.
8.
9.
10.
11.
12.
13.
14.
15.
16.
17.
18.
19.
20.
21.
22.
23.
24.

D

Distance problems: Distance (d) in terms of rate (r) and time (t) is given by the formula $d = rt$. Equations for solving these problems usually indicate that the sum (or difference) of two distances is equal to the total distance. Solve the following.

25. Two cyclists start in opposite directions from the same point, one at 8 mph and the other at 10 mph. How long will it take until they are 54 miles apart? Hint: the simpler word form is

(distance of rider one) + (distance of rider two) is 54.

26. One automobile starts out from town at 6 A.M. and travels at an average speed of 45 mph. Two hours later a second automobile starts out to overtake the first, traveling at an average speed of 60 mph. How long will it take the second to overtake the first?

27. Two cars, 580 miles apart, start out toward each other at 1 P.M. One car is averaging 40 mph and the other 60 mph. The one traveling at 60 mph takes a one-hour lunch break. At what time will the two cars meet?

28. One car starts out at 3 P.M. traveling at 40 mph. If a second car starts out to overtake the first at 5 P.M., how fast will it have to travel in order to overtake the first by 10 P.M.?

E   *Maintain Your Skills*

Find the products:

29. $(-8)(-12)(-4)$

30. $(-14)(12)(-7)$

31. $(12 - 18)(6 - 9)$

32. $(-5.2)(-3.6)(2.5)(-5)$

33. $(-1)(-1)(-1)(-1)(-1)(-1)$

6.7 Dividing Polynomials

OBJECTIVE

67. Find the quotient of two polynomials.

APPLICATION

A cylindrical tank is to be painted two colors, the top half red and the bottom half blue. In order to purchase the correct amount of paint, the surface area for half the tank must be found. If $S = 2\pi r^2 + 2\pi rh$ is the formula for the surface area of the tank, what is the formula for half the area?

VOCABULARY

No new vocabulary.

HOW AND WHY

To divide any polynomial by a monomial, change the subtraction signs in the polynomial to addition of opposites and then divide each term by the monomial divisor.

For example, a polynomial divided by a monomial may be written as a fraction or as a long division problem.

$$(3x^4 - 6x^3 + 12x^2 - 9x) \div (-3x)$$

$$\frac{3x^4 + -6x^3 + 12x^2 + -9x}{-3x} = \frac{3x^4}{-3x} + \frac{-6x^3}{-3x} + \frac{12x^2}{-3x} + \frac{-9x}{-3x}$$

$$= -x^3 + 2x^2 - 4x + 3$$

or

$$
\begin{array}{r}
-x^3 + 2x^2 - 4x + 3 \\
-3x{\overline{\smash{\big)}\,3x^4 + -6x^3 + 12x^2 + -9x}} \\
\underline{3x^4} \\
0 + -6x^3 \\
\underline{-6x^3} \\
0 + 12x^2 \\
\underline{+ 12x^2} \\
0 + -9x \\
\underline{-9x} \\
0
\end{array}
$$

Division of polynomials by a polynomial follows the same pattern as division of whole numbers.

$x + 3{\overline{\smash{\big)}\,x^2 + 7x + 12}}$ *Think: What times x will give a product of x^2? $(x)(x) = x^2$*

$$
\begin{array}{r}
x \\
x + 3{\overline{\smash{\big)}\,x^2 + 7x + 12}} \\
\underline{x^2 + 3x} = x(x + 3) \\
4x + 12
\end{array}
$$

Place x over the x column, multiply x by $x + 3$, then subtract.

Think: What times x will give a product of 4x? $4(x) = 4x$

$$\begin{array}{r} x +\ \ 4 \\ x+3\overline{)x^2+7x+12} \\ \underline{x^2+3x} \\ 4x+12 \\ \underline{4x+12} \\ 0 \end{array}$$

Place 4 *over the last column, multiply* 4 *by* $x+3$, *then subtract.*

Since there is no remainder, the quotient is $x+4$.

Let's try another one.

$$\begin{array}{r} x^3+x^2\ \ \ \ \ \ \ -3 \\ x+1\overline{)x^4+2x^3+x^2-3x-3} \\ \underline{x^4+\ \ x^3} \ \ \ \ \ \ \ \ \ \ \ = x^3(x+1) \\ x^3+x^2-3x-3 \\ \underline{x^3+x^2} \ \ \ \ \ \ \ \ \ \ \ = x^2(x+1) \\ 0\ -3x-3 \\ \underline{-3x-3} = -3(x+1) \\ 0 \end{array}$$

The quotient is x^3+x^2-3.

If the division has a remainder, we will show the remainder as $\dfrac{\text{remainder}}{\text{divisor}}$. A remainder is identified when the exponent of the variable in the remainder is less than the exponent of the variable in the divisor.

$$\begin{array}{r} x +\ \ 4 \\ x+3\overline{)x^2+7x-\ \ 8} \\ \underline{x^2+3x} \\ 4x-\ \ 8 \\ \underline{4x+12} \\ -20 \end{array}$$

The quotient is $x+4+\dfrac{-20}{x+3}$.

EXAMPLES

a. $(8y^2-4y)\div(2y)$

$$\frac{8y^2+-4y}{2y}=\frac{8y^2}{2y}+\frac{-4y}{2y}=4y-2$$

b. $(36a^3b-27a^2b^2+18ab^3)\div(-9ab)$

$$\frac{36a^3b+-27a^2b^2+18ab^3}{-9ab}=\frac{36a^3b}{-9ab}+\frac{-27a^2b^2}{-9ab}+\frac{18ab^3}{-9ab}$$

$$=-4a^2+3ab-2b^2$$

c. $(25x^3-20x^2-55x)\div(-5x)$

$$\begin{array}{r} -5x^2+\ \ \ \ 4x\ +\ \ \ 11 \\ -5x\overline{)25x^3+-20x^2+-55x} \\ \underline{25x^3} \\ 0\ +-20x^2 \\ \underline{-20x^2} \\ 0\ \ +-55x \\ \underline{-55x} \\ 0 \end{array}$$

d. $(x^2 - 5x - 14) \div (x - 7)$

$$
\begin{array}{r}
x + 2 \\
x - 7 \overline{\smash{)}x^2 - 5x - 14} \\
\underline{x^2 - 7x} \\
2x - 14 \\
\underline{2x - 14} \\
0
\end{array}
$$

e. $2x - 3\overline{\smash{)}4x^4 + x^2 - x + 9}$

$$
\begin{array}{r}
2x^3 + 3x^2 + 5x + 7 + \dfrac{30}{2x - 3} \\
2x - 3 \overline{\smash{)}4x^4 + 0 \cdot x^3 + x^2 - x + 9} \\
\underline{4x^4 - 6x^3} \\
6x^3 + x^2 - x + 9 \\
\underline{6x^3 - 9x^2} \\
10x^2 - x + 9 \\
\underline{10x^2 - 15x} \\
14x + 9 \\
\underline{14x - 21} \\
30
\end{array}
$$

Insert $0x^3$ as a placeholder for the missing x^3 term

APPLICATION SOLUTION

To find one-half of the surface area $\left(\dfrac{S}{2}\right)$, we do the following.

Formula: $S = 2\pi r^2 + 2\pi rh$

Solve: $\dfrac{S}{2} = \dfrac{2\pi r^2 + 2\pi rh}{2}$

$\dfrac{S}{2} = \dfrac{2\pi r^2}{2} + \dfrac{2\pi rh}{2}$

Answer: $\dfrac{S}{2} = \pi r^2 + \pi rh$

WARM UPS

1. $(3x - 6) \div 3$

2. $(8x - 24) \div 8$

3. $(4x + 12) \div 4$

4. $(12x - 16) \div 4$

5. $(4x^2 + 8x + 12) \div 2$

6. $(8y^3 - 16y^2 + 8) \div 8$

7. $(4x^2y - 16xy^2) \div 4xy$

8. $(15abc + 25ab) \div 5ab$

9. $5x\overline{\smash{)}15x^2 + 10x}$

10. $50a\overline{\smash{)}100a^2 + 50a}$

11. $x + 2\overline{\smash{)}x^2 + 4x + 4}$

12. $x - 2\overline{\smash{)}x^2 - 4x + 4}$

Exercises 6.7

A

Divide:

1. $(3x^2 - 6x - 9) \div 3$

2. $(x^3 - x^2 + x) \div x$

3. $(3a^2b - 3ab^2) \div 3ab$

4. $5x\overline{)10x^2 - 5x}$

5. $(12ab^2c - 24a^2b^2c^2) \div 6ab$

6. $(36p^2q - 18p^2q^2 - 24p^2q^3) \div 6p^2$

B

7. $(5x^2y^2 + 10xy^2 - 15xy) \div 5xy$

8. $\dfrac{12r^2s^2t^2 - 16r^2st^2 - 24rs^2t}{-4rst}$

9. $-15ab\overline{)-30abc - 45a^2b^2c^2 - 15ab^2c^2}$

10. $(12x^5 - 15x^4 + 18x^3 - 6x^2) \div 3x$

11. $(14y^2z + 16yz^2 + 24yz - 30y^2z^2) \div 24z$

12. $(a^3b^2c^4 + a^2bc^3 + ab^2c^2 - a^2bc - abc) \div (-abc)$

13. $x + 1\overline{)x^2 + 3x + 2}$

14. $x - 1\overline{)x^2 - 5x + 4}$

C

15. $x - 5\overline{)x^2 - 7x + 10}$

16. $a + 3\overline{)a^2 + 10a + 21}$

17. $3x + 1\overline{)6x^2 - 13x - 5}$

18. $2y - 3\overline{)10y^2 - 3y - 18}$

19. $(8a^2 + 6ab - 9b^2) \div (4a - 3b)$

20. $(x^2 + 8x + 10) \div (x - 5)$

21. $(x^3 + 3x^2 + 3x + 1) \div (x + 1)$

22. $(x^3 - 3x^2 + 3x - 1) \div (x - 1)$

23. $(x^3 + 0x^2 + 0x + 1) \div (x + 1)$

24. $(x^4 - 1) \div (x - 1)$

1. _____
2. _____
3. _____
4. _____
5. _____
6. _____
7. _____
8. _____
9. _____
10. _____
11. _____
12. _____
13. _____
14. _____
15. _____
16. _____
17. _____
18. _____
19. _____
20. _____
21. _____
22. _____
23. _____
24. _____

D

25. The formula for the total surface area of a rectangular solid is $A = 2\ell w + 2h\ell + 2hw$. Write a formula that expresses one-half of the total surface area $\left(\dfrac{A}{2}\right)$.

Division of polynomials is used to help locate roots in theory of equations, which is a topic for a higher mathematics course. If a polynomial is divisible by $x - a$ (with 0 remainder), then a is a root of the equation in which the polynomial is equal to 0. For example, 3 is a root of the equation $x^2 - 9x + 18 = 0$ since $(x - 3)$ divides $x^2 - 9x + 18$. Also, -3 is a root of $x^2 + 8x + 15 = 0$ since $(x + 3)$ divides $x^2 + 8x + 15$.

Use division of polynomials to test whether the given value of x is a root of the given equation.

26. Is -2 a root of $x^2 - 6x - 16 = 0$?

27. Is 4 a root of $x^2 - 3x - 28 = 0$?

28. Is 3 a root of $x^3 + x^2 - 8x - 12 = 0$?

29. Is -3 a root of $x^3 + 4x^2 - 3x - 18 = 0$?

30. Is 5 a root of $x^3 - 4x^2 - x - 15 = 0$?

E   *Maintain Your Skills*

Divide:

31. $360 \div [(-8)(-3)]$

32. $(9 - 12 - 17) \div (-4)$

33. $[(-12)(15)] \div (-45)$

34. $\left(-\dfrac{25}{72}\right) \div \left(\dfrac{25}{12}\right)$

35. $[(-15.4)(2.8)] \div (-1.4)$

6.8 Factoring Polynomials: Common Monomial Factor

OBJECTIVE

68. Factor a polynomial by writing it as the product of a monomial and another polynomial.

APPLICATION

Equations that have a polynomial on one side and zero on the other can often be solved by factoring. An example of such an equation is $s = 144t - 16t^2$. When s (height of falling object) is known, t (the time the object falls) can be found. If $s = 0$, t is the time it takes the object to hit the ground. (See Section 6.11.)

VOCABULARY

To *factor* a number or polynomial means to write it in the form of a multiplication problem.

HOW AND WHY

In an earlier section the distributive law was used to multiply monomials and polynomials.

$$3x(x^2 - 4y) = 3x^3 - 12xy$$

To factor a polynomial is to reverse this process.

$$4y^3 - 12xy = 4y(y^2 - 3x)$$

To *factor a polynomial* we look for all of the common factors of the terms. The common monomial factor contains every factor that each of the terms of the polynomial has in common. This monomial factor is called the HCF (Highest Common Factor) of the terms of the polynomial. The HCF may often be seen by inspecting the polynomial. If it is not easily discovered, it can also be found by writing each term in factored form.

$$34x^2y - 51xy^2 - 68x^2y^2$$
$$= (2 \cdot 17 \cdot x \cdot x \cdot y) - (3 \cdot 17 \cdot x \cdot y \cdot y) - (2 \cdot 2 \cdot 17 \cdot x \cdot x \cdot y \cdot y)$$
$$= 17xy(2x) - 17xy(3y) - 17xy(4xy)$$

By rearranging the terms so that the common factors are written first, you can see that the HCF is $17xy$. The polynomial can then be written as a product (using the distributive law).

$$17xy(2x - 3y - 4xy)$$

If you can find the HCF by inspection, you can short-cut the factoring steps by doing them mentally.

To factor a polynomial as the product of a monomial and another polynomial,

1. **Determine the HCF of the terms of the polynomial.**
2. **If the HCF is not one (1), use the distributive law to write it as a product.**
3. **If the HCF is one (1), we say that the polynomial cannot be factored.**

EXAMPLES

a. $4x + 16y = 4(x + 4y)$

b. $4x + 17xy = x(4 + 17y)$

c. $5x + 6y = 5x + 6y$ *Since HCF = 1, the polynomial cannot be factored.*

d. $2xyz - 2wxy + 4wyz = 2y(xz - wx + 2wz)$

e. $15a^2b - 35b^2c - 25abc = 5b(3a^2 - 7bc - 5ac)$

f. $-4r^2st - 20rs^2t - 36rst^2 = -4rst(r + 5s + 9t)$

WARM UPS

Factor:

1. $2x + 6$

2. $8x - 12$

3. $6x + 12$

4. $3x - 9$

5. $12a - 16$

6. $18ab - 27b$

7. $36x + 48y$

8. $6ab - 2b$

9. $4x - xy$

10. $2ab + b$

11. $4xy + 5xz$

12. $12abc + 3bc$

Exercises 6.8

A

Factor each of the following:

1. $6xy - 5y$ **2.** $2ab - 4a$ **3.** $2x^2 - 3x$

4. $3ab - 15b$ **5.** $15xyz - 7wxz$ **6.** $12bc - 16bd$

B

7. $\pi r^2 h - \pi R^2 h$ **8.** $2\pi rh + 2\pi r^2$

9. $4ab + 8bc - 6cd$ **10.** $25rst - 35st + 5stx$

11. $\pi r^2 h + \pi r^2 + 2\pi r$ **12.** $15x^3 y - 25x^2 y + 40xy$

13. $3xy + 4yz - 5xz$ **14.** $36ac - 8ab - 12ac$

C

15. $x^2 y^2 + x^3 y^3 - x^4 y^4$

16. $3m^2 n^2 - 6mn - 12m^2 n - 15mn^2$

17. $8x^3 y^2 + 9x^2 y^3 - 7x^2 y^2$

18. $14ax + 10ay + 6az$

19. $24x^4 - 12x^3 - 36x^2 - 48x$

20. $16x^2 y^2 z + 24x^2 yz^2 + 32x^2 yz$

21. $24a^3 b^3 + 18a^2 bc^2 + 30a^2 bc$

22. $3m^2 n^2 - 6mn^2 - 42m^2 n^2 - 30m^3 n^2$

23. $4x^2 y^2 + 8xy - 12x^3 y - 16xy^2$

24. $24x^4 - 48x^3 - 36x^2 - 12x$

E   *Maintain Your Skills*

Solve:

25. $15x - 12 = 18$ **26.** $\dfrac{4}{5}y - \dfrac{2}{3} = \dfrac{14}{15}$

27. $2.5y + 1.8 = -3.2$ **28.** Solve for x: $x + 3y = 5$

29. Solve for y: $3x + y = 4$

ANSWERS

1. _____
2. _____
3. _____
4. _____
5. _____
6. _____
7. _____
8. _____
9. _____
10. _____
11. _____
12. _____
13. _____
14. _____
15. _____
16. _____
17. _____
18. _____
19. _____
20. _____
21. _____
22. _____
23. _____
24. _____
25. _____
26. _____
27. _____
28. _____
29. _____

ANSWERS TO WARM UPS (6.8)

1. $2(x + 3)$	**2.** $4(2x - 3)$	**3.** $6(x + 2)$	**4.** $3(x - 3)$
5. $4(3a - 4)$	**6.** $9b(2a - 3)$	**7.** $12(3x + 4y)$	**8.** $2b(3a - 1)$
9. $x(4 - y)$	**10.** $b(2a + 1)$	**11.** $x(4y + 5z)$	**12.** $3bc(4a + 1)$

6.9 Factoring Polynomials: Trinomials

OBJECTIVE

69. Factor a trinomial.

APPLICATION

The time it takes a rock to fall to the bottom of a gully 176 feet deep if it is thrown *upward* from the rim at the rate of 56 ft/sec is given by the equation $16t^2 - 56t - 176 = 0$. (See Section 6.11.)

VOCABULARY

No new vocabulary.

HOW AND WHY

A trinomial (with no common factor) that can be factored using whole numbers and/or their opposites is the product of two binomials. The factors can be found by looking at all possible products of the numerical coefficients of the first and last terms (F and L from FOIL). For instance, study the polynomial

$x^2 + 6x + 8$

We would like to find two numbers a and b such that

$(x + a)(x + b) = x^2 + 6x + 8$

that is,

$(x + a)(x + b) = x^2 + (a + b)x + ab$

The first term has a coefficient of 1. So we look for two numbers such that their sum is 6 (the coefficient of the middle term) and whose product is 8 (the last term). The product is positive and the sum is positive. Thus the two numbers we are searching for are positive. Both (2)(4) and (8)(1) yield a product of 8, but only $2 + 4$ yields a sum of 6. Thus 2 and 4 are the numbers we want. So

$x^2 + 6x + 8 = (x + 2)(x + 4)$
  $2 + 4$ $(2)(4)$

To factor $x^2 + x - 30$: Again the coefficient of x^2 is 1, so we want two numbers whose product is -30 and whose sum is 1. A product of -30 indicates that one number is positive and the other is negative. We have

$(-1)(30)$ or $1(-30)$ sum is 29 or -29

$(-2)(15)$ or $2(-15)$ sum is 13 or -13

$(-3)(10)$ or $3(-10)$ sum is 7 or -7

$(-5)(6)$ or $5(-6)$ sum is 1 or -1

Therefore -5 and 6 are the two numbers we want. Thus

$x^2 + x - 30 = (x - 5)(x + 6)$

If the coefficient of the first term is not 1, there are more possibilities. For instance, look at

$6x^2 - 19x + 10$

The first term has coefficient 6 and has factors of

(1)(6)   or   (2)(3)

The last term is 10 and has factors

$(-1)(-10), (-2)(-5), (1)(10),$ or $(2)(5)$

Since the middle term is -19, we ignore the products $(1)(10)$ and $(2)(5)$; the sum would be positive if those factors were used. The remaining possibilities are

$(x - 10)(6x - 1)$

$(x - 1)(6x - 10)$

$(x - 2)(6x - 5)$

$(x - 5)(6x - 2)$

$(2x - 10)(3x - 1)$

$(2x - 1)(3x - 10)$

$(2x - 2)(3x - 5)$

$(2x - 5)(3x - 2)$

Multiply each pair of binomials until the correct factoring is found.

$6x^2 - 19x + 10 = (2x - 5)(3x - 2)$

In some cases there is a common monomial factor in the trinomial, and it should be factored first. This will reduce the number of possibilities in factoring the remaining trinomial. (See Example e.)

EXAMPLES

a. $m^2 + 20m + 64$

Since the coefficient of m^2 is 1, we look for two integers whose product is 64 and whose sum is 20. The product and sum are positive, so both integers are positive.

(1)(64)	$1 + 64 = 65$	no
(2)(32)	$2 + 32 = 34$	no
(4)(16)	$4 + 16 = 20$	yes

Therefore $m^2 + 20m + 64 = (m + 4)(m + 16)$.

b. $c^2 + 5c - 50$

The coefficient of c^2 is 1, so we need two integers whose product is -50 and whose sum is 5. The product is negative and the sum is positive, so one integer is positive and one is negative. Since the sum is positive, the positive number has the larger absolute value.

(−1)(50)	$-1 + 50 = 49$	no
(−2)(25)	$-2 + 25 = 23$	no
(−5)(10)	$-5 + 10 = 5$	yes

So $c^2 + 5c - 50 = (c - 5)(c + 10)$

c. $a^2 - 15a + 48$

The coefficient of a^2 is 1, so we need two integers whose product is 48 and whose sum is -15. Since the product is positive, both of the integers are positive or both are negative. The sum is negative, so each is negative.

$$(-1)(-48) \qquad -1 + (-48) = -49 \qquad \text{no}$$
$$(-2)(-24) \qquad -2 + (-24) = -26 \qquad \text{no}$$
$$(-3)(-16) \qquad -3 + (-16) = -19 \qquad \text{no}$$
$$(-4)(-12) \qquad -4 + (-12) = -16 \qquad \text{no}$$
$$(-6)(-8) \qquad -6 + (-8) \ = -14 \qquad \text{no}$$

Since there are no more different pairs of negative factors of 48, we conclude that $a^2 - 15a + 48$ cannot be factored at this time.

d. $4y^2 - 3y - 10$

coefficient of the first term: $4 = (1)(4)$ or $(2)(2)$
last term: $-10 = (-1)(10), (1)(-10), (-2)(5),$ or $(2)(-5)$

$$(y - 1)(4y + 10) = 4y^2 + 6y - 10 \qquad \text{no}$$
$$(y + 1)(4y - 10) = 4y^2 - 6y - 10 \qquad \text{no}$$
$$(y - 10)(4y + 1) = 4y^2 - 39y - 10 \qquad \text{no}$$
$$(y + 10)(4y - 1) = 4y^2 + 49y - 10 \qquad \text{no}$$
$$(y - 2)(4y + 5) \ = 4y^2 - 3y - 10 \qquad \text{yes}$$

e. $18w^2 + 66w + 36 = 6(3w^2 + 11w + 6)$

In $3w^2 + 11w + 6$,

coefficient of first term: $3 = (1)(3)$
last term: $6 = (1)(6)$ or $(2)(3)$

$$(w + 1)(3w + 6) = 3w^2 + 9w + 6 \qquad \text{no}$$
$$(w + 6)(3w + 1) = 3w^2 + 19w + 6 \qquad \text{no}$$
$$(w + 2)(3w + 3) = 3w^2 + 9w + 6 \qquad \text{no}$$
$$(w + 3)(3w + 2) = 3w^2 + 11w + 6 \qquad \text{yes}$$

So $18w^2 + 66w + 36 = 6(w + 3)(3w + 2)$.

WARM UPS

Factor:

1. $a^2 + a - 6$ **2.** $b^2 + 10b + 24$

3. $x^2 + 5x + 6$ **4.** $y^2 + 12y + 35$

5. $x^2 + x - 2$ **6.** $x^2 + 6x + 9$

7. $x^2 + x - 90$ **8.** $y^2 - y - 12$

9. $b^2 - 3b - 18$ **10.** $a^2 - 5a - 24$

11. $m^2 + 9m + 20$ **12.** $q^2 + 8q + 12$

Exercises 6.9

A

Factor if possible.

1. $x^2 - x - 42$ **2.** $y^2 - 6y + 8$ **3.** $x^2 - 2x - 80$

4. $x^2 + 7x + 12$ **5.** $a^2 - 5a + 4$ **6.** $c^2 + 4c - 21$

B

7. $bx^2 - 8bx - 20b$ **8.** $ay^2 + 6ay - 16a$

9. $2y^2 - 3y - 5$ **10.** $3x^2 + 5x + 2$

11. $3a^2 - 10a + 8$ **12.** $2y^2 + 7y - 20$

13. $3x^2 - 10xy + 3y^2$ **14.** $6x^2 + 13x + 7$

C

15. $6x^2 + 3x - 7$ **16.** $4y^2 - 6y - 10$

17. $25x^2 - 40xy + 16y^2$ **18.** $3a^2 - 13ab + 4b^2$

19. $6x^2 + 2x + 1$ **20.** $7a^2 - 17a - 12$

21. $12x^2 + 4x - 5$ **22.** $12x^2 - 3x - 1$

23. $15x^2 + x - 6$ **24.** $15x^2 - x - 6$

25. $6x^2 + 7x + 2$ **26.** $6x^2 - x - 2$

27. $15x^2 + 19x + 6$ **28.** $14x^2 + x - 3$

29. $35x^2 + 42x - 24$ **30.** $17x^2 - 50xy - 3y^2$

E   *Maintain Your Skills*

Simplify:

31. $(x^{10})(x^{12})$ **32.** $(2a^3b^2)^3$ **33.** $(4a^5b^2c^3)^2$

34. $\dfrac{(x^2y^3)^3}{x^3y^2}$ **35.** $\left(\dfrac{x^2}{y^4}\right)^3$

315

ANSWERS

1. _____
2. _____
3. _____
4. _____
5. _____
6. _____
7. _____
8. _____
9. _____
10. _____
11. _____
12. _____
13. _____
14. _____
15. _____
16. _____
17. _____
18. _____
19. _____
20. _____
21. _____
22. _____
23. _____
24. _____
25. _____
26. _____
27. _____
28. _____
29. _____
30. _____
31. _____
32. _____
33. _____
34. _____
35. _____

ANSWERS TO WARM UPS (6.9)

ANSWERS TO WARM UPS (6.9)

1. $(a + 3)(a - 2)$	**2.** $(b + 6)(b + 4)$	**3.** $(x + 2)(x + 3)$
4. $(y + 7)(y + 5)$	**5.** $(x + 2)(x - 1)$	**6.** $(x + 3)(x + 3)$
7. $(x + 10)(x - 9)$	**8.** $(y - 4)(y + 3)$	**9.** $(b - 6)(b + 3)$
10. $(a - 8)(a + 3)$	**11.** $(m + 5)(m + 4)$	**12.** $(q + 6)(q + 2)$

6.10 Factoring Polynomials: Difference of Two Squares

OBJECTIVE

70. Factor a binomial that is the difference of two squares.

APPLICATION

In arithmetic $(50 - 2)(50 + 2) = 2500 - 4 = 2496$ is a shortcut for multiplying $48 \cdot 52$. (Note that $50 - 2 = 48$ and $50 + 2 = 52$.) Similarly $1600 - 49 = 1551$ is a short cut for multiplying what two numbers?

VOCABULARY

No new vocabulary.

HOW AND WHY

Binomials that are the difference of two squares are factored by observation (they can also be factored by the method discussed in the last section). By multiplication of the sum and difference of the terms x and 5

$(x + 5)(x - 5)$

we obtain

$x^2 - 25$

which is the difference of two squares. Also, since

$(y + 5)(y - 5) = y^2 - 25$

$(m + 5)(m - 5) = m^2 - 25$

we judge that, given $r^2 - 25$, we may factor as follows:

$r^2 - 25 = (r + 5)(r - 5)$

The factoring is correct as we can check by multiplying. Similarly,

$9x^2 - 100$

is the difference of two squares, since $3x \cdot 3x = (3x)^2 = 9x^2$ and $10 \cdot 10 = 10^2 = 100$. We try the factors that are the sum and difference of $3x$ and 10. Does

$9x^2 - 100 = (3x + 10)(3x - 10)$?

After multiplying (to check) we can answer yes.

To factor a binomial that is the difference of two squares, write the product of the sum and difference of the terms that are squared.

$a^2 - b^2 = (a + b)(a - b)$

EXAMPLES

a. $a^2 - 144 = (a + 12)(a - 12)$

b. $16t^2 - 1 = (4t + 1)(4t - 1)$

c. $x^2 - 10$ *Not factored at this time, since* 10 *is not the square of a whole number.*

d. $196y^2 - 81 = (14y + 9)(14y - 9)$

e. $4x^2 - 36 = 4(x^2 - 9) = 4(x + 3)(x - 3)$

APPLICATION SOLUTION

Since 1600 is the square of 40, and 49 is the square of 7, we can write

$$1600 - 49 = 40^2 - 7^2$$
$$= (40 + 7)(40 - 7) \quad \textit{Factor the difference of two squares}$$
$$= (47)(33)$$

Answer: $1600 - 49$ is shortcut for multiplying 47 times 33.

WARM UPS

Factor:

1. $x^2 - 1$

2. $x^2 - 4$

3. $x^2 - 9$

4. $a^2 - 100$

5. $z^2 - 64$

6. $x^2 - 36$

7. $a^2 - 49$

8. $81c^2 - 1$

9. $64y^2 - 1$

10. $x^2 - 121$

11. $y^2 - 144$

12. $16x^2 - 9$

Exercises 6.10

ANSWERS

1. _____
2. _____
3. _____
4. _____
5. _____
6. _____
7. _____
8. _____
9. _____
10. _____
11. _____
12. _____
13. _____
14. _____
15. _____
16. _____
17. _____
18. _____
19. _____
20. _____
21. _____
22. _____
23. _____
24. _____
25. _____
26. _____
27. _____
28. _____
29. _____
30. _____
31. _____
32. _____
33. _____

A

Factor each of the following if possible.

1. $a^2 - 16$ **2.** $y^2 - 25$ **3.** $4x^2 - 9$

4. $9a^2 - 1$ **5.** $a^2 - 64$ **6.** $x^2 - 36$

B

7. $9a^2 - 64b^2$ **8.** $81x^2 - 25$ **9.** $25x^2 - 36y^2$

10. $100x^2 - 81y^2$ **11.** $144a^2 - 121b^2$ **12.** $36a^2b^2 - 169$

13. $400x^2 - 9$ **14.** $225x^2 - 64$

C

15. $48a^2 - 75$ **16.** $100y^2 - 64z^2$ **17.** $9x^2 - 36$

18. $4ax^2 - 4a$ **19.** $36a^3 - 64a$ **20.** $-36 + 25a^2$

21. $25b^2 - 16$ **22.** $2\pi r^2 - 2\pi R^2$ **23.** $14a^3 - 56ab^2$

24. $72a^2b - 200b^3$

D

25. $2500 - 9$ is a shortcut for multiplying what two numbers?

26. $4900 - 4$ is a shortcut for multiplying what two numbers?

27. $16900 - 25$ is a shortcut for multiplying what two numbers?

28. $19600 - 81$ is a shortcut for multiplying what two numbers?

E   *Maintain Your Skills*

Do the indicated operations:

29. $6x^2 + 8x^2 - 6x^2 + 7x$ **30.** $(4x)^3 + 2x^3 + (2x)^3$

Solve:

31. $-8x + 7 = 2x - 3$ **32.** $2x + 7 - x = 5x + 19$

33. Solve for a: $2a + 3b = a + c$

319

ANSWERS TO WARM UPS (6.10)

1. $(x + 1)(x - 1)$ 2. $(x + 2)(x - 2)$ 3. $(x + 3)(x - 3)$

4. $(a + 10)(a - 10)$ 5. $(z + 8)(z - 8)$ 6. $(x + 6)(x - 6)$

7. $(a + 7)(a - 7)$ 8. $(9c + 1)(9c - 1)$ 9. $(8y + 1)(8y - 1)$

10. $(x + 11)(x - 11)$ 11. $(y + 12)(y - 12)$ 12. $(4x + 3)(4x - 3)$

6.11 Quadratic Equations — Solved by Factoring

OBJECTIVE

71. Solve a quadratic equation by factoring.

APPLICATION

The time it takes a rock to fall to the bottom of a gully 176 ft deep if it is thrown upward from the rim at the rate of 56 ft/sec is given by the equation $16t^2 - 56t - 176 = 0$. How long will it take the rock to reach the bottom? (The negative value of t has no physical meaning.)

VOCABULARY

In a *quadratic equation* the variable has the number 2 as the largest exponent ($2x^2 + x - 3 = 0$ or $4x^2 - 1 = 0$, but not $x^3 - x^2 + 1 = 7x$). The *solutions* (there are usually two) are the number replacements for the variable that make the equation true.

HOW AND WHY

By the use of another fundamental law of equations, we can write a series of equations to find the solution of a quadratic equation (recall the two laws used for linear equations).

Third law: If two expressions have a product of 0, then one or both of them must equal 0.

Examples: If $ab = 0$, then $a = 0$ or $b = 0$.
If $(x - c)(x - d) = 0$, then $x - c = 0$ or $x - d = 0$.

To solve the equation $2x^2 - 9x + 4 = 0$, we can write this series of equations:

$2x^2 - 9x + 4 = 0$ *Original equation*

$(2x - 1)(x - 4) = 0$ *Factor the left side so the product of the two factors equals 0*

$2x - 1 = 0$ or $x - 4 = 0$ *One or both factors must equal 0*

$2x = 1$ or $x = 4$ *Solve the two linear equations as usual*

$x = \dfrac{1}{2}$ or $x = 4$

Check: if $x = \dfrac{1}{2}$,

$$2\left(\frac{1}{2}\right)^2 - 9\left(\frac{1}{2}\right) + 4 = 0$$

$$2\left(\frac{1}{4}\right) - \frac{9}{2} + 4 = 0$$

$$\frac{1}{2} - \frac{9}{2} + 4 = 0$$

$$-4 + 4 = 0$$

$$0 = 0$$

If $x = 4$,

$$2(4)^2 - 9(4) + 4 = 0$$

$$2(16) - 36 + 4 = 0$$

$$32 - 36 + 4 = 0$$

$$-4 + 4 = 0$$

$$0 = 0$$

If the original quadratic equation does not have zero as its right member, the fundamental laws of equations are used to change it. (See Example c.)

> **To solve a quadratic equation by factoring,**
>
> 1. **Rewrite the equation so the right member equals zero.**
> 2. **Factor the left member.**
> 3. **Set both factors equal to zero.**
> 4. **Solve both linear equations.**

Note: Not all quadratic equations can be solved by this method. (See Chapter 10.)

EXAMPLES

a. $x^2 - 4x - 77 = 0$
$(x + 7)(x - 11) = 0$
$x + 7 = 0$ or $x - 11 = 0$
$x = -7$ or $x = 11$

b. $15x^2 - 16x + 4 = 0$
$(3x - 2)(5x - 2) = 0$
$3x - 2 = 0$ or $5x - 2 = 0$
$3x = 2$ or $5x = 2$
$x = \dfrac{2}{3}$ or $x = \dfrac{2}{5}$

c. $4x + 12 = x^2$
$-x^2 + 4x + 12 = -x^2 + x^2$
$-x^2 + 4x + 12 = 0$
$-1(-x^2 + 4x + 12) = -1(0)$
$x^2 - 4x - 12 = 0$
$(x + 2)(x - 6) = 0$
$x + 2 = 0$ or $x - 6 = 0$
$x = -2$ or $x = 6$

d. The product of two consecutive positive numbers is 56. What are the numbers?

Let x represent the first number; then
$x + 1$ represents the second number.

(first number)(second number) is 56

$$(x)(x + 1) = 56$$

$$x^2 + x = 56$$

$$x^2 + x - 56 = 0$$

$$(x + 8)(x - 7) = 0$$

$$x + 8 = 0 \quad \text{or } x - 7 = 0$$

$$x = -8 \text{ or} \quad x = 7$$

$$x + 1 = -7 \quad x + 1 = 8$$

Check: The product is 56. Now, $(7)(8) = 56$. Also, $(-8)(-7) = 56$, but the problem statement says that the numbers are positive. So the numbers are 7 and 8.

APPLICATION SOLUTION

To find out how long it takes the rock to reach the bottom of the gully, we solve the given equation.

Formula: $16t^2 - 56t - 176 = 0$

Solve: $8(2t^2 - 7t - 22) = 0$ *Common factor*

$8(2t - 11)(t + 2) = 0$ *Factor the trinomial*

$2t - 11 = 0 \text{ or } t + 2 = 0$ *Solve the quadratic equation*

$2t = 11 \qquad t = -2$

$t = \dfrac{11}{2} = 5\dfrac{1}{2}$

Answer: $t = -2$ has no physical meaning, so we discard that root. So $t = 5\dfrac{1}{2}$ is the accepted root. Therefore it will take $5\dfrac{1}{2}$ seconds for the rock to reach the bottom of the gully.

WARM UPS

Solve:

1. $x^2 + 4x + 3 = 0$

2. $x^2 + 5x + 6 = 0$

3. $x^2 - 2x - 15 = 0$

4. $y^2 + 3y - 28 = 0$

5. $y^2 - 2y + 1 = 0$

6. $y^2 + 5y - 14 = 0$

7. $x^2 + 10x + 16 = 0$

8. $x^2 - 11x + 30 = 0$

9. $x^2 - 12x + 11 = 0$

10. $y^2 - y - 12 = 0$

11. $p^2 + 6p + 5 = 0$

12. $q^2 + 8q + 12 = 0$

Exercises 6.11

A

Solve the following:

1. $(x + 7)(x - 8) = 0$

2. $(x + 5)(x + 6) = 0$

3. $(x - 9)(x - 4) = 0$

4. $(x + 8)(x - 11) = 0$

5. $x(x - 1)(x - 3) = 0$

6. $x^2 - x - 72 = 0$

B

7. $x^2 - 9 = 8x$

8. $10a = 24 - a^2$

9. $0 = b^2 + 3b - 10$

10. $y^2 = 49$

11. $4x^2 = 36$

12. $x^2 + 4x = 32$

13. $x^2 + 144 = -30x$

14. $x^2 = 7x + 144$

C

15. $3x^2 + 11x + 6 = 0$

16. $7x^2 - 17x - 12 = 0$

17. $3x^2 - 10x + 3 = 0$

18. $6x^2 + 1 = -5x$

19. $4x^2 + 4x + 1 = 0$

20. $-90 = -e^2 - e$

21. $x^2 + 3x = 0$

22. $10x^2 + 11x - 6 = 0$

23. $2x^2 + 19x = -35$

24. $6x^2 = -x + 5$

D

25. If the height (s) of a falling object is given by the formula $s = 144t - 16t^2$, where t is the time the object is falling, how long will it take the object to hit the ground ($s = 0$)?

26. The area of a triangle whose altitude is x and whose base is $x + 7$ is 60. What is the measure of each, if the formula for the area of a triangle is $\frac{1}{2}ba = A$?

27. The product of two consecutive positive odd numbers is 63. Find the numbers.

28. One positive number is six more than another number. The product of the two numbers is 72. Find the numbers.

29. A rectangular lot is 3 ft longer than it is wide. It contains 88 sq ft in area. Find the dimensions.

30. The perimeter of a rectangle is 40 in. Its area is 96 sq in. Find its dimensions.

E *Maintain Your Skills*

Simplify:

31. $(6x + y) - (-2x - 3y)$

32. $x - \{2 - [2x - (x - 4)]\}$

Solve:

33. $8 - (6x + 18) = 2 - (3x - 6)$

34. $6a - (2a - 6) + (a - 5) = 26$

35. Solve for b: $4a + 2b - (3a - 3b) = 3 - (a - b)$

ANSWERS

1. _____
2. _____
3. _____
4. _____
5. _____
6. _____
7. _____
8. _____
9. _____
10. _____
11. _____
12. _____
13. _____
14. _____
15. _____
16. _____
17. _____
18. _____
19. _____
20. _____
21. _____
22. _____
23. _____
24. _____
25. _____
26. _____
27. _____
28. _____
29. _____
30. _____
31. _____
32. _____
33. _____
34. _____
35. _____

Chapter 6
POST-TEST

1. _____

2. _____

3. _____

4. _____

5. _____

6. _____

7. _____

8. _____

9. _____

10. _____

11. _____

12. _____

13. _____

14. _____

15. _____

16. _____

17. _____

18. _____

19. _____

20. _____

21. _____

22. _____

23. _____

24. _____

25. _____

1. **(Obj. 59)** Solve for x: $4x + 5 = 33$

2. **(Obj. 67)** _Divide:_ $(-9ab^2c + 15a^2bc - 3abc^2) \div 3abc$

3. **(Obj. 62)** _Add:_ $4x + 2x + 5x$

4. **(Obj. 63)** _Simplify:_ $(6a + 2b - 7) - (3a - 2b + 4)$

5. **(Obj. 64)** Solve for x: $(4x + 5) - (2x + 6) = 18$

6. **(Obj. 71)** Solve for x: $x^2 - 13x + 42 = 0$

7. **(Obj. 66)** _Multiply:_ $(x - 7)(x + 5)$

8. **(Obj. 60)** Express as a power of x: $\dfrac{(x^7)^2}{x^3}$

9. **(Obj. 61)** _Evaluate:_ $(4^3)^2$

10. **(Obj. 62)** Do the indicated operations:

$$3x + \frac{6x^3}{2x^2} - 2x$$

11. **(Obj. 67)** _Divide:_ $(4x^2 - 7x - 2) \div (x - 2)$

12. **(Obj. 63)** _Add:_ $3x + 3y + 6$, $7x - 5y + 9$, and $2x + 4y - 16$

13. **(Obj. 63)** Solve for y: $4y + 8 = 2y - 10$

14. **(Obj. 69)** _Factor:_ $2y^2 - 15y + 7$

15. **(Obj. 70)** _Factor:_ $16x^2 - 49$

16. **(Obj. 66)** _Multiply:_ $(2x + 5)(3x + 2)$

17. **(Obj. 60)** Express as a power of a: $a^3 \cdot a^5 \cdot a^6$

18. **(Obj. 68)** _Factor:_ $8xy + 12xyz - 4xyz^2$

19. **(Obj. 66)** Solve for x: $3(x - 7) + 2(x + 5) = 27$

20. **(Obj. 69)** _Factor:_ $x^2 - 7x - 30$

21. **(Obj. 71)** Solve for x: $x^2 - 43x + 42 = 0$

22. **(Obj. 66)** Solve for x: $3(x + 2) - 5(2x + 5) = 7(3 - 2x)$

23. **(Obj. 71)** One positive number is three less than twice another. The product of the two numbers is 20. Find the numbers.

24. **(Obj. 65)** The sum of four consecutive even numbers is 100. Find the numbers.

25. **(Obj. 65)** A landscaper planted six more mugho pines than he did azaleas. He also planted twice as many roses as he did mugho pines. If he planted 62 plants in all, how many of each did he plant?

ANSWERS TO WARM UPS (6.11)

1. $x = -3$ or $x = -1$
2. $x = -2$ or $x = -3$
3. $x = -3$ or $x = 5$
4. $y = -7$ or $y = 4$
5. $y = 1$
6. $y = -7$ or $y = 2$
7. $x = -8$ or $x = -2$
8. $x = 5$ or $x = 6$
9. $x = 1$ or $x = 11$
10. $y = -3$ or $y = 4$
11. $p = -5$ or $p = -1$
12. $q = -6$ or $q = -2$

7

Rational Expressions

1. _____

2. _____

3. _____

4. _____

5. _____

6. _____

7. _____

8. _____

9. _____

10. _____

11. _____

12. _____

13. _____

14. _____

Chapter 7
PRE-TEST

The problems in the following pre-test are a sample of the material in the chapter. You may already know how to work some of these. If so, this will allow you to spend less time on those parts. As a result, you will have more time to give to the sections that you find more difficult or that are new to you. The answers are in the back of the text.

1. (Obj. 72) *Multiply:* $\dfrac{7x}{5} \cdot \dfrac{8x-3}{9}$

2. (Obj. 72) *Multiply:* $\dfrac{17p}{5q} \cdot \dfrac{7p^2}{9q}$

3. (Obj. 73) *Reduce:* $\dfrac{n^2-49}{3n^2+9n-84}$

4. (Obj. 73) *Reduce:* $\dfrac{90x^4y^3z^2}{120x^3y^5z^2}$

5. (Obj. 74) Find the missing numerator:

$$\frac{x+3}{x+7} = \frac{?}{x^2-2x-63}$$

6. (Obj. 74) Find the missing numerator:

$$\frac{15ab}{2c} = \frac{?}{24b^2c^2}$$

7. (Obj. 75) Multiply and reduce to lowest terms:

$$\frac{a^2b-b}{ab^2} \cdot \frac{5abc}{a^2-2a+1}$$

8. (Obj. 76) Divide and reduce:

$$\frac{14mn^3}{9p^3} \div \frac{7m^3n^2}{18p^4}$$

9. (Obj. 76) Divide and reduce:

$$\frac{y-6}{y^2+3y-40} \div \frac{y^2-y-30}{y^2+13y+40}$$

10. (Obj. 77) Find the LCM of the following:

$$6b^2, 2a^2b, 5a^2b^3c$$

11. (Obj. 77) Find the LCM of the following:

$$5a+10b, 3a+6b, 27$$

12. (Obj. 78) *Add:* $\dfrac{7y-8}{4y^2} + \dfrac{3y+8}{4y^2}$

13. (Obj. 78) *Add:* $\dfrac{a}{a-3} + \dfrac{3}{a-4}$

14. (Obj. 79) Subtract and reduce:

$$\frac{5x}{x^2-25} - \frac{25}{x^2-25}$$

15. _____

16. _____

17. _____

18. _____

19. _____

20. _____

21. _____

22. _____

23. _____

24. _____

25. _____

15. **(Obj. 79)** *Subtract:* $\dfrac{y}{y-5} - \dfrac{3}{y+3}$

16. **(Obj. 80)** *Simplify:* $\dfrac{\dfrac{a}{5} + \dfrac{b}{2}}{\dfrac{a}{10} - \dfrac{b}{5}}$

17. **(Obj. 80)** *Simplify:* $\dfrac{\dfrac{1}{x^2} - \dfrac{1}{x}}{\dfrac{1}{x^2} + \dfrac{1}{x}}$

18. **(Obj. 81)** *Solve:* $\dfrac{a}{6} - \dfrac{4}{5} = \dfrac{2}{3}$

19. **(Obj. 81)** *Solve:* $\dfrac{3}{a-1} + \dfrac{2}{3} = \dfrac{1}{4}$

20. **(Obj. 82)** If $V = \ell wh$, find h if $\ell = 2.8$, $w = 2.2$, and $V = 61.6$.

21. **(Obj. 82)** If $A = \dfrac{h}{2}(B + b)$, find b if $h = 12$, $B = 30$, and $A = 300$.

22. **(Obj. 83)** If $I = \dfrac{E - e}{R}$, solve for e.

23. **(Obj. 83)** If $S = 2ab + 2ah + 2bh$, solve for a.

24. **(Obj. 81)** A painter and her apprentice can paint a house in 8 days. If the painter could do the painting alone in 12 days, how long would it take the apprentice alone?

25. **(Obj. 81)** Bill drove 400 miles on Wednesday. On Thursday he averaged 10 mph more than on Wednesday and traveled 480 miles. If Bill drove the same number of hours each day, what was his average speed each day?

7.1 Multiplication of Rational Expressions

OBJECTIVE	**72. Multiply two or more rational expressions.**

APPLICATION	The main application of this procedure is in the solution of fractional equations.

VOCABULARY	A *rational expression* is a fraction that has a polynomial for its numerator and/or denominator. A rational expression is also an indicated quotient of two polynomials. Some examples are

$$\frac{1}{x}, \frac{3x}{4y}, \text{ and } \frac{2a+b}{x+3}$$

For this chapter it will be assumed that no variable will have a value that will make any denominator zero.

HOW AND WHY	Since rational expressions look like fractions in arithmetic, one might expect that the operations performed on fractions would be used in a similar manner when working with rational expressions. This is exactly the case.

To multiply two rational expressions, multiply the numerators and multiply the denominators.

$$\frac{A}{B} \cdot \frac{C}{D} = \frac{AC}{BD}, BD \neq 0$$

So, $\dfrac{3}{x} \cdot \dfrac{1}{2} = \dfrac{3 \cdot 1}{x \cdot 2} = \dfrac{3}{2x}$.

EXAMPLES	**a.** $\dfrac{2}{a} \cdot \dfrac{3}{b} = \dfrac{2 \cdot 3}{a \cdot b} = \dfrac{6}{ab}$
	b. $\dfrac{4x}{5} \cdot \dfrac{3a}{5b} = \dfrac{4x \cdot 3a}{5 \cdot 5b} = \dfrac{12ax}{25b}$
	c. $\dfrac{2x+y}{4a} \cdot \dfrac{b}{x} = \dfrac{(2x+y)b}{4ax} = \dfrac{2bx+by}{4ax}$

WARM UPS	*Multiply:*

1. $\dfrac{3}{2} \cdot \dfrac{w}{z}$

2. $\dfrac{6}{m} \cdot \dfrac{p}{7}$

3. $\dfrac{12}{x} \cdot \dfrac{2}{y}$

4. $\dfrac{8a}{5b} \cdot \dfrac{9c}{7d}$

5. $\dfrac{-5x}{7u} \cdot \dfrac{12xy}{11uv}$

6. $\dfrac{-2}{x} \cdot \dfrac{-y}{7x}$

7. $\dfrac{3x}{5} \cdot \dfrac{2a}{5}$

8. $\dfrac{6ab}{7x} \cdot \dfrac{3ad}{5y}$

9. $\dfrac{x}{4} \cdot \dfrac{3x-2}{5}$

10. $\dfrac{a}{2} \cdot \dfrac{a+3}{3}$

11. $\dfrac{2x}{3} \cdot \dfrac{x-1}{5}$

12. $\dfrac{4}{y} \cdot \dfrac{3}{2y-5}$

333

NAME _____ CLASS _____

ANSWERS

Exercises 7.1

A

Multiply:

1. $\dfrac{5}{7} \cdot \dfrac{s}{t}$

2. $\dfrac{4}{m} \cdot \dfrac{n}{7}$

3. $\dfrac{3m}{2p} \cdot \dfrac{5n}{p}$

4. $\dfrac{-6y}{w} \cdot \dfrac{3y}{5w}$

5. $\dfrac{3ab}{2c} \cdot \dfrac{-7ab}{c^2}$

6. $\dfrac{-7cd}{3xy} \cdot \dfrac{-5ce}{2yz}$

B

7. $\dfrac{-2}{m} \cdot \dfrac{a+b}{m-n}$

8. $\dfrac{b}{3} \cdot \dfrac{2b-3}{-2a^2}$

9. $\dfrac{5-3c}{p} \cdot \dfrac{-c}{p-4}$

10. $\dfrac{26+z}{m} \cdot \dfrac{2y}{7m}$

11. $\dfrac{19x}{13y} \cdot \dfrac{-8a}{15b}$

12. $\dfrac{15rs}{7t} \cdot \dfrac{18r}{7v}$

13. $\dfrac{5a+3b}{x} \cdot \dfrac{9ab}{3a+c}$

14. $\dfrac{4x-2y}{3a-b} \cdot \dfrac{7}{4a}$

C

15. $\dfrac{z+y}{z} \cdot \dfrac{y}{z-y}$

16. $\dfrac{x-4}{x+3} \cdot \dfrac{x-4}{x+3}$

17. $\dfrac{y+3}{2t+1} \cdot \dfrac{y-7}{3t-2}$

18. $\dfrac{7p-2q}{2r} \cdot \dfrac{q}{6r+7s}$

19. $\dfrac{x}{-5y} \cdot \dfrac{-11x}{15y} \cdot \dfrac{7}{-2y}$

20. $\dfrac{a}{-3b} \cdot \dfrac{-5a^2c}{4bd^2} \cdot \dfrac{7ac^2}{-2b^2d}$

21. $\dfrac{3}{x} \cdot \dfrac{x+4}{2x} \cdot \dfrac{x-1}{x+1}$

22. $\dfrac{5}{b} \cdot \dfrac{b-3}{b+3} \cdot \dfrac{b+2}{4b}$

23. $\dfrac{c}{a} \cdot \dfrac{c+6}{4a^2} \cdot \dfrac{c-6}{a+2}$

24. $\dfrac{s}{3t} \cdot \dfrac{s-5}{2t+1} \cdot \dfrac{2s+3}{2t-1}$

E   *Maintain Your Skills*

Multiply:

25. $(x-8)(x+12)$

26. $(2x+5)(3x-7)$

27. $(2x+3)(2x^2-3x+5)$

28. $4x-3$
    $\underline{5x-7}$

29. $5x^2+3x-8$
    $\underline{3x-4}$

335

1. $\dfrac{3w}{2z}$ 2. $\dfrac{6p}{7m}$ 3. $\dfrac{24}{xy}$ 4. $\dfrac{72ac}{35bd}$ 5. $\dfrac{-60x^2y}{77u^2v}$

6. $\dfrac{2y}{7x^2}$ 7. $\dfrac{6ax}{25}$ 8. $\dfrac{18a^2bd}{35xy}$ 9. $\dfrac{3x^2-2x}{20}$ 10. $\dfrac{a^2+3a}{6}$

11. $\dfrac{2x^2-2x}{15}$ 12. $\dfrac{12}{2y^2-5y}$

7.2 Renaming Rational Expressions

OBJECTIVES

73. Reduce rational expressions.
74. Build rational expressions.

APPLICATION

The main application of this procedure is in adding and subtracting rational expressions.

VOCABULARY

Reducing rational expressions has the same meaning as reducing fractions in arithmetic. Eliminate like factors from the numerator and denominator. The reduced and the given rational expressions are said to be *equivalent*. For instance, $\dfrac{4xy}{6xz}$ and $\dfrac{2y}{3z}$ are equivalent.

To build a rational expression is to write an equivalent rational expression by introducing like factors in the numerator and denominator.

HOW AND WHY

The process of building or reducing algebraic expressions is similar to that used with fractions in arithmetic.

To build an algebraic expression, introduce a common factor in both the numerator and the denominator.

$$\frac{A}{B} \cdot \frac{C}{C} = \frac{AC}{BC}, \; BC \neq 0$$

To reduce an algebraic expression, eliminate a common factor in both the numerator and the denominator.

$$\frac{AC}{BC} = \frac{A}{B} \cdot \frac{C}{C} = \frac{A}{B} \cdot 1 = \frac{A}{B}, \; BC \neq 0$$

Reduce:

$$\frac{2}{4} = \frac{2 \cdot 1}{2 \cdot 2} = \frac{2}{2} \cdot \frac{1}{2} = 1 \cdot \frac{1}{2} = \frac{1}{2}$$

We can use the same method for rational expressions.

$$\frac{4x}{12y} = \frac{4 \cdot x}{4 \cdot 3y} = \frac{4}{4} \cdot \frac{x}{3y} = 1 \cdot \frac{x}{3y} = \frac{x}{3y}$$

EXAMPLES

a. $\dfrac{7a}{21b} = \dfrac{7 \cdot a}{7 \cdot 3b} = \dfrac{7}{7} \cdot \dfrac{a}{3b} = 1 \cdot \dfrac{a}{3b} = \dfrac{a}{3b}$

b. $\dfrac{4x}{5x^2} = \dfrac{4 \cdot x}{5x \cdot x} = \dfrac{4}{5x} \cdot \dfrac{x}{x} = \dfrac{4}{5x} \cdot 1 = \dfrac{4}{5x}$

c. $\dfrac{2x + 4}{x^2 + 5x + 6} = \dfrac{2(x + 2)}{(x + 3)(x + 2)} = \dfrac{2}{x + 3} \cdot \dfrac{x + 2}{x + 2} = \dfrac{2}{x + 3}$

d. $\dfrac{3}{4} = \dfrac{?}{8}$ *Multiply by* $\dfrac{2}{2}$ *since* $8 \div 4 = 2$

$\dfrac{3}{4} = \dfrac{3}{4} \cdot \dfrac{2}{2} = \dfrac{6}{8}$

e. $\dfrac{4x}{5y} = \dfrac{?}{20y^2}$ *Multiply by* $\dfrac{4y}{4y}$ *since* $20y^2 \div 5y = 4y$

$\dfrac{4x}{5y} = \dfrac{4x}{5y} \cdot \dfrac{4y}{4y} = \dfrac{16xy}{20y^2}$

f. $\dfrac{x + y}{5} = \dfrac{?}{10x + 15y}$

$\dfrac{x + y}{5} \cdot \dfrac{2x + 3y}{2x + 3y} = \dfrac{2x^2 + 5xy + 3y^2}{10x + 15y}$

WARM UPS

Reduce:

1. $\dfrac{6a}{18b}$

2. $\dfrac{5xyz}{10xz}$

3. $\dfrac{8ab^2}{12ab}$

4. $\dfrac{a(x - 3)}{b(x - 3)}$

5. $\dfrac{9(2x + 3)}{12(2x + 3)}$

6. $\dfrac{2(x - y)^2}{4(x - y)}$

Find the missing numerator.

7. $\dfrac{8y}{9x} = \dfrac{?}{9x^2}$

8. $\dfrac{7p}{13q} = \dfrac{?}{26pq}$

9. $\dfrac{a}{x} = \dfrac{?}{x^2y}$

10. $\dfrac{3}{7x} = \dfrac{?}{14x}$

11. $\dfrac{3x}{a + b} = \dfrac{?}{2a + 2b}$

12. $\dfrac{5}{x + 3} = \dfrac{?}{x^2 + 5x + 6}$

Exercises 7.2

A

Reduce:

1. $\dfrac{16a^2}{8a}$

2. $\dfrac{3ab^2}{9ab}$

3. $\dfrac{5ax^2}{10x^2}$

Find the missing numerator:

4. $\dfrac{3b}{18c} = \dfrac{?}{36c}$

5. $\dfrac{7y}{6x} = \dfrac{?}{12x^2}$

6. $\dfrac{3y-1}{8} = \dfrac{?}{32x}$

B

Reduce:

7. $\dfrac{16x^2y^2}{24x}$

8. $\dfrac{21b^2d}{35bd^2}$

9. $\dfrac{x+2}{2x+4}$

10. $\dfrac{2a+6b}{a+3b}$

Find the missing numerator:

11. $\dfrac{5x}{12y} = \dfrac{?}{24xy}$

12. $\dfrac{a+b}{6} = \dfrac{?}{6a+6b}$

13. $\dfrac{4x+2y}{x+y} = \dfrac{?}{(x+y)^2}$

14. $\dfrac{x+2}{x+3} = \dfrac{?}{x^2+5x+6}$

C

Reduce:

15. $\dfrac{28a^3b^2c^5}{21a^4b^5c}$

16. $\dfrac{-96x^7y^6z^4}{60x^4y^6z^6}$

17. $\dfrac{x^2+6x}{x^2+10x+24}$

18. $\dfrac{6x+12}{x^2+5x+6}$

19. $\dfrac{x^2-9}{x^2+5x-24}$

ANSWERS

1. _____

2. _____

3. _____

4. _____

5. _____

6. _____

7. _____

8. _____

9. _____

10. _____

11. _____

12. _____

13. _____

14. _____

15. _____

16. _____

17. _____

18. _____

19. _____

Find the missing numerators:

20. $\dfrac{5ac}{9xz} = \dfrac{?}{45bx^2z^3}$

21. $\dfrac{-3xy}{2mn} = \dfrac{?}{18m^2nx^2}$

22. $\dfrac{2x-7}{3x+1} = \dfrac{?}{6x^2-13x-5}$

23. $\dfrac{4x+5}{7x-5} = \dfrac{?}{21x^2-64x+35}$

24. $\dfrac{x^2+2x+3}{3x-4} = \dfrac{?}{9x^2-16}$

E   *Maintain Your Skills*

Divide:

25. $(16x^4 + 24x^3 - 8x^2 + 32x) \div 8x$

26. $6xy\overline{)36x^3y + 48x^2y^2 - 12xy^3}$

27. $(8x^2 - 6x - 35) \div (2x - 5)$

28. $x - 1\overline{)15x^2 - 8x + 5}$

29. $2x + 3\overline{)2x^3 + 17x^2 + 11x - 15}$

ANSWERS TO WARM UPS (7.2)

1. $\dfrac{a}{3b}$ 2. $\dfrac{y}{2}$ 3. $\dfrac{2b}{3}$ 4. $\dfrac{a}{b}$ 5. $\dfrac{3}{4}$ 6. $\dfrac{x-y}{2}$

7. $8xy$ 8. $14p^2$ 9. axy 10. 6 11. $6x$ 12. $5x + 10$

7.3 A Second Look at Multiplying Rational Expressions

OBJECTIVE

75. **Find the product of two rational expressions and reduce the product to lowest terms.**

APPLICATION

Floyd can mow his lawn in 3 hours and his grandson Norm can mow the same lawn in 2 hours. If they work together, the equation for the time (t) it takes them together is given by $\dfrac{1}{3} + \dfrac{1}{2} = \dfrac{1}{t}$. Simplify this equation by multiplying each term of the equation by $\dfrac{6t}{1}$. Find the time (t) it takes for them to mow the lawn together.

VOCABULARY

No new vocabulary.

HOW AND WHY

Recall from multiplication of fractions in arithmetic that there were times that it was possible to do the work more quickly by reducing before performing the multiplication. For example,

$$\frac{3}{4} \cdot \frac{4}{5} = \frac{3}{\overset{}{\underset{1}{\cancel{4}}}} \cdot \frac{\overset{1}{\cancel{4}}}{5} = \frac{3}{5}$$

The same holds true when multiplying rational expressions. If a factor in a numerator is the same as a factor in a denominator, they may be eliminated (divided out) and the multiplication performed as before. Following the example in arithmetic just used, we can proceed as follows:

$$\frac{5}{x} \cdot \frac{x}{7} = \frac{5}{\underset{1}{\cancel{x}}} \cdot \frac{\overset{1}{\cancel{x}}}{7} = \frac{5}{7}$$

To multiply two or more rational expressions,

1. **Factor the numerators and the denominators.**
2. **Reduce, if possible.**
3. **Multiply.**

EXAMPLES

a. $\dfrac{x}{2a + 2b} \cdot \dfrac{a + b}{y} = \dfrac{x}{2\underset{1}{\cancel{(a+b)}}} \cdot \dfrac{\overset{1}{\cancel{(a+b)}}}{y} = \dfrac{x}{2y}$

b. $\dfrac{ax}{x+y} \cdot \dfrac{x^2-y^2}{bx} = \dfrac{a\overset{1}{\cancel{x}}}{\cancel{x+y}} \cdot \dfrac{\overset{1}{\cancel{(x+y)}}(x-y)}{b\cancel{x}} = \dfrac{a(x-y)}{b} = \dfrac{ax-ay}{b}$

APPLICATION SOLUTION

First simplify the equation by multiplying each term by $\dfrac{6t}{1}$.

Equation: $\dfrac{1}{3} + \dfrac{1}{2} = \dfrac{1}{t}$

Solve: $\dfrac{6t}{1} \cdot \dfrac{1}{3} + \dfrac{6t}{1} \cdot \dfrac{1}{2} = \dfrac{6t}{1} \cdot \dfrac{1}{t}$

$$2t + 3t = 6$$
$$5t = 6$$
$$t = \dfrac{6}{5}$$

Answer: Floyd and Norm can mow the lawn together in $1\dfrac{1}{5}$ hours or 1 hour 12 minutes.

WARM UPS

Multiply and reduce to lowest terms:

1. $\dfrac{3y}{8} \cdot \dfrac{4}{y^2}$

2. $\dfrac{5a}{11b} \cdot \dfrac{22}{a}$

3. $\dfrac{9c}{2d} \cdot \dfrac{d^2}{3c}$

4. $\dfrac{12mn}{7} \cdot \dfrac{21}{m^2}$

5. $\dfrac{8a^2}{5b^2} \cdot \dfrac{3b}{16a}$

6. $\dfrac{14x^2y}{9z^2} \cdot \dfrac{18z}{7xy^2}$

7. $\dfrac{3(x+7)}{x-9} \cdot \dfrac{x-9}{4(x+7)}$

8. $\dfrac{2x(x+6)}{3y(x+4)} \cdot \dfrac{2(x+4)}{3(x+6)}$

9. $\dfrac{5m(m-3)}{10n(m+3)} \cdot \dfrac{2(m+3)}{m-3}$

10. $\dfrac{(x+7)(x-8)}{x+5} \cdot \dfrac{x+5}{(x-6)(x+7)}$

11. $\dfrac{x^2-4}{3x} \cdot \dfrac{x}{x-2}$

12. $\dfrac{x^2-25}{5x^2} \cdot \dfrac{5x}{x+5}$

Exercises 7.3

A

Multiply and reduce to lowest terms:

1. $\dfrac{2x}{3y} \cdot \dfrac{12}{5x}$

2. $\dfrac{10u^3}{12z^2} \cdot \dfrac{6z}{5u}$

3. $\dfrac{3(x+4)}{6(x-2)} \cdot \dfrac{14(x-2)}{4(x+4)}$

4. $\dfrac{2x+2}{5x} \cdot \dfrac{15x}{x+1}$

5. $\dfrac{33}{x^2-9} \cdot \dfrac{x+3}{3}$

6. $\dfrac{x^2+5x+6}{10y} \cdot \dfrac{2y}{x+2}$

B

7. $\dfrac{6(x+4)}{3(x-4)} \cdot \dfrac{4y}{8y}$

8. $\dfrac{y}{x+z} \cdot \dfrac{x+z}{z}$

9. $\dfrac{a+b}{5} \cdot \dfrac{6a}{a+b}$

10. $\dfrac{x+y}{x+3} \cdot \dfrac{x+3}{x+y}$

11. $\dfrac{ax}{y^2} \cdot \dfrac{y^2+2y}{a^2+a}$

12. $\dfrac{2a+3b}{x+y} \cdot \dfrac{x^2+xy}{2ax+3bx}$

13. $\dfrac{(x+2)(x+3)}{x+5} \cdot \dfrac{x+7}{(x+3)(x+7)}$

14. $\dfrac{(x+9)(x-9)}{(x-6)(x+9)} \cdot \dfrac{(x-6)(x+4)}{(x-4)(x-9)}$

C

15. $\dfrac{x^2+2x+3}{x+1} \cdot \dfrac{1}{x+2}$

16. $\dfrac{a^2x^2y^2}{a+b} \cdot \dfrac{a^2+ab}{ab}$

17. $\dfrac{x^2-y^2}{15} \cdot \dfrac{40}{x-y}$

18. $\dfrac{b^2+4b}{ab} \cdot \dfrac{a}{b^2-16}$

19. $\dfrac{4bc^2}{x^2+4x+4} \cdot \dfrac{x^2+6x+8}{4x+16}$

20. $\dfrac{12a^2-48}{a^2-6a-8} \cdot \dfrac{a^2-a-12}{a+3}$

21. $\dfrac{x^2-7x+12}{x^2+x-30} \cdot \dfrac{x^2+8x+12}{x^2-x-6}$

22. $\dfrac{y^2+12y+36}{2y^2+9y-18} \cdot \dfrac{6y^2-5y-6}{3y^2-13y-10}$

23. $\dfrac{6a^2-21a-12}{20a^2-10a-10} \cdot \dfrac{10a^2+5a}{6a^3-96a}$

24. $\dfrac{2x^3-11x^2+15x}{2x^2y-5xy-3y} \cdot \dfrac{2x^2y+11xy+5y}{2x^3-5x^2}$

ANSWERS

1. _____
2. _____
3. _____
4. _____
5. _____
6. _____
7. _____
8. _____
9. _____
10. _____
11. _____
12. _____
13. _____
14. _____
15. _____
16. _____
17. _____
18. _____
19. _____
20. _____
21. _____
22. _____
23. _____
24. _____

ANSWERS

25.

26.

27.

28.

29.

30.

31.

33.

33.

D

25. At Acme Electronics, assembly line A can assemble 1000 circuit boards in 6 days and assembly line B can assemble 1000 circuit boards in 4 days. The formula for finding how many days (d) it would take to produce 1000 circuit boards using both lines is

$$\frac{1}{6} + \frac{1}{4} = \frac{1}{d}$$

Simplify the formula by multiplying each term by $\frac{12d}{1}$. Find the number of days (d).

26. Crew A at the Green Thumb Nursery can plant 15000 seedlings in 30 hours. Crew B can plant the 15000 seedlings in 40 hours. The formula for the number of hours (t) it would take the crews working together to plant the seedlings is

$$\frac{1}{30} + \frac{1}{40} = \frac{1}{t}$$

Simplify the formula by multiplying each term by $\frac{120t}{1}$ and find the number of hours (t).

27. The total resistance (R) of three resistors, 10 ohms, 8 ohms, and 12 ohms, in parallel is given by the formula

$$\frac{1}{10} + \frac{1}{8} + \frac{1}{12} = \frac{1}{R}$$

Simplify the formula by multiplying each term by $\frac{120R}{1}$ and find R.

28. The total resistance (R) of three resistors, 21 ohms, 14 ohms, and 42 ohms, in parallel is given by

$$\frac{1}{21} + \frac{1}{14} + \frac{1}{42} = \frac{1}{R}$$

Simplify the formula by multiplying each term by $\frac{42R}{1}$ and solve for R.

E   *Maintain Your Skills*

Factor:

29. $36x + 48y$

30. $75ab - 15a$

31. $48x^2y^2 - 72xy^3 + 36x^3y$

32. $28a^3bc^3 - 14ab^3c^2 + 21a^2b^2c^2$

33. $25x^2yz + 35xz - 49yz^3$

344

7.4 Division of Rational Expressions

OBJECTIVE

76. Divide rational expressions.

APPLICATION

Expressions involving division of fractions can always be expressed as a single fraction. It is sometimes easier to work with the single fraction.

VOCABULARY

Two rational expressions whose product is 1 are said to be *reciprocals* of each other. For instance, $\dfrac{3}{x} \cdot \dfrac{x}{3} = 1$.

HOW AND WHY

Recall from division of fractions that an equivalent multiplication problem was solved.

To divide rational expressions, multiply by the reciprocal of the divisor.

$$\frac{A}{B} \div \frac{C}{D} = \frac{A}{B} \cdot \frac{D}{C}, \; BCD \neq 0$$

$$\frac{5a^2b}{a-b} \div \frac{15ab^2}{a-b}$$

$$\frac{\overset{a}{\cancel{5a^2b}}}{\underset{1}{\cancel{a-b}}} \cdot \frac{\overset{1}{\cancel{a-b}}}{\underset{3b}{\cancel{15ab^2}}} \qquad \textit{Multiply by the reciprocal}$$

$$\frac{a}{3b} \qquad \textit{Simplify}$$

EXAMPLES

a. $\dfrac{2x}{3y} \div \dfrac{3}{4} = \dfrac{2x}{3y} \cdot \dfrac{4}{3} = \dfrac{8x}{9y}$

b. $\dfrac{4a}{5b} \div \dfrac{10a^2}{9b} = \dfrac{\overset{}{\cancel{4a}}}{\underset{1}{\cancel{5b}}} \cdot \dfrac{\overset{1}{\cancel{9b}}}{\underset{5a}{\cancel{10a^2}}} = \dfrac{18}{25a}$

c. $\dfrac{ab}{2a + 4b} \div \dfrac{a^2 - ab}{a^2 + 2ab} = \dfrac{ab}{2a + 4b} \cdot \dfrac{a^2 + 2ab}{a^2 - ab} = \dfrac{ab}{2\cancel{(a + 2b)}} \cdot \dfrac{\overset{1}{\cancel{a}}\overset{1}{\cancel{(a + 2b)}}}{\cancel{a}(a - b)}$

$$= \dfrac{ab}{2(a - b)}$$

WARM UPS

Divide and reduce where possible:

1. $\dfrac{x}{y} \div \dfrac{a}{b}$

2. $\dfrac{3}{4} \div \dfrac{p}{q}$

3. $\dfrac{r}{s} \div \dfrac{4}{7}$

4. $\dfrac{2x}{3} \div \dfrac{1}{3}$

5. $\dfrac{6}{5w} \div \dfrac{3}{5}$

6. $\dfrac{x + 4}{6} \div \dfrac{x + 4}{5}$

7. $\dfrac{x + 3}{7x^2} \div \dfrac{x + 3}{14x}$

8. $\dfrac{x - 2}{x + 3} \div \dfrac{x}{x + 3}$

9. $\dfrac{a - 5}{a + 7} \div \dfrac{a - 5}{6a}$

10. $\dfrac{a - 10}{(a - 3)(a + 2)} \div \dfrac{a - 10}{a + 2}$

11. $\dfrac{x + 7}{(x + 1)(x - 1)} \div \dfrac{x + 7}{x - 1}$

12. $\dfrac{4xy}{x + y} \div \dfrac{y(x + 1)}{x + y}$

Exercises 7.4

A

Divide and reduce where possible:

1. $\dfrac{2}{x} \div \dfrac{6}{x^2}$

2. $\dfrac{5a}{14b} \div \dfrac{10a^2}{7b^2}$

3. $\dfrac{x-6}{18y} \div \dfrac{x-6}{24xy^2}$

4. $\dfrac{20xy^2}{y+x} \div \dfrac{10x^2y}{y+x}$

5. $\dfrac{x(x+1)}{x+3} \div \dfrac{y(x+1)}{x+3}$

6. $\dfrac{5a(a-1)}{a+5} \div \dfrac{10a(a-1)}{b(a+5)}$

B

7. $\dfrac{x^2-9}{2} \div \dfrac{x+3}{1}$

8. $\dfrac{1}{x^2+5x+6} \div \dfrac{1}{x+2}$

9. $\dfrac{2x+6}{x-5} \div \dfrac{3x+9}{x-5}$

10. $\dfrac{ax-x}{36} \div \dfrac{bx-x}{12}$

11. $\dfrac{4a}{5b} \div \dfrac{1}{2}$

12. $\dfrac{6x}{5y} \div \dfrac{3}{4}$

13. $\dfrac{2a}{3b} \div \dfrac{6a^2}{5b^3}$

14. $\dfrac{25xy}{3a} \div \dfrac{5x}{9a^2}$

C

15. $\dfrac{a+2}{15} \div \dfrac{a-2}{10a}$

16. $\dfrac{4x-6y}{3ay} \div \dfrac{2x-3y}{15y^2}$

17. $\dfrac{a+b}{x+y} \div \dfrac{a-b}{x-y}$

18. $\dfrac{x+2}{x^2+6x+5} \div \dfrac{x^2-4}{x+5}$

19. $\dfrac{3x+9}{6a^2-3a} \div \dfrac{x+3}{8ax-4x}$

20. $\dfrac{25ab^2}{b^2-16} \div \dfrac{15abx}{b^2-b-12}$

21. $\dfrac{x+4y}{2a+1} \div \dfrac{x^2-16y^2}{4a^2-1}$

22. $\dfrac{3a+4b}{5x+2} \div \dfrac{9a^2-16b^2}{25x^2-4}$

23. $\dfrac{b^2-b-12}{b^2+5b+6} \div \dfrac{b^2-4}{b^2-5b-6}$

24. $\dfrac{c^2-c-12}{c^2+c-6} \div \dfrac{c^2-2c-8}{c^2-4}$

E *Maintain Your Skills*

Factor:

25. $x^2-x-156$

26. $x^2+16x+63$

27. $4a^2+19a-63$

28. $16a^2+40ab+25b^2$

29. x^2+x-7

ANSWERS

1. _____
2. _____
3. _____
4. _____
5. _____
6. _____
7. _____
8. _____
9. _____
10. _____
11. _____
12. _____
13. _____
14. _____
15. _____
16. _____
17. _____
18. _____
19. _____
20. _____
21. _____
22. _____
23. _____
24. _____
25. _____
26. _____
27. _____
28. _____
29. _____

7.5 Least Common Multiple

OBJECTIVE

77. Find the least common multiple of two or more polynomials.

APPLICATION

Find the least common multiple of the denominators of the following fractions: $\dfrac{3}{x}, \dfrac{4}{x^2}, \dfrac{5}{2x}$. The least common multiple will be used to add and subtract such fractions.

VOCABULARY

Recall that the *least common multiple* (LCM) of two or more whole numbers is the smallest number other than zero that is a multiple of each of the given numbers. The *LCM of two or more polynomials* is similar in that it is the polynomial with the least number of factors that is a multiple of each of the given polynomials. The LCM of $6x$ and $9x^2$ is $18x^2$.

HOW AND WHY

Recall that when we found the LCM of two or more whole numbers, each of the whole numbers was prime factored. The LCM was then constructed from these prime factors. To find the LCM of 15, 50, and 75, we did the following:

$$15 = 3 \cdot 5$$

$$50 = 2 \cdot 5 \cdot 5 = 2 \cdot 5^2$$

$$75 = 3 \cdot 5 \cdot 5 = 3 \cdot 5^2$$

The LCM is the product of 2, 3, and 5^2 ($2 \cdot 3 \cdot 25 = 150$). We chose 5^2 since the largest exponent of 5 is 2. (See Chapter 2.)

> **To find the LCM of two or more polynomials,**
>
> **1. Factor each polynomial completely.**
> **2. The LCM is the product of the highest power of each factor.**

To find the LCM of x^2, y^2, and x^3y^3 we have:

$$x^2 = x^2$$

$$y^2 = y^2$$

$$x^3y^3 = x^3 \cdot y^3$$

The different factors are x and y, and the highest power of each is 3; therefore, the LCM is x^3y^3.

EXAMPLES

a. Find the LCM of $4x^2$, $3y^2$, and $8xy$.

$4x^2 = 2^2x^2$

$3y^2 = 3y^2$

$8xy = 2^3xy$

$\text{LCM} = 2^3 \cdot 3x^2y^2 = 24x^2y^2$

b. Find the LCM of x, $x + y$, and $x^2 + xy$.

$x = x$

$x + y = x + y$

$x^2 + xy = x(x + y)$

$\text{LCM} = x(x + y) = x^2 + xy$

c. Find the LCM of $x^2 - y^2$, $x + y$, and $x - y$.

$x^2 - y^2 = (x + y)(x - y)$

$x + y = x + y$

$x - y = x - y$

$\text{LCM} = (x + y)(x - y) = x^2 - y^2$

APPLICATION SOLUTION

To find the LCM of the denominators, factor each completely.

$x = x$

$x^2 = x^2$

$2x = 2 \cdot x$

$\text{LCM} = 2 \cdot x^2 = 2x^2$

The least common multiple of the denominators is $2x^2$.

WARM UPS

Find the LCM of the following. LCM may be written in factored form.

1. $2x$, $3x$ **2.** $3x^2$, $2x$

3. $5x^2$, $3x^2$ **4.** $6x^2$, $10x$, 15

5. $2a$, $3b$, $5c$ **6.** $2(x + 3)$, $3(x + 3)$

7. $6(3a + 7y)$, $8(3a + 7y)$ **8.** $4x(a + 1)$, $5y(a + 1)$

9. $3x + 3y$, $2x + 2y$ **10.** $x + 5$, $x^2 - 25$

11. $x + 3$, $x^2 + 5x + 6$ **12.** $x - 1$, $x^2 + 7x - 8$

Exercises 7.5

ANSWERS

1. _____

2. _____

3. _____

4. _____

5. _____

6. _____

7. _____

8. _____

9. _____

10. _____

11. _____

12. _____

13. _____

14. _____

15. _____

16. _____

17. _____

18. _____

19. _____

20. _____

21. _____

22. _____

23. _____

24. _____

A

Find the LCM of each of the following. Leave in factored form.

1. $2ab$, $4a^2$, $8b^2$ **2.** $12z$, $16xy$, $4x$

3. $6a^2b$, $8ab$, $12ab^2$ **4.** $x + 4$, $x + 6$

5. $x - y$, $x^2 - y^2$ **6.** $a + b$, $a - b$

B

7. $4a^2 - 9$, $14a + 21$ **8.** $y^2 - 9$, $(y - 3)^2$

9. $x + y$, $x^2 + 2xy + y^2$ **10.** x, $x + y$

11. $b + c$, $b - c$, $b^2 - c^2$ **12.** $3x$, $12x^2$, $x + y$

13. $y + 5$, $y + 2$, $y + 1$ **14.** $2y + 6$, $2y - 8$

C

15. $x^2 - 25$, $x^2 + 7x + 10$

16. $x^2 - 7x$, $x^3 - 7x^2$

17. $x^2 + 6x + 9$, $x^2 - 9$

18. $x^3 - 3x^2 - 40x$, $x^3 - 8x^2$

19. $12x^3$, $8x^2 - 8x$, $18x^3 - 18x^2$

20. $15abc$, $10a^2 + 30a$, $9ab + 27b$

21. $x^2 + 9x + 14$, $x^2 + 10x + 21$, $x^2 + 5x + 6$

22. $x^2 - 36$, $x^2 + 12x + 36$, $x^2 - 12x + 36$

23. $4abc - 4ab$, $6c^2 - 6$, $18a^3bc - 18a^3b$

24. $36a^3bc^2$, $48a^2b^3c$, $80a^5b^4$, $15a^4b - 15a^4$

351

ANSWERS

25.

26.

27.

28.

29.

30.

31.

32.

33.

D

Find the least common multiple of the denominators of the following groups of fractions.

25. $\dfrac{1}{4x^2y}, \dfrac{3}{6xy^3}, \dfrac{-a}{10x^2y^2}$

26. $\dfrac{-x^5}{m^3n}, \dfrac{5a}{4mn^3}, \dfrac{-7b}{20m^2n^2}, \dfrac{16}{15m^4n}$

27. $\dfrac{1}{x-3}, \dfrac{5x}{x^2-9}, \dfrac{2y-1}{x^2-6x+9}$

28. $\dfrac{1}{2x+3}, \dfrac{x+2}{4x^2-9}, \dfrac{x-1}{4x^2+12x+9}$

E *Maintain Your Skills*

Factor:

29. $16y^2 - 49z^2$ **30.** $49b^2 - 36c^2$

31. $121a^2b^2 - 144c^2$ **32.** $36a^2 - 144b^2$

33. $48ab^2 - 75ac^2$

7.6 Addition of Rational Expressions

OBJECTIVE

78. Add rational expressions.

APPLICATION

Expressions involving addition of one or more fractions can always be expressed as a single fraction. Sometimes it is easier to work with the single fraction.

VOCABULARY

No new vocabulary.

HOW AND WHY

The sum of two or more rational expressions is found by using the same process as was used when adding fractions in arithmetic.

To add two rational expressions that have a common denominator,

1. Add the numerators and put the sum over the common denominator.
2. Reduce, if possible.

$$\frac{2}{x} + \frac{5}{x} = \frac{7}{x}$$

To add two rational expressions that do not have a common denominator,

1. Build each as an equivalent expression with a common denominator.
2. Add.
3. Reduce, if possible.

To add $\dfrac{3}{2x} + \dfrac{5}{x^2}$, we pick the LCM of x^2 and $2x$, which is $2x^2$, for the common denominator. We then have

$$\frac{3}{2x} \cdot \frac{x}{x} + \frac{5}{x^2} \cdot \frac{2}{2} = \frac{3x}{2x^2} + \frac{10}{2x^2} = \frac{3x+10}{2x^2}$$

EXAMPLES

a. $\dfrac{4}{x} + \dfrac{5}{x} = \dfrac{4+5}{x} = \dfrac{9}{x}$

b. $\dfrac{3x}{a+b} + \dfrac{4y}{a+b} = \dfrac{3x+4y}{a+b}$

c. $\dfrac{3}{a} + \dfrac{4}{b}$ *ab is the LCM of a and b*

$\dfrac{3}{a} + \dfrac{4}{b} = \dfrac{3}{a} \cdot \dfrac{b}{b} + \dfrac{4}{b} \cdot \dfrac{a}{a} = \dfrac{3b}{ab} + \dfrac{4a}{ab} = \dfrac{4a+3b}{ab}$

d. $\dfrac{x}{a} + \dfrac{y}{a+b}$ *a(a + b) is the LCM of a and a + b*

$$\frac{x}{a} + \frac{y}{a+b} = \frac{x}{a} \cdot \frac{a+b}{a+b} + \frac{y}{a+b} \cdot \frac{a}{a}$$

$$= \frac{ax + bx}{a(a+b)} + \frac{ay}{a(a+b)}$$

$$= \frac{ax + bx + ay}{a(a+b)} = \frac{ax + bx + ay}{a^2 + ab}$$

e. $\dfrac{3}{x} + \dfrac{4}{x^2 + x}$ *x(x + 1) or x² + x is the LCM of x and x² + x*

$$\frac{3}{x} + \frac{4}{x^2 + x} = \frac{3}{x} + \frac{4}{x(x+1)}$$

$$\frac{3}{x} + \frac{4}{x(x+1)} = \frac{3}{x} \cdot \frac{(x+1)}{(x+1)} + \frac{4}{x(x+1)}$$

$$= \frac{3x+3}{x(x+1)} + \frac{4}{x(x+1)} = \frac{3x+7}{x(x+1)} = \frac{3x+7}{x^2+x}$$

WARM UPS

Add and reduce where possible:

1. $\dfrac{y}{5} + \dfrac{y}{5}$

2. $\dfrac{z}{6} + \dfrac{z}{6}$

3. $\dfrac{3}{14w} + \dfrac{4}{14w}$

4. $\dfrac{3}{15a} + \dfrac{5}{15a}$

5. $\dfrac{3x}{x+2} + \dfrac{7}{x+2}$

6. $\dfrac{5y}{x-3} + \dfrac{2y}{x-3}$

7. $\dfrac{a+3}{b} + \dfrac{2a-5}{b}$

8. $\dfrac{2x-3}{4y} + \dfrac{2x+3}{4y}$

9. $\dfrac{4x-3}{7} + \dfrac{2x-1}{7}$

10. $\dfrac{x}{2} + \dfrac{x}{6}$

11. $\dfrac{m}{2y} + \dfrac{3m}{8y}$

12. $\dfrac{3}{p^2} + \dfrac{4}{3p}$

Exercises 7.6

ANSWERS

1. _____

2. _____

3. _____

4. _____

5. _____

6. _____

7. _____

8. _____

9. _____

10. _____

11. _____

12. _____

13. _____

14. _____

15. _____

16. _____

17. _____

18. _____

19. _____

20. _____

21. _____

22. _____

23. _____

24. _____

25. _____

26. _____

27. _____

28. _____

29. _____

A

Add and reduce where possible:

1. $\dfrac{2a}{3} + \dfrac{5a}{3}$

2. $\dfrac{7}{y} + \dfrac{11}{y}$

3. $\dfrac{5a}{12} + \dfrac{a}{12}$

4. $\dfrac{2y-3}{7} + \dfrac{y+3}{7}$

5. $\dfrac{2a+7}{10b} + \dfrac{2a-7}{10b}$

6. $\dfrac{3y}{4x} + \dfrac{y}{2x}$

B

7. $\dfrac{x+2}{5} + \dfrac{x+3}{10}$

8. $\dfrac{7}{y} + \dfrac{3}{z}$

9. $\dfrac{x}{y} + \dfrac{x}{3y} + \dfrac{x}{3}$

10. $\dfrac{1}{4ab} + \dfrac{b}{6a} + \dfrac{a}{2b}$

11. $\dfrac{3}{xy} + \dfrac{5}{x^2} + \dfrac{8}{y^2}$

12. $\dfrac{8a}{9b} + \dfrac{a}{6b}$

13. $\dfrac{3}{5a} + 7$

14. $\dfrac{2}{3a} + \dfrac{5}{a} + \dfrac{3}{6a}$

C

15. $x + 3 + \dfrac{1}{x}$

16. $\dfrac{4}{a+b} + \dfrac{7}{a-b}$

17. $\dfrac{2}{x+5} + \dfrac{7}{x-1}$

18. $\dfrac{4}{y} + \dfrac{3}{y+1}$

19. $\dfrac{x}{x+y} + \dfrac{y}{x-y} + \dfrac{y}{x^2-y^2}$

20. $\dfrac{a}{a+b} + \dfrac{a-b}{a}$

21. $\dfrac{1}{x} + (x+3)$

22. $\dfrac{a}{b} + \dfrac{a+b}{a} + \dfrac{b}{a+b}$

23. $\dfrac{x+1}{x^2-x-20} + \dfrac{x-3}{x^2-25}$

24. $\dfrac{-x}{x^2+10x+24} + \dfrac{x+2}{x^2+5x-6}$

E   *Maintain Your Skills*

Solve:

25. $(x-5)(x+4) = 0$

26. $x^2 - 12x - 64 = 0$

27. $2x^2 + x - 3 = 0$

28. $x^2 + 7x - 18 = 0$

29. $3x^2 = 8x + 35$

ANSWERS TO WARM UPS (7.6)

1. $\dfrac{2y}{5}$ 2. $\dfrac{z}{3}$ 3. $\dfrac{1}{2w}$ 4. $\dfrac{8}{15a}$ 5. $\dfrac{3x+7}{x+2}$

6. $\dfrac{7y}{x-3}$ 7. $\dfrac{3a-2}{b}$ 8. $\dfrac{x}{y}$ 9. $\dfrac{6x-4}{7}$ 10. $\dfrac{2x}{3}$

11. $\dfrac{7m}{8y}$ 12. $\dfrac{9+4p}{3p^2}$

7.7 Subtraction of Rational Expressions

OBJECTIVE

79. **Subtract rational expressions.**

APPLICATION

Expressions involving subtraction of fractions can always be expressed as a single fraction. Sometimes it is easier to work with the single fraction.

VOCABULARY

No new vocabulary.

HOW AND WHY

The difference of two rational expressions is found using the same process as was used when subtracting fractions in arithmetic.

> To subtract two rational expressions that have a common denominator,
>
> 1. **Subtract the numerators and put the difference over the common denominator.**
> 2. **Reduce, if possible.**

$$\frac{x}{a+b} - \frac{3}{a+b} = \frac{x-3}{a+b}$$

> To subtract two rational expressions that do not have a common denominator,
>
> 1. **Build each one as an equivalent expression with a common denominator.**
> 2. **Subtract.**
> 3. **Reduce, if possible.**

$$\frac{4}{3} - \frac{5}{x} = \frac{4x}{3x} - \frac{15}{3x} = \frac{4x-15}{3x}$$

EXAMPLES

a. $\dfrac{5}{x} - \dfrac{2}{x} = \dfrac{5-2}{x} = \dfrac{3}{x}$

b. $\dfrac{9a}{a+b} - \dfrac{2a}{a+b} = \dfrac{9a-2a}{a+b} = \dfrac{7a}{a+b}$

c. $\dfrac{8y}{x} - \dfrac{6b}{a}$ *ax is the LCM of a and x*

$\dfrac{8y}{x} - \dfrac{6b}{a} = \dfrac{8y}{x} \cdot \dfrac{a}{a} - \dfrac{6b}{a} \cdot \dfrac{x}{x} = \dfrac{8ay}{ax} - \dfrac{6bx}{ax} = \dfrac{8ay-6bx}{ax}$

d. $\dfrac{5x}{x+y} - \dfrac{4y}{x-y}$ *(x + y)(x − y) is the LCM of x + y and x − y*

$$\dfrac{5x}{x+y} - \dfrac{4y}{x-y} = \dfrac{5x}{x+y} \cdot \dfrac{x-y}{x-y} - \dfrac{4y}{x-y} \cdot \dfrac{x+y}{x+y}$$

$$= \dfrac{5x(x-y)}{(x+y)(x-y)} - \dfrac{4y(x+y)}{(x-y)(x+y)}$$

$$= \dfrac{5x^2 - 5xy - (4xy + 4y^2)}{(x+y)(x-y)}$$

$$= \dfrac{5x^2 - 5xy - 4xy - 4y^2}{(x+y)(x-y)}$$

$$= \dfrac{5x^2 - 9xy - 4y^2}{x^2 - y^2}$$

e. $\dfrac{3}{a-b} - \dfrac{3b}{a(a-b)}$ *The LCM of a − b and a(a − b) is a a(a − b)*

$$\dfrac{3}{a-b} - \dfrac{3b}{a(a-b)} = \dfrac{3}{a-b} \cdot \dfrac{a}{a} - \dfrac{3b}{a(a-b)}$$

$$= \dfrac{3a}{a(a-b)} - \dfrac{3b}{a(a-b)}$$

$$= \dfrac{3a - 3b}{a(a-b)}$$

$$= \dfrac{3\cancel{(a-b)}}{a\cancel{(a-b)}}$$

$$= \dfrac{3}{a}$$

WARM UPS

Subtract and reduce where possible:

1. $\dfrac{5a}{12} - \dfrac{a}{12}$ **2.** $\dfrac{4}{a^2} - \dfrac{7}{a^2}$

3. $\dfrac{7a}{x-y} - \dfrac{3}{x-y}$ **4.** $\dfrac{6b}{y+3} - \dfrac{4b}{y+3}$

5. $\dfrac{9}{x+2} - \dfrac{5}{x+2}$ **6.** $\dfrac{5x}{3} - \dfrac{2x}{3}$

7. $\dfrac{4}{b} - \dfrac{3}{b^2}$ **8.** $\dfrac{3}{x} - \dfrac{5}{x^3}$

9. $\dfrac{x}{y} - \dfrac{x}{3y}$ **10.** $\dfrac{3}{xy} - \dfrac{5}{x^2}$

11. $\dfrac{7x-4}{6} - \dfrac{x}{2}$ **12.** $\dfrac{8x+2}{4y} - \dfrac{3x}{2y}$

Exercises 7.7

ANSWERS

1. _____
2. _____
3. _____
4. _____
5. _____
6. _____
7. _____
8. _____
9. _____
10. _____
11. _____
12. _____
13. _____
14. _____
15. _____
16. _____
17. _____
18. _____
19. _____
20. _____
21. _____
22. _____
23. _____
24. _____
25. _____
26. _____
27. _____
28. _____
29. _____

A

Subtract and reduce where possible:

1. $\dfrac{7x}{8} - \dfrac{3x}{8}$

2. $\dfrac{3}{14w} - \dfrac{1}{14w}$

3. $\dfrac{2x}{y-1} - \dfrac{3}{y-1}$

4. $\dfrac{p}{q+1} - \dfrac{2}{q+1}$

5. $\dfrac{8}{15a} - \dfrac{1}{5a}$

6. $\dfrac{2x}{x-1} - \dfrac{2}{x-1}$

B

7. $\dfrac{1}{4a} - \dfrac{2}{3b}$

8. $\dfrac{6}{5y} - \dfrac{3}{10y}$

9. $\dfrac{x}{2y} - \dfrac{z}{5y}$

10. $\dfrac{8a}{9b} - \dfrac{a}{6b}$

11. $\dfrac{3}{5a} - 7$

12. $\dfrac{2}{3a} - \dfrac{5}{a} - \dfrac{3}{6a}$

13. $x - 3 - \dfrac{1}{x}$

14. $y + 2 - \dfrac{3}{y}$

C

15. $\dfrac{4}{a+b} - \dfrac{7}{a-b}$

16. $\dfrac{2}{x+5} - \dfrac{7}{x-1}$

17. $\dfrac{4}{y} - \dfrac{3}{y+1}$

18. $\dfrac{x}{x+y} - \dfrac{y}{x-y}$

19. $\dfrac{a}{a+b} - \dfrac{a+b}{a}$

20. $\dfrac{a}{a+b} - \dfrac{a-b}{a}$

21. $\dfrac{1}{x} - (x-3)$

22. $\dfrac{a}{b} - \dfrac{a+b}{a} - \dfrac{b}{a+b}$

23. $\dfrac{x-2}{x^2-3x-4} - \dfrac{x+3}{x^2-16}$

24. $\dfrac{x+1}{x^2+7x+10} - \dfrac{x-1}{x^2+3x-10}$

E　　*Maintain Your Skills*

Multiply:

25. $\left(\dfrac{5}{x}\right)\left(\dfrac{3}{y}\right)\left(\dfrac{w}{z}\right)$

26. $\left(\dfrac{-5}{a}\right)\left(\dfrac{b}{c}\right)\left(\dfrac{d}{-e}\right)$

27. $\left(\dfrac{4}{x}\right)\left(\dfrac{x-2}{3x}\right)\left(\dfrac{x+2}{5x}\right)$

28. $\left(\dfrac{6x-5y}{25a}\right)\left(\dfrac{-8x}{5y}\right)$

29. $\left(\dfrac{y+z}{a+b}\right)\left(\dfrac{y-z}{a-b}\right)\left(\dfrac{-7y}{4b}\right)$

ANSWERS TO WARM UPS (7.7)

1. $\dfrac{a}{3}$ 2. $\dfrac{-3}{a^2}$ 3. $\dfrac{7a-3}{x-y}$ 4. $\dfrac{2b}{y+3}$ 5. $\dfrac{4}{x+2}$

6. x 7. $\dfrac{4b-3}{b^2}$ 8. $\dfrac{3x^2-5}{x^3}$ 9. $\dfrac{2x}{3y}$ 10. $\dfrac{3x-5y}{x^2y}$

11. $\dfrac{2(x-1)}{3}$ 12. $\dfrac{x+1}{2y}$

7.8 Complex Rational Expressions

OBJECTIVE

80. Simplify complex rational expressions.

APPLICATION

Complex rational expressions can always be expressed as a single fraction. It is sometimes easier to work with the single fraction.

VOCABULARY

Complex rational expressions are those rational expressions that contain rational expressions within their numerator and/or denominator.

$$\dfrac{\dfrac{1}{x}+2}{\dfrac{2}{x}+3} \text{ is a complex fraction.}$$

HOW AND WHY

To simplify a complex rational expression,

1. Find the LCM of the denominators that are within the numerator and/or denominator of the complex fraction.

2. Multiply the numerator *and* denominator of the complex fraction by the LCM and simplify.

To simplify $\dfrac{\dfrac{1}{x}}{\dfrac{1}{y}}$, find the LCM of x and y, which is xy. Now multiply both the numerator and denominator by xy and simplify.

$$\dfrac{\dfrac{1}{x}\cdot\dfrac{xy}{1}}{\dfrac{1}{y}\cdot\dfrac{xy}{1}}=\dfrac{\dfrac{xy}{x}}{\dfrac{xy}{y}}=\dfrac{y}{x}$$

EXAMPLES

a. Simplify $\dfrac{\dfrac{1}{2}+\dfrac{1}{y}}{5}$.

The LCM of 2 and y is $2y$; therefore, we have

$$\frac{\left(\dfrac{1}{2}+\dfrac{1}{y}\right)\cdot\dfrac{2y}{1}}{\dfrac{5}{1}\cdot\dfrac{2y}{1}}=\frac{\dfrac{2y}{2}+\dfrac{2y}{y}}{10y}$$

$$=\frac{y+2}{10y}$$

b. Simplify $\dfrac{x+\dfrac{1}{y}}{\dfrac{x}{y}}$.

The LCM of y and y is y; therefore, we get

$$\frac{\left(x+\dfrac{1}{y}\right)\cdot\dfrac{y}{1}}{\dfrac{x}{y}\cdot\dfrac{y}{1}}=\frac{\dfrac{xy}{1}+\dfrac{y}{y}}{\dfrac{xy}{y}}$$

$$=\frac{xy+1}{x}$$

c. Simplify $\dfrac{\dfrac{1}{x}-\dfrac{3}{2}}{\dfrac{1}{2x}+\dfrac{1}{3x}}$.

The LCM of x, 2, $2x$, and $3x$ is $6x$; therefore, we get

$$\frac{\left(\dfrac{1}{x}-\dfrac{3}{2}\right)\cdot\dfrac{6x}{1}}{\left(\dfrac{1}{2x}+\dfrac{1}{3x}\right)\cdot\dfrac{6x}{1}}=\frac{6-9x}{3+2}=\frac{6-9x}{5}$$

WARM UPS   *Simplify:*

1. $\dfrac{\dfrac{1}{t}}{\dfrac{1}{w}}$ 2. $\dfrac{\dfrac{3}{b}}{\dfrac{2}{b}}$ 3. $\dfrac{\dfrac{a}{x-1}}{\dfrac{b}{x-1}}$

4. $\dfrac{\dfrac{3}{x+1}}{\dfrac{y}{x+1}}$ 5. $\dfrac{\dfrac{3}{x}}{\dfrac{3}{x}}$ 6. $\dfrac{\dfrac{1}{y}}{\dfrac{4}{y}}$

7. $\dfrac{\dfrac{1}{3}}{\dfrac{1}{x}}$ 8. $\dfrac{\dfrac{1}{y}}{\dfrac{1}{6}}$ 9. $\dfrac{\dfrac{2}{y+6}}{\dfrac{3}{y+6}}$

10. $\dfrac{\dfrac{y-3}{7}}{\dfrac{y-3}{8}}$ 11. $\dfrac{\dfrac{1}{a}}{b}$ 12. $\dfrac{m}{\dfrac{1}{n}}$

Exercises 7.8

A

Simplify:

1. $\dfrac{\frac{1}{4}}{\frac{3}{5}}$

2. $\dfrac{\frac{1}{a}}{\frac{1}{b}}$

3. $\dfrac{\frac{2}{x}}{\frac{3}{x^2}}$

4. $\dfrac{\frac{1}{x}+\frac{1}{y}}{6}$

5. $\dfrac{\frac{2}{3}+\frac{3}{4}}{\frac{1}{2}}$

6. $\dfrac{x+\frac{1}{2}}{\frac{2}{3}}$

B

7. $\dfrac{a+\frac{b}{c}}{\frac{1}{a}}$

8. $\dfrac{\frac{1}{x}+\frac{1}{y}}{\frac{1}{x}-\frac{1}{y}}$

9. $\dfrac{\frac{1}{x}+\frac{1}{y}}{\frac{1}{w}}$

10. $\dfrac{a+b}{\frac{1}{x}+\frac{1}{y}}$

11. $\dfrac{a+\frac{1}{b}}{a-\frac{1}{b}}$

12. $\dfrac{\frac{a}{b}-\frac{b}{a}}{a-b}$

13. $\dfrac{\frac{1}{y}-\frac{1}{z}}{\frac{y-z}{z}}$

14. $\dfrac{1+\frac{1}{b}}{b+\frac{1}{b}}$

C

15. $\dfrac{\frac{x}{y}+\frac{x}{z}}{\frac{y}{x}+z}$

16. $\dfrac{\frac{1}{a}-\frac{2}{b}}{\frac{3}{a}+4}$

17. $\dfrac{\frac{1}{5}+\frac{x^2}{2}}{\frac{x}{4}}$

18. $\dfrac{\frac{x}{5}}{\frac{1}{4}+\frac{x^2}{2}}$

19. $\dfrac{\frac{5}{x}+\frac{2}{x^2}}{\frac{x}{4}}$

20. $\dfrac{\frac{5}{x}}{\frac{2}{x^2}-\frac{x}{4}}$

363

ANSWERS

1. _____

2. _____

3. _____

4. _____

5. _____

6. _____

7. _____

8. _____

9. _____

10. _____

11. _____

12. _____

13. _____

14. _____

15. _____

16. _____

17. _____

18. _____

19. _____

20. _____

ANSWERS

21. _____

22. _____

23. _____

24. _____

25. _____

26. _____

27. _____

28. _____

29. _____

21. $\dfrac{\dfrac{1+\dfrac{1}{x}}{x}}{\dfrac{1}{x}-1}$

22. $\dfrac{\dfrac{1}{x-1}+\dfrac{1}{x+1}}{\dfrac{1}{x-1}-\dfrac{1}{x+1}}$

23. $\dfrac{\dfrac{1}{a+3}-\dfrac{1}{a-3}}{\dfrac{1}{a+3}+\dfrac{1}{a-3}}$

24. $\dfrac{2y-1+\dfrac{1}{y-1}}{3y+2+\dfrac{1}{y-1}}$

E   *Maintain Your Skills*

Reduce:

25. $\dfrac{5x+10}{x^2+5x+6}$

26. $\dfrac{15a^2-15b^2}{3a+3b}$

27. $\dfrac{x^2-8x}{x^2-6x-16}$

Find the missing numerator:

28. $\dfrac{3x}{x+y}=\dfrac{?}{(x+y)^2}$

29. $\dfrac{x+3}{x-3}=\dfrac{?}{x^2-9}$

ANSWERS TO WARM UPS (7.8)

1. $\dfrac{w}{t}$ 2. $\dfrac{3}{2}$ 3. $\dfrac{a}{b}$ 4. $\dfrac{3}{y}$ 5. 1 6. $\dfrac{1}{4}$

7. $\dfrac{x}{3}$ 8. $\dfrac{6}{y}$ 9. $\dfrac{2}{3}$ 10. $\dfrac{8}{7}$ 11. $\dfrac{1}{ab}$ 12. mn

7.9 Fractional Equations

OBJECTIVE

81. Solve equations involving rational expressions (fractions).

APPLICATION

The formula for finding the total resistance (R) of two resistors wired in parallel is $\dfrac{1}{R} = \dfrac{1}{r_1} + \dfrac{1}{r_2}$. If $r_1 = 5$ ohms and $r_2 = 8$ ohms, what is the total resistance?

VOCABULARY

No new vocabulary.

HOW AND WHY

To solve an equation such as $\dfrac{x}{9} + \dfrac{x}{3} = 5$, we can perform an operation so that it will have the same form as equations previously encountered. Using the multiplication law of equality and multiplying each member of the equation by 9 (the LCM of 3 and 9), the equation becomes:

$$9\left(\frac{x}{9} + \frac{x}{3}\right) = 9 \cdot 5$$

$$9 \cdot \frac{x}{9} + 9 \cdot \frac{x}{3} = 45$$

$$x + 3x = 45$$

$$4x = 45$$

$$x = \frac{45}{4}$$

To solve an equation that contains rational expressions, multiply both members of the equation by the LCM of the denominators. This will eliminate all of the denominators, and the resulting equation can then be solved by methods previously discussed.

EXAMPLES

a. $\dfrac{x}{3} + \dfrac{x}{4} = \dfrac{1}{2}$

Since 12 is the LCM of 2, 3, and 4, multiply both members by 12.

$$12\left(\dfrac{x}{3} + \dfrac{x}{4}\right) = 12 \cdot \dfrac{1}{2}$$

$$12 \cdot \dfrac{x}{3} + 12 \cdot \dfrac{x}{4} = 6$$

$$4x + 3x = 6$$

$$7x = 6$$

$$x = \dfrac{6}{7}$$

b. $\dfrac{18}{x} + \dfrac{6}{x} = 12$

x is the LCM; therefore, multiply both members by x.

$$x\left(\dfrac{18}{x} + \dfrac{6}{x}\right) = x \cdot 12$$

$$x \cdot \dfrac{18}{x} + x \cdot \dfrac{6}{x} = 12x$$

$$18 + 6 = 12x$$

$$24 = 12x$$

$$2 = x$$

c. $\dfrac{5}{x} = \dfrac{3}{x + 2}$

$x(x + 2)$ is the LCM; therefore, multiply both members by $x(x + 2)$.

$$x(x + 2) \cdot \dfrac{5}{x} = x(x + 2) \cdot \dfrac{3}{x + 2}$$

$$5(x + 2) = x \cdot 3$$

$$5x + 10 = 3x$$

$$5x - 3x = -10$$

$$2x = -10$$

$$x = -5$$

APPLICATION SOLUTION

To find the total resistance, substitute into the formula and solve.

Formula: $\quad \dfrac{1}{R} = \dfrac{1}{r_1} + \dfrac{1}{r_2}, r_1 = 5, r_2 = 8$

Substitute: $\quad \dfrac{1}{R} = \dfrac{1}{5} + \dfrac{1}{8}$

Continued on next page.

Solve: $40R\left(\dfrac{1}{R}\right) = 40R\left(\dfrac{1}{5}+\dfrac{1}{8}\right)$

$$40 = 8R + 5R$$

$$40 = 13R$$

$$R = \dfrac{40}{13}$$

Answer: The total resistance of the system is $\dfrac{40}{13}$ ohms or $3\dfrac{1}{13}$ ohms.

WARM UPS

Solve:

1. $\dfrac{3}{x} + \dfrac{5}{x} = 2$

2. $\dfrac{7}{x} - \dfrac{3}{x} = 4$

3. $\dfrac{1}{a} + \dfrac{6}{a} = 7$

4. $\dfrac{3}{b} - \dfrac{2}{b} = 5$

5. $\dfrac{1}{3a} + \dfrac{1}{a} = \dfrac{1}{3}$

6. $\dfrac{1}{4y} - \dfrac{1}{y} = \dfrac{1}{4}$

7. $\dfrac{1}{x} + \dfrac{2}{x} = \dfrac{1}{4}$

8. $\dfrac{2}{b} - \dfrac{1}{3b} = \dfrac{1}{6}$

9. $\dfrac{1}{3} + \dfrac{1}{6} = \dfrac{1}{x}$

10. $\dfrac{1}{8} + \dfrac{1}{2} = \dfrac{1}{y}$

11. $\dfrac{2}{y} - \dfrac{5}{3} = \dfrac{3}{6y}$

12. $\dfrac{5}{x} - \dfrac{3}{4} = \dfrac{7}{8x}$

Exercises 7.9

A

Solve:

1. $\dfrac{8}{x} + \dfrac{6}{x} = 7$

2. $\dfrac{4}{y} - \dfrac{3}{y} = 2$

3. $\dfrac{1}{x} - \dfrac{1}{2x} = \dfrac{1}{2}$

4. $\dfrac{5}{3y} + \dfrac{2}{6y} = \dfrac{1}{6}$

5. $\dfrac{1}{4} + \dfrac{1}{2} = \dfrac{1}{x}$

6. $\dfrac{1}{5} + \dfrac{1}{10} = \dfrac{1}{x}$

B

7. $\dfrac{a}{3} + \dfrac{a}{4} = \dfrac{7}{2}$

8. $\dfrac{b}{14} + \dfrac{b}{7} + \dfrac{b}{21} = 10$

9. $\dfrac{1}{8a} - \dfrac{1}{6a} = \dfrac{1}{4}$

10. $\dfrac{2}{x} + \dfrac{1}{2x} = 6$

11. $\dfrac{3}{x-2} = \dfrac{2}{x+2}$

12. $\dfrac{2}{a+3} = \dfrac{5}{a-4}$

13. $\dfrac{2}{a} + 1 = \dfrac{4}{a}$

14. $\dfrac{2}{x-5} = \dfrac{5}{x-2}$

C

15. $\dfrac{3}{x+2} = \dfrac{6}{x+4}$

16. $\dfrac{5}{x} + \dfrac{6}{x} = 33$

17. $\dfrac{x}{4} = x - 3$

18. $\dfrac{4a-3}{6} + 2 = \dfrac{a-7}{4} + a$

19. $\dfrac{4}{b+2} + \dfrac{3}{b} = \dfrac{5}{b^2+2b}$

20. $\dfrac{10}{x+4} - \dfrac{3}{x-2} = 0$

21. $\dfrac{9}{a-5} - \dfrac{3}{a-2} = \dfrac{5}{a-2}$

22. $\dfrac{a}{a+3} - \dfrac{2}{a-3} = 1$

23. $\dfrac{x}{x^2-9} + \dfrac{2}{x^2+x-12} = \dfrac{x-1}{x^2+7x+12}$

24. $\dfrac{5}{a^2-8a+15} + \dfrac{a}{a^2-10a+21} = \dfrac{a+4}{a^2-12a+35}$

D

25. The formula for finding the total resistance (R) of two resistors wired in parallel is $\dfrac{1}{R} = \dfrac{1}{r_1} + \dfrac{1}{r_2}$. If r_1 has resistance that is 15 ohms greater than r_2, and the total resistance is 4 ohms, find the resistance of each resistor.

26. Jim drove 500 miles on Wednesday at a certain speed. On Thursday he drove 15 mph faster and went 180 miles farther. If Jim drove the same number of hours each day, find the number of hours he drove each day.

27. Flight 182A flew 800 miles in the same time it took Flight 291B to fly 625 miles. If Flight 291B averaged 75 mph less than Flight 182A, what was the rate of both planes?

1. _____
2. _____
3. _____
4. _____
5. _____
6. _____
7. _____
8. _____
9. _____
10. _____
11. _____
12. _____
13. _____
14. _____
15. _____
16. _____
17. _____
18. _____
19. _____
20. _____
21. _____
22. _____
23. _____
24. _____
25. _____
26. _____
27. _____

28. Dan can row his rubber life raft 4 mph in still water. He rows up a stream 4 miles and then rows downstream 6 miles. The trip upstream takes the same time as the trip downstream. Find the rate of the stream.

Work Problems: Work problems can be set up by determining the part of the job accomplished per day, per hour, or during some other unit of time (rate of work). If it takes 6 hours to do a certain job, the rate of work would be $\frac{1}{6}$ of the job per hour. The sum of all the rates is the rate of those working together.

29. Mildred can mow her lawn in 2 hours and her granddaughter Pat can mow the same lawn in 3 hours. If they work together, how long will it take to mow the lawn?

Let $x =$ the time it takes to mow the lawn together.

(Mildred's rate) + (Pat's rate) is (rate working together)

$$\frac{1}{2} \quad + \quad \frac{1}{3} \quad = \quad \frac{1}{x}$$

30. Larry and Greg can hoe an acre of strawberries in 3 hours working together. Working alone Larry can hoe the acre in 5 hours. How long would it take Greg to hoe the acre by himself?

31. A water tank for a small city can be filled in four days. During the summer months, the average rate of water use drains the tank in six days. If the tank is empty at the beginning of the summer, how many days will it take the tank to fill?

32. A tank can be filled through two pipes in 2 hours. One pipe alone can fill the tank in 3 hours less than the other can fill the tank alone. How long does it take each pipe to fill the tank?

E *Maintain Your Skills*

Multiply and reduce to lowest terms:

33. $\left(\dfrac{4x}{5y}\right)\left(\dfrac{15}{32}\right)$

34. $\left(\dfrac{5a^3}{18b^2}\right)\left(\dfrac{6b}{10a}\right)$

35. $\left(\dfrac{x^2 - 5x + 6}{5a}\right)\left(\dfrac{10a}{x - 2}\right)$

36. $\left(\dfrac{x^2 + 5x + 6}{x^2 - x - 20}\right)\left(\dfrac{x^2 + x - 12}{x^2 - x - 6}\right)$

37. $\dfrac{x^2 + 4x - 21}{x^2 + 6x - 27} \cdot \dfrac{x^2 + 10x + 9}{x^2 + 8x + 7}$

370

ANSWERS TO WARM UPS (7.9)

1. $x = 4$ 2. $x = 1$ 3. $a = 1$ 4. $b = \dfrac{1}{5}$ 5. $a = 4$

6. $y = -3$ 7. $x = 12$ 8. $b = 10$ 9. $x = 2$ 10. $y = \dfrac{8}{5}$

11. $y = \dfrac{9}{10}$ 12. $x = \dfrac{11}{2}$

7.10 Solving Formulas

OBJECTIVES

82. Evaluate a formula for any of its variables when given the value of the others.

83. Solve or rewrite a formula for any one of its variables.

APPLICATION

Fred wants to make a horespower table or chart for use in his diesel shop. The formula for engine torque is $T = \dfrac{5252H}{S}$, where H is horsepower and S is engine speed in revolutions per minute. What formula can Fred use for horsepower in terms of torque and engine speed?

VOCABULARY

Recall that variables are often called unknowns when working with formulas.

HOW AND WHY

Formulas can be evaluated by substituting the known values for the variables in the formula and using the techniques of previous sections.

Find r if $A = p + prt$, given $p = 20,000$, $A = 45,500$, and $t = 15$.

$45,500 = 20,000 + 20,000 \cdot r \cdot 15$ *Substitute the known values*

$45,500 = 20,000 + 300,000r$ *Simplify*

$45,500 + -20,000 = 20,000 + -20,000 + 300,000r$ *Add* $-20,000$ *to both sides*

$25,500 = 300,000r$ *Simplify*

$\dfrac{17}{200} = .085 = r$ *Divide both sides by 300,000*

When a formula is to be rewritten to solve for a different variable or unknown, the general steps are similar. Now, however, there are no known values to substitute, so we treat the variables as if they were numbers.

Solve for r if $A = p + prt$.

$A + -p = p + -p + prt$ *Add* $-p$ *to both sides*

$A + -p = prt$ *Simplify*

$\dfrac{A + -p}{pt} = \dfrac{A - p}{pt} = r$ *Divide both sides by p and t*

The *new* formula, $r = \dfrac{A - p}{pt}$, is handier when using a calculator to find r when the values of A, p, and t are known.

EXAMPLES

a. $C = \dfrac{5}{9}(F - 32)$ Find F if $C = 60$.

$$60 = \dfrac{5}{9}(F - 32)$$

$$60 = \dfrac{5}{9}F - \dfrac{160}{9}$$

$$540 = 5F - 160$$

$$700 = 5F$$

$$140 = F$$

b. $\dfrac{1}{T} = \dfrac{1}{a} + \dfrac{1}{b} + \dfrac{1}{c}$ Find b if $T = .2$, $a = 7$, and $c = 3$.

$$\dfrac{1}{.2} = \dfrac{1}{7} + \dfrac{1}{b} + \dfrac{1}{3}$$

$$4.2b\left(\dfrac{1}{.2}\right) = 4.2b\left(\dfrac{1}{7} + \dfrac{1}{b} + \dfrac{1}{3}\right)$$

$$21b = .6b + 4.2 + 1.4b$$

$$21b = 2b + 4.2$$

$$19b = 4.2$$

$$b = \dfrac{4.2}{19} \approx .22$$

c. Calculator example

$A = p + prt$ Find p if $A = 19950$, $r = .11$, and $t = 3$.
First solve for p.

$A = p + prt$

$A = p(1 + rt)$

$$p = \dfrac{A}{1 + rt} = A \div (1 + rt)$$

$$= 19950 \div [1 + (.11)(3)]$$

ENTER	DISPLAY
19950	19950.
÷	19950.
(	19950.
1	1.
+	1.
.11	0.11
×	0.11
3	3.
)	1.33
=	15000.

So $p = 15000$.

d. $I = \dfrac{E}{R + r}$ 　　Solve for r.

$$(R + r)I = (R + r)\dfrac{E}{R + r}$$

$$RI + rI = E$$

$$RI + -RI + rI = E + -RI$$

$$rI = E - RI$$

$$r = \dfrac{E - RI}{I} = \dfrac{E}{I} - R$$

e. $\dfrac{1}{T} = \dfrac{1}{a} + \dfrac{1}{b} + \dfrac{1}{c}$ 　　Solve for a.

$$abcT \cdot \dfrac{1}{T} = abcT \cdot \left(\dfrac{1}{a} + \dfrac{1}{b} + \dfrac{1}{c}\right)$$

$$abc = bcT + acT + abT$$

$$abc - acT - abT = bcT + acT - acT + abT - abT$$

$$abc - acT - abT = bcT$$

$$a(bc - cT - bT) = bcT$$

$$a = \dfrac{bcT}{bc - cT - bT}$$

APPLICATION SOLUTION

To find the horsepower in terms of the torque and engine speed, solve the formula for H.

Formula: 　　$T = \dfrac{5252H}{S}$

Solve: 　$S(T) = S\left(\dfrac{5252H}{S}\right)$

$$ST = 5252H$$

$$H = \dfrac{ST}{5252}$$

Answer: 　Fred can use the formula $H = \dfrac{ST}{5252}$.

WARM UPS

1. $P = 4s$ 　　Find P if $s = 10$.

2. $A = s^2$ 　　Find A if $s = 10$.

3. $P = 4s$ 　　Find s if $P = 80$.

4. $P = 2w + 2\ell$ 　　Find w if $P = 90$ and $\ell = 30$.

5. $S - M = C$ 　　Find M if $S = 20$ and $C = 10$.

6. $V = \ell wh$ 　　Find ℓ if $V = 250$, $w = 5$, and $h = 5$.

7. $C = pn$ Find n if $C = 150$ and $p = 5$.

8. $A = \dfrac{bh}{2}$ Find b if $A = 48$ and $h = 8$.

9. $P = 4s$ Solve for s.

10. $V = \ell wh$ Solve for h.

11. $A = p + prt$ Solve for r.

12. $S - M = C$ Solve for S.

Chapter 7
POST-TEST

1. _____

2. _____

3. _____

4. _____

5. _____

6. _____

7. _____

8. _____

9. _____

10. _____

11. _____

12. _____

13. _____

14. _____

15. _____

16. _____

1. **(Obj. 82)** If $A = p(1 + r)^n$, find p if $r = .05$, $n = 2$, and $A = 882$.

2. **(Obj. 77)** Find the LCM of the following:

$$3x^2, 5xy, \text{ and } 10xy^2$$

3. **(Obj. 72)** *Multiply:* $\dfrac{23w}{4p} \cdot \dfrac{5w}{37p}$

4. **(Obj. 75)** Multiply and reduce to lowest terms:

$$\frac{x^2 - x}{10x^2} \cdot \frac{6x}{x^2 - 1}$$

5. **(Obj. 73)** *Reduce:* $\dfrac{32a^2b^3}{40a^3b^2}$

6. **(Obj. 76)** Divide and reduce:

$$\frac{4a^2b}{6a} \div \frac{8ab}{3ab^2}$$

7. **(Obj. 83)** If $I = \dfrac{E - e}{R}$, solve for E.

8. **(Obj. 74)** Find the missing numerator:

$$\frac{x - 1}{x + 3} = \frac{?}{x^2 - 9}$$

9. **(Obj. 79)** Subtract and reduce:

$$\frac{3x}{x^2 - 4} - \frac{6}{x^2 - 4}$$

10. **(Obj. 72)** *Multiply:* $\dfrac{3x}{5} \cdot \dfrac{12x + 5}{17}$

11. **(Obj. 79)** *Subtract:* $\dfrac{a}{a + 4} - \dfrac{2}{a + 7}$

12. **(Obj. 81)** *Solve:* $\dfrac{x}{5} - \dfrac{3}{4} = \dfrac{x}{2}$

13. **(Obj. 80)** *Simplify:* $\dfrac{\dfrac{3}{x + 3} + 1}{\dfrac{4}{x - 3} - 2}$

14. **(Obj. 78)** *Add:* $\dfrac{x}{x + 2} + \dfrac{3}{x - 4}$

15. **(Obj. 80)** *Simplify:* $\dfrac{\dfrac{x}{3} + \dfrac{y}{2}}{\dfrac{x}{4} - \dfrac{y}{6}}$

16. **(Obj. 74)** Find the missing numerator:

$$\frac{7xy}{6w} = \frac{?}{72w^2z}$$

17. _____

18. _____

19. _____

20. _____

21. _____

22. _____

23. _____

24. _____

25. _____

17. **(Obj. 76)** Divide and reduce:

$$\frac{y+5}{y^2-4} \div \frac{y^2+6y+5}{y+2}$$

18. **(Obj. 73)** *Reduce:* $\dfrac{m^2-25}{2m^2-6m-20}$

19. **(Obj. 78)** *Add:* $\dfrac{3y-5}{xy} + \dfrac{5y-3}{xy}$

20. **(Obj. 83)** If $S = 2ab + 2ah + 2bh$, solve for h.

21. **(Obj. 77)** Find the LCM of the following:

$$3a + 6b,\ 2a + 4b,\ 4$$

22. **(Obj. 81)** *Solve:* $\dfrac{2}{x+3} + \dfrac{1}{2} = \dfrac{2}{3}$

23. **(Obj. 82)** If $A = \dfrac{bh}{2}$, find h if $b = 2.1$ and $A = 7.56$.

24. **(Obj. 81)** A carpenter and his apprentice can frame a house in 10 days. If the carpenter could do the framing alone in 16 days, how long would it take the apprentice alone?

25. **(Obj. 81)** Three fifths of a certain number is two thirds more than half of the same number. Find the number.

ANSWERS TO WARM UPS (7.10)   **1.** $P = 40$ **2.** $A = 100$ **3.** $s = 20$ **4.** $w = 15$ **5.** $M = 10$

6. $\ell = 10$ **7.** $n = 30$ **8.** $b = 12$ **9.** $s = \dfrac{P}{4}$ **10.** $h = \dfrac{V}{\ell w}$

11. $r = \dfrac{A - p}{pt}$ **12.** $S = C + M$

8

Proportion, Variation, and Percent

ANSWERS

1. _____

2. _____

3. _____

4. _____

5. _____

6. _____

7. _____

8. _____

9. _____

10. _____

11. _____

12. _____

13. _____

14. _____

15. _____

16. _____

Chapter 8
PRE-TEST

The problems in the following pre-test are a sample of the material in the chapter. You may already know how to work some of these. If so, this will allow you to spend less time on those parts. As a result, you will have more time to give to the sections that you find more difficult or that are new to you. The answers are in the back of the text.

1. **(Obj. 84)** Write a ratio to compare 3 quarters to 6 dollars (in quarters). Reduce to lowest terms.

2. **(Obj. 85)** Is the following proportion true or false?

$$\frac{4 \text{ feet}}{4 \text{ yards}} = \frac{12 \text{ inches}}{3 \text{ feet}}$$

3. **(Obj. 85)** Is the following proportion true or false?

$$\frac{45}{72} = \frac{25}{36}$$

4. **(Obj. 86)** Solve the following proportion:

$$\frac{18}{32} = \frac{z}{160}$$

5. **(Obj. 86)** Solve the following proportion:

$$\frac{\frac{2}{3}}{5} = \frac{\frac{8}{3}}{x}$$

6. **(Obj. 87)** If 30 lb of beef contains 6 lb of bones, how many pounds of bones may be expected in 100 lb of beef?

7. **(Obj. 87)** There is a canned food sale at the supermarket. A case of 24 cans of peas is priced at $10.50. To the nearest cent, what is the price of 10 cans of peas?

8. **(Obj. 87)** If Mary is paid $46.06 for 7 hours of work, how much should she expect to earn for 12 hours of work?

9. **(Obj. 88)** A real estate salesperson's salary varies directly as her total sales. If she receives $1450 for sales of $25000, how much will she receive for sales of $62500?

10. **(Obj. 89)** The time it takes to fly between two cities varies inversely as the speed. If it takes 6 hours at 400 mph, how long will it take at 600 mph?

11. **(Obj. 91)** Write $.27\frac{3}{4}$ as a percent.

12. **(Obj. 91)** Write 3.5 as a percent.

13. **(Obj. 92)** Write 325% as a decimal.

14. **(Obj. 92)** Write .2% as a decimal.

15. **(Obj. 93)** Change $\frac{3}{4}$ to a percent.

16. **(Obj. 93)** Change $2\frac{5}{9}$ to a percent (give answer to nearest tenth of a percent).

17. _____

18. _____

19. _____

20. _____

21. _____

22. _____

23. _____

24. _____

25. _____

17. (Obj. 94) Change 275% to a mixed number.

18. (Obj. 94) Change $13\frac{5}{6}\%$ to a fraction.

19. (Obj. 98) 48% of what number is 36?

20. (Obj. 98) What percent of 86 is 38.7?

21. (Obj. 98) What number is 13% of 830?

22. (Obj. 99) 72 of 90 people enrolled in a class received a grade of C or better. What percent received a C or better?

23. (Obj. 99) A student answered 48 out of 64 questions correctly. What percent did the student answer correctly?

24. (Obj. 99) On the first day of her P.E. class, Julia did 25 sit-ups. At the end of the term she did 85 sit-ups. What was the percent of increase during the term?

25. (Obj. 99) The cost of an item is discounted 20%. If the item originally sold for $10.80, what is the discounted price?

8.1 Ratio and Proportion

OBJECTIVES

84. Write a fraction that shows a ratio comparison of two numbers or two measurements.

85. Determine whether or not a given proportion is true.

APPLICATION

In the parking lot at the Rural Community Center there are 48 parking spaces for compact cars and 32 parking spaces for larger cars.

a. What is the ratio of the number of compact spaces to the number of larger spaces?

b. What is the ratio of the number of compact spaces to the total number of spaces?

VOCABULARY

A *ratio* is a comparison of a pair of numbers or quantities by division. When two ratios are written with an equal sign between them, the resulting equation is called a *proportion*. A *measurement* is written with a number and a unit of measure.

HOW AND WHY

Two numbers can be compared by subtraction or by division. Suppose we wish to compare the measurements 12 dollars and 3 dollars. Since $12 - 3 = 9$, we can say that

$12 is nine dollars more than $3.

And since $3\overline{)12}$ or $12 \div 3$ is 4, we can say that

$12 is four times larger than $3.

The indicated division is called a ratio. These are common ways to write the ratio comparison of 12 and 3:

$$12 : 3 \qquad 12 \div 3 \qquad 12 \text{ to } 3 \qquad \frac{12}{3}$$

Here we shall treat ratios as fractions.

If a car runs 208 miles on 8 gallons of gas, we can compare the unlike measurements "280 miles" and "8 gallons" by writing $\frac{208 \text{ miles}}{8 \text{ gallons}}$. This symbol can be reduced in the same way as a fraction, as long as the units are stated:

$$\frac{208 \text{ miles}}{8 \text{ gallons}} = \frac{104 \text{ miles}}{4 \text{ gallons}} = \frac{26 \text{ miles}}{1 \text{ gallon}} = 26 \text{ miles per gallon}$$

It is this process that leads to statements such as "There are 3.1 children to a family," since

$$\frac{31 \text{ children}}{10 \text{ families}} = \frac{3.1 \text{ children}}{1 \text{ family}}$$

The last ratio must be understood as a comparison and not as a fact, since no family has 3.1 children.

If two measurements have the same units, such as $\frac{\$3}{\$100}$, then the unit labels or words may be dropped:

$$\frac{\$3}{\$100} = \frac{3}{100}$$

If two measurements do not have the same units, as in $\dfrac{26 \text{ miles}}{1 \text{ gallon}}$, the unit labels must be written.

Equations with two ratios set equal, such as

$$\frac{5 \text{ inches}}{2 \text{ feet}} = \frac{20 \text{ inches}}{8 \text{ feet}} \quad \text{and} \quad \frac{62 \text{ miles}}{1 \text{ hour}} = \frac{100 \text{ kilometers}}{1 \text{ hour}},$$

are called proportions. These are read as "5 inches is to 2 feet as 20 inches is to 8 feet" and "62 miles is to 1 hour as 100 kilometers is to 1 hour." To be true, the ratios must represent equivalent fractions when the units are the same.

The test to determine whether a proportion is true is often called "cross multiplication." $\dfrac{14}{8} = \dfrac{35}{20}$ is a true proportion because

$$\frac{14}{8} = \frac{35}{20}$$

$$20 \cdot 14 = 8 \cdot 35$$
$$280 = 280$$

The test is based on the multiplication law of equality. It is equivalent to multiplying both sides of the equation by the product of the denominators.

The proportion $\dfrac{4}{9} = \dfrac{2}{3}$ is not true because $3 \cdot 4 \neq 9 \cdot 2$.

To tell whether a proportion is true or false,

1. **Check that the ratios have the same units.**
2. **Cross-multiply.**
3. **If the products are equal, the proportion is true.**

$\dfrac{a}{b} = \dfrac{c}{d}$ **is true provided** $ad = bc$.

EXAMPLES

a. The ratio of 5 chairs to 6 chairs is

$$\frac{5 \text{ chairs}}{6 \text{ chairs}} = \frac{5}{6}.$$

Since the units are the same they may be dropped.

b. The ratio of 10 chairs to 8 people is

$$\frac{10 \text{ chairs}}{8 \text{ people}} \quad \text{or} \quad \frac{5 \text{ chairs}}{4 \text{ people}} \quad \text{or} \quad \frac{1.25 \text{ chairs}}{1 \text{ person}}.$$

The units must be stated since they are unlike.

c. The ratio of the length of a room to its width, if it measures 24 feet by 18 feet, is

$$\frac{24 \text{ feet}}{18 \text{ feet}} = \frac{24}{18} = \frac{4}{3}.$$

d. The ratio of six dimes to fourteen nickels is

$$\frac{6 \text{ dimes}}{14 \text{ nickels}} = \frac{6 \text{ dimes}}{7 \text{ dimes}} = \frac{6}{7},$$

$$\text{or } \frac{6 \text{ dimes}}{14 \text{ nickels}} = \frac{12 \text{ nickels}}{14 \text{ nickels}} = \frac{12}{14} = \frac{6}{7},$$

$$\text{or } \frac{6 \text{ dimes}}{14 \text{ nickels}} = \frac{60 \text{ cents}}{70 \text{ cents}} = \frac{60}{70} = \frac{6}{7}.$$

e. The proportion $\dfrac{6}{5} = \dfrac{72}{60}$ is true, since $6 \cdot 60 = 5 \cdot 72$.

f. The proportion $\dfrac{2.1}{3.1} = \dfrac{2}{3}$ is false since $(2.1)(3)$ is not equal to $(3.1)(2)$.

g. The proportion $\dfrac{1 \text{ dollar}}{3 \text{ quarters}} = \dfrac{8 \text{ dimes}}{12 \text{ nickels}}$ is true, since we may write it as

$$\frac{20 \text{ nickels}}{15 \text{ nickels}} = \frac{16 \text{ nickels}}{12 \text{ nickels}} \quad \text{or} \quad \frac{20}{15} = \frac{16}{12}$$

and $(20)(12) = (15)(16)$.

h. Calculator example

The population density of a region is the ratio of the number of people to the number of square miles of area. Find the population density of Skarf County if the population is 12550 and the area is 1700 square miles. Reduce to a 1 square mile comparison, rounded to the nearest tenth.

$$\text{density} = \frac{12550 \text{ people}}{1700 \text{ square miles}}$$

ENTER	DISPLAY
12550	12550.
÷	12550.
1700	1700.
=	7.3823529

The density to the nearest tenth is 7.4 people per square mile.

APPLICATION SOLUTION

To find the ratio of the number of compact spaces to the number of larger spaces, we write the fraction

$$\frac{48}{32} = \frac{3}{2}$$

To find the ratio of the number of compact spaces to the total number of spaces, we first find the total number of spaces $(48 + 32 = 80)$. Then we write the fraction

$$\frac{48}{80} = \frac{3}{5}$$

Therefore,

a. The ratio of compact spaces to larger spaces is $\dfrac{3}{2}$.

b. The ratio of compact spaces to the total number of spaces is $\dfrac{3}{5}$.

Write as a ratio and reduce:

1. 8 dollars to 3 dollars

2. 8 dollars to 3 records

3. 22 dimes to 2 dimes

4. 22 logs to 2 trucks

5. 9 families to 27 children

6. 8 families to 12 children

7. 5 dimes to 15 cents (in cents)

8. 6 dimes to 6 nickels (in cents)

Are the following proportions true or false?

9. $\dfrac{7}{9} = \dfrac{5}{7}$ 10. $\dfrac{3}{4} = \dfrac{9}{12}$

11. $\dfrac{4}{6} = \dfrac{6}{9}$ 12. $\dfrac{7}{5} = \dfrac{10}{7}$

Exercises 8.1

A

Write a ratio to compare the following pairs of numbers. Reduce to lowest terms where possible.

1. 8 people to 11 chairs

2. 3 inches to 12 inches

3. 1 nickel to 1 dime (compare in cents)

4. 4 cm to 8 cm

5. 80 cm to 1 m

6. 16 families to 48 children

B

Tell whether the following proportions are true or false.

7. $\dfrac{1}{2} = \dfrac{5}{10}$ 8. $\dfrac{2}{3} = \dfrac{18}{24}$ 9. $\dfrac{6}{4} = \dfrac{9}{10}$

10. $\dfrac{\frac{1}{2}}{5} = \dfrac{1}{10}$ 11. $\dfrac{7}{8} = \dfrac{11}{12}$ 12. $\dfrac{\frac{1}{3}}{\frac{7}{9}} = \dfrac{\frac{1}{2}}{1\frac{1}{6}}$

Write a ratio to compare the following pairs of numbers. Reduce to lowest terms where possible.

13. 3 dimes to 7 nickels (compare in nickels)   14. 12 meters to 10 meters

C

15. 600 miles to (per) 8 hours

16. $2.37 to (per) 3 pounds of potatoes

17. 1300 television sets to 1000 houses (reduce to a 1-house comparison)

18. 32 games won to 20 games lost

19. The low gear ratio in a truck's transmission, if the large gear has 189 teeth and the small gear has 14 teeth

Tell whether the following proportions are true or false.

20. $\dfrac{2}{4} = \dfrac{14}{28}$ 21. $\dfrac{4\frac{1}{2}}{3} = \dfrac{9}{6}$

22. $\dfrac{2.1}{3.2} = \dfrac{1.2}{2.3}$ 23. $\dfrac{2.6}{4.8} = \dfrac{3.9}{7.2}$

24. $\dfrac{5 \text{ pounds}}{\$1.30} = \dfrac{8 \text{ ounces}}{14 \text{ cents}}$

1. _____
2. _____
3. _____
4. _____
5. _____
6. _____
7. _____
8. _____
9. _____
10. _____
11. _____
12. _____
13. _____
14. _____
15. _____
16. _____
17. _____
18. _____
19. _____
20. _____
21. _____
22. _____
23. _____
24. _____

D

25. The parking lot in the lower level of the Knew Office Building has 18 spaces for compact cars and 24 spaces for larger cars.
   a. What is the ratio of compact spaces to larger spaces?
   b. What is the ratio of compact spaces to the total number of spaces?

26. The Reliable Auto Repair Service building has eight stalls for repairing automobiles and four stalls for repairing small trucks.
   a. What is the ratio of the number of stalls for small trucks to the number of stalls for automobiles?
   b. What is the ratio of the number of stalls for small trucks to the total?

27. One section of the country has 3500 TV sets per 1000 houses. A second section has 500 TV sets per 150 houses. Are the ratios of the TV sets to the number of houses the same in both parts of the country?

28. In City A there are 5000 automobiles per 3000 households. In City B there are 8000 automobiles per 4800 households. Are the ratios of the number of automobiles to the number of households the same?

29. What is the population density of the city of Dryton if 22,450 people live there and the area is 230 square miles? Reduce to a 1 square mile comparison, rounded to the nearest tenth.

30. What is the population density of Struvaria if there are 950,000 people and the area is 18000 square miles? Reduce to a 1 square mile comparison, rounded to the nearest tenth.

E   *Maintain Your Skills*

Add, reduce if possible:

31. $\dfrac{5a-b}{12} + \dfrac{3a-b}{12}$

32. $\dfrac{9}{x} + \dfrac{12}{z}$

33. $\dfrac{4x}{3y} + \dfrac{2x}{9y}$

34. $\dfrac{3}{x+1} + \dfrac{4}{x-2}$

35. $\dfrac{1}{x} + (x-2)$

ANSWERS TO WARM UPS (8.1)

1. $\dfrac{8}{3}$ 2. $\dfrac{8 \text{ dollars}}{3 \text{ records}}$ 3. $\dfrac{11}{1}$ 4. $\dfrac{11 \text{ logs}}{1 \text{ truck}}$

5. $\dfrac{1 \text{ family}}{3 \text{ children}}$ 6. $\dfrac{2 \text{ families}}{3 \text{ children}}$ 7. $\dfrac{10}{3}$ 8. $\dfrac{2}{1}$

9. false 10. true 11. true 12. false

8.2 Solving Proportions

OBJECTIVE

86. Solve a proportion.

APPLICATION

Applications of this objective come from a variety of fields. These applications are so important that an entire section (8.3) is given to them.

VOCABULARY

In proportions we use a letter to hold the place of a missing number. The letter that is used is called an unknown or a *variable*. Finding that replacement for the missing number which will make the proportion true is called *solving the proportion*.

HOW AND WHY

Proportions are used to solve many problems in science and technology which involve ratios in some manner. There are four members in a proportion, and if any three of them are known, it is possible to solve the proportion to find the missing member. The following method of solving a proportion is based upon the fact that the cross products must be equal. Consider the following question: "What number is to 5 as 15 is to 25?" To answer this, we will use a variable (x) to represent the missing number, and we can write the following: x is to 5 as 15 is to 25,

$$\frac{x}{5} = \frac{15}{25}$$

Since the cross products are equal, we have

$$\frac{x}{5} = \frac{15}{25}$$

$$25 \cdot x = 75$$

The missing number is 3; therefore,

$$\frac{3}{5} = \frac{15}{25}$$

To solve a proportion,

1. **Cross multiply.**
2. **Set the cross products equal to each other.**
3. **Solve the equation.**

EXAMPLES

Solve the following proportions.

a. $\dfrac{4}{9} = \dfrac{8}{x}$

$4x = 72$

$x = 18$

b. $\dfrac{.6}{x} = \dfrac{1.2}{.84}$

$1.2x = .504$

$x = .42$

c. $\dfrac{\frac{3}{4}}{1\frac{2}{3}} = \dfrac{\frac{1}{2}}{x}$

$\dfrac{3}{4}x = \dfrac{5}{6}$

$x = 1\dfrac{1}{9}$

d. **Calculator example**
Proportions with whole numbers or decimals can be solved with a calculator. For instance, consider

$\dfrac{3}{x} = \dfrac{9.6}{7.32}$

STEPS	ENTER	DISPLAY
	$\boxed{3}$	3.
$(3)(7.32) = 9.6(x)$	$\boxed{\times}$	3.
	$\boxed{7.32}$	7.32
	$\boxed{=}$	21.96
$21.96 \div 9.6 = x$	$\boxed{\div}$	21.96
	$\boxed{9.6}$	9.6
	$\boxed{=}$	2.2875

So $x \approx 2.29$ to the nearest hundredth.

WARM UPS

Solve each of the following proportions:

1. $\dfrac{a}{1} = \dfrac{6}{3}$

2. $\dfrac{1}{b} = \dfrac{5}{15}$

3. $\dfrac{56}{7} = \dfrac{c}{1}$

4. $\dfrac{8}{4} = \dfrac{1}{d}$

5. $\dfrac{u}{72} = \dfrac{1}{9}$

6. $\dfrac{72}{v} = \dfrac{4}{1}$

7. $\dfrac{3}{2} = \dfrac{x}{6}$

8. $\dfrac{6}{y} = \dfrac{4}{6}$

9. $\dfrac{9}{18} = \dfrac{w}{2}$

10. $\dfrac{x}{10} = \dfrac{2}{4}$

11. $\dfrac{2}{8} = \dfrac{3}{m}$

12. $\dfrac{9}{3} = \dfrac{n}{2}$

Exercises 8.2

ANSWERS

1. _____
2. _____
3. _____
4. _____
5. _____
6. _____
7. _____
8. _____
9. _____
10. _____
11. _____
12. _____
13. _____
14. _____
15. _____
16. _____
17. _____
18. _____
19. _____
20. _____
21. _____
22. _____
23. _____
24. _____
25. _____

A

Solve the following proportions.

1. $\dfrac{6}{8} = \dfrac{12}{x}$ **2.** $\dfrac{x}{42} = \dfrac{5}{7}$ **3.** $\dfrac{6}{y} = \dfrac{9}{5}$

4. $\dfrac{y}{25} = \dfrac{3}{5}$ **5.** $\dfrac{5}{2} = \dfrac{w}{9}$ **6.** $\dfrac{14}{8} = \dfrac{7}{w}$

B

7. $\dfrac{3}{2} = \dfrac{R}{100}$ **8.** $\dfrac{5}{4} = \dfrac{R}{100}$ **9.** $\dfrac{.2}{.3} = \dfrac{8}{x}$

10. $\dfrac{x}{40} = \dfrac{\frac{3}{4}}{5}$ **11.** $\dfrac{\frac{1}{2}}{y} = \dfrac{\frac{2}{3}}{10}$ **12.** $\dfrac{8}{9} = \dfrac{\frac{1}{3}}{y}$

13. $\dfrac{w}{2.5} = \dfrac{3}{5}$ **14.** $\dfrac{418}{154} = \dfrac{w}{7}$

C

15. $\dfrac{6.5}{26} = \dfrac{y}{.04}$ **16.** $\dfrac{3}{5} = \dfrac{R}{100}$ **17.** $\dfrac{.05\frac{1}{2}}{1} = \dfrac{R}{100}$

18. $\dfrac{.014}{x} = \dfrac{7}{50}$ **19.** $\dfrac{3\frac{1}{2}}{10\frac{1}{2}} = \dfrac{8}{x}$ **20.** $\dfrac{.05}{.9} = \dfrac{y}{4.5}$

21. $\dfrac{y}{3} = \dfrac{15}{16}$ **22.** $\dfrac{2\frac{1}{2}}{3\frac{1}{3}} = \dfrac{4\frac{1}{4}}{x}$ **23.** $\dfrac{A}{35} = \dfrac{6.2}{100}$

24. $\dfrac{1.2}{2.7} = \dfrac{3.4}{w}$ **25.** $\dfrac{2.5}{x} = \dfrac{3.2}{300}$

ANSWERS

26. _____

27. _____

28. _____

29. _____

30. _____

31. _____

32. _____

33. _____

34. _____

35. _____

D

Solve these proportions. Round to the nearest hundredth.

26. $\dfrac{7}{11} = \dfrac{y}{15}$

27. $\dfrac{6}{z} = \dfrac{7}{17}$

28. $\dfrac{4}{3.7} = \dfrac{a}{.53}$

29. $\dfrac{4.4}{.23} = \dfrac{9}{b}$

30. $\dfrac{341}{202} = \dfrac{c}{8}$

E   *Maintain Your Skills*

Subtract, reduce if possible:

31. $\dfrac{16a}{13} - \dfrac{3a}{13}$

32. $\dfrac{7}{4y} - \dfrac{9}{12y}$

33. $\dfrac{5}{x + y} - \dfrac{7}{x - y}$

34. $\dfrac{5}{x + 2} - \dfrac{2}{x - 5}$

35. $\dfrac{1}{x} - (x - 2)$

ANSWERS TO WARM UPS 8.2

1. $a = 2$ **2.** $b = 3$ **3.** $c = 8$ **4.** $d = .5$ or $d = \dfrac{1}{2}$

5. $u = 8$ **6.** $v = 18$ **7.** $x = 9$ **8.** $y = 9$

9. $w = 1$ **10.** $x = 5$ **11.** $m = 12$ **12.** $n = 6$

8.3 Word Problems (Proportion)

OBJECTIVE

87. Solve word problems using proportions.

APPLICATION

Paula decides to start a savings account. Her weekly take-home pay (without overtime) is $278. She decides to save $13.90 of this each week. One week she works overtime and her take-home pay is $320. If she wants to save from this check in the same ratio, how much should she save?

VOCABULARY

No new words.

HOW AND WHY

Many common situations occur in which two quantities are compared by ratios: cost (price per pound), map scale (miles per inch), geometry (length of a shadow to the height of the object), time to do a certain amount of work, and profit to a specific amount of investment, to name a few.

Assuming that the comparison of the two quantities is constant, the given comparison can be used to discover the missing part of a second one. For instance, if 2 pounds of bananas cost $.36, what will 12 pounds of bananas cost?

	Pounds of Bananas	Cost in Dollars
Case 1	2	.36
Case 2	12	

In the preceding table, it is seen that the cost in Case 2 is missing. Assign this missing value the letter y.

	Pounds of Bananas	Cost in Dollars
Case 1	2	.36
Case 2	12	y

Now form the ratios of the like quantities, Case 1 to Case 2. These must be equal if the relationship is constant.

$$\frac{2 \text{ lb of bananas}}{12 \text{ lb of bananas}} = \frac{\$.36}{\$y}$$

Since the units are the same, they may be dropped.

$$\frac{2}{12} = \frac{.36}{y}$$

Solving, we have:

$2y = 4.32$

$y = 2.16$

The conclusion is that 12 pounds of bananas will cost $2.16.

Although different ratios also remain constant $\left(\dfrac{\text{dollars}}{\text{pounds}}\right)$ in the relationship, we prefer to form the ones that compare the same units $\left(\dfrac{\text{pounds}}{\text{pounds}} = \dfrac{\text{dollars}}{\text{dollars}}\right)$. The denominator of each ratio must come from the same case. The table helps us to keep this in mind because it simulates the ratios we use.

EXAMPLE

On a road map of Oregon, $\dfrac{1}{4}$ inch represents 50 miles. How many miles are represented by $1\dfrac{1}{2}$ inches?

	Inches	Miles
Case 1	$\dfrac{1}{4}$	50
Case 2	$1\dfrac{1}{2}$	N

$$\frac{\frac{1}{4}}{1\frac{1}{2}} = \frac{50}{N}$$

$$\frac{1}{4} \cdot N = 1\frac{1}{2} \cdot 50$$

$$N = 300$$

Therefore, $1\dfrac{1}{2}$ inches on the map represents 300 miles.

APPLICATION SOLUTION

To find out how much Paula should save, we make the following chart:

	Pay	Savings
Case 1	278	13.90
Case 2	320	x

The proportion we get is

$$\frac{278}{320} = \frac{13.90}{x}$$

Continued on next page.

$$278 \cdot x = 13.90(320)$$

$$278 \cdot x = 4448$$

$$x = 4448 \div 278$$

$$x = 16$$

If she wants to save from this check in the same ratio, she should save $16.

WARM UPS

If a fir tree 20 ft tall casts a shadow of 12 ft, how tall is a tree that casts a shadow of 18 feet?

	Height	Shadow
First tree	(1)	(2)
Second tree	(3)	(4)

1. What goes in box 1? **2.** What goes in box 2?

3. What goes in box 3? **4.** What goes in box 4?

5. What is the proportion for the problem?

6. How tall is the second tree?

Three women can assemble 17 television sets in 4 days. How many days will it take them to assemble 68 sets?

	TV sets	Number of Days
Case 1	(5)	(6)
Case 2	(7)	(8)

7. What goes in box 5? **8.** What goes in box 6?

9. What goes in box 7? **10.** What goes in box 8?

11. What is the proportion for the problem?

12. How many days does it take to assemble 68 sets?

Exercises 8.3

A

A photograph that measures 3 inches wide and 5 inches high is to be enlarged so that the height will be 10 inches. What will be the width of the enlargement?

	Width	Height
Case 1	(a)	(b)
Case 2	(c)	(d)

1. What goes in box a?

2. What goes in box b?

3. What goes in box c?

4. What goes in box d?

5. What is the proportion for the problem?

6. What is the width of the enlargement?

B

7. Grains and drams are units of weight used in pharmacy. (8 drams = 480 grains.) If a doctor's prescription calls for 24 grains of a drug, what part of a dram is required?

8. A photograph that measures 8 cm wide and 13 cm high is to be enlarged so that the height will be 26 cm. What will be the width of the enlargement?

9. Merle is knitting a sweater. The knitting gauge is 6 rows to one inch. How many rows must she knit to complete $9\frac{1}{2}$ inches of the sweater?

10. At a certain time of the day, a tree 15 m tall casts a shadow of 12 m, while a second tree casts a shadow of 20 m. How tall is the second tree?

11. If 16 lb of fertilizer will cover 1500 sq ft of lawn, how much fertilizer is needed to cover 2500 sq ft?

12. The Mudville Elementary School expects an enrollment of 980 students. The district assigns teachers at the rate of 3 teachers for every 70 students. The district now employs 32 teachers. How many additional teachers does the district need to hire?

13. The Mitchells pay $810 taxes on their home, which has an assessed value of $28,000. How much will the taxes be on a $42,000 home in the same district?

ANSWERS

14.

15.

16.

17.

18.

19.

20.

21.

22.

23.

24.

14. One can of frozen orange juice concentrate mixed with three cans of water makes one liter of juice. At the same rate, how many cans of water are needed to make four liters of juice?

C

15. A room that contains 24 square yards was carpeted at a cost of $204. If the same kind of carpet is used in a room that contains 16 square yards, what will be the cost?

16. The counter on a tape recorder registers 520 after the recorder has been running for 20 minutes. What would the counter register after half an hour?

17. During the first 320 miles of their vacation trip, the Scaberys used 35 gallons of gasoline. At this rate, to the nearest tenth of a gallon how many gallons of gasoline will be needed to finish the remaining 580 miles of their trip?

18. If it takes 4 men 12 hours to do a certain job, how many of these jobs could they do in 66 hours?

19. In the first 12 games of an 18 game schedule, Dave's basketball team scored a total of 974 points. At this rate, how many points can they expect to score in their remaining games?

20. A 16-ounce can of pears cost $.57 and a 29-ounce can costs $.89. Is the price per ounce the same in both cases? If not, then to the nearest cent, what would the price of the 29-ounce can need to be to make the prices per ounce the same?

21. A map of the western United States is scaled so that $\frac{3}{4}$ inch represents 100 miles. How many miles is it between San Diego and Seattle if the distance on the map is 9 inches?

22. If Wayne receives $650 for $\frac{3}{4}$ of a ton of strawberries, how much will he receive for $1\frac{4}{5}$ tons?

23. The Utah Construction Co. has a job that takes 5 men 18 hours to do. How many of these jobs could they do in 90 hours?

24. If a 20-foot beam of structural steel contracts .0053 inch for each drop of 5 degrees in temperature, then to the nearest ten-thousandth of an inch, how much would a 13-foot beam of structural steel contract for a drop of 5 degrees in temperature?

25. The ratio of boys to girls taking chemistry is 4 to 3. How many boys are there in a chemistry class of 84 students? (Hint: Fill in the rest of the table.)

	Number of boys	Number of students
Case 1		7
Case 2		84

26. Betty prepares a mixture of nuts that has cashews and peanuts in a ratio of 3 to 5. How many pounds of each will she need to make 48 pounds of the mixture?

27. The estate of the late Mr. John Redgrave is to be divided among his 3 nephews in the ratio of 5 to 2 to 2. How much of the $72,081 in the estate did each nephew receive?

28. A brass alloy is 4 parts copper and 3 parts zinc. How many kilograms of copper are needed to make 200 kilograms of the alloy (to the nearest tenth of a kg)?

29. A yard (cubic yard) of concrete will make a 4-inch-thick slab with an area of 81 ft². How many yards of concrete (to the nearest tenth of a yard) are needed to pour a garage floor that is to be 440 ft²?

30. A doctor requires that the nurse, Ida, give 10 milligrams of a certain drug to his patient. The drug is in a solution that contains 25 milligrams in one cubic centimeter. How many cc's should Ida use for the injection?

E   *Maintain Your Skills*

Divide, reduce if possible:

31. $\dfrac{c}{d} \div \dfrac{e}{f}$

32. $\dfrac{2x+3}{5} \div \dfrac{2x+3}{10}$

33. $\dfrac{1}{x^2+8x+15} \div \dfrac{1}{x+5}$

34. $\dfrac{4x+8}{14x+21} \div \dfrac{x+2}{6x+9}$

35. $\dfrac{x^2+4x+3}{x^2+6x+5} \div \dfrac{x^2+7x+12}{x^2+2x-15}$

25. _____

26. _____

27. _____

28. _____

29. _____

30. _____

31. _____

32. _____

33. _____

34. _____

35. _____

ANSWERS TO WARM UPS (8.3)

ANSWERS TO WARM UPS (8.3)   **1.** 20   **2.** 12   **3.** unknown h **4.** 18

5. $\dfrac{20}{h} = \dfrac{12}{18}$ **6.** 30 ft   **7.** 17   **8.** 4

9. 68   **10.** unknown x **11.** $\dfrac{17}{68} = \dfrac{4}{x}$ **12.** 16 days

8.4 Variation

OBJECTIVES

88. Solve problems involving direct variation.
89. Solve problems involving inverse variation.

APPLICATION

The weight (w) of a gold ingot varies directly as the volume (v). If an ingot containing $\dfrac{3}{4}$ cubic foot weighs 906 pounds, what will an ingot of $1\dfrac{1}{2}$ cubic feet weigh?

VOCABULARY

Two quantities *vary directly* when one is a multiple of the other (their quotient is a constant). In the formula

$$D = 55t \left(\text{or} \quad \frac{D}{t} = 55 \right)$$

(distance equals 55 mph times time), *D varies directly* as *t*. Note that as time increases, distance increases, and as time decreases, so does distance.

Two quantities *vary inversely* when their product is constant. In the formula

$$PV = 50$$

(pressure times volume is constant if the temperature remains the same), *P varies inversely* as *V*. Note that pressure increases as volume decreases and that pressure decreases as volume increases.

HOW AND WHY

In our physical world most things are in a state of change or variation. Measurements of changes in temperature, rainfall, light, heat, and fuel supplies are recorded. Changes in a person's height, weight, and blood pressure can be important to health. Many relationships between these can be expressed in formulas. Here we are concerned about only two types of variation, direct and inverse.

Some examples of direct variation are:

Distance traveled at a constant speed varies directly as time:

$$d = kt \quad \text{or} \quad \frac{d}{t} = k \quad \text{(quotient is a constant).}$$

The area of a circle varies directly as the square of its radius:

$$A = \pi R^2 \quad \text{or} \quad \frac{A}{R^2} = \pi.$$

The total cost of a certain number of articles varies directly as the price:

$$C = nP \quad \text{or} \quad \frac{C}{P} = n.$$

In solving a problem involving direct variation, a constant must be identified. Suppose a variable S varies directly as a variable P. If $S = 40$ when $P = 5$, find S when $P = 8$. Since the variables vary directly, we know that $\frac{S}{P} = k$. So

$$\frac{S}{P} = k$$

$$\frac{40}{5} = k$$

$$8 = k$$

We now have the formula $\frac{S}{P} = 8$ or $S = 8P$ and can solve for S when $P = 8$.

$$S = 8P = 8(8) = 64$$

Some examples of inverse variation are:

The time t it takes to cover a certain distance d varies inversely as the speed R:

$$R \cdot t = d \quad \text{(product is a constant distance)}$$

The number of vibrations n in a musical string varies inversely as its length ℓ:

$$n \cdot \ell = k$$

When two pulleys are connected by a belt, the revolutions per minute each makes varies inversely as their respective diameters:

$$d \cdot r = k \quad \text{and} \quad D \cdot R = k$$

If the small pulley has a diameter of 10 inches and revolves 80 times per minute, how many rpm will the longer one make if it has a diameter of 16 inches?

SMALLER PULLEY	LARGER PULLEY
$d \cdot r = k$	$D \cdot R = k$
$(10)(80) = k$	$16 \cdot R = 800$
$800 = k$	$R = 50$

So the larger pulley will make 50 revolutions per minute.

Some formulas show variation between three or more variables. (See Example c.)

EXAMPLES

a. The weight w of a piece of aluminum varies directly as the volume v. If a piece of aluminum containing $1\frac{1}{2}$ cubic feet weighs 254 pounds, find the weight of a piece containing 5 cubic feet.

$$\frac{w}{v} = k \qquad \textit{Direct variation}$$

$$\frac{254}{1\frac{1}{2}} = k$$

$$k = \frac{508}{3}$$

So

$$\frac{w}{v} = \frac{508}{3}$$

$$\frac{w}{5} = \frac{508}{3}$$

$$w = \frac{2540}{3} = 846\frac{2}{3}$$

So the weight of a piece containing 5 cubic feet is $846\frac{2}{3}$ pounds.

b. The base b of a rectangle with constant area k varies inversely with its height. One rectangle has a base of 10 and a height of 6. Find the height of another rectangle with the same area whose base is 12.

$$b \cdot h = k \qquad \textit{Inverse variation}$$

$$(10)(6) = k$$

$$k = 60$$

So

$$b \cdot h = 60$$

$$12(h) = 60$$

$$h = 5$$

So the height of the second rectangle is 5.

c. The interest i paid on borrowed money varies directly as the principle P and the time t. If \$80 interest is earned on a loan of \$600 in 2 years, how much interest is earned on \$1000 borrowed for 3 years?

$$\frac{i}{P \cdot t} = k \qquad \begin{array}{l}\textit{i varies directly as P and t} \\ \textit{(sometimes read: i varies jointly as P and t)}\end{array}$$

$$\frac{80}{(600)(2)} = k$$

$$k = \frac{1}{15}$$

So

$$\frac{i}{(1000)(3)} = \frac{1}{15}$$

$$i = 200$$

So the interest earned is \$200.

APPLICATION SOLUTION

In the application, we know that the weight of the gold ingot varies directly as the volume $\left(\dfrac{w}{v} = k\right)$. The constant k can be found by replacing w and v with the given values.

Formula: $\dfrac{w}{v} = k$, $w = 906$ and $v = \dfrac{3}{4}$ ft^3

Substitute: $\dfrac{906}{\frac{3}{4}} = k$

$1208 = k$

We can now write a formula for w in terms of v.

$\dfrac{w}{v} = k$ or $w = kv$ *w varies directly as v*

$w = 1208 \cdot 1\dfrac{1}{2}$ *Replace k by 1208 and v by $1\dfrac{1}{2}$*

$= 1812$

Answer: The $1\dfrac{1}{2}$ cubic foot ingot will weigh 1812 pounds.

Exercises 8.4

1. The weight of a metal ingot varies directly as the volume v. If an ingot containing $1\frac{2}{3}$ cubic feet weighs 350 pounds, what will an ingot of $2\frac{1}{3}$ cubic feet weigh?

2. The force of attraction between two magnetic poles of opposite polarity varies inversely as the square of the distance between them. What is the force when they are 5 cm apart if it is 36 dynes at a distance of 9 cm?

3. The time it takes to make a certain trip varies inversely as the speed. If it takes 5 hours at 50 mph, how long will it take at 60 mph?

4. A car salesman's salary varies directly as his total sales. If he receives $372 for sales of $3100, how much will he receive for sales of $4500?

5. The amount of quarterly income one receives varies directly as the amount of money invested. If Dan earns $115 on a $2000 investment, how much would he earn on a $4300 investment?

6. The number of amperes varies directly as the number of watts. For a reading of 50 watts, the number of amperes is $\frac{5}{11}$. What are the amperes when the watts are 75?

7. The weight of wire varies directly as its length. If 1000 ft of wire weighs 45 lb, what will one mile of wire weigh?

8. The length of a rectangle with a constant area varies inversely as the width. If one rectangle has a length of 12 and a width of 8, what will be the length of a rectangle with a width of 6?

ANSWERS

9. _____

10. _____

11. _____

12. _____

13. _____

14. _____

15. _____

16. _____

17. _____

9. The force needed to raise an object with a crowbar varies inversely with the length of the crowbar. If it takes 40 lb of force to lift a certain object with a 2-ft long crowbar, what force will be necessary if you use a 3-ft crowbar?

10. As a rule of thumb, realtors suggest that the price you can afford to pay for a house varies directly with your annual salary. If a person earning $18,500 can purchase a $46,250 home, what price home can a person earning $24,000 annually afford?

11. Assuming that each person works at the same rate, the time it takes to complete a job varies inversely as the number of people assigned to it. If it takes 5 people 12 hours to do a job, how long will it take 3 people?

12. The weight of wire varies directly as its length. If 850 ft of wire weighs 68 lb, what will one mile of wire weigh?

E   *Maintain Your Skills*

Simplify:

13. $\dfrac{\dfrac{4}{5}+\dfrac{2}{3}}{\dfrac{1}{2}}$

14. $\dfrac{x+\dfrac{1}{3}}{\dfrac{3}{4}}$

15. $\dfrac{x-\dfrac{1}{y}}{x+\dfrac{1}{y}}$

16. $\dfrac{\dfrac{y}{8}}{\dfrac{1}{2}+\dfrac{y}{4}}$

17. $\dfrac{\dfrac{1-\dfrac{1}{y}}{y}}{\dfrac{1}{y}-1}$

8.5 What is Percent?

| OBJECTIVE | **90. Write a percent to express a comparison of two numbers.** |

| APPLICATION | Donna has \$300 in her savings account. Last year she received a total of \$24 in interest. What percent of the total in her account is the interest? |

| VOCABULARY | When using ratios to compare numbers, the denominator is called the *base unit*. In comparing 80 to 100 $\left(\text{as the ratio } \dfrac{80}{100}\right)$, 100 is the base unit. The *percent comparison,* or just the *percent,* is a ratio with a base unit of 100. The percent, $\dfrac{80}{100} = (80)\left(\dfrac{1}{100}\right)$, is usually written 80%. The symbol "%" is read "percent," and $\% = \dfrac{1}{100} = .01$. Also $100\% = 1$. |

| HOW AND WHY | Consider the comparison of 24 to 100 as a percent. The base unit is 100. Figure 8.1 shows the base unit divided into 100 equal parts, each of size one. We see that 24 of the 100 equal parts are shaded. Since each part represents $\dfrac{1}{100}$ of the 100, the 24 (shaded part) is $(24)\left(\dfrac{1}{100}\right)$ or 24% of 100. |

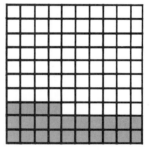

Figure 8.1

Figure 8.1 also illustrates that if the first number is smaller than the base unit, then not all of the base unit will be shaded and hence the comparison will be less than 100%. If the first number equals the base unit, the entire unit will be shaded and the comparison will be 100%. If the first number is larger than the base unit, then additional parts will be needed to represent it and the comparison will be more than 100%.

The percent comparison of two numbers can also be obtained from their ratio. The ratio of 24 to 100 is $\dfrac{24}{100} = 24 \div 100 = 24 \cdot \dfrac{1}{100} = 24\%$. This gives rise to the interpretation of percent as "so many per hundred."

The ratio of two numbers can be used to find the percent when the base unit is not 100. Compare 7 to 20. The ratio is $\dfrac{7}{20}$. Now find the equivalent ratio with denominator 100.

$$\frac{7}{20} = \frac{R}{100}$$ *Write the two ratios as a proportion.*

$$20 \cdot R = 7 \cdot 100$$ *Set the cross products equal.*

$$R = 35$$

So, $\dfrac{7}{20} = \dfrac{35}{100} = 35 \cdot \dfrac{1}{100} = 35\%$.

To find the percent comparison of two numbers:

1. Write the ratio of the first number to the base number.
2. Find the equivalent ratio with denominator 100.
3. $\dfrac{\text{numerator}}{100} = \text{numerator} \cdot \dfrac{1}{100} = \text{numerator} \%$.

EXAMPLES

a. The shaded portion represents 55 of the 100 equal parts or $55 \cdot \dfrac{1}{100} = 55\%$ of the region.

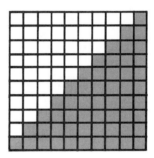

b. The shaded portion represents $\dfrac{4}{4}$ of one unit.

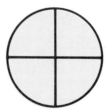

$$\frac{4}{4} = \frac{100}{100} = 100\left(\frac{1}{100}\right) = 100\%$$

c. The shaded portion represents $\dfrac{5}{4}$ of one unit.

 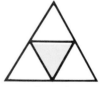

One unit   One unit

$$\frac{5}{4} = \frac{125}{100} = 125\left(\frac{1}{100}\right) = 125\%$$

d. At a recent sporting event, there were 30 women among the first 100 people to enter. What percent were women? The ratio of women to people is $\dfrac{30}{100} = 30 \cdot \dfrac{1}{100}$ or 30% who were women.

e. The comparison of 42 to 25 as a ratio is $\dfrac{42}{25}$. As a percent it is 168%, since

$$\frac{42}{25} = \frac{168}{100} = 168 \cdot \frac{1}{100} = 168\%.$$

f. Calculator example
   Percents can be found by proportions, so the calculator can be used.
      Find the percent comparison of 35 to 280.

$$\frac{35}{280} = \frac{R}{100}$$

ENTER	DISPLAY
35	35.
×	35.
100	100.
÷	3500.
280	280.
=	12.5

or

ENTER	DISPLAY
35	35.
÷	35.
280	280.
%	12.5

So 35 is 12.5% of 280.

APPLICATION SOLUTION

We want to compare $24 to $300, so we write the following proportion.

$$\frac{24}{300} = \frac{R}{100}$$

$$300(R) = 100(24)$$

$$300(R) = 2400$$

$$R = 2400 \div 300$$

$$R = 8$$

So we have

$$\frac{24}{300} = \frac{8}{100} = 8\left(\frac{1}{100}\right) = 8\%$$

Therefore, the interest was 8% of the total in Donna's account.

1. What percent of the following region is shaded?

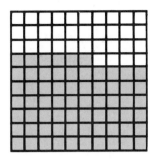

What percent of each of the following is shaded?

2. 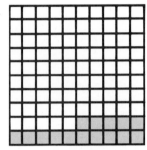   **3.**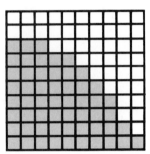

4.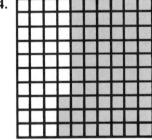

Write a percent for these comparisons:

5. 20 out of 100   **6.** 33 out of 100

7. 22 to 50   **8.** 40 to 50

9. 10 parts per 25 parts   **10.** 11 parts per 25 parts

11. 70 parts per 200 parts   **12.** 120 parts per 200 parts

Exercises 8.5

A

Write a percent to express each of the following comparisons.

1. 27 out of 100   **2.** 68 parts per 200

3. 16 to 10   **4.** 215 to 100

5. 40 parts per 25 parts   **6.** 33 per 20

B

7. 12 to 25   **8.** 62 to 62   **9.** 53 to 500

10. 93 to 80   **11.** 7 to 9   **12.** 18 per 300

13. 515 to 100   **14.** 8 to 4

C

15. 63 to 500   **16.** 73 to 80

17. 7 parts per 11 parts   **18.** 9 per 15

19. 11 to 12   **20.** 11 to 15

21. The ratio of 2 to 25 is the same as _____ to 100 or _____ %.

22. The ratio of 4 to 32 is the same as _____ to 100 or _____ %.

23. The ratio of 8 to 36 is the same as _____ to 100 or _____ %.

24. The ratio of 16 to 9 is the same as _____ to 100 or _____ %.

ANSWERS

1. _____
2. _____
3. _____
4. _____
5. _____
6. _____
7. _____
8. _____
9. _____
10. _____
11. _____
12. _____
13. _____
14. _____
15. _____
16. _____
17. _____
18. _____
19. _____
20. _____
21. _____
22. _____
23. _____
24. _____

25. _____

26. _____

27. _____

28. _____

29. _____

30. _____

31. _____

32. _____

33. _____

34. _____

35. _____

36. _____

37. _____

D

25. The fact that 10% of all women are blond indicates that _____ out of 100 women are blond.

26. In a recent election, out of every 100 eligible voters, 62 cast their ballots. What percent of the eligible voters exercised the right to vote?

27. Of the people who use Shiny toothpaste, 37 out of 100 report that they have fewer cavities. Out of every 100 people who report, what percent do not report fewer cavities?

28. If the telephone tax rate is 5 cents per dollar, what percent is this?

29. For every $100 spent on gasoline, the state receives $6 tax. What percent of the price of gasoline is the state tax?

30. A bank pays $5.25 interest per year for every $100 in savings. What is the annual interest rate?

31. An electronic calculator originally priced at $100 is on sale for $93. What is the percent of discount? (Discount is the difference between the original price and the sale price.)

32. A filbert grower grades the quality of the crop by cracking a sample of 100. He finds that 7 nuts are wormy or blanks. What is the estimated percent of rejects for the entire crop?

E   *Maintain Your Skills*

Solve:

33. $\dfrac{5}{x} + \dfrac{3}{x} = 16$

34. $\dfrac{a}{12} - \dfrac{a}{8} + \dfrac{a}{16} = 2$

35. $\dfrac{1}{3x} + \dfrac{5}{6x} = 2$

36. $\dfrac{3x-4}{5} + 2 = \dfrac{x+2}{10} + x$

37. $\dfrac{8}{x+2} - \dfrac{2}{x-2} = 0$

ANSWERS TO WARM UPS (8.5)   **1.** 66%   **2.** 15%   **3.** 56%   **4.** 63%   **5.** 20%

6. 33%   **7.** 44%   **8.** 80%   **9.** 40%   **10.** 44%

11. 35%   **12.** 60%

8.6 Changing Decimals to Percents

| OBJECTIVE | **91. Write a given decimal or decimal with fraction as a percent.** |

| APPLICATION | To pass a job entrance examination, an applicant must get a minimun of 136 of the 200 problems correct. This is .68 of the problems correct. Express this as a percent. |

| VOCABULARY | No new words. |

HOW AND WHY

In multiplication, where one factor is $\frac{1}{100}$, the indicated multiplication can be read as percent. That is, $75\left(\frac{1}{100}\right) = 75\%$, $.8\left(\frac{1}{100}\right) = .8\%$, and $\frac{3}{4}\left(\frac{1}{100}\right) = \frac{3}{4}\%$.

To write a number as a percent, multiply by $100 \cdot \frac{1}{100}$, a name for one. This is shown in the following table.

Number	Multiply by 1 $100\left(\dfrac{1}{100}\right) = 1$	Multiply by 100	Percent
.45	$.45(100)\left(\dfrac{1}{100}\right)$	$45.\left(\dfrac{1}{100}\right)$	45%
.2	$.2(100)\left(\dfrac{1}{100}\right)$	$20.\left(\dfrac{1}{100}\right)$	20%
5	$5(100)\left(\dfrac{1}{100}\right)$	$500.\left(\dfrac{1}{100}\right)$	500%

In each case the decimal point is moved two places to the right and the percent symbol (%) is inserted.

To change a decimal or decimal-with-fraction to a percent:

1. Move the decimal point two places to the right. (Insert zeros if necessary.)
2. Insert the percent symbol (%).

EXAMPLES

Change from a decimal to a percent.

a. .35 = 35%   **b.** .04 = 4%   **c.** .217 = 21.7%

d. .003 = .3%   **e.** 9 = 900%   **f.** .7 = 70%

g. $.25\frac{1}{3} = 25\frac{1}{3}\%$ (recall that $5\frac{1}{3}$ is in the hundredths place)

APPLICATION SOLUTION

To change .68 to a percent, we move the decimal point two places to the right and insert the percent symbol.

.68 = .68% = 68%

Therefore, the applicant must score at least 68% to pass the test.

WARM UPS

Write each decimal as a percent:

1. .24   **2.** .17   **3.** 1   **4.** 3

5. 1.23   **6.** 2.84   **7.** 0.06   **8.** 0.09

9. .8   **10.** .1   **11.** .007   **12.** .005

Exercises 8.6

A

Write each of the following as a percent.

1. .36 **2.** 5.95 **3.** .08 **4.** 8.33

5. 1.6 **6.** .007

B

7. .214 **8.** .083 **9.** 7 **10.** 27

11. 13.21 **12.** .005 **13.** 1.27 **14.** .745

C

15. .0256 **16.** $.03\frac{1}{3}$ **17.** 5.75 **18.** $.74\frac{1}{6}$

19. .1025 **20.** $.10\frac{1}{4}$ **21.** $.03\frac{1}{2}$ **22.** $.07\frac{1}{3}$

23. $.27\frac{2}{3}$ **24.** $.27\frac{5}{9}$

D

25. If the tax rate on a building lot is given as .03, what is the tax rate expressed as a percent?

26. If the tax rate on a person's income is .22, what is that rate expressed as a percent?

Answers: 1. 2. 3. 4. 5. 6. 7. 8. 9. 10. 11. 12. 13. 14. 15. 16. 17. 18. 19. 20. 21. 22. 23. 24. 25. 26.

27. The completion rate in a certain math class is .85. What is this rate as a percent?

28. The sales tax in a certain state is .055. Express this as a percent.

29. If .375 of the contestants in a race withdraw, what is this expressed as a percent?

30. To pass an entrance examination for a private college, a student must get a minimum of 164 of the 200 questions correct. This is .82 of the problems correct. Express this as a percent.

E   *Maintain Your Skills*

31. $I = prt$. Find r if $I = 1500$, $p = 10,000$, and $t = 2$.

32. $V = hwd$. Find h if $V = 120$, $d = 6$, and $w = 8$.

33. $\dfrac{c}{d} = \pi$. Find d if $\pi \approx 3.14$ and $c = 78.5$.

34. $I = \dfrac{E - e}{R}$. Solve for E.

35. $F = \dfrac{kmM}{r^2}$. Solve for m.

8.7 Changing Percents to Decimals

OBJECTIVE

92. **Write a decimal or decimal with fraction that is equivalent to a given percent.**

APPLICATION

When ordering bricks, a contractor orders 3% more than is needed, to allow for breakage. What decimal is entered into the computer to calculate the extra number of bricks to be added to the order?

VOCABULARY

No new words. Recall:

1. The decimal point for a whole number follows the ones digit.
2. In a decimal with fraction, the fraction part is attached to the digit preceding it and has the same place value as the digit.
3. Any fraction can be thought of as a decimal with fraction, since

$$\frac{1}{2} = 0\frac{1}{2}, \frac{5}{8} = 0\frac{5}{8}, \text{ and so on.}$$

HOW AND WHY

The definition of percent gives us a clue that we can use to change from a percent to a decimal. The percent symbol indicates multiplication by $\frac{1}{100}$; hence 17% can be written as $17 \cdot \frac{1}{100}$. Since multiplying by the reciprocal of a number is the same as dividing by that number, $17 \cdot \frac{1}{100}$ can be thought of as $17 \div 100$. To divide a number by 100, move the decimal point two places to the left; hence 17% = .17.

To change a percent to a decimal or a decimal-with-fraction:

1. **Move the decimal point two places to the left. (Insert zeros if necessary.)**
2. **Drop the percent symbol (%).**

EXAMPLES

a. $14.5\% = .145$, since $14.5\% = 14.5\left(\frac{1}{100}\right) = 14.5 \div 100 = .145$

b. $29\frac{1}{3}\% = 29\frac{1}{3}\left(\frac{1}{100}\right) = 29\frac{1}{3} \div 100 = .29\frac{1}{3}$

c. $35\% = .35$

d. $295\% = 2.95$

e. $83\frac{1}{2}\% = .83\frac{1}{2}$ or $83\frac{1}{2}\% = 83.5\% = .835$

f. $.5\% = .005$

g. $\dfrac{3}{4}\% = .75\% = .0075$ or $.00\dfrac{3}{4}$

h. $\dfrac{1}{3}\% = 0\dfrac{1}{3}\% = .00\dfrac{1}{3}$

i. $\dfrac{1}{3}\% = .3\dfrac{1}{3}\% = .003\dfrac{1}{3}$ $\left.\rule{0pt}{60pt}\right\}$ Same problem

j. $\dfrac{1}{3}\% = .33\dfrac{1}{3}\% = .0033\dfrac{1}{3}$

APPLICATION SOLUTION

To change 3% to a decimal, we move the decimal point two places to the left and drop the percent sign.

$3\% = .03 = .03$

Therefore, .03 is the number that must be entered into the computer.

WARM UPS

Write each of the following as a decimal:

1. 68%	**2.** 127%	**3.** 3.7%
4. 24%	**5.** .35%	**6.** 14%
7. 27%	**8.** 92.3%	**9.** 2.79%
10. .8%	**11.** 3%	**12.** 3.6%

Exercises 8.7

A

Write each of the following as a decimal.

1. 14%   **2.** 27%   **3.** 92.3%   **4.** 2.79%

5. 3%   **6.** .81%

B

7. 16%   **8.** 5.9%   **9.** 82%   **10.** 36%

11. 2.15%   **12.** 312%   **13.** 563%   **14.** 100%

C

15. 53.7%   **16.** .04%   **17.** $\frac{1}{8}$%   **18.** $\frac{4}{5}$%

19. 314.7%   **20.** 261.3%   **21.** $35\frac{1}{6}$%

22. $4.8\frac{1}{3}$%   **23.** $\frac{1}{4}$%   **24.** $475\frac{1}{2}$%

D

25. A contractor orders 5% more sand than is needed, to allow for shrinkage. What decimal part is this?

26. When bidding for a job, an estimator adds 10% to cover unexpected expenses. What decimal part is this?

27. A merchant advertises merchandise for sale at cost plus 20%. What decimal must be multiplied by the price to find the amount to add to the cost? (This will find the selling price.)

28. If the merchant in Exercise 27 advertised cost plus 33%, what decimal would be used to multiply by the cost?

ANSWERS

1. _____
2. _____
3. _____
4. _____
5. _____
6. _____
7. _____
8. _____
9. _____
10. _____
11. _____
12. _____
13. _____
14. _____
15. _____
16. _____
17. _____
18. _____
19. _____
20. _____
21. _____
22. _____
23. _____
24. _____
25. _____
26. _____
27. _____
28. _____

E   *Maintain Your Skills*

Tell whether the proportions are true or false:

29. $\dfrac{16}{36} = \dfrac{24}{45}$

30. $\dfrac{98}{72} = \dfrac{245}{180}$

31. $\dfrac{7.5}{25} = \dfrac{.75}{2.5}$

32. $\dfrac{12 \text{ inches}}{3 \text{ feet}} = \dfrac{8 \text{ inches}}{2 \text{ feet}}$

33. $\dfrac{\$12}{4 \text{ days}} = \dfrac{\$18}{1 \text{ week}}$

ANSWERS TO WARM UPS (8.7)   **1.** .68   **2.** 1.27   **3.** .037   **4.** .24   **5.** .0035

6. .14   **7.** .27   **8.** .923   **9.** .0279   **10.** .008

11. .03   **12.** .036

8.8 Fractions to Percent

OBJECTIVE

93. Change a given fraction or mixed number to percent.

APPLICATION

In a survey it was determined that $\frac{5}{8}$ of the residents of Hillsboro, Oregon, own their homes. What percent does this represent?

VOCABULARY

No new words.

HOW AND WHY

Since we already know how to change fractions to decimals and decimals to percent, we can combine the two ideas to change fractions to percent.

To change a fraction or mixed number to a percent:

1. **Change to a decimal with two decimal places. The decimal is rounded off or carried out as directed.**
2. **Change the decimal to percent.**

EXAMPLES

a. $\frac{5}{8} = .62\frac{1}{2} = 62\frac{1}{2}\%$ or

$= .625 = 62.5\%$

b. $\frac{5}{6} = .83\frac{1}{3} = 83\frac{1}{3}\%$

c. $2\frac{1}{2} = 2.5 = 250\%$

d. $\frac{3}{7} = .42\frac{6}{7} = 42\frac{6}{7}\%$ or, to the nearest tenth of a percent,

$\approx .429 = 42.9\%$

e. $\frac{1}{40} = .025 = 2.5\%$

f. $\frac{7}{320} = .02\frac{3}{16} = 2\frac{3}{16}\%$ or

$= .0218\frac{3}{4} = 2.18\frac{3}{4}\%$ or, to the nearest tenth of a percent,

$\approx .022 = 2.2\%$

g. Calculator example: $\dfrac{3}{80}$

ENTER	DISPLAY
3	3.
$\div$	3.
80	80.
$\%$	3.75

$$\dfrac{3}{80} = 3.75\%$$

h. Calculator example: $2\dfrac{9}{40}$

$$2\dfrac{9}{40} = \dfrac{89}{40}$$

ENTER	DISPLAY
89	89.
$\div$	89.
40	40.
$\%$	222.5

$$2\dfrac{9}{40} = 222.5\%$$

APPLICATION SOLUTION

Since $\dfrac{5}{8}$ of the residents of Hillsboro, Oregon own their homes, we need to change $\dfrac{5}{8}$ to a percent. We first change the fraction to a decimal and then change the decimal to a percent:

$$\dfrac{5}{8} = .625 = 62.5\%$$

Therefore, 62.5% of the residents of Hillsboro, Oregon own their homes.

WARM UPS

Change the following fractions to decimals and then change the decimals to percent:

1. $\dfrac{3}{4}$ 2. $\dfrac{3}{5}$ 3. $\dfrac{7}{50}$

4. $\dfrac{7}{25}$ 5. $\dfrac{3}{25}$ 6. $\dfrac{3}{20}$

Change each fraction to a two-place decimal with fraction. Then change the decimal with fraction to percent:

7. $\dfrac{1}{8}$ 8. $\dfrac{1}{3}$ 9. $\dfrac{1}{6}$ 10. $\dfrac{1}{9}$

Change each fraction to a decimal rounded to the nearest thousandth. Then change the decimal to a percent:

11. $\dfrac{1}{8}$ 12. $\dfrac{1}{3}$

Exercises 8.8

A

Change the following fractions to decimals and then change the decimals to percent.

1. $\dfrac{7}{100}$ **2.** $\dfrac{50}{100}$ **3.** $\dfrac{11}{50}$

4. $\dfrac{3}{10}$ **5.** $\dfrac{17}{20}$ **6.** $\dfrac{4}{25}$

B

Change each fraction to a two-digit decimal with fraction. Then change the decimal with fraction to percent.

7. $\dfrac{2}{3}$ **8.** $\dfrac{1}{6}$ **9.** $1\dfrac{7}{8}$ **10.** $3\dfrac{7}{9}$

11. $\dfrac{21}{400}$ **12.** $\dfrac{14}{15}$ **13.** $\dfrac{17}{200}$ **14.** $2\dfrac{5}{7}$

C

Change each fraction to a decimal rounded to the nearest thousandth. Then change the decimal to a percent. (The result is the fraction expressed as a percent to the nearest tenth.)

15. $\dfrac{1}{3}$ **16.** $\dfrac{7}{500}$ **17.** $\dfrac{1}{80}$ **18.** $2\dfrac{5}{9}$

19. $\dfrac{8}{21}$ **20.** $\dfrac{5}{6}$ **21.** $3\dfrac{11}{16}$ **22.** $\dfrac{17}{10,000}$

23. $1\dfrac{1}{80}$ **24.** $\dfrac{15}{28}$

1. _____
2. _____
3. _____
4. _____
5. _____
6. _____
7. _____
8. _____
9. _____
10. _____
11. _____
12. _____
13. _____
14. _____
15. _____
16. _____
17. _____
18. _____
19. _____
20. _____
21. _____
22. _____
23. _____
24. _____

25. _____

26. _____

27. _____

28. _____

29. _____

30. _____

31. _____

32. _____

33. _____

34. _____

35. _____

D

25. In a survey it was determined that $\frac{3}{5}$ of the automobile tires sold are radials. What percent does this represent?

26. Steve applied for a loan to buy a motorcycle. The annual interest charged was 23¢ for each $2 of the loan $\left(\frac{23}{200}\right)$. What was the annual rate (percent) of interest?

27. Four-sevenths of the eligible voters turned out for the recent Hillsboro city elections. What percent of the voters turned out, to the nearest tenth of a percent?

28. In a supermarket, 2 eggs out of 9 dozen are lost because of cracks. What percent of eggs must be discarded, to the nearest tenth of a percent?

29. The owner of a small business pays 17 cents for insurance for every $100 of insurance coverage. What is the percent rate of the insurance?

30. The tachometer on an automobile can register 6000 revolutions per minute (rpm). If the reading on the tachometer when the motor is idling is 900, what percent of the maximum reading is the motor turning?

E   *Maintain Your Skills*

Solve the following proportions:

31. $\dfrac{4}{x} = \dfrac{16}{5}$

32. $\dfrac{6}{8} = \dfrac{x}{18}$

33. $\dfrac{x}{30} = \dfrac{16}{20}$

34. $\dfrac{3.5}{x} = \dfrac{28}{8}$

35. $\dfrac{15}{27} = \dfrac{4.5}{x}$

ANSWERS TO WARM UPS (8.8)   1. 75%   2. 60%   3. 14%   4. 28%   5. 12%

6. 15%   7. $12\frac{1}{2}$%   8. $33\frac{1}{3}$%   9. $16\frac{2}{3}$%   10. $11\frac{1}{9}$%

11. 12.5%   12. 33.3%

8.9 Percents to Fractions

OBJECTIVE

94. Change percents to fractions or mixed numbers.

APPLICATION

It is estimated that the spraying for the gypsy moth in Salemtown was 92% successful. What fraction (ratio comparison) of the moths were eliminated?

VOCABULARY

No new words.

HOW AND WHY

$6.5\% = 6.5 \times \dfrac{1}{100}$. This gives a very efficient method for changing a percent to a fraction. Change 6.5 to a fraction and do the multiplication. See Example a.

To change a percent to a fraction or a mixed number:

1. **Drop the percent symbol (%).**

2. **Multiply by $\dfrac{1}{100}$. Before multiplying, rewrite the other factor (if necessary) as a fraction.**

EXAMPLES

a. $6.5\% = 6.5 \times \dfrac{1}{100} = 6\dfrac{5}{10} \times \dfrac{1}{100} = \dfrac{65}{10} \times \dfrac{1}{100} = \dfrac{65}{1000} = \dfrac{13}{200}$

b. $312\% = 312 \times \dfrac{1}{100} = \dfrac{312}{100} = 3\dfrac{12}{100} = 3\dfrac{3}{25}$

c. $12\dfrac{2}{3}\% = 12\dfrac{2}{3} \times \dfrac{1}{100} = \dfrac{38}{3} \times \dfrac{1}{100} = \dfrac{38}{300} = \dfrac{19}{150}$

d. $.05\dfrac{1}{3}\% = .05\dfrac{1}{3} \times \dfrac{1}{100} = 5\dfrac{1}{3} \times \dfrac{1}{100} \times \dfrac{1}{100} = \dfrac{16}{3} \times \dfrac{1}{100} \times \dfrac{1}{100}$

$= \dfrac{4 \times 4 \times 1 \times 1}{3 \times 4 \times 25 \times 4 \times 25} = \dfrac{1}{1875}$

APPLICATION SOLUTION

To write 92% as a fraction we do the following:

$$92\% = 92 \cdot \frac{1}{100}$$

$$= \frac{92}{100}$$

$$= \frac{23}{25}$$

Therefore, $\frac{23}{25}$ of the moths were destroyed.

WARM UPS

Change each of the following percents to fractions or mixed numbers:

1. 25%   **2.** 40%   **3.** 75%   **4.** 80%

5. 90%   **6.** 100%   **7.** 3%   **8.** 9%

9. 15%   **10.** 36%   **11.** 120%   **12.** 150%

Exercises 8.9

ANSWERS

1. _____
2. _____
3. _____
4. _____
5. _____
6. _____
7. _____
8. _____
9. _____
10. _____
11. _____
12. _____
13. _____
14. _____
15. _____
16. _____
17. _____
18. _____
19. _____
20. _____
21. _____
22. _____
23. _____
24. _____
25. _____
26. _____
27. _____

A

Change each of the following to a fraction or mixed number.

1. 5%   **2.** 35%   **3.** 125%

4. 20%   **5.** 70%   **6.** 400%

B

7. 56%   **8.** 82%   **9.** 112%   **10.** 6.5%

11. $\frac{3}{4}$%   **12.** $16\frac{1}{4}$%   **13.** $9\frac{1}{5}$%   **14.** 32.6%

C

15. $16\frac{1}{2}$%   **16.** 285%   **17.** $\frac{1}{3}$%   **18.** .0016%

19. $11\frac{1}{9}$%   **20.** .25%   **21.** 2.5%   **22.** 20.5%

23. $16\frac{5}{7}$%   **24.** $19\frac{1}{5}$%

D

25. A census determined that $37\frac{1}{2}$% of the residents of a certain city were of age 40 or over and 50% were of age 25 or under. What fraction of the residents are between the ages of 25 and 40?

26. George and Ethel paid 16% of their annual income in taxes last year. What fractional part of their income went to taxes?

27. The enrollment at City Community College this year is 118% of last year's enrollment. What fractional increase in enrollment took place this year?

ANSWERS

28. _____

29. _____

30. _____

31. _____

32. _____

33. _____

34. _____

35. _____

28. The enrollment at Carver Community College this year is 123% of last year's enrollment. What fractional increase in enrollment took place this year?

29. A survey determined that 38% of the residents of a certain city were in favor of a sales tax and that 53% were opposed. What fraction of the residents did the survey not account for?

30. Brooks and Foster form a partnership. If Brooks' investment is $56\frac{1}{4}$% of the total, what fraction of the total is Foster's share?

E *Maintain Your Skills*

31. If 8 lb of oranges cost $1, at the same rate what is the cost of 20 lb of oranges?

32. A school assigns teachers at the rate of 2 teachers for every 47 students. How many teachers will be needed if the school expects an enrollment of 893 students?

33. A counter on a tape recorder registers 420 after running for 12 minutes. What would the counter register after running for an hour?

34. If 36 lb of fertilizer will cover 1500 ft² of lawn, how much is needed to cover 3500 ft²?

35. On a map $\frac{7}{8}$ inch represents 100 miles. What distance would be represented by a length of $3\frac{1}{2}$ inches?

ANSWERS TO WARM UPS (8.9)   1. $\dfrac{1}{4}$ 2. $\dfrac{2}{5}$ 3. $\dfrac{3}{4}$ 4. $\dfrac{4}{5}$ 5. $\dfrac{9}{10}$ 6. 1

7. $\dfrac{3}{100}$ 8. $\dfrac{9}{100}$ 9. $\dfrac{3}{20}$ 10. $\dfrac{9}{25}$ 11. $1\dfrac{1}{5}$ 12. $1\dfrac{1}{2}$

8.10 Fractions, Percents, and Decimals

OBJECTIVES

95. Write a percent as a decimal and as a fraction.
96. Write a fraction as a percent and as a decimal.
97. Write a decimal as a percent and as a fraction.

VOCABULARY

No new vocabulary.

HOW AND WHY

Decimals, fractions, and percents can each be expressed in terms of the others. This can be shown as follows:

$$50\% = 50 \cdot \frac{1}{100} = \frac{50}{100} = \frac{1}{2} \quad \text{and} \quad 50\% = 50(.01) = .50$$

$$\frac{3}{4} = 3 \div 4 = .75 \quad \text{and} \quad \frac{3}{4} = .75 = 75\%$$

$$.65 = 65(.01) = 65\% \quad \text{and} \quad .65 = \frac{65}{100} = \frac{13}{20}$$

EXAMPLES

Fill in the empty spaces with the related fraction, decimal, or percent.

Percent	Decimal	Fraction
30%		
		$\dfrac{7}{8}$
	.62	

We find 30% = .30, and 30% = $\dfrac{3}{10}$;

$\dfrac{7}{8} = .87\dfrac{1}{2}$ (also .875), and $\dfrac{7}{8} = 87\dfrac{1}{2}\%$ (also 87.5%);

.62 = 62%, and .62 = $\dfrac{31}{50}$.

The filled-in chart is:

Percent	Decimal	Fraction
30%	.30 or .3	$\dfrac{3}{10}$
$.87\dfrac{1}{2}\%$ or 87.5%	$.87\dfrac{1}{2}$ or .875	$\dfrac{7}{8}$
62%	.62	$\dfrac{31}{50}$

WARM UPS

Fill in the empty spaces with the related percent, decimal, or fraction:

Percent	Decimal	Fraction
		$\dfrac{2}{3}$
27%		
	.72	
		$\dfrac{73}{100}$
160%		
	1.3	

Exercises 8.10

Fill in the empty spaces with the related percent, decimal, or fraction:

	Fraction	Decimal	Percent
*	$\frac{1}{10}$		
*			30%
*		.75	
*	$\frac{9}{10}$		
			145%
*	$\frac{3}{8}$		
		.001	
		1	
	$2\frac{1}{4}$		
*		.8	
			$5\frac{1}{2}\%$
*		.875	
			$\frac{1}{2}\%$
*		.6	
*			$62\frac{1}{2}\%$
			$\frac{3}{4}\%$
	$\frac{2}{3}$		
*			25%
*		.20	
*			40%
*			$33\frac{1}{3}\%$
*		.125	
*	$\frac{7}{10}$		

* If you are going to be working with problems that will involve percent in your job or in your personal finances (loans, savings, insurance, etc.), it is advisable to know (memorize) these special relationships.

1. _____

2. _____

3. _____

4. _____

5. _____

1. The illumination, I (in foot candles) of a light source varies inversely as the square of the distance (d) from the source. Find the constant of variation if $I = 80$ and $d = 40$.

2. Using the formula from Exercise 1, what is the distance if $I = 45$ and the constant of variation (K) is 112500?

3. If the speed is constant, the distance traveled varies directly as time. An auto traveled 300 miles in 8 hours. At that rate, how long would it take to travel 125 miles?

4. Using the information from Exercise 3, how long would it take the auto to travel 575 miles?

5. Juanita's salary varies directly as her total sales. If she receives $78 for sales of $975, how much will she receive for sales of $1350?

ANSWERS TO WARM UPS (8.10)

Percent	Decimal	Fraction
$66\frac{2}{3}\%$	$.66\frac{2}{3}$	$\frac{2}{3}$
27%	.27	$\frac{27}{100}$
72%	.72	$\frac{18}{25}$
73%	.73	$\frac{73}{100}$
160%	1.6	$1\frac{3}{5}$
130%	1.3	$1\frac{3}{10}$

8.11 Solving Percent Problems

OBJECTIVE

98. Solve problems that are written in the form "*A* is *R*% of *B*" or "*R*% of *B* is *A*."

APPLICATION

Applications of this objective come from a variety of fields. These applications are so important that an entire section (8.12) is given to them.

VOCABULARY

To *solve* a percent problem means to do one of the following:

1. Find *A*, given *R* and *B*.
2. Find *B*, given *R* and *A*.
3. Find *R*, given *A* and *B*.

In the statement *R*% of *B* is *A* ($R\% \times B = A$),
  R is the *rate* of percent and is followed by the "%" symbol.
  B is the *base* unit and follows the word "of", also called the *whole*.
  A is the *amount* that is compared to *B*, also called the *part*.

HOW AND WHY

We show two methods for solving percent problems. It is recommended that you look over both methods, pick your favorite, and use it for doing percent problems.

METHOD ONE: FORMULA	METHOD TWO: PROPORTION
The word "of" in the statement above and in other places in mathematics indicates multiplication. The word "is" describes the relationship "is equal to" or "=". Thus we may write $R\%$ of B is A in the form:	Since $R\%$ is a comparison of A to B and we have seen that this comparison can be thought of as a ratio, the following proportion can be formed:
$R\% \cdot B = A$.	$$\frac{R}{100} = \frac{A}{B}$$
When any two of R, B, and A are known, we can find the other value. 25% of 60 is what number? $25\% \cdot 60 = A$ Since $25\% = .25$,	Now if anyone of A, B, or R is missing, it can be found by solving the proportion. 25% of 60 is what number? Here $R = 25$, $B = 60$, and $A = ?$ So
$.25(60) = A$ $15.00 = A$	$$\frac{25}{100} = \frac{A}{60}$$
So 25% of 60 is 15.	$(25)(60) = 100(A)$ $1500 = 100(A)$
	$15 = A$
	So 25% of 60 is 15.

EXAMPLES

a. 45% of what number is 9?

METHOD ONE	METHOD TWO
R is 45 (45 tells what percent), B is unknown (B follows the word "of"), and A is 9. Now we write the problem	$R\%$ of B is A $R = 45$, $B = ?$, $A = 9$
$45\% (B) = 9$	$$\frac{45}{100} = \frac{9}{B}$$
$.45(B) = 9$	$45(B) = (9)(100)$
$B = 9 \div .45$	$45(B) = 900$
$B = 20$	$B = 900 \div 45$
So 45% of 20 is 9.	$B = 20$
	So 45% of 20 is 9.

b. 5 is what percent of 20?

METHOD ONE	METHOD TWO
R is unknown.	A is $R\%$ of B
$5 = (R\%)(20)$ $\% = .01$	$A = 5$, $R = ?$, $B = 20$
$5 = (R)(.01)(20)$	$$\frac{R}{100} = \frac{5}{20}$$
$5 = (R)(.20)$	$20(R) = (5)(100)$
$5 \div .20 = R$	$20(R) = 500$
$25 = R$	$R = 500 \div 20$
So 5 is 25% of 20.	$R = 25$
	So 5 is 25% of 20.

c. 135% of ___?___ is 54.

METHOD ONE

$(135\%)(B) = 54$

$(1.35)(B) = 54$

$B = 54 \div 1.35$

$B = 40$

So 135% of 40 is 54.

METHOD TWO

$R = 135, B = ?, A = 54$

$$\frac{135}{100} = \frac{54}{B}$$

$(135)(B) = 5400$

$B = 5400 \div 135$

$B = 40$

So 135% of 40 is 54.

d. 78% of 36 is ___?___ (to the nearest tenth).

METHOD ONE

$(78\%)(36) = A$

$(.78)(36) = A$

$28.08 = A$

So 78% of 36 is 28.1 to the nearest tenth.

METHOD TWO

$R = 78, B = 36, A = ?$

$$\frac{78}{100} = \frac{A}{36}$$

$(78)(36) = (100)(A)$

$2808 = (100)(A)$

$2808 \div 100 = A$

$28.08 = A$

So 78% of 36 is 28.1 to the nearest tenth.

e. 50 is ___?___ % of 180 (to the nearest tenth of one percent).

METHOD ONE

$50 = (R\%)(180)$

$50 = (R)(.01)(180)$

$50 = (R)(1.80)$

$50 \div 1.80 = R$

$27.777 \approx R$

So 50 is 27.8% of 180 to the nearest tenth of one percent.

METHOD TWO

$A = 50, R = ?, B = 180$

$$\frac{R}{100} = \frac{50}{180}$$

$(180)(R) = (100)(50)$

$(180)(R) = 5000$

$R = 5000 \div 180$

$R \approx 27.777$

So 50 is 27.8% of 180 to the nearest tenth of one percent.

f. $27\frac{2}{3}\%$ of 60 is ___?___ .

METHOD ONE	METHOD TWO

METHOD ONE

$$\left(27\frac{2}{3}\%\right)(60) = A$$

$$\left(27\frac{2}{3}\right)\left(\frac{1}{100}\right)(60) = A$$

$$\left(\frac{83}{3}\right)\left(\frac{1}{100}\right)(60) = A$$

$$\frac{83}{5} = A$$

$$16\frac{3}{5} = A$$

METHOD TWO

$$\frac{27\frac{2}{3}}{100} = \frac{A}{60}$$

$$\left(27\frac{2}{3}\right)(60) = (100)(A)$$

$$1660 = (100)(A)$$

$$1660 \div 100 = A$$

$$16.6 = A$$

Since $16\frac{3}{5} = 16.6$, $27\frac{2}{3}\%$ of 60 is $16\frac{3}{5}$ or 16.6.

WARM UPS

Solve:

1. 20% of 10 is ___?___ .

2. 8% of 50 is ___?___ .

3. 9 is 50% of ___?___ .

4. 9 is 75% of ___?___ .

5. ___?___ is 3% of 200.

6. ___?___ is 6% of 300.

7. 3 is ___?___ % of 5.

8. 3 is ___?___ % of 4.

9. 9 is 60% of ___?___ .

10. 9 is 90% of ___?___ .

11. 6% of 30 is ___?___ .

12. 600% of 30 is ___?___ .

Exercises 8.11

ANSWERS

1. _____

2. _____

3. _____

4. _____

5. _____

6. _____

7. _____

8. _____

9. _____

10. _____

11. _____

12. _____

13. _____

14. _____

15. _____

16. _____

17. _____

A

Solve:

1. __?__ % of 60 is 30.   **2.** __?__ % of 65 is 13.

3. 70% of __?__ is 28.   **4.** 80% of __?__ is 28.

5. 80% of 45 is __?__ .   **6.** __?__ is 80% of 25.

B

7. 67% of 40 is __?__ .   **8.** 38% of 70 is __?__ .

9. 48 is __?__ % of 36.   **10.** 56 is __?__ % of 48.

11. 140% of __?__ is 35.   **12.** 175% of __?__ is 52.5.

13. 9.3% of 60 is __?__ .   **14.** 36.75% of 28 is __?__ .

C

15. 3.2% of .7 is __?__ .   **16.** __?__ is $\frac{1}{2}$% of .5.

17. 28% of __?__ is 36. (To the nearest tenth.)

18. 219 is 12% of ___?___ .

19. 47 is ___?___ % of 30. (To the nearest tenth of one percent.)

20. 82.5 is ___?___ % of 37.6. (To the nearest whole number percent.)

21. ___?___ is 31.6% of 57.8. (To the nearest tenth.)

22. $5\frac{1}{3}$% of $6\frac{1}{2}$ is ___?___ . (As a fraction.)

23. ___?___ is 53% of $15\frac{2}{3}$. (As a two-place decimal with fraction.)

24. 1.25% of 1250 is ___?___ .

E   *Maintain Your Skills*

Write the following as a percent:

25. .85   **26.** .023   **27.** .0003   **28.** $.00\frac{1}{2}$

29. 18.5

ANSWERS TO WARM UPS (8.11)   **1.** 2   **2.** 4   **3.** 18   **4.** 12   **5.** 6   **6.** 18
7. 60%   **8.** 75%   **9.** 15   **10.** 10   **11.** 1.8   **12.** 180

8.12 Word Problems (Percent)

OBJECTIVE

99. Solve word problems involving percent.

APPLICATION

In a recent survey it was found that, of the 285 people surveyed, 114 people preferred eating whole wheat bread. What percent of the people surveyed preferred eating whole wheat bread?

VOCABULARY

No new words.

HOW AND WHY

When a word problem is translated into the simpler word form,

"What percent of what is what?"

the answers can be found by the procedures of the last section. Either Method One or Method Two can be used. The two methods are used alternately in the examples.

EXAMPLES

a. Rod bought a car with a 12% annual loan. If the interest payment was $54 per year, how much was his loan? (Method One.)

The $54 interest is 12% of the loan, so we can translate the problem into the simpler word form

12% of what is 54?

Using the formula $R\% \cdot B = A$, we write the equation

$$12\% \,(B) = 54$$

$$.12 \, B = 54$$

$$B = 450$$

so Rod's loan was for $450.

b. The population of Century County is now 130% of its population ten years ago. The population ten years ago was 117,000. What is the present population? (Method Two.)

The simpler word form is

130% of 117,000 is what number?

and, using a proportion,

$$\frac{130}{100} = \frac{A}{117,000}$$

$$15,210,000 = (100)(A)$$

$$152,100 = A$$

The population is now 152,100.

c. A bike has depreciated to 65% of its original cost. If the value of the bike is presently $1469, what did it cost originally? (Method One.)

65% of what is 1469?   *(simpler word form)*

$.65B = 1469$ *(equation)*

$B = 2260$

The original cost was $2260.

d. The Goliath Bakery has 500 loaves of day-old bread they want to sell. If the price was originally 52¢ a loaf and they sell it for 36¢ a loaf, what percent discount, based on the original price, should they advertise? (Method Two.)

The discount is 16 cents so:

What percent of 52 is 16?   *(simpler word form)*

$$\frac{R}{100} = \frac{16}{52}$$ *(equation)*

$(52)(R) = 1600$

$R \approx 30.8$ (about 30.8%)

The bakery will probably advertise "over 30% off."

e. In a poll taken among a group of students, 2 said they walked to their school, 7 said they rode the bus, 10 drove in car pools, and 3 drove their own cars. What percent of the group rode the bus? (Method One.)

There are 22 students in the group, so the problem can be translated to

What percent of 22 is 7?   *(simpler word form)*

$R(.01)(22) = 7$ *(equation)*

$.22R = 7$

$R = 31\dfrac{9}{11} \approx 32$

$31\dfrac{9}{11}\%$ or approximately 32% of the students take the bus.

APPLICATION SOLUTION

The simpler word form of the problem is: 114 is what percent of 285?

$$\frac{114}{285} = \frac{R}{100}$$ *Method Two*

$285(R) = 114(100)$

$285(R) = 11400$

$R = 11400 \div 285$

$R = 40$

Therefore, 40% of the people surveyed preferred eating whole wheat bread.

Exercises 8.12

A

1. In a survey it was found that of the 324 people surveyed, 81 people preferred eating chocolate ice cream. What percent of the people surveyed preferred eating chocolate ice cream?

2. If there is a 4% sales tax on a television set costing $119.95, how much is the tax?

3. Dan bought a used motorcycle for $955. He made a down payment of 18%. How much cash did he pay as a down payment?

4. Frank is selling magazine subscriptions. He keeps 16% of the cost of each subscription he sells as his salary. How many dollars' worth of subscriptions must he sell to earn $125?

5. Carol pays $270, or 30% of her monthly income, for rent. What is her monthly income?

6. Last year Joan had 14% of her salary withheld for taxes. If the total amount withheld was $2193.10, what is Joan's yearly salary?

B

7. John got 26 problems correct on a 30-problem test. What was his percent score (to the nearest whole number percent)?

8. The manager of a fruit stand lost $16\frac{2}{3}$% of his bananas to spoilage and sold the rest. He discarded 3 boxes of bananas in two weeks. How many boxes did he have in stock?

9. To pass a test to qualify for a job interview, Carol must score at least 70%. If there are 40 questions on the test, how many must Carol get correct to score 70%?

10. Vera's house is valued at $65,000 and rents for $4875 per year. What percent of the value of the house is the annual income from rent? (To the nearest tenth of a percent.)

11. Eddie and his family went to a restaurant for dinner. The dinner check was $19.75. He left the waiter a tip of $3. What percent of the check was the tip (to the nearest whole number percent)?

12. Floyd earns a monthly salary of $805. He spends $130 a month at the supermarket. What percent of his salary is spent at the supermarket (to the nearest whole number percent)?

C

13. Adams High School's basketball team finished the season with a record of 15 wins and 9 losses. What percent of the games played were won?

14. A furniture store has a sale on sofas. Every sofa is marked 20% off. What is the sale price of a sofa that was priced at $549.95?

15. A state charges a gasoline tax of 9% of the cost. The federal tax is 4¢ per gallon. If gasoline costs $1.30 per gallon before taxes, what is the total price per gallon including both taxes?

2. _____
3. _____
4. _____
5. _____
6. _____
7. _____
8. _____
9. _____
10. _____
11. _____
12. _____
13. _____
14. _____
15. _____

16. The local police department set up a vehicle inspection station at the high school parking lot. Of the 128 cars inspected on a particular day, 6.25% were found to have faulty brakes. How many vehicles did not have faulty brakes?

17. The cost of dairy items increased an average of 2.3% during the month of February. If the price of eggs on February 1 was 94¢ per dozen, what was the price on March 1? (Assume that the rate of increase for eggs was close to the average increase.)

18. Gene bought 125 shares of IWW Inc. for $21.75 a share and sold them for $29.33 a share. If the brokerage fee on both transactions together totaled $39.50, what percent profit did he make on his investment? (To the nearest whole percent.)

19. The town of Verboort has a population of 15,560, which is 45% male. Of the men, 32% are of age 40 or older. How many men are there in Verboort who are younger than 40?

D

20. The population of Port City increased 15% since the last census. If the former population was 124,000, what is the present population?

21. For customers who use a bank's credit card, there is a $1\frac{3}{4}$% finance charge on monthly accounts that have a balance of $400 or less. Merle's finance charge for August was $2.80. What was the amount of her account for that month?

22. In preparing a mixture of concrete, Susan uses 300 pounds of gravel, 100 pounds of cement, and 200 pounds of sand. What percent of the mixture is gravel?

23. St. Joseph's Hospital has eight three-bed wards, twenty four-bed wards, twelve two-bed wards, and ten private rooms. What percent of the capacity of St. Joseph's Hospital is in private rooms? (To the nearest tenth of a percent.)

24. A discount house advertises that they sell all appliances at cost plus 10%. If Kathy buys a TV set for $605, what is the profit for the store?

25. It is claimed that in 15,000 hours, or six years, a gasoline engine will be down 32 days for routine maintenance, whereas a diesel engine will be down only 13 days. What percent less time is the diesel engine down compared with the gasoline engine? (To the nearest whole percent.)

26. During a six-year period, the cost of maintaining a diesel engine averages $2650 and the cost of maintaining a gasoline engine averages $4600. What is the percent of saving of the diesel compared with the gasoline engine? (To the nearest whole percent.)

27. Carol's baby weighed $7\frac{1}{2}$ lb when he was born. On his first birthday he weighed $23\frac{3}{4}$ lb. What was the percent of increase during the year? (To the nearest whole percent.)

E   *Maintain Your Skills*

Write the percent comparisons:

28. 38 out of 50

29. 48 per 200

30. 115 per 100

31. 56 to 80

32. 120 to 60

444

1. _____

2. _____

3. _____

4. _____

5. _____

6. _____

7. _____

8. _____

9. _____

10. _____

11. _____

12. _____

13. _____

14. _____

15. _____

16. _____

17. _____

18. _____

19. _____

20. _____

Chapter 8
POST-TEST

1. **(Obj. 85)** Is the following proportion true or false?

$$\frac{3 \text{ feet}}{2 \text{ yards}} = \frac{12 \text{ inches}}{2 \text{ feet}}$$

2. **(Obj. 91)** Write $.13\frac{1}{8}$ as a percent.

3. **(Obj. 92)** Write 119% as a decimal.

4. **(Obj. 94)** Change 145% to a mixed number.

5. **(Obj. 98)** 63% of what number is 75.6?

6. **(Obj. 98)** What percent of 64 is 72?

7. **(Obj. 93)** Change $\frac{7}{8}$ to a percent.

8. **(Obj. 92)** Write 1.8% as a decimal.

9. **(Obj. 98)** What number is 11.6% of 210?

10. **(Obj. 86)** Solve the following proportion. $\frac{8}{27} = \frac{y}{81}$

11. **(Obj. 85)** Is the following proportion true or false?

$$\frac{15}{36} = \frac{20}{52}$$

12. **(Obj. 93)** Change $1\frac{3}{11}$ to a percent. (Give answer to the nearest tenth of a percent.)

13. **(Obj. 84)** Write a ratio to compare 8 hours to 2 days and reduce. (Compare in hours)

14. **(Obj. 94)** Change $26\frac{1}{3}$% to a fraction.

15. **(Obj. 86)** Solve the following proportion. $\frac{\frac{1}{3}}{2} = \frac{2\frac{3}{4}}{x}$

16. **(Obj. 91)** Write 1.6 as a percent.

17. **(Obj. 87)** On the assembly line at the local electronics firm, John can make 45 welds in 3 hours. At this rate, how many welds will he make in an 8-hour day?

18. **(Obj. 87)** If Kathy is paid $24.85 for 7 hours of work, how much would she expect to earn for 16 hours work?

19. **(Obj. 88)** The weight of a piece of metal varies directly as the volume. If a piece of the metal containing $2\frac{1}{3}$ cubic feet weighs 420 lb, find the weight of a piece containing 4 cubic feet.

20. **(Obj. 89)** The time it takes to travel a certain distance varies inversely with the speed. If it takes 4 hours to cover the distance at 45 mph, how long will it take at 60 mph?

ANSWERS

21. _____

22. _____

23. _____

24. _____

25. _____

21. (Obj. 99) If 44 out of every 50 people are right-handed, what percent of the population is right-handed?

22. (Obj. 99) The local nurseryman sells an average of 112 dozen flowering plants a day during the spring planting season. If 38% of what he sells are petunias, how many dozen petunias does he sell in a day? (Nearest dozen.)

23. (Obj. 99) Carol's baby weighed $6\frac{3}{4}$ lb when he was born. On his first birthday he weighed 22 lb. What was the percent of increase during the year to the nearest tenth of a percent?

24. (Obj. 99) A discount store advertises 30% off the suggested retail price. What does the store charge for an item that has a suggested price of $38.50?

25. (Obj. 87) In preparing a mixture of cement, Charlie uses 300 lb of gravel and 100 lb of cement. How many pounds of cement should he use for a mixture containing 720 lb of gravel?

9

**Roots and
Exponents**

ANSWERS

1. _____

2. _____

3. _____

4. _____

5. _____

6. _____

7. _____

8. _____

9. _____

10. _____

11. _____

12. _____

13. _____

14. _____

15. _____

16. _____

17. _____

18. _____

19. _____

20. _____

21. _____

22. _____

Chapter 9
PRE-TEST

The problems in the following pre-test are a sample of the material in the chapter. You may already know how to work some of these. If so, this will allow you to spend less time on those parts. As a result, you will have more time to give to the sections that you find more difficult or that are new to you. The answers are in the back of the text.

1. **(Obj. 100)** Compute the root: $\sqrt{169}$

2. **(Obj. 102)** Compute the root: $\sqrt[3]{\dfrac{1}{8}}$

3. **(Obj. 103)** In a right triangle, the lengths of the legs are $a = 2.5$ and $b = 6$. Find the length of side c (the hypotenuse).

4. **(Obj. 104)** *Multiply:* $\sqrt{11} \cdot \sqrt{13}$

5. **(Obj. 104)** *Multiply:* $\sqrt{7}(3 + \sqrt{5})$

6. **(Obj. 104)** Multiply and simplify: $(\sqrt{2} + \sqrt{7})(\sqrt{2} - \sqrt{7})$

7. **(Obj. 105)** *Simplify:* $\sqrt{252}$

8. **(Obj. 105)** *Simplify:* $\sqrt{117xy^2}$

9. **(Obj. 105)** *Simplify:* $\sqrt[3]{54a^2b^3}$

10. **(Obj. 106)** Rationalize the denominator: $\dfrac{\sqrt{7}}{\sqrt{10}}$

11. **(Obj. 106)** Rationalize the denominator and simplify: $\dfrac{\sqrt{50}}{\sqrt{7}}$

12. **(Obj. 106)** Rationalize the denominator and simplify: $\dfrac{3}{\sqrt{11} - \sqrt{7}}$

13. **(Obj. 107)** *Combine:* $\sqrt{75} + \sqrt{27}$

14. **(Obj. 107)** *Combine:* $\sqrt{50} - 2\sqrt{8} + \sqrt{98}$

15. **(Obj. 107)** *Combine:* $\sqrt[3]{81} - \sqrt[3]{3}$

16. **(Obj. 108)** Write $\sqrt{6a^5}$ in exponential form.

17. **(Obj. 109)** Write $5^{1/4}xy^{3/4}$ in radical form.

18. **(Obj. 110)** Write without fractions, using negative exponents:

$$\frac{6}{x^2} - \frac{1}{y^3}$$

19. **(Obj. 111)** Write $\dfrac{3^{-1}a^3}{2b^{-1}}$ with positive exponents and simplify.

20. **(Obj. 112)** Write .000039 in scientific notation.

21. **(Obj. 113)** Write 2.71×10^8 in place value notation.

22. **(Obj. 103)** A rope fastened to the top of a flagpole is 30 feet long. If the rope is pulled tight, away from the bottom of the pole, the end will reach the ground 16 feet from the bottom of the flagpole. How tall is the pole? (Write the answer as a simplified radical or correct to the nearest tenth of a foot.)

23. (Obj. 112) The planet Venus is approximately 67,000,000 miles from the Sun. Write the distance in scientific notation.

24. (Obj. 109) The formula for the radius of a circle in terms of its area is

$$r \approx \sqrt{.32A}$$

Write the right side of the equation in exponential form (with fraction exponent).

25. (Obj. 111) The light (illumination) from a light source is measured in foot-candles. The formula for the illumination I is

$$I = kd^{-2}$$

Write the formula without negative exponents.

9.1 Square Roots and Cube Roots

100. Find the square roots of perfect squares.
101. Find the approximate positive square root of a positive real number.
102. Find the cube roots of perfect cubes.

APPLICATION

What is the length of the side of a square whose area is 64 cm²? (The formula for the length of the side of a square is $S = \sqrt{A}$.)

VOCABULARY

The words *perfect square* and *perfect cube* used in this section refer to whole numbers and fractions that are the squares or cubes of other whole numbers or fractions. For instance, $16 = 4^2$ and $\frac{1}{9} = \left(\frac{1}{3}\right)^2$, so 16 and $\frac{1}{9}$ are perfect *squares*. However, 7, 8, $\frac{1}{5}$, and $\frac{4}{11}$ are not perfect squares.

The square roots of a given number are the positive number and the negative number that when squared (used as a factor two times) yield that given number as a product. The symbol "$\sqrt{a}$" means the positive square root of a, and the symbol "$-\sqrt{a}$" means the negative square root of a. So, $\sqrt{25} = 5$ and $-\sqrt{25} = -5$.

The square roots of non-negative numbers can be located on the number line. All numbers on the number line are called *real numbers*. ($\sqrt{9}$, $\sqrt{14}$, $\sqrt{3.14}$, and $\sqrt{\frac{7}{8}}$ are real numbers.) Square roots of negative numbers cannot be located on the number line and are not real numbers. Note that neither 5^2 nor $(-5)^2$ is equal to -25. We say that $\sqrt{-25}$ is not a real number. Such numbers are studied in a later course.

The cube root of a given number is the number (positive or negative) that when cubed (used as a factor three times) yields that given number as a product. The symbol "$\sqrt[3]{a}$" means the cube root of a. Note that a can be either positive or negative. So, $\sqrt[3]{27} = 3$, $-\sqrt[3]{27} = -3$, and $\sqrt[3]{-27} = -3$.

The symbols $\sqrt{}$ and $\sqrt[3]{}$ are called *radicals*.

HOW AND WHY

Some whole numbers or fractions are squares of other whole numbers or fractions. For example, the whole number 4 is the square of 2 or -2 since $2 \cdot 2 = 4$ and $(-2)(-2) = 4$. Therefore, we can say

$\sqrt{4} = 2$ *The positive square root of 4 is 2*

$-\sqrt{4} = -2$ *The negative square root of 4 is -2*

$\sqrt{-4} = ?$ *The square root of a negative number is not a real number*

$\sqrt{9} = 3$ *The positive square root of 9 is 3*

$-\sqrt{9} = -3$ *The negative square root of 9 is -3*

$\sqrt{\frac{4}{9}} = \frac{2}{3}$ *The positive square root of $\frac{4}{9}$ is $\frac{2}{3}$*

$\sqrt[3]{8} = 2$ *The cube root of 8 is 2*

$\sqrt[3]{-8} = -2$ *The cube root of -8 is -2*

Some numbers (such as 2, 3, and 5) are not perfect squares, but their square roots can be approximated. Also, some squares of whole numbers are large and it is difficult to determine the square root by observation. These roots and approximations can be found by using a calculator. It will be assumed from this point on that you have a calculator.

The following roots were found by calculator.

$$\sqrt{2} \approx 1.41 \qquad \sqrt{35} \approx 5.92 \qquad \sqrt{571536} = 756$$

EXAMPLES

a. $\sqrt{16} = 4$ *since* $4^2 = 16$

b. $\sqrt{25} = 5$ *since* $5^2 = 25$

c. $-\sqrt{64} = -8$ *since* $(-8)^2 = 64$

d. $\sqrt{\dfrac{9}{25}} = \dfrac{3}{5}$ *since* $\left(\dfrac{3}{5}\right)^2 = \dfrac{9}{25}$

e. $\sqrt[3]{27} = 3$ *since* $3^3 = 27$

f. $\sqrt[3]{-64} = -4$ *since* $(-4)^3 = -64$

g. $\sqrt[3]{\dfrac{8}{27}} = \dfrac{2}{3}$ *since* $\left(\dfrac{2}{3}\right)^3 = \dfrac{8}{27}$

h. **Calculator example**

$\sqrt{115}$

ENTER	DISPLAY
115	115.
$\sqrt{}$	10.723805

So $\sqrt{115} \approx 10.72$ rounded to two decimal places.

i. $\sqrt{88} \approx 9.381$ *(approximation by calculator, rounded to three decimal places)*

j. $\sqrt{412} \approx 20.30$ *(approximation by calculator, rounded to two decimal places)*

k. $\sqrt{3025} = 55$ *(root found by calculator)*

APPLICATION SOLUTION

Formula:	$S = \sqrt{A}, A = 64$
Substitute:	$S = \sqrt{64}$
Simplify:	$S = 8$ since $8(8) = 64$
Answer:	The length of one side of the square is 8 centimetres.

WARM UPS

Find the roots:

1. $\sqrt{1}$ 2. $-\sqrt{1}$ 3. $\sqrt{49}$

4. $-\sqrt{49}$ 5. $\sqrt{81}$ 6. $-\sqrt{9}$

7. $\sqrt[3]{-1}$ 8. $\sqrt[3]{8}$ 9. $\sqrt[3]{27}$

10. $\sqrt{144}$ 11. $\sqrt{.25}$ 12. $\sqrt{.09}$

Exercises 9.1

A

Find the roots:

1. $\sqrt{100}$ **2.** $-\sqrt{36}$ **3.** $-\sqrt{25}$

4. $\sqrt{\dfrac{64}{81}}$ **5.** $\sqrt[3]{-8}$ **6.** $\sqrt[3]{64}$

B

Find the roots:

7. $\sqrt{121}$ **8.** $\sqrt{169}$ **9.** $\sqrt{\dfrac{121}{169}}$

10. $\sqrt{\dfrac{9}{196}}$ **11.** $\sqrt[3]{\dfrac{8}{27}}$ **12.** $-\sqrt[3]{\dfrac{64}{125}}$

13. $-\sqrt[3]{-27}$ **14.** $\sqrt[3]{\dfrac{1}{216}}$

C

Find the approximate root rounded to two decimal places (use a calculator or square root table).

15. $\sqrt{8}$ **16.** $\sqrt{12}$ **17.** $\sqrt{24}$

18. $\sqrt{124}$ **19.** $\sqrt{200}$ **20.** $\sqrt{1215}$

21. $\sqrt{892}$ **22.** $\sqrt{596}$ **23.** $\sqrt{10.6}$

24. $\sqrt{3.48}$

1. _____
2. _____
3. _____
4. _____
5. _____
6. _____
7. _____
8. _____
9. _____
10. _____
11. _____
12. _____
13. _____
14. _____
15. _____
16. _____
17. _____
18. _____
19. _____
20. _____
21. _____
22. _____
23. _____
24. _____

ANSWERS

25. _____

26. _____

27. _____

28. _____

29. _____

30. _____

31. _____

32. _____

33. _____

34. _____

35. _____

36. _____

D

25. What is the length of the side of a square whose area is 225 ft²? (Use the formula in the application.)

26. What is the length of the side of a square whose area is 1.44 yd²?

27. Evaluate the formula $t = 2\pi\sqrt{\dfrac{\ell}{g}}$ for t if $\ell = 36$, $g = 49$, and $\pi \approx \dfrac{22}{7}$.

28. A formula for computing the area of a triangle is

$$A = \sqrt{s(s - a)(s - b)(s - c)},$$

where the sides of the triangle are a, b, and c, and $s = \dfrac{1}{2}(a + b + c)$. What is the area of a triangle whose sides are 6′, 8′, and 10′?

29. The lens setting of a camera varies inversely as the square root of the shutter speed. If the speed is $\dfrac{1}{100}$ sec for a setting of 20, what should the lens setting be for a speed of $\dfrac{1}{25}$ sec?

30. The number of vibrations of a pendulum varies inversely as the square root of the length. If the number of vibrations is 28 when the length is 9, what will be the number of vibrations when the length is 16?

31. To meet the city code the attic of a new house needs a vent with a minimum area of 706 square inches. What is the radius of a circular vent that will meet the code requirement? The radius of a circle in terms of the area is given by $r = \sqrt{.32A}$. Compute the radius to one decimal place.

E *Maintain Your Skills*

Change each of the following fractions to a decimal rounded to the nearest thousandth; then change the decimal to a percent.

32. $\dfrac{81}{80}$ **33.** $\dfrac{13}{15}$ **34.** $\dfrac{19}{24}$ **35.** $\dfrac{38}{45}$

36. $\dfrac{13}{44}$

9.2 The Pythagorean Theorem

OBJECTIVE	**103.** Solve right triangles using the Pythagorean theorem.

APPLICATION	What is the length of a rafter that has a rise of 6′ and a run of 12′?

VOCABULARY	The *Pythagorean theorem* (theorem of Pythagoras) describes a relationship between the sides of a right triangle.

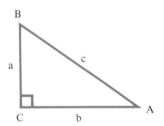   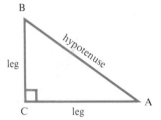

In the above right triangle the legs are a and b and the hypotenuse is c. It is always true for a right triangle that $a^2 + b^2 = c^2$. That is, the square of the hypotenuse is equal to the sum of the squares of the legs.

It is also true that if the square of one side is equal to the sum of the squares of the other two sides, the triangle is a right triangle.

Using the laws for solving equations, we can write

HOW AND WHY

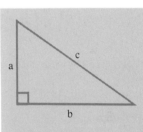

$$a^2 + b^2 = c^2$$

or

$$b^2 = c^2 - a^2$$

or

$$a^2 = c^2 - b^2$$

If any two sides of a right triangle are known, the third can be found by substituting the known values into one of the proceding formulas.

Consider a right triangle with one leg 6″ and the hypotenuse 10″. To determine leg b we use the formula $b^2 = c^2 - a^2$ and we have

$b^2 = 10^2 - 6^2$

$b^2 = 100 - 36$

$b^2 = 64$

$b = \sqrt{64}$

$b = 8$

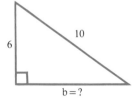

EXAMPLES

a. Find the length of the leg of a right triangle whose second leg is 12 cm and hypotenuse is 15 cm.

$$a^2 = c^2 - b^2$$

$$a^2 = 15^2 - 12^2$$

$$a^2 = 225 - 144$$

$$a^2 = 81$$

$$a = 9$$

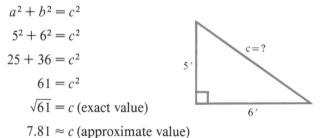

b. Given a right triangle whose legs are 5′ and 6′, respectively, what is the length of the hypotenuse?

$$a^2 + b^2 = c^2$$

$$5^2 + 6^2 = c^2$$

$$25 + 36 = c^2$$

$$61 = c^2$$

$$\sqrt{61} = c \text{ (exact value)}$$

$$7.81 \approx c \text{ (approximate value)}$$

c. Is the triangle whose sides are 12′, 18′, and 24′ a right triangle?

Does $12^2 + 18^2 = 24^2$?

$$12^2 + 18^2 \overset{?}{=} 24^2$$

$$144 + 324 \overset{?}{=} 576$$

$$468 \neq 576$$

Therefore the triangle is not a right triangle.

d. Calculator example
Find the length of the leg of a right triangle that has a second leg of 14 inches and a hypotenuse of 15 inches. Round answer to the nearest hundredth:

$$b^2 = c^2 - a^2$$

$$b^2 = 15^2 - 14^2$$

$$b = \sqrt{15^2 - 14^2}$$

Using a calculator with algebraic logic:

ENTER	DISPLAY	or	ENTER	DISPLAY
15	15.		15	15.
×	15.		x^2	225.
15	15.		−	225.
−	225.		14	14.
14	14.		x^2	196.
×	14.		=	29.
14	14.		√	5.3851648
=	29.			
√	5.3851648			

The leg is approximately 5.39 inches long (to the nearest hundredth).

APPLICATION SOLUTION

Formula:	$a^2 + b^2 = c^2$
Substitute:	$6^2 + 12^2 = c^2$
Simplify:	$36 + 144 = c^2$
	$180 = c^2$
Solve:	$\sqrt{180} = c$ (exact value)
	$13.42 \approx c$ (approximate value)
Answer:	The length of the rafter is approximately 13.4 feet.

Exercises 9.2

ANSWERS

1. _____
2. _____
3. _____
4. _____
5. _____
6. _____
7. _____
8. _____
9. _____
10. _____
11. _____
12. _____
13. _____
14. _____
15. _____
16. _____
17. _____
18. _____
19. _____
20. _____
21. _____
22. _____
23. _____
24. _____
25. _____
26. _____

A

Find the missing side of each of the following right triangles.

1. $a = 3, b = 4, c = ?$

2. $a = 6, b = ?, c = 10$

3. $a = ?, b = 12, c = 13$

4. $a = 8, b = 15, c = ?$

5. $a = 9, b = 12, c = ?$

6. $a = 12, b = ?, c = 13$

B

Find the missing side of each of the following right triangles.

7. $a = 4.5, b = 6, c = ?$

8. $a = 3, b = ?, c = 3.25$

9. $a = 8, b = ?, c = 17$

10. $a = 2, b = ?, c = 4.25$

11. $a = ?, b = 27, c = 45$

12. $a = ?, b = 16, c = 34$

13. $a = 18, b = 24, c = ?$

14. $a = 16, b = ?, c = 34$

C

Find the missing side of each of the following right triangles. Round answers to the nearest hundredth.

15. $a = 1, b = 2, c = ?$

16. $a = 1, b = ?, c = 2$

17. $a = 2, b = 3, c = ?$

18. $a = 2, b = ?, c = 3$

19. $a = ?, b = 10, c = 15$

20. $a = 4, b = ?, c = 11$

21. $a = 4, b = 11, c = ?$

22. $a = 5, b = ?, c = 20$

23. $a = 17, b = ?, c = 20$

24. $a = 12, b = 22, c = ?$

D

25. What is the length of a rafter with a rise of 5′ and a run of 13′ (to the nearest hundredth)?

26. What is the length of a rafter with a rise of 7′ and a run of 14′ (to the nearest hundredth)?

459

27. What is the length of a rafter with a rise of 8′ and a run of 12′ (to the nearest hundredth)?

28. What is the rise of a rafter that is 20 feet long and has a run of 16 feet?

29. What is the length of cable needed to replace a brace that connects a 50′ power pole to a ground level anchor that is 35′ from the base of the pole? (To the nearest tenth of a foot).

30. A plane is flying south at a speed of 200 miles per hour. A wind is blowing from the west at a rate of 50 miles per hour. To the nearest tenth of a mile, how many miles does the plane actually fly in a southeasterly direction during one hour? (The following figure illustrates the problem.)

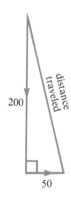

31. A baseball "diamond" is actually a square that is 90 feet on each side (between the bases). To the nearest tenth of a foot, what is the distance the catcher must throw when attempting to put out a runner who is stealing second base? (The following figure illustrates the problem.)

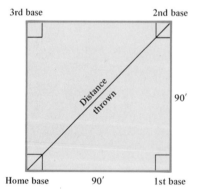

E *Maintain Your Skills*

Change each of the following to a fraction or a mixed number.

32. 38%

33. 148%

34. $27\frac{1}{2}\%$

35. 3.5%

36. $10\frac{1}{9}\%$

9.3 Multiplying Radical Expressions (Square Roots)

OBJECTIVE	**104. Multiply two or more radical expressions (square roots).**

APPLICATION	The height of an equilateral triangle is given by the formula $h = \dfrac{s}{2}\sqrt{3}$ where h represents the height of the triangle and s represents one of the three equal sides of the triangle. Find the height of a triangular medallion, in simplified radical form and to the nearest hundredth, if each side is $\sqrt{7}$ cm long.

VOCABULARY	No new vocabulary.

HOW AND WHY	Two square roots can be multiplied using the procedures for multiplying whole numbers and fractions, together with the following rule. **The product of two square root expressions is the square root of the product of the two numbers. In other words, for all positive numbers a and b,** $\sqrt{a} \cdot \sqrt{b} = \sqrt{a \cdot b}$

EXAMPLES	**a.** $\sqrt{5} \cdot \sqrt{6} = \sqrt{5 \cdot 6} = \sqrt{30}$
	
	b. $\sqrt{2} \cdot \sqrt{8} = \sqrt{2 \cdot 8} = \sqrt{16} = 4$
	
	c. $\sqrt{3} \cdot \sqrt{5} \cdot \sqrt{2} = \sqrt{30}$
	
	d. $\sqrt{3}(2 - \sqrt{2}) = \sqrt{3} \cdot 2 - \sqrt{3}\sqrt{2}$ *Monomial times binomial*
	                 $= 2\sqrt{3} - \sqrt{6}$ *Simplify*
	
	e. $(3 + \sqrt{2})(3 - \sqrt{2}) = 3^2 - (\sqrt{2})^2$ *Use FOIL shortcut*
	                            $= 9 - 2$ *Simplify*
	                            $= 7$

APPLICATION SOLUTION	To find the height of the medallion, we solve the following problem: *Formula:* $h = \dfrac{s}{2}\sqrt{3}$ where $s = \sqrt{7}$ *Substitute:* $h = \dfrac{\sqrt{7}}{2} \cdot \sqrt{3}$ *Solve:* $h = \dfrac{\sqrt{21}}{2}$ $h \approx \dfrac{4.58}{2} = 2.29$ cm *Nearest hundredth* *Answer:* The medallion is $\dfrac{\sqrt{21}}{2}$ cm or approximately 2.29 cm in height.

WARM UPS

Multiply and simplify. Assume all variables represent positive numbers.

1. $\sqrt{2}\,\sqrt{3}$

2. $\sqrt{3}\,\sqrt{5}$

3. $\sqrt{5}\,\sqrt{2}$

4. $\sqrt{8}\,\sqrt{2}$

5. $\sqrt{7}\,\sqrt{7}$

6. $\sqrt{3}\,\sqrt{12}$

7. $\sqrt{a}\,\sqrt{b}$

8. $\sqrt{x}\,\sqrt{yz}$

9. $\sqrt{5}\,\sqrt{5}$

10. $\sqrt{2}\,\sqrt{18}$

11. $\sqrt{2}(\sqrt{3}+\sqrt{5})$

12. $\sqrt{5}\,\sqrt{x}$

Exercises 9.3

A

Multiply and simplify.

1. $\sqrt{3}\,\sqrt{7}$ **2.** $\sqrt{10}\,\sqrt{7}$ **3.** $\sqrt{5}\,\sqrt{20}$

4. $\sqrt{2}\,\sqrt{32}$ **5.** $\sqrt{6}\,\sqrt{7}$ **6.** $\sqrt{2}\,\sqrt{3}\,\sqrt{11}$

B

Multiply and simplify. Assume all variables represent positive numbers.

7. $\sqrt{3}\,\sqrt{3}$ **8.** $\sqrt{11}\,\sqrt{11}$

9. $\sqrt{2}\,\sqrt{72}$ **10.** $\sqrt{2}\,\sqrt{50}$

11. $\sqrt{a}\,\sqrt{b}\,\sqrt{c}$ **12.** $\sqrt{a}\,\sqrt{ab}\,\sqrt{b}$

13. $\sqrt{7}\,\sqrt{14}\,\sqrt{2}$ **14.** $\sqrt{45}\,\sqrt{5}$

C

Multiply and simplify.

15. $\sqrt{3}(\sqrt{5}+2)$ **16.** $\sqrt{2}(3+\sqrt{3})$

17. $\sqrt{6}(3+\sqrt{5})$ **18.** $2(3+\sqrt{7})$

19. $\sqrt{6}(\sqrt{6}-\sqrt{5})$ **20.** $2(\sqrt{3}-\sqrt{7})$

21. $\sqrt{15}(\sqrt{2}+\sqrt{7})$ **22.** $(4-\sqrt{2})(4+\sqrt{2})$

23. $(\sqrt{5}-3)(\sqrt{5}+3)$ **24.** $(\sqrt{5}-\sqrt{3})(\sqrt{5}+\sqrt{3})$

1.
2.
3.
4.
5.
6.
7.
8.
9.
10.
11.
12.
13.
14.
15.
16.
17.
18.
19.
20.
21.
22.
23.
24.

25. _____

26. _____

27. _____

28. _____

29. _____

30. _____

31. _____

D

25. Each side of an equilateral triangle is $\sqrt{8}$ feet. Use the formula in the application to find the height of the triangle to the nearest tenth.

26. Each side of an equilateral triangle is $\sqrt{15}$ metres. Use the formula in the application to find the height of the triangle to the nearest tenth.

27. Find the area of a rectangle with base $\sqrt{12}$ inches and height $\sqrt{15}$ inches to the nearest tenth.

28. Find the area of a triangle with base $\sqrt{10}$ cm and height $\sqrt{15}$ cm, to the nearest tenth.

E   *Maintain Your Skills*

Fill in the empty spaces with a related fraction, decimal, or percent.

	Fraction	Decimal	Percent
29.	$\frac{7}{8}$		
30.		.354	
31.			136%

9.4 Simplifying Radical Expressions

OBJECTIVE	**105. Simplify a radical expression.**

APPLICATION	A city lot has an area of 5600 square feet. If the lot is a square, what is the length of one side? The formula is $s = \sqrt{A}$ where s represents the length of one side and A represents the area.

VOCABULARY	The number or expression under a radical sign is called the *radicand*. In $\sqrt{5}$, the radicand is 5. In $\sqrt[3]{19w}$, the radicand is $19w$.

HOW AND WHY

Irrational numbers such as $\sqrt{15}$ and $\sqrt[3]{14}$ are usually written in two ways.

1. We can find an approximation, which is the usual procedure in a practical problem.

$$\sqrt{15} \approx 3.87 \qquad \sqrt[3]{14} \approx 2.41$$

2. We can leave the number in radical form, which is the only way to represent the exact value.

A number that is left in radical form is usually simplified. To simplify a radical such as $\sqrt{1400}$, we use the following basic principle of radicals:

The square root of a product is the product of the square roots, if neither radicand is negative.

$\sqrt{ab} = \sqrt{a}\,\sqrt{b}$, if a and b are not negative

$\sqrt{1400} = \sqrt{100} \cdot \sqrt{41} = 10\sqrt{41}$

Similarly, the cube root of a product is the product of the cube roots.

$\sqrt[3]{54} = \sqrt[3]{27} \cdot \sqrt[3]{2} = 3\sqrt[3]{2}$

A radical indicating a positive square root is in simplified form when the radicand is a whole number that has no perfect-square factors except 1. A radical indicating a cube root is in simplified form when the radicand is a whole number that has no perfect-cube factors except 1. Radicals with fractional radicands are covered in Section 9.5.

To simplify a radical with a whole-number radicand,

1. **If possible, factor the radicand so that at least one factor is a perfect square (or cube.) If the radicand has no perfect-square (or perfect-cube) factors other than 1, it is already in simplest form.**
2. **Find the square root (or cube root) of the perfect square (or perfect cube).**
3. **Repeat steps 1 and 2 until the radicand is simplified.**

When a radical indicating a positive square root contains a variable factor, we must be careful that it represents a positive number (or zero) so we can use the previously stated principle for radicals.

$\sqrt{x^3} = \sqrt{x^2} \cdot \sqrt{x} = x\sqrt{x}$, if x represents a non-negative number.

Throughout the remainder of this text, we will assume that variables in the radicand are restricted so that the radicand does not represent a negative number.

EXAMPLES

a. $\sqrt{8} = \sqrt{4} \cdot \sqrt{2} = 2\sqrt{2}$

b. $\sqrt{x^6} = \sqrt{(x^3)^2} = x^3$

c. $\sqrt{50y^2} = \sqrt{25} \cdot \sqrt{y^2} \cdot \sqrt{2} = 5y\sqrt{2}$

d. $\sqrt{150} = \sqrt{25} \cdot \sqrt{6} = 5\sqrt{6}$

e. $\sqrt{288y^7} = \sqrt{144} \cdot \sqrt{y^6} \cdot \sqrt{2y} = 12y^3\sqrt{2y}$

f. $\sqrt{14} = \sqrt{2} \cdot \sqrt{7} = \sqrt{14}$ *Already in simplest form*

g. $\sqrt[3]{108} = \sqrt[3]{27} \cdot \sqrt[3]{3} = 3\sqrt[3]{3}$

h. $\sqrt[3]{40} = \sqrt[3]{8} \cdot \sqrt[3]{5} = 2\sqrt[3]{5}$

APPLICATION SOLUTION

To find the length of one side of the square, we substitute $A = 5600$ in the formula and solve:

Formula:	$s = \sqrt{A}$		
Substitute:	$s = \sqrt{5600}$	or	$s = \sqrt{5600}$
Solve:	$s = \sqrt{100 \cdot 56}$		$s = \sqrt{400 \cdot 14}$
	$s = \sqrt{100} \cdot \sqrt{56}$		$s = \sqrt{400} \cdot \sqrt{14}$
	$s = 10\sqrt{56}$		$s = 20\sqrt{14}$
	$s = 10\sqrt{4 \cdot 14}$		
	$s = 10\sqrt{4}\sqrt{14}$		
	$s = 10 \cdot 2\sqrt{14}$		
	$s = 20\sqrt{14}$		

Answer: The square is $20\sqrt{14}$ ft on each side.
Using a calculator or a square root table, we get 74.83 ft (rounded to the nearest hundredth of a foot).

WARM UPS

Simplify; assume all variables represent positive numbers:

1. $\sqrt{8}$ 2. $\sqrt{24}$ 3. $\sqrt{28}$ 4. $\sqrt{54}$

5. $\sqrt{48}$ 6. $\sqrt{12}$ 7. $\sqrt{20}$ 8. $\sqrt{18}$

9. $\sqrt{45}$ 10. $\sqrt{x^2y}$ 11. $\sqrt{pq^2}$ 12. $\sqrt[3]{32}$

Exercises 9.4

A

Simplify. Assume all variables represent positive numbers.

1. $\sqrt{32}$ **2.** $\sqrt{52}$ **3.** $\sqrt{40}$ **4.** $\sqrt{50}$

5. $\sqrt{26}$ **6.** $\sqrt{30}$

B

7. $\sqrt{16c^2}$ **8.** $\sqrt{100x^2}$ **9.** $\sqrt{80}$ **10.** $\sqrt{125}$

11. $\sqrt{84}$ **12.** $\sqrt{128}$ **13.** $\sqrt[3]{56}$ **14.** $\sqrt[3]{40}$

C

15. $\sqrt{176}$ **16.** $\sqrt{224}$ **17.** $\sqrt{245}$

18. $\sqrt{1800}$ **19.** $\sqrt{1125}$ **20.** $\sqrt{396}$

21. $\sqrt{150w^2}$ **22.** $\sqrt{80b^2}$ **23.** $\sqrt[3]{81}$

24. $\sqrt[3]{128a^3}$

D

25. A city lot has an area of 4500 square feet. If the lot is a square, represent the length of one side as a radical in simplest form. Find the approximate value to the nearest tenth.

26. A sandlot ball diamond (square shape) has an area of 7500 square feet. Express the distance between home plate and first base as a radical in simplest form. (Use the formula in the application.)

1. _____
2. _____
3. _____
4. _____
5. _____
6. _____
7. _____
8. _____
9. _____
10. _____
11. _____
12. _____
13. _____
14. _____
15. _____
16. _____
17. _____
18. _____
19. _____
20. _____
21. _____
22. _____
23. _____
24. _____
25. _____
26. _____

ANSWERS

27. _____

28. _____

29. _____

30. _____

31. _____

32. _____

33. _____

27. An electric space heater requires 600 watts of power. If the resistance is 18 ohms, what is the voltage requirement expressed as a simplified radical? The formula is

$$V = \sqrt{(\text{watts})(\text{ohms})} = \sqrt{ER}$$

where V represents the number of volts, E represents the number of watts of power, and R represents the number of ohms of resistance.

28. Express the voltage needed for a circuit with 30 watts of power and a resistance of 40 ohms in simplest radical form.

E   *Maintain Your Skills*

29. 78% of 36 is ___?___ .

30. ___?___ % of 75 is 12.

31. 48% of ___?___ is 21.

32. .8% of 150 is ___?___ .

33. 5.3% of ___?___ is 20.14.

ANSWERS TO WARM UPS (9.4)   **1.** $2\sqrt{2}$ **2.** $2\sqrt{6}$ **3.** $2\sqrt{7}$ **4.** $3\sqrt{6}$ **5.** $4\sqrt{3}$

6. $2\sqrt{3}$ **7.** $2\sqrt{5}$ **8.** $3\sqrt{2}$ **9.** $3\sqrt{5}$ **10.** $x\sqrt{y}$

11. $q\sqrt{p}$ **12.** $2\sqrt[3]{4}$

9.5 Dividing Radical Expressions

OBJECTIVE	**106. Divide two radical expressions.**

APPLICATION

The clock at the Amtrak Station has a pendulum with a length of 30 feet. The time it takes the pendulum to make one cycle is given by

$$T = 2\pi \frac{\sqrt{L}}{\sqrt{32}}$$

where L is the length of the pendulum in feet and T is the time in seconds. Find T as a product of π and an expression without a radical in the denominator.

VOCABULARY

Rewriting a radical expression containing a radical in the denominator so that the denominator is a whole number is called *rationalizing the denominator*. For example, $\sqrt{\dfrac{3}{2}} = \dfrac{\sqrt{6}}{2}$. Conjugate binomials are the sum and difference of the same two terms. For example, $17x + \sqrt{5}$ and $17x - \sqrt{5}$ are conjugate binomials.

HOW AND WHY

The quotient of two radicals (square roots) is equal to the square root of the quotient of the two numbers.

$$\frac{\sqrt{a}}{\sqrt{b}} = \sqrt{\frac{a}{b}}$$

This quotient can be approximated by dividing the numbers and approximating the square root on a calculator.

$$\frac{\sqrt{3}}{\sqrt{2}} = \sqrt{\frac{3}{2}} = \sqrt{1.5} \approx 1.225$$

The common way of stating the exact value of the quotient is to write an indicated division with a whole number for the denominator. This is called rationalizing the denominator.

To rationalize the denominator,

1. **Write the quotient as a fraction, if necessary.**
2. **Multiply by 1 ($a \cdot 1 = a$). The name for 1 that is chosen will eliminate the radical in the denominator.**
3. **Simplify.**

For instance, when we divide $\sqrt{6}$ by $\sqrt{5}$ we get $\dfrac{\sqrt{6}}{\sqrt{5}}$. To rationalize the denominator, we multiply by $\dfrac{\sqrt{5}}{\sqrt{5}}$ (name for 1) so the denominator will be $\sqrt{25}$ or 5.

$$\sqrt{6} \div \sqrt{5} = \frac{\sqrt{6}}{\sqrt{5}} = \frac{\sqrt{6}}{\sqrt{5}} \cdot \frac{\sqrt{5}}{\sqrt{5}} = \frac{\sqrt{30}}{5}$$

If the denominator of the fraction is a binomial whose terms contain one or more radicals, the denominator can be rationalized by multiplying by the *conjugate* of the denominator.

$$\frac{5}{\sqrt{7} + \sqrt{5}}$$

Can be rationalized by multiplying both numerator and denominator by $\sqrt{7} - \sqrt{5}$, which is the conjugate of $\sqrt{7} + \sqrt{5}$.

$$= \frac{5}{\sqrt{7} + \sqrt{5}} \cdot \frac{\sqrt{7} - \sqrt{5}}{\sqrt{7} - \sqrt{5}}$$

$$= \frac{5(\sqrt{7} - \sqrt{5})}{(\sqrt{7} + \sqrt{5})(\sqrt{7} - \sqrt{5})} = \frac{5\sqrt{7} - 5\sqrt{5}}{7 - 5}$$

$$= \frac{5\sqrt{7} - 5\sqrt{5}}{2}$$

$$\frac{\sqrt{x}}{\sqrt{x} - \sqrt{y}}$$

Can be rationalized by multiplying both numerator and denominator by $\sqrt{x} + \sqrt{y}$, which is the conjugate of $\sqrt{x} - \sqrt{y}$.

$$= \frac{\sqrt{x}}{\sqrt{x} - \sqrt{y}} \cdot \frac{\sqrt{x} + \sqrt{y}}{\sqrt{x} + \sqrt{y}}$$

$$= \frac{x - \sqrt{xy}}{x - y}$$

EXAMPLES

a. $\sqrt{\dfrac{5}{3}} = \dfrac{\sqrt{5}}{\sqrt{3}} = \dfrac{\sqrt{5}}{\sqrt{3}} \cdot \dfrac{\sqrt{3}}{\sqrt{3}} = \dfrac{\sqrt{15}}{3}$

b. $\sqrt{15} \div \sqrt{18} = \dfrac{\sqrt{15}}{\sqrt{18}} = \dfrac{\sqrt{15}}{\sqrt{18}} \cdot \dfrac{\sqrt{2}}{\sqrt{2}} = \dfrac{\sqrt{30}}{\sqrt{36}} = \dfrac{\sqrt{30}}{6}$

c. $4\sqrt{3} \div 2\sqrt{8} = \dfrac{4\sqrt{3}}{2\sqrt{8}} = \dfrac{4\sqrt{3}}{2\sqrt{8}} \cdot \dfrac{\sqrt{2}}{\sqrt{2}} = \dfrac{4\sqrt{6}}{2\sqrt{16}}$

$$= \frac{4\sqrt{6}}{2 \cdot 4} = \frac{4\sqrt{6}}{8} = \frac{\sqrt{6}}{2}$$

d. $\dfrac{3}{4 + \sqrt{3}} = \dfrac{3}{4 + \sqrt{3}} \cdot \dfrac{4 - \sqrt{3}}{4 - \sqrt{3}} = \dfrac{3(4 - \sqrt{3})}{(4 + \sqrt{3})(4 - \sqrt{3})}$

$$= \frac{12 - 3\sqrt{3}}{16 - 3} = \frac{12 - 3\sqrt{3}}{13}$$

APPLICATION SOLUTION

To find the time it takes the pendulum in the clock to make one cycle, we substitute $L = 30$ in the formula and solve:

Formula: $T = 2\pi \dfrac{\sqrt{L}}{\sqrt{32}}$

Substitute: $T = 2\pi \dfrac{\sqrt{30}}{\sqrt{32}}$

Solve: $= 2\pi \dfrac{\sqrt{30}}{\sqrt{32}} \cdot \dfrac{\sqrt{2}}{\sqrt{2}}$

$= 2\pi \dfrac{\sqrt{60}}{\sqrt{64}}$

$= 2\pi \dfrac{\sqrt{4}\,\sqrt{15}}{8}$

$= \dfrac{4\pi\sqrt{15}}{8}$

$= \dfrac{\pi\sqrt{15}}{2}$

Answer: The pendulum takes $\dfrac{\pi\sqrt{15}}{2}$ sec to complete a cycle, or approximately 6.08 sec.

WARM UPS

Divide; rationalize the denominator:

1. $\dfrac{1}{\sqrt{3}}$ 2. $\dfrac{1}{\sqrt{7}}$ 3. $\dfrac{\sqrt{3}}{\sqrt{5}}$ 4. $\dfrac{\sqrt{5}}{\sqrt{7}}$

5. $\dfrac{1}{\sqrt{2}}$ 6. $\dfrac{3}{\sqrt{3}}$ 7. $\dfrac{2}{\sqrt{5}}$ 8. $\dfrac{3}{\sqrt{11}}$

9. $\dfrac{6}{\sqrt{13}}$ 10. $\dfrac{2}{\sqrt{6}}$ 11. $\dfrac{7}{\sqrt{7}}$ 12. $\dfrac{\sqrt{11}}{\sqrt{3}}$

Exercises 9.5

A

Divide; rationalize the denominator:

1. $\dfrac{1}{\sqrt{5}}$ 　　　**2.** $\dfrac{1}{\sqrt{11}}$ 　　　**3.** $\sqrt{5} \div \sqrt{13}$ 　　　**4.** $\sqrt{2} \div \sqrt{7}$

5. $\dfrac{\sqrt{7}}{\sqrt{6}}$ 　　　**6.** $\dfrac{\sqrt{5}}{\sqrt{11}}$

B

7. $\dfrac{3}{\sqrt{6}}$ 　　　**8.** $\dfrac{4}{\sqrt{12}}$ 　　　**9.** $\dfrac{\sqrt{6}}{\sqrt{8}}$

10. $\dfrac{\sqrt{10}}{\sqrt{5}}$ 　　　**11.** $\dfrac{\sqrt{12}}{\sqrt{18}}$ 　　　**12.** $\dfrac{\sqrt{20}}{\sqrt{15}}$

13. $\sqrt{24} \div \sqrt{30}$ 　　　**14.** $\sqrt{15} \div \sqrt{40}$

C

15. $\dfrac{6\sqrt{15}}{\sqrt{75}}$ 　　　**16.** $\dfrac{5\sqrt{40}}{\sqrt{80}}$ 　　　**17.** $\dfrac{\sqrt{32}}{3\sqrt{96}}$

18. $\dfrac{6\sqrt{50}}{5\sqrt{75}}$ 　　　**19.** $\dfrac{3\sqrt{18}}{\sqrt{120}}$ 　　　**20.** $\dfrac{7\sqrt{20}}{\sqrt{27}}$

21. $\dfrac{11}{\sqrt{6} + \sqrt{3}}$ 　　　**22.** $\dfrac{13}{\sqrt{7} + \sqrt{2}}$ 　　　**23.** $\dfrac{2}{4 - \sqrt{3}}$

24. $\dfrac{9}{1 - \sqrt{3}}$

ANSWERS

1. _____
2. _____
3. _____
4. _____
5. _____
6. _____
7. _____
8. _____
9. _____
10. _____
11. _____
12. _____
13. _____
14. _____
15. _____
16. _____
17. _____
18. _____
19. _____
20. _____
21. _____
22. _____
23. _____
24. _____

25. _____

D

25. Using the formula in the application, find the time it takes a pendulum to complete a cycle if the pendulum is 2 ft long. (Write answer in simplest radical form and rounded to the nearest hundredth.)

26. _____

26. Using the formula in the application, find the time it takes a pendulum 12 inches long to complete a cycle. (Write answer in simplest radical form and rounded to the nearest hundredth.)

27. _____

E *Maintain Your Skills*

27. A car has depreciated to 38% of its original value. If the value of the car is presently $1900, what was the original value?

28. _____

28. Betty wanted to buy a calculator. She saw an advertisement for 15% off the regular price. What was the regular price if the advertised price was $20.40?

29. _____

29. Last year the enrollment at Grand College was 8920. This year there has been an increase of 10%. What is the enrollment this year?

30. _____

30. The Local High School team finished the season with a record of 20 wins and 5 defeats. What percent of the games played did they lose?

31. Nancy earned $872 from which $156.96 was withheld for taxes. What percent of her earnings was withheld for taxes?

31. _____

9.6 Combining Radical Expressions

OBJECTIVE

107. Combine radical expressions.

APPLICATION

Ethyl plans to heat her living room with two portable heaters. The first heater uses 605 watts and has a resistance of 25 ohms. The second heater uses 845 watts and has a resistance of 36 ohms. If the voltage (V) required to operate each heater is given by

$$V = \sqrt{\text{watts} \cdot \text{ohms}}$$

how many volts are needed to operate the heaters? (Simplest radical form.)

VOCABULARY

No new vocabulary.

HOW AND WHY

Radical expressions are combined using the same laws of algebra that are used to combine algebraic terms.

If two radicals are the same and have the same radicand ($\sqrt{7}$ and $\sqrt{7}$, but not $\sqrt{7}$ and $\sqrt[3]{7}$ and not $\sqrt{7}$ and $\sqrt{17}$), we can call them *like radical numbers* (similar to *like terms*).

To combine like radical numbers, add (or subtract) the numerical coefficients and annex the like radical number. It is sometimes necessary to simplify the radicals in the radical expression before it is possible to combine the expressions.

EXAMPLES

(Assume all variables represent positive numbers.)

a. $\sqrt{2} + \sqrt{2} = (1+1)\sqrt{2} = 2\sqrt{2}$

b. $3\sqrt{5} + 4\sqrt{5} = (3+4)\sqrt{5} = 7\sqrt{5}$

c. $8\sqrt[3]{2} - 4\sqrt[3]{2} = (8-4)\sqrt[3]{2} = 4\sqrt[3]{2}$

d. $4\sqrt{x} + 3\sqrt{y} + 7\sqrt{x} = 4\sqrt{x} + 7\sqrt{x} + 3\sqrt{y}$
$$= (4+7)\sqrt{x} + 3\sqrt{y}$$
$$= 11\sqrt{x} + 3\sqrt{y}$$

e. $\sqrt{75} + \sqrt{20} - \sqrt{\dfrac{1}{3}} = \sqrt{25} \cdot \sqrt{3} + \sqrt{4} \cdot \sqrt{5} - \dfrac{\sqrt{1}}{\sqrt{3}}$
$$= 5\sqrt{3} + 2\sqrt{5} - \dfrac{\sqrt{3}}{3}$$
$$= \left(5 - \dfrac{1}{3}\right)\sqrt{3} + 2\sqrt{5}$$
$$= \dfrac{14}{3}\sqrt{3} + 2\sqrt{5}$$

f. $\sqrt{8x^3} - \sqrt{50x^3} + \sqrt[3]{27x^3} = \sqrt{4x^2} \cdot \sqrt{2x} - \sqrt{25x^2} \cdot \sqrt{2x} + 3x$
$= 2x\sqrt{2x} - 5x\sqrt{2x} + 3x$
$= -3x\sqrt{2x} + 3x$

g. $\sqrt{24} - 6\sqrt{6} + \sqrt{50} \cdot \sqrt{3} = \sqrt{4} \cdot \sqrt{6} - 6\sqrt{6} \quad b\sqrt{25} \cdot \sqrt{2} \cdot \sqrt{3}$
$= 2\sqrt{6} - 6\sqrt{6} + 5\sqrt{2} \cdot \sqrt{3}$
$= 2\sqrt{6} - 6\sqrt{6} + 5\sqrt{6}$
$= \sqrt{6}$

APPLICATION SOLUTION

The voltage needed to operate the two heaters is the sum of the voltages of both heaters. Let V_1 be the voltage of the first heater and V_2 the voltage of the second heater. We then have

Simpler word form: total voltage = sum of voltages

Translate to algebra: $V = V_1 + V_2$

Solve: Voltage of the first heater:

$V_1 = \sqrt{605 \cdot 25}$
$= \sqrt{121 \cdot 5 \cdot 25}$
$= 55\sqrt{5}$

Voltage of the second heater:

$V_2 = \sqrt{845 \cdot 36}$
$= \sqrt{169 \cdot 5 \cdot 36}$
$= 78\sqrt{5}$

Total voltage:

$V = 55\sqrt{5} + 78\sqrt{5}$
$V = 133\sqrt{5}$

Answer: The total voltage needed to operate the heaters is $113\sqrt{5}$ volts.

WARM UPS

Combine:

1. $\sqrt{2} + 2\sqrt{2}$

2. $\sqrt{3} + \sqrt{3}$

3. $3\sqrt{5} + 2\sqrt{5}$

4. $4\sqrt{2} + 6\sqrt{2}$

5. $3\sqrt{6} - \sqrt{6}$

6. $8\sqrt{5} - 2\sqrt{5}$

7. $11\sqrt{7} - 7\sqrt{7}$

8. $8\sqrt{13} - 5\sqrt{13}$

9. $2\sqrt{x} + 3\sqrt{x}$

10. $\sqrt{3} + 2\sqrt{3} + 3\sqrt{3}$

11. $4\sqrt{2} + 8\sqrt{2} + \sqrt{2}$

12. $4\sqrt{y} + 5\sqrt{y}$

Exercises 9.6

ANSWERS

1. _____

2. _____

3. _____

4. _____

5. _____

6. _____

7. _____

8. _____

9. _____

10. _____

11. _____

12. _____

13. _____

14. _____

15. _____

16. _____

17. _____

18. _____

19. _____

20. _____

21. _____

22. _____

23. _____

24. _____

A

Combine:

1. $3\sqrt{5} + 6\sqrt{5}$

2. $8\sqrt{3} + 6\sqrt{3}$

3. $\sqrt{7} + 3\sqrt{7}$

4. $5\sqrt{6} + \sqrt{6}$

5. $9\sqrt{2} - 7\sqrt{2}$

6. $10\sqrt{3} - 5\sqrt{3}$

B

7. $\sqrt{5} + 3\sqrt{5} + 2\sqrt{5}$

8. $8\sqrt{6} + 11\sqrt{6} + \sqrt{6}$

9. $3\sqrt{2} + \sqrt{8}$

10. $5\sqrt{3} + \sqrt{12}$

11. $2\sqrt{6} + \sqrt{24}$

12. $3\sqrt{5} + \sqrt{20}$

13. $\sqrt{24x} - \sqrt{6x}$

14. $\sqrt{32x} - \sqrt{18x}$

C

15. $\sqrt{18} + \sqrt{32}$

16. $\sqrt{108} - \sqrt{48}$

17. $\sqrt{72} + \sqrt{50} - \sqrt{128}$

18. $\sqrt{27} - \sqrt{75} + \sqrt{192}$

19. $\sqrt{72} - \sqrt{75} + \sqrt{98} - \sqrt{27}$

20. $\sqrt{24} + \sqrt{32} - \sqrt{128} - \sqrt{150}$

21. $\sqrt{28} - \sqrt{700} + \sqrt{63}$

22. $-\sqrt{98} - 2\sqrt{32} + \sqrt{18}$

23. $3\sqrt{75} - 14\sqrt{12} - 2\sqrt{48}$

24. $\sqrt{500} - 8\sqrt{45} - \sqrt{180}$

ANSWERS

25.

26.

27.

28.

29.

30.

31.

D

25. Using the formula in the application, find the voltage needed to operate three heaters. The first uses 200 watts and has a resistance of 48 ohms. The second uses 300 watts and has a resistance of 50 ohms. The third uses 625 watts and has a resistance of 24 ohms. (Give answer in simplest radical form.)

26. A plot of ground is surveyed as two equilateral triangles. One has sides 25 yd long, and the other has sides 20 yd long. What is the area of the entire plot of ground? (*Hint:* The area of an equilateral triangle is given by the formula

$$A = \frac{1}{4} s^2 \sqrt{3}$$

where s is a length of a side.) (Give answer in simplest radical form and rounded to the nearest tenth.)

E   *Maintain Your Skills*

Compute the following:

27. $\sqrt{1024}$ 28. $-\sqrt{625}$ 29. $\sqrt[3]{125}$

30. $-\sqrt[3]{216}$ 31. $\sqrt{28.09}$

9.7 Fractional Exponents

OBJECTIVES

108. Write radical expressions in exponential form.
109. Write expressions written in exponential form as radical expressions.

APPLICATION

Two lines of a computer program read

LINE NUMBER	COMMAND
275	$y = 9$
280	$x = y \wedge 1.5$

The symbol "$\wedge$" is used to show that the number following is an exponent. In algebra we write this as

$$x = y^{1.5} = 9^{1.5} = 9^{3/2}$$

where the exponent is a rational number $\left(\dfrac{3}{2}\right)$ but not an integer. What is the value of this expression?

VOCABULARY

No new vocabulary.

HOW AND WHY

It is possible to write radical expressions in exponential form. If the laws of exponents for squaring a number are accepted, we can consider the following:

$$(a^{1/2})^2 = a^{1/2} \cdot a^{1/2} = a^{1/2+1/2} = a^1 = a$$

or

$$(a^{1/2})^2 = a^{1/2 \cdot 2} = a^1 = a$$

We also know that the following is true:

$$(\sqrt{a})^2 = \sqrt{a} \cdot \sqrt{a} = a$$

Since $(a^{1/2})^2$ and $(\sqrt{a})^2$ both yield an answer of a, we make the following definition:

$\sqrt{a} = a^{1/2}$, **where** a **represents a positive number**

A similar argument with cube root leads to the following definition:

$$\sqrt[3]{a} = a^{1/3}, \qquad \sqrt[3]{a^2} = a^{2/3}$$

Using the laws of exponents adopted earlier we can see that

$$x^{2/3} = x^{2 \cdot 1/3} = (x^2)^{1/3} = \sqrt[3]{x^2}$$

and

$$\sqrt{a^3} = (a^3)^{1/2} = a^{3 \cdot 1/2} = a^{3/2}$$

In general, if x **is positive we can write:**
$$x^{m/n} = \sqrt[n]{x^m} = (\sqrt[n]{x})^m$$

EXAMPLES

Write the following in exponential form.

a. $\sqrt{5} = 5^{1/2}$

b. $\sqrt[3]{x} = x^{1/3}$

c. $\sqrt{xy^3} = (xy^3)^{1/2} = x^{1/2} \cdot y^{3 \cdot 1/2} = x^{1/2}y^{3/2}$

Write the following as radical expressions.

d. $7^{1/2} = \sqrt{7}$

e. $9^{2/3} = \sqrt[3]{9^2} = \sqrt[3]{81} = 3\sqrt[3]{3}$

f. $x^{1/3}y^{1/3} = (xy)^{1/3} = \sqrt[3]{xy}$

g. $(8y^5)^{1/2} = \sqrt{8y^5} = 2y^2\sqrt{2y}$

APPLICATION SOLUTION

In the application, the value of x is written

$x = 9 \wedge 1.5$ *Computer statement*

$x = 9^{1.5} = 9^{3/2}$ *Algebra statement*

Using the Third Law of Exponents, (See Section 6.2), we can write

$x = (9^{1/2})^3$

and since $9^{1/2} = 3$,

$x = (3)^3$ *Then multiply*

$x = 27$

WARM UPS

Write the following in exponential form.

1. $\sqrt{2}$ **2.** $\sqrt[3]{5}$ **3.** $\sqrt{6}$

4. $\sqrt[4]{7}$ **5.** $\sqrt[3]{a^2}$ **6.** $\sqrt{a^3}$

Write the following as radical expressions.

7. $a^{1/3}$ **8.** $9^{1/3}$ **9.** $x^{1/2}$

10. $5^{1/2}$ **11.** $5^{3/2}$ **12.** $5^{2/3}$

Exercises 9.7

A

Write the following in exponential form.

1. $\sqrt{7}$ **2.** $\sqrt[3]{2}$ **3.** $\sqrt{x}$

Write the following as radical expressions.

4. $11^{1/2}$ **5.** $12^{1/3}$ **6.** $x^{1/3}$

B

Write the following in exponential form.

7. $\sqrt{57}$ **8.** $\sqrt[3]{xy}$ **9.** $\sqrt{ab}$ **10.** $\sqrt[3]{x^2}$

Write the following as radical expressions.

11. $(7b)^{1/2}$ **12.** $xy^{1/3}$ **13.** $(xy)^{1/3}$ **14.** $(xy)^{2/3}$

C

Write the following in exponential form. Simplify.

15. $\sqrt[3]{8x^2}$ **16.** $\sqrt[3]{5x^3}$ **17.** $\sqrt{49x^3}$

18. $\sqrt{7y^4}$ **19.** $\sqrt[4]{5x^3}$

Write the following in radical form. Simplify.

20. $(8x^2)^{1/2}$ **21.** $(32x^3)^{1/2}$ **22.** $(27x^3)^{1/3}$

23. $(16x^2)^{1/3}$ **24.** $(3y^3)^{2/3}$

ANSWERS

1. _____
2. _____
3. _____
4. _____
5. _____
6. _____
7. _____
8. _____
9. _____
10. _____
11. _____
12. _____
13. _____
14. _____
15. _____
16. _____
17. _____
18. _____
19. _____
20. _____
21. _____
22. _____
23. _____
24. _____

481

ANSWERS

25. _____

26. _____

27. _____

28. _____

29. _____

30. _____

31. _____

32. _____

33. _____

D

25. Two lines of a computer program read

LINE NUMBER	COMMAND
310	$y = 16$
320	$x = y \wedge 1.5$

What is the numerical value of x?

26. Two lines of a computer program read

1130	$w = 4$
1140	$x = w \wedge 2.5$

What is the numerical value of x?

27. A line of a computer program reads

960    $x = 5 + 64 \wedge .5$

What is the numerical value of x? (Recall that the order of operations in Chapter 1 states that exponential expressions are calculated first.)

28. A line of a computer program reads

960    $x = 18 - 81 \wedge .25$

What is the numerical value of x?

E   *Maintain Your Skills*

Find the missing side of each of the following right triangles. Round decimal answers to the nearest hundredth.

29. $a = 5, b = 12, c = ?$

30. $a = ?, b = 12, c = 17$

31. $a = 8, b = ?, c = 20$

Are the following right triangles?

32. $a = 24, b = 32, c = 40$

33. $a = 12, b = 48, c = 50$

9.8 Negative Exponents

OBJECTIVES

110. Write expressions in an equivalent form using negative exponents.
111. Write an expression written with negative exponents as an expression without negative exponents.

APPLICATION

The radius of a red corpuscle is about 3.8×10^{-5} centimetre. Write this measurement in decimal form.

VOCABULARY

An expression such as $x^3 + 3x + x^{-1}$ has a *negative exponent*. In this case it is -1. The expression is read "x cubed plus three x plus x to the negative one."

HOW AND WHY

In order to find a meaning for the expression 5^{-2}, study the following chart:

EXPRESSION	REDUCE	OBSERVATION
$\dfrac{5^3}{5}$	$\dfrac{5 \cdot 5 \cdot 5}{5} = 5 \cdot 5 = 5^2$	$\dfrac{5^3}{5^1} = 5^{3-1} = 5^2$
$\dfrac{5^3}{5^2}$	$\dfrac{5 \cdot 5 \cdot 5}{5 \cdot 5} = 5 = 5^1$	$\dfrac{5^3}{5^2} = 5^{3-2} = 5^1$
$\dfrac{5^3}{5^3}$	$\dfrac{5 \cdot 5 \cdot 5}{5 \cdot 5 \cdot 5} = 1 = 5^0$	$\dfrac{5^3}{5^3} = 5^{3-3} = 5^0$
$\dfrac{5^3}{5^4}$	$\dfrac{5 \cdot 5 \cdot 5}{5 \cdot 5 \cdot 5 \cdot 5} = \dfrac{1}{5} = 5^?$	$\dfrac{5^3}{5^4} = 5^{3-4} = 5^{-1}$
$\dfrac{5^3}{5^5}$	$\dfrac{5 \cdot 5 \cdot 5}{5 \cdot 5 \cdot 5 \cdot 5 \cdot 5} = \dfrac{1}{5 \cdot 5} = 5^?$	$\dfrac{5^3}{5^5} = 5^{3-5} = 5^{-2}$

In the second column we see that the exponent is decreasing by one from top to bottom. If this pattern is continued, we see that

$$5^{-1} = \frac{1}{5}$$

and

$$5^{-2} = \frac{1}{5^2}$$

This pattern leads us to the following definition:

If x is any non-zero rational number and n is a positive integer, then

$$x^{-n} = \frac{1}{x^n}, \quad x \neq 0$$

Zero is excluded as the base, since division by zero is not defined. The negative sign in the exponent indicates "the reciprocal of."

EXAMPLES

(Assume all variables represent positive numbers.)

a. $3^{-1} = \dfrac{1}{3}$

b. $x^{-1/2} = \dfrac{1}{x^{1/2}}$

c. $(a + b)^{-1} = \dfrac{1}{a + b}$

d. $(x + y)^{-4} = \dfrac{1}{(x + y)^4}$

e. $\dfrac{x}{a^4} = xa^{-4}$

f. $\dfrac{1}{x} + \dfrac{1}{y} = x^{-1} + y^{-1}$

g. $\dfrac{1}{(x + y)^{1/3}} = (x + y)^{-1/3}$

h. $\dfrac{1}{(a + 2b + c)^3} = (a + 2b + c)^{-3}$

APPLICATION SOLUTION

To find the radius of the red corpuscle in decimal form, rewrite 10^{-5} without exponents and perform the indicated multiplication.

3.8×10^{-5}

$3.8 \times \dfrac{1}{10^5}$

$3.8 \times \dfrac{1}{100,000}$

$3.8 \times .00001$

$.000038$

So the radius is .000038 centimetre.

WARM UPS

Write the following using positive exponents.

1. 4^{-1} **2.** a^{-1} **3.** x^{-2}

4. $4a^{-1}$ **5.** $3y^{-1}$ **6.** $6z^{-2}$

Write the following without fractions, using negative exponents.

7. $\dfrac{1}{w}$ **8.** $\dfrac{2}{a^2}$ **9.** $\dfrac{5}{a^3}$

10. $\dfrac{2}{x^3}$ **11.** $\dfrac{xy}{z}$ **12.** $\dfrac{x}{yz}$

Exercises 9.8

ANSWERS

1. _____

2. _____

3. _____

4. _____

5. _____

6. _____

7. _____

8. _____

9. _____

10. _____

11. _____

12. _____

13. _____

14. _____

15. _____

16. _____

17. _____

18. _____

19. _____

20. _____

21. _____

22. _____

23. _____

24. _____

A

Write the following using positive exponents.

1. t^{-1} **2.** 7^{-1} **3.** $3^{-1}x^{-2}$

Write the following without fractions, using negative exponents.

4. $\dfrac{1}{9}$ **5.** $\dfrac{2}{b^3}$ **6.** $\dfrac{4}{x^5}$

B

Write the following using positive exponents.

7. $a^{-1}b^{-2}$ **8.** rs^{-2} **9.** $3^{-1}x^2$ **10.** x^2y^{-2}

Write the following without fractions, using negative exponents.

11. $\dfrac{7}{x^2}$ **12.** $\dfrac{1}{x+y}$ **13.** $\dfrac{4}{m+n}$ **14.** $\dfrac{x+y}{2}$

C

Write the following using positive exponents.

15. $2x^{-1}$ **16.** $(2x)^{-2}$ **17.** $\dfrac{1}{x^{-3}}$

18. $x + y^{-2/3}$ **19.** $(x+y)^{-2/3}$

Write the following without fractions, using negative exponents.

20. $\dfrac{x^2}{y^2}$ **21.** $\dfrac{x^{-2}}{y^2}$ **22.** $\dfrac{a^{1/2}c}{b^3}$ **23.** $\dfrac{1}{(a+b)^{1/2}}$

24. $\dfrac{1}{ab^{1/2}}$

ANSWERS

25. _____

26. _____

27. _____

28. _____

29. _____

30. _____

31. _____

D

25. In photography, the time for taking a picture (exposure time) is given by the formula

$$t = kd^{-2}$$

where d is the diameter of the camera lens and k is a constant. Write the formula for t without the negative exponent.

26. An equation of a probability curve is $y = e^{-x^2}$. Some teachers use this curve when "grading on the curve." Write the right-hand member with all exponents positive.

E   *Maintain Your Skills*

Multiply and simplify. Assume all variables represent positive numbers.

27. $(\sqrt{5})(\sqrt{20})$ **28.** $(\sqrt{3})(\sqrt{12})$ **29.** $(\sqrt{6})(\sqrt{2})(\sqrt{3})$

30. $(\sqrt{x})(\sqrt{y})(\sqrt{z})$ **31.** $(\sqrt{5})(\sqrt{6} - \sqrt{2})$

ANSWERS TO WARM UPS (9.8)

1. $\dfrac{1}{4}$ 2. $\dfrac{1}{a}$ 3. $\dfrac{1}{x^2}$ 4. $\dfrac{4}{a}$ 5. $\dfrac{3}{y}$

6. $\dfrac{6}{z^2}$ 7. w^{-1} 8. $2a^{-2}$ 9. $5a^{-3}$ 10. $2x^{-3}$

11. xyz^{-1} 12. $x(yz)^{-1}$ or $xy^{-1}z^{-1}$

9.9 Scientific Notation

OBJECTIVES

112. Write a number in scientific notation.
113. Change a number written in scientific notation to place value notation.

APPLICATION

A *light year* is a measure of distance. It is the distance that light travels in one year. Rounded to the nearest ten billion, a light year is approximately 5,870,000,000,000 miles. Write this number in scientific notation.

VOCABULARY

No new words.

HOW AND WHY

Very large numbers and very small numbers arise in science, technology, and industry. A way of writing such numbers, called "scientific notation," has two big advantages. First, we can write such numbers in less space, and second, the degree of accuracy (number of significant figures) is easier to display.

A number written in scientific notation has the following form:

$a \times 10^m$

where

1. a is written as a decimal,
2. a is 1 or a number between 1 and 10, and
3. m is an integer.

These numbers are written in scientific notation:

5.4×10^4 7.23×10^{-6} 9.89×10^{15} 1.03×10^{-12}

To rewrite these numbers in place value notation, we have only to do the indicated multiplications.

$5.4 \times 10^4 = 5.4(10)(10)(10)(10) = 54000$

$7.23 \times 10^{-6} = 7.23 \left(\dfrac{1}{10}\right)\left(\dfrac{1}{10}\right)\left(\dfrac{1}{10}\right)\left(\dfrac{1}{10}\right)\left(\dfrac{1}{10}\right)\left(\dfrac{1}{10}\right) = .00000723$

Since calculators are limited in the number of digits they can display, scientific calculators automatically display numbers in scientific notation when the result is too large or too small for the calculator to show as a decimal. For instance, 3.69×10^{23} is displayed as 3.69 23. On a computer the space between 3.69 and 23 is replaced by an E, 3.69 E 23.

The shortcut for multiplying a number by 10 is to move the decimal point to the right. The inverse, dividing a number by 10 (or multiplying the number by $\dfrac{1}{10}$ or 10^{-1}), is to move the decimal point to the left. We can use these shortcuts to quickly write a number in scientific notation.

To write a number in scientific notation,

1. Move the decimal point right or left so that the result is 1 or a number between 1 and 10.
2. Multiply the number in step 1 by a power of ten. The appropriate exponent of 10 to use is the one that will make the new product equal to the original number. Note that this exponent shows the number of decimal places moved in step 1, and it is positive if the decimal point was moved to the left and negative if the decimal point was moved to the right.

To change a number from scientific notation to place value notation, do the indicated multiplication.

Study the examples below to see how to use the rule.

EXAMPLES

a. 593,000,000

   Step 1. 5.93 is between 1 and 10

   Step 2. 5.93 times 100,000,000 (or 10^8) is 593,000,000

   Notice that the decimal point in 593,000,000. was moved eight places to the left in step 1, and the exponent of 10 is 8.

   So 593,000,000 $= 5.93 \times 10^8$

b. .000000011

   Step 1. 1.1 is between 1 and 10

   Step 2. $1.1 \times \frac{1}{10} \times \frac{1}{10} \times \frac{1}{10} \times \frac{1}{10} \times \frac{1}{10} \times \frac{1}{10} \times \frac{1}{10} \times \frac{1}{10}$ $\left(\text{or } \frac{1}{10^8}\right)$ is .000000011

   Notice that the decimal point in .000000011 was moved eight places to the right in step 1, and the exponent of 10 is −8.

   So .000000011 $= 1.1 \times 10^{-8}$

c. Write 1×10^{12} in place value notation.

   $1 \times 10^{12} = 1,000,000,000,000$ *Move the decimal point 12 places to the right*

   $= $ one trillion

d. Write 5.62×10^{-11} in place value notation.

   $5.62 \times 10^{-11} = .0000000000562$ *Move the decimal point 11 places to the left*

APPLICATION SOLUTION

To write the distance, in miles, of a light year in scientific notation, we do these steps: 5,870,000,000,000

Step 1. 5.87 is between 1 and 10

Step 2. 5.87×10^{12} is 5,870,000,000,000

so 5,870,000,000,000 $= 5.87 \times 10^{12}$

WARM UPS

Write in scientific notation:

1. 47   **2.** .4   **3.** 370   **4.** .37

5. .002   **6.** 5900

Change to place value notation:

7. 7.0×10^3 **8.** 7.0×10^{-3} **9.** 3.1×10^{-2}

10. 3.1×10^2 **11.** 7.28×10^3 **12.** 3.62×10^{-4}

Exercises 9.9

A

Write in scientific notation:

1. 30,000   **2.** 34,000   **3.** .0003

Change to place value notation:

4. 8.8×10^2 **5.** 7.1×10^2 **6.** 6.3×10^{-2}

B

Write in scientific notation:

7. 692,000   **8.** 127,000   **9.** .00348

10. .00671

Change to place value notation:

11. 4.68×10^{-1} **12.** 9.77×10^{-2} **13.** 3.29×10^1

14. 6.04×10^3

C

Write in scientific notation:

15. 8446   **16.** 4641   **17.** .776   **18.** .324

19. .000067

Change to place value notation:

20. 5.6×10^6 **21.** 6.9×10^7 **22.** 1.44×10^{-8}

23. 5.03×10^{-7} **24.** 2.337×10^{-3}

491

ANSWERS

1. _____
2. _____
3. _____
4. _____
5. _____
6. _____
7. _____
8. _____
9. _____
10. _____
11. _____
12. _____
13. _____
14. _____
15. _____
16. _____
17. _____
18. _____
19. _____
20. _____
21. _____
22. _____
23. _____
24. _____

25. _____

26. _____

27. _____

28. _____

29. _____

30. _____

31. _____

32. _____

33. _____

34. _____

35. _____

36. _____

37. _____

38. _____

39. _____

D

25. The total land area of the earth is approximately 52,400,000 square miles. What is the total area written in scientific notation?

26. The speed of light is approximately 11,160,000 miles per minute. Write the speed in scientific notation.

27. Gamma rays are 2×10^{-8} cm in length. Write this measurement in place value notation.

28. A wave of red light is 7.2×10^{-8} cm in length. Write this number in place value notation.

29. The earth is approximately 1.72×10^8 kilometres from the sun. Write this distance in place value notation.

30. By the end of 1982, the total investment in real estate tax shelters had grown to $4,700,000,000 in the United States. Write this value in scientific notation.

31. The assets of 260 money funds listed in a local report was 1.86×10^{11} on April 30. Write the value of the assets in place value notation.

32. In a recent seven hour day, 100,000,000 shares were traded on the New York Stock Exchange. This amounted to selling one share every .000252 seconds. Write the time per share in scientific notation.

33. The local computer shop offers a home computer with 128k (128,000) bytes of memory. Write the number of bytes in scientific notation.

34. A business computer is advertised as having 165 MB (1.65×10^8 bytes) of memory. Write the number of bytes in place value notation.

E   *Maintain Your Skills*

Simplify. Assume all variables represent positive numbers.

35. $\sqrt{48}$

36. $\sqrt{120}$

37. $\sqrt[3]{81}$

38. $\sqrt{27x}$

39. $\sqrt{16x^2 y}$

1.

2.

3.

4.

5.

6.

7.

8.

9.

10.

11.

12.

13.

14.

15.

16.

17.

18.

19.

20.

21.

22.

23.

24.

25.

Chapter 9
POST-TEST

1. **(Obj. 104)** *Multiply:* $\sqrt{7} \cdot \sqrt{6}$

2. **(Obj. 105)** *Simplify:* $\sqrt{12}$

3. **(Obj. 100)** Find the root: $\sqrt{49}$

4. **(Obj. 106)** Rationalize the denominator: $\dfrac{\sqrt{2}}{\sqrt{7}}$

5. **(Obj. 111)** Write $\dfrac{4x^{-2}}{y^{-2}}$ with positive exponents.

6. **(Obj. 106)** Rationalize the denominator and simplify: $\sqrt{16} \div \sqrt{6}$

7. **(Obj. 105)** *Simplify:* $\sqrt{128x^3y^2}$

8. **(Obj. 106)** Rationalize the denominator: $\dfrac{2}{5 - \sqrt{3}}$

9. **(Obj. 107)** *Add:* $\sqrt{8} + \sqrt{2}$

10. **(Obj. 109)** Write $3x^{1/2}y^{1/2}$ in radical form.

11. **(Obj. 113)** Write 2.31×10^{-5} in place value notation.

12. **(Obj. 104)** *Multiply:* $(\sqrt{6} + \sqrt{3})(\sqrt{5} - \sqrt{7})$

13. **(Obj. 108)** Write $\sqrt{ab^3}$ in exponential form.

14. **(Obj. 102)** *Evaluate:* $\sqrt[3]{64}$

15. **(Obj. 107)** *Subtract:* $\sqrt[3]{16} - \sqrt[3]{2}$

16. **(Obj. 110)** Write $\dfrac{1}{a + b}$ without fractions using negative exponents.

17. **(Obj. 104)** *Multiply:* $\sqrt{5}(\sqrt{2} + \sqrt{3})$

18. **(Obj. 107)** *Add:* $7\sqrt{6} + \sqrt{54} + 3\sqrt{24}$

19. **(Obj. 105)** *Simplify:* $\sqrt[3]{8x^4}$

20. **(Obj. 112)** Write 6,230,000 in scientific notation.

21. **(Obj. 103)** In a right triangle, if side $b = 15$ and hypotenuse $c = 17$, what is the length of side a?

22. **(Obj. 103)** A fresh air duct will have a cross section area of 256.25 square inches. If the duct will be a square, what is the length of a side? (To the nearest tenth of an inch.)

23. **(Obj. 103)** What is the length of the hypotenuse of a right triangle whose legs are 6 and 12, respectively? (Leave answer in simplified radical form.)

24. **(Obj. 103)** What is the length of the side of a square food crib that has a floor area of 1089 sq ft?

25. **(Obj. 103)** What is the length of the diagonal of a rectangle whose length is 15 m and whose width is 8 m?

ANSWERS TO WARM UPS (9.9)   **1.** 4.7×10^1 **2.** 4×10^{-1} **3.** 3.7×10^2 **4.** 3.7×10^{-1} **5.** 2×10^{-3}

6. 5.9×10^3 **7.** 7000   **8.** .007   **9.** .031   **10.** 310

11. 7280   **12.** .000362

10

Quadratic
Equations

Chapter 10
PRE-TEST

The problems in the following pre-test are a sample of the material in the chapter. You may already know how to work some of these. If so, this will allow you to spend less time on those parts. As a result, you will have more time to give to the sections that you find more difficult or that are new to you. The answers are in the back of the text.

Solve:

1. (Obj. 114) $x^2 = 45$

2. (Obj. 114) $x^2 - 64 = 0$

3. (Obj. 115) Solve using the quadratic formula:

$$x^2 - 8x + 15 = 0$$

4. (Obj. 115) Solve using the quadratic formula:

$$2x^2 - 7x + 5 = 0$$

5. (Obj. 115) Solve using the quadratic formula:

$$12x^2 - 23x + 10 = 0$$

6. (Obj. 116) $6x^2 + 15x = -6$

7. (Obj. 116) $\dfrac{1}{x+1} - \dfrac{4}{x} = 1$

8. (Obj. 117) $\sqrt{5x - 1} = 7$

9. (Obj. 117) $\sqrt{x + 2} - 8 = 3$

10. (Obj. 117) $\sqrt{3x + 17} - 2 = \sqrt{3x + 5}$

11. (Obj. 114) A square plot of land has an area of 2116 ft². What is the length of one side?

12. (Obj. 116) An arrow is shot directly upward from a height of 6 feet. If its distance from the ground is given by the formula $S = 6 + 150t - 16t^2$, how long will it take to reach a height of 300 feet? (S in feet, t in seconds). Give answer to the nearest hundredth of a second.

2.

3.

4.

5.

6.

7.

8.

9.

10.

11.

12.

10.1 Quadratic Equations — Solved by Square Roots

OBJECTIVE	**114.** Solve a quadratic equation that can be written in the form $x^2 = a$, where a is not negative.

APPLICATION	A fluid moving quickly through a pipe can actually erode the inner surface of the pipe and reduce its life. Therefore, it is important to determine the velocity through a pipe of a given size for a required volumetric flow rate. The flow rate through the pipe is

$$F = \frac{\pi}{4} D^2 V*$$

where F is the volumetric flow rate, ft^3/sec
     D is the diameter of the pipe, ft
     V is the velocity of the fluid, ft/sec

Find the diameter of the pipe (to the nearest hundredth) when $F = 3.5$ ft³/sec and $V = 40$ ft/sec. Let $\pi \approx 3.14$.

VOCABULARY	Recall that every positive number has two square roots, one positive and one negative, and that the square of any real number is not negative.

The symbol $\pm$ indicates a positive or a negative number. So $x = \pm 6$ means $x = +6$ or $x = -6$.

HOW AND WHY	**Given a quadratic equation written in the form $x^2 = a$, where a is not negative, the roots of the equation are found by taking the square root of both sides. (The $\pm$ is omitted in front of the x since it does not affect the solution.)**

So if $x^2 = a$, then
    $x = \pm\sqrt{a}$

Consider the equation

$x^2 - 25 = 0$

    $x^2 = 25$ *Form of $x^2 = a$*

    $x = \pm\sqrt{25}$ *Square root of both sides*

    $x = \pm 5$

$x^2 = -16$ has no real solution, since the square of any positive number is positive and the square of any negative number is positive.

Therefore, $x^2 = a$, when a is negative, has no real solutions. If $x^2 = 0$, then $x = 0$ is the only solution.

EXAMPLES	**a.** $x^2 = 16$

    $x = \pm\sqrt{16}$

    $x = \pm 4$

    Check: $(+4)^2 = 16$ $(-4)^2 = 16$

               $16 = 16$ $16 = 16$

* Formula courtesy of Exxon Co.

b. $x^2 = 32$

$$x = \pm\sqrt{32} = \pm\sqrt{16 \cdot 2}$$

$$x = \pm 4\sqrt{2}$$

Check: $(4\sqrt{2})^2 = 16 \cdot 2 = 32$

$(-4\sqrt{2})^2 = 16 \cdot 2 = 32$

c. $4x^2 = 25$

$$x^2 = \frac{25}{4}$$

$$x = \pm\sqrt{\frac{25}{4}}$$

$$x = \pm\frac{5}{2}$$

Check: $4\left(+\dfrac{5}{2}\right)^2 = 25$ $4\left(-\dfrac{5}{2}\right)^2 = 25$

$4\left(\dfrac{25}{4}\right) = 25$ $4\left(\dfrac{25}{4}\right) = 25$

$25 = 25$ $25 = 25$

d. $6 + 2x^2 = x^2 + 9$

$$x^2 = 3$$

$$x = \pm\sqrt{3}$$

e. $3x^2 + 15 = x^2 + 9$

$$2x^2 = -6$$

$$x^2 = -3$$

No real solution because x^2 cannot be negative.

APPLICATION SOLUTION

To find the diameter of the pipe, we substitute into the given formula.

Formula: $F = \dfrac{\pi}{4}(D^2)(V)$, $F = 3.5$ ft³/sec

$V = 40$ ft/sec

$\pi \approx 3.14$

Substitute: $3.5 \approx \dfrac{3.14}{4}(D^2)(40)$

Solve: $\dfrac{4(3.5)}{(3.14)(40)} \approx D^2$

$\pm\sqrt{\dfrac{14}{125.6}} \approx D$

$\pm.33 \approx D$

Answer: Since distance is not measured with negative numbers, we reject $-.33$. Therefore, the diameter of the pipe is .33 ft to the nearest hundredth.

WARM UPS

Solve:

1. $x^2 = 16$ **2.** $x^2 = 36$ **3.** $y^2 = 49$

4. $y^2 = 81$ **5.** $a^2 = \dfrac{4}{9}$ **6.** $a^2 = \dfrac{9}{25}$

7. $b^2 = .09$ **8.** $b^2 = .81$ **9.** $m^2 = 10$

10. $m^2 = 11$ **11.** $t^2 = -9$ **12.** $t^2 = -16$

Exercises 10.1

ANSWERS
1. _____
2. _____
3. _____
4. _____
5. _____
6. _____
7. _____
8. _____
9. _____
10. _____
11. _____
12. _____
13. _____
14. _____
15. _____
16. _____
17. _____
18. _____
19. _____
20. _____
21. _____
22. _____
23. _____
24. _____

A

Solve:

1. $x^2 = 25$ **2.** $x^2 = 81$ **3.** $x^2 = 121$

4. $y^2 = 1600$ **5.** $x^2 = -16$ **6.** $x^2 = -36$

B

7. $x^2 = 5$ **8.** $x^2 = 7$ **9.** $x^2 = 8$

10. $x^2 = 27$ **11.** $36 = x^2$ **12.** $225 = y^2$

13. $2x^2 = 32$ **14.** $5x^2 = 25$

C

15. $x^2 - 10 = 49$ **16.** $x^2 + 10 = 49$

17. $x^2 + 15 = 6$ **18.** $x^2 + 25 = 0$

19. $2x^2 - 35 = x^2 - 19$ **20.** $3 - 2x - x^2 = 5 - 2x - 2x^2$

21. $4(x^2 - 5) = 3(x^2 + 6)$ **22.** $3x^2 - 5x^2 = 2x^2 - 7x^2$

23. $x^4 = 256$ (*Hint:* Take the square root twice.)

24. $x^4 - 625 = 0$

D

25. Using the formula in the application, find the diameter of a pipe to the nearest hundredth when $F = 18$ ft³/sec and $V = 50$ ft/sec. (Let $\pi \approx 3.14$.)

26. $d = 16t^2$ is a formula to compute the distance that an object has fallen (d is distance in feet, t is time in seconds). How long will it take an object to fall a distance of 3600 feet? Negative numbers have no meaning in this case.

27. Using the formula in Exercise 26, how long will it take an object to fall a distance of 1444 feet?

28. Using the formula in Exercise 26, how long will it take an object to fall 800 feet? Leave answer in simplest radical form.

29. The illumination (in foot candles) of a light source varies inversely as the square of the distance from the source. The formula is $Id^2 = k$ (I is the illumination and d is the distance in feet). If the constant of variation (k) is 16,000 and the illumination is 80 foot candles, what is the distance from the light source?

30. The illumination (in foot candles) of a light source varies inversely as the square of the distance from the light source. The formula is $Id^2 = k$. (I is the illumination and d is the distance in feet.) If the constant of variation is 36, what is the distance from the light source if the illumination is 2 foot candles?

E   *Maintain Your Skills*

Divide by rationalizing the denominator.

31. $\dfrac{5}{\sqrt{8}}$ **32.** $\dfrac{\sqrt{2}}{\sqrt{12}}$ **33.** $\sqrt{\dfrac{3}{5}}$

34. $\sqrt[3]{\dfrac{7}{4}}$ **35.** $\dfrac{1}{1 - \sqrt{3}}$

ANSWERS TO WARM UPS (10.1)

1. $x = 4$ or $x = -4$ **2.** $x = 6$ or $x = -6$

3. $y = 7$ or $y = -7$ **4.** $y = 9$ or $y = -9$

5. $a = \dfrac{2}{3}$ or $a = -\dfrac{2}{3}$ **6.** $a = \dfrac{3}{5}$ or $a = -\dfrac{3}{5}$

7. $b = .3$ or $b = -.3$ **8.** $b = .9$ or $b = -.9$

9. $m = \sqrt{10}$ or $m = -\sqrt{10}$ **10.** $m = \sqrt{11}$ or $m = -\sqrt{11}$

11. No real roots   **12.** No real roots

10.2 Quadratic Equations—Solved by Quadratic Formula

OBJECTIVE

115. **Find the solutions of a quadratic equation by using the quadratic formula.**

APPLICATION

A flat piece of aluminum that is 12 in. wide and 40 ft long is to be formed into a rectangular gutter by bending up an equal amount on each side. What will be the depth of the gutter if the area of the cross section is 18 in.²?

VOCABULARY

A quadratic equation is an equation that can be written in the form $ax^2 + bx + c = 0$, with $a \neq 0$. The quadratic formula, $x = \dfrac{-b \pm \sqrt{b^2 - 4ac}}{2a}$, gives the roots of the quadratic equation.

HOW AND WHY

In Section 6.11, we solved quadratic equations by factoring. It is not always practical to solve by factoring. The following method can be used in all cases.

To solve a quadratic by using the formula, first write the equation in the form $ax^2 + bx + c = 0$. Then identify the coefficients a, b, and c. Substitute these in the formula and evaluate

$$x = \frac{-b \pm \sqrt{b^2 - 4ac}}{2a}$$

Consider

$x^2 - 3x - 10 = 0$	*Form $ax^2 + bx + c = 0$*
$a = 1, b = -3, c = -10$	*Identify a, b, and c*
$x = \dfrac{-b \pm \sqrt{b^2 - 4ac}}{2a}$	*Quadratic formula*
$x = \dfrac{-(-3) \pm \sqrt{(-3)^2 - 4(1)(-10)}}{2(1)}$	*Substitute*
$x = \dfrac{3 \pm \sqrt{49}}{2}$	*Evaluate*
$x = \dfrac{10}{2}$ or $x = \dfrac{-4}{2}$	
$x = 5$ or -2	*Solutions*

The part of the formula under the radical, $b^2 - 4ac$ (called the discriminant), identifies whether the equation has real roots. Recall that the square root of a negative number is not a real number. So if $b^2 - 4ac$ is negative, the equation has no real roots.

EXAMPLES

a. $4x^2 + 7x - 1 = 0$

$ax^2 + bx + c = 0$

$a = 4, b = 7, c = -1$

$$x = \frac{-b \pm \sqrt{b^2 - 4ac}}{2a}$$

$$x = \frac{-(7) \pm \sqrt{(7)^2 - 4(4)(-1)}}{2(4)}$$

$$x = \frac{-7 \pm \sqrt{49 + 16}}{8}$$

$$x = \frac{-7 \pm \sqrt{65}}{8}$$

b. $x^2 = 2x + 11$

$x^2 - 2x - 11 = 0$

$ax^2 + bx + c = 0$

$a = 1, b = -2, c = -11$

$$x = \frac{-b \pm \sqrt{b^2 - 4ac}}{2a}$$

$$x = \frac{-(-2) \pm \sqrt{(-2)^2 - 4(1)(-11)}}{2(1)}$$

$$x = \frac{2 \pm \sqrt{4 + 44}}{2}$$

$$x = \frac{2 \pm \sqrt{48}}{2}$$

$$x = \frac{2 \pm 4\sqrt{3}}{2}$$

$$x = 1 \pm 2\sqrt{3}$$

c. $x^2 - 3x + 5 = 0$

$a = 1, b = -3, c = 5$

$$x = \frac{-(-3) \pm \sqrt{(-3)^2 - 4(1)(5)}}{2(1)}$$

$$x = \frac{3 \pm \sqrt{9 - 20}}{2}$$

$$x = \frac{3 \pm \sqrt{-11}}{2}$$

Since $\sqrt{-11}$ is not real, the equation has no real roots.

d. If a sum of I dollars is invested at r percent compounded annually, at the end of two years it will amount to $A = I(1 + r)^2$. What is the interest rate if $1000 grew to $1440 in two years?

Formula: $A = I(1 + r)^2, A = 1440, I = 1000$

Substitute: $1440 = 1000(1 + r)^2$

Solve: $1440 = 1000(1 + 2r + r^2)$

$1440 = 1000 + 2000r + 1000r^2$

$1000r^2 + 2000r - 440 = 0$ *Write in standard form*

$25r^2 + 50r - 11 = 0$ *Divide both sides by 40*

$a = 25, b = 50, c = -11$

$$x = \frac{-b \pm \sqrt{b^2 - 4ac}}{2a}$$

$$x = \frac{-50 \pm \sqrt{50^2 - 4 \cdot 25(-11)}}{2 \cdot 25}$$

$$x = \frac{-50 \pm \sqrt{3600}}{50}$$

$$x = \frac{-50 \pm 60}{50}$$

$$x = \frac{-50 + 60}{50} \quad \text{or} \quad x = \frac{-50 - 60}{50}$$

$$x = \frac{10}{50} \quad\quad\quad\quad\quad x = -\frac{110}{50}$$

$x = .2$ or $x = -2.2$ is the solution of the equation.

Answer: We reject -2.2 as an answer to the problem, since rate of interest is not measured with negative numbers. Therefore, the interest rate was .2, which is expressed as 20%.

e. Calculator example

$3x^2 + 11x - 35 = 0$

$a = 3, b = 11, c = -35$

$$x = \frac{-11 \pm \sqrt{(11)^2 - 4(3)(-35)}}{2(3)}$$

First find the value of $\sqrt{(11)^2 - 4(3)(-35)}$

ENTER	DISPLAY
11	11.
x^2	121.
$-$	121.
4	4.
$\times$	4.
3	3.
$\times$	12.
35	35.

$\boxed{+/-}$	$-35.$
$\boxed{=}$	$541.$
$\boxed{\sqrt{}}$	23.25941

$$x \approx \frac{-11 + 23.25941}{6} \qquad x \approx \frac{-11 - 23.25941}{6}$$

ENTER	DISPLAY	ENTER	DISPLAY
$\boxed{11}$	$11.$	$\boxed{11}$	$11.$
$\boxed{+/-}$	$-11.$	$\boxed{+/-}$	$-11.$
$\boxed{+}$	$-11.$	$\boxed{-}$	$-11.$
$\boxed{23.25941}$	23.25941	$\boxed{23.25941}$	23.25941
$\boxed{=}$	12.25941	$\boxed{=}$	-34.25941
$\boxed{\div}$	12.25941	$\boxed{\div}$	-34.25941
$\boxed{6}$	$6.$	$\boxed{6}$	$6.$
$\boxed{=}$	2.043235	$\boxed{=}$	-5.709902

So $x \approx 2.04$ or $x \approx -5.71$.

APPLICATION SOLUTION

To solve the application, we need to substitute in the formula $A = \ell \cdot w$. As seen in the following drawing, the depth of the gutter is x and the width is $12 - 2x$. The area is 18 in.².

x x Equal amount bent up on each edge

12 in. wide $12 - 2x$

Formula: $A = \ell w$

Substitute: $18 = (12 - 2x)(x)$

Solve: $18 = 12x - 2x^2$

$2x^2 - 12x + 18 = 0$

$x^2 - 6x + 9 = 0$ *Divide both sides by 2*

$a = 1, b = -6, c = 9$

$x = \dfrac{-(-6) \pm \sqrt{(-6)^2 - 4(1)(9)}}{2 \cdot 1}$

$x = \dfrac{6 \pm \sqrt{36 - 36}}{2}$

$x = \dfrac{6 \pm \sqrt{0}}{2}$

$x = \dfrac{6}{2}$

$x = 3$

Answer: Therefore, the depth of the gutter is 3 in.

Exercises 10.2

A

Solve:

1. $x^2 - 3x - 18 = 0$ **2.** $x^2 + 6x - 40 = 0$

3. $y^2 + 12y + 35 = 0$ **4.** $x^2 + 11x + 18 = 0$

5. $x^2 + 9x + 14 = 0$ **6.** $x^2 - 9x + 14 = 0$

B

7. $6x^2 - 5x - 6 = 0$ **8.** $4x^2 - 27x - 7 = 0$

9. $20x^2 - 23x - 21 = 0$ **10.** $2x^2 + 11x - 6 = 0$

11. $12x^2 - 7x + 1 = 0$ **12.** $4x^2 - 5x - 21 = 0$

13. $4x^2 - 4x + 5 = 0$ **14.** $y^2 - 2y + 16 = 0$

C

15. $x^2 - 5x - 10 = 0$ **16.** $x^2 + 5x - 10 = 0$

17. $y^2 - y - 14 = 0$ **18.** $x(2x - 4) - 3 = 0$

19. $5x^2 - 3x - 1 = 0$ **20.** $-x^2 + 6x - 10 = 0$

21. $17x^2 + 2x - 13 = 0$ **22.** $26x + 45 = 7x^2$

23. $8x^2 - 10x - 20 = 7x - 2x^2$ **24.** $8x^2 + 5x + 10 = 3x^2 - 4x + 4$

ANSWERS

1. _____
2. _____
3. _____
4. _____
5. _____
6. _____
7. _____
8. _____
9. _____
10. _____
11. _____
12. _____
13. _____
14. _____
15. _____
16. _____
17. _____
18. _____
19. _____
20. _____
21. _____
22. _____
23. _____
24. _____

ANSWERS

25. _____

26. _____

27. _____

28. _____

29. _____

30. _____

31. _____

32. _____

33. _____

34. _____

35. _____

D

25. A piece of metal that is 16 in. wide is to be formed into a gutter of rectangular shape. Both sides are to be bent up an equal amount. What will be the depth of the gutter if the cross-sectional area is 32 in.²?

26. What would be the depth of the gutter in Exercise 25 if the piece of metal is 20 in. wide and the cross-sectional area is 50 in.²?

27. Elna wants to put her $1000 into an account that will be worth $1210 in two years. What rate of interest must she look for? (*Hint:* Use the formula in Example d, with $A = 1210$ and $I = 1000$.)

28. Jimmy wants to buy a microcomputer whose cost is $1690. If he has $1000, what interest rate will he need to find so that he will have $1690 in two years? (Use the formula in Example d.)

29. A rectangular piece of cardboard is 3 inches longer than it is wide. A 2-inch square is cut out of each corner and the edges turned up to form a container. If the volume of the container is 496 in.³, what are the dimensions of the piece of cardboard? (To the nearest hundredth.)

30. A variable electrical current is given by $i = t^2 - 8t + 15$. If t is in seconds, at what time is the current (i) equal to 16 amperes? (To the nearest hundredth.)

E   *Maintain Your Skills*

Combine the following radical expressions.

31. $4\sqrt{2} - \sqrt{18}$

32. $8\sqrt{6} - \sqrt{24}$

33. $\sqrt{48} + \sqrt{75} + \sqrt{27}$

34. $\sqrt{5} + \dfrac{1}{\sqrt{5}}$

35. $\sqrt{72} - \sqrt{8} - \sqrt{18}$

508

10.3 Quadratic Equations: A Review

OBJECTIVE

116. Solve quadratic equations by factoring, by taking the square root of each member, or by using the quadratic formula.

APPLICATION

A 4×6 in. photo is mounted in a frame of uniform width. What are the outside dimensions of the frame if the area of the frame and the area of the photo are equal?

VOCABULARY

No new vocabulary.

HOW AND WHY

Although all quadratic equations can be solved by using the quadratic formula, the other two methods are sometimes easier.

When given a quadratic equation, use the easiest of the following:

1. Check to see if it can be written in the form $x^2 = a$. If so, solve by taking the square root of each member.
2. Write the quadratic equation in the form $ax^2 + bx + c = 0$ and attempt to factor the left member. If you succeed, solve by setting each factor equal to zero.
3. Solve by using the quadratic formula.

$$x = \frac{-b \pm \sqrt{b^2 - 4ac}}{2a}$$

EXAMPLES

a. $3x^2 - 4x + 17 = 25 - 4x$

$$3x^2 = 8$$

$$x = \pm\sqrt{\frac{8}{3}} = \pm\frac{2}{3}\sqrt{6}$$

b. $\dfrac{x}{x+1} + \dfrac{10x}{(x+1)(x+3)} - \dfrac{15}{x+3} = 0$ *Note that $x \neq -1$ and $x \neq -3$ because division by 0 is not defined.*

$$x(x+3) + 10x - 15(x+1) = 0$$

$$x^2 + 3x + 10x - 15x - 15 = 0$$

$$x^2 - 2x - 15 = 0$$

$$(x-5)(x+3) = 0$$

$$x - 5 = 0 \quad \text{or} \quad x + 3 = 0$$

$$x = 5 \quad \text{or} \quad x = -3$$

Since $x \neq -3$ and $x \neq -1$, the solution is $x = 5$.

c. $(x-5)(x+4) = x(3x-5) + 3$

$$x^2 - x - 20 = 3x^2 - 5x + 3$$

$$2x^2 - 4x + 23 = 0$$

$$x = \frac{-(-4) \pm \sqrt{(-4)^2 - 4(2)(23)}}{2(2)}$$

$$x = \frac{4 \pm \sqrt{-168}}{4}$$

No real roots.

APPLICATION SOLUTION

If the area of the frame and the area of the photo are the same, we can conclude that the combined area is twice the area of the photo.

Simpler word form: $\left(\begin{array}{c}\text{area of frame}\\ \text{and photo}\end{array}\right) = 2\left(\begin{array}{c}\text{area of}\\ \text{photo}\end{array}\right)$

Select variable: Let x represent the width of the frame.

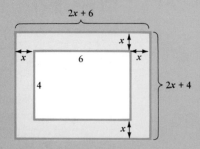

Translate to algebra: $(2x + 6)(2x + 4) = 2(4 \cdot 6)$

Solve: $4x^2 + 20x + 24 = 48$

$4x^2 + 20x - 24 = 0$

$x^2 + 5x - 6 = 0$

$(x + 6)(x - 1) = 0$

$x = -6 \quad \text{or} \quad x = 1$

Answer: Since the width of a picture frame cannot be negative, the root -6 is discarded. So the frame is 1 in. wide, and the outside dimensions of the frame are 6 in. × 8 in.

Exercises 10.3

ANSWERS

1. _____

2. _____

3. _____

4. _____

5. _____

6. _____

7. _____

8. _____

9. _____

10. _____

11. _____

12. _____

13. _____

14. _____

15. _____

16. _____

17. _____

18. _____

19. _____

20. _____

21. _____

A

Solve:

1. $x^2 = 4x + 5$

2. $16 = 5x - 2x^2$

3. $3x^2 - 8x - 60 = 0$

4. $x(x - 5) = 5(4 - x)$

5. $16x^2 + 4x = 8x^2$

6. $7x = 5 - 6x^2$

B

7. $3x^2 - x - 9 = 0$

8. $3x^2 + 8 = -12x$

9. $(x - 4)(2x + 3) = x^2 - 4x + 8$

10. $3x(x - 6) + 2 = (2x - 4)(x - 7) + 3$

11. $2(x - 1)^2 = (x + 3)^2 - 4$

12. $(x + 5)(x + 6) = (2x - 3)(x + 7)$

13. $(2x + 6)(3x - 5) + 32 = (2x - 3)(x + 7)$

14. $x(2x - 1) + 3x(1 - 4x) + 8 = 0$

C

15. $3 + \dfrac{2}{x} - \dfrac{5}{x^2} = 0$

16. $1 - \dfrac{3}{x} - \dfrac{40}{x^2} = 0$

17. $\dfrac{3}{x - 2} + 2 = \dfrac{2}{x + 5}$

18. $\dfrac{x}{x - 9} + \dfrac{3}{x - 4} = \dfrac{3}{(x - 9)(x - 4)}$

19. $\dfrac{6}{x - 10} = \dfrac{x + 2}{8}$

20. $(x + 3)^2 - 39 = (2x - 5)^2$

21. $\dfrac{x}{x - 7} - \dfrac{2x}{(x - 7)(x - 5)} + \dfrac{6}{x - 5} = 0$

22. $x - \dfrac{5x}{x+2} = \dfrac{10}{x+2}$

23. $\dfrac{x^2}{x-4} = 9 + \dfrac{56}{x-4}$

24. $\dfrac{10}{x+3} = x + 6$

D

25. A landscaper has a rectangular plot of ground that measures 18 ft by 14 ft. He wishes to put in a petunia bed as a border around the entire plot. What width should the bed be if the landscaper wants 140 ft² of the plot remaining for grass?

26. A metalworker must cut a rectangle with an area of 180 cm² from a flat piece of iron. If the rectangle is to be 3 cm longer than it is wide, what are the desired dimensions?

27. A metalworker must cut a rectangle with an area of 216 sq cm out of a flat piece of iron. If the rectangle is to be 6 cm longer than it is wide, what are the desired dimensions?

28. The radius of a circular arch of a certain height (h) and span (b) is given by

$$r = \frac{b^2 + 4h^2}{8h}$$

Find h when $r = 30$ ft and $b = 24$ ft.

29. A sheet metal worker needs to construct a rectangular duct with a cross section area of 340 sq cm. If the width is 3 cm less than the length, what are the needed dimensions?

30. A variable electrical current is given by $i = t^2 - 4t + 16$. If t is in seconds, at what time is the current (i) equal to 61 amperes?

E   *Maintain Your Skills*

Write the following in exponential form:

31. $\sqrt[3]{7}$ **32.** $\sqrt{8}$ **33.** $\sqrt{15}$

Write the following in radical form:

34. $(ab)^{1/3}$ **35.** $ab^{1/3}$

512

10.4 Equations Involving Radicals

OBJECTIVE

117. **Find the solution of an equation that contains the variable within a radical expression.**

APPLICATION

The pressure of the stream from a fire hydrant can be calculated from the formula

$$G = 26.8d^2\sqrt{p}$$

where G represents the discharge in gallons per minute (gpm), d represents the diameter of the outlet in inches, and p represents the pressure in pounds per square inch (psi). Find the pressure (psi) of water from an outlet that is 3.5 inches in diameter and that discharges at 680 gallons per minute (round to the nearest tenth).

VOCABULARY

An *extraneous root* is a root derived from an equation by multiplying each member by zero or raising each member to a power. The extraneous root is not a root of the original equation and can be eliminated by checking. For instance, an extraneous root is obtained from $x = 3$ by squaring both members: $(x)^2 = (3)^2$ or $x^2 = 9$. The roots of $x^2 = 9$ are $x = \pm 3$. However, -3 does not check in the original equation, $x = 3$, because $-3 \neq 3$.

HOW AND WHY

In order to solve an equation containing the variable in a radical expression, we eliminate the radical. Squaring a radical expression eliminates the radical. Assume that all expressions under the radical represent nonnegative numbers. $(\sqrt{3})^2 = 3$, $(\sqrt{x})^2 = x$, $(\sqrt{x-2})^2 = x - 2$, and $(\sqrt{x^2-5})^2 = x^2 - 5$.

To solve a radical equation that contains a single radical:

1. **Isolate the radical on one side of the equation.**
2. **Square both sides.**
3. **Solve for x.**
4. **Check for "extraneous solutions" (necessary since the equations may not be equivalent).**

Consider

$$\sqrt{x} = 9$$

$(\sqrt{x})^2 = (9)^2$ *Square both members*

$\quad x = 81$ *The solution is possibly 81*

Check:

$\sqrt{81} = 9$

$\quad 9 = 9$

Therefore, $x = 81$ is the solution.

If the radical is not isolated, it will not be eliminated by squaring.

$$\sqrt{x-1} + 1 = 7$$

$$(\sqrt{x-1} + 1)^2 = (7)^2 \qquad \text{\textit{Square both members}}$$

$$x - 1 + 2\sqrt{x-1} + 1 = 49 \qquad \text{\textit{Radical not eliminated}}$$

For this reason, if the radical is not isolated, it is necessary to isolate it prior to squaring.

$$\sqrt{x-1} + 1 = 7$$

$$\sqrt{x-1} = 6 \qquad \textit{Isolate the radical}$$

$$(\sqrt{x-1})^2 = (6)^2 \qquad \textit{Square both members}$$

$$x - 1 = 36$$

$$x = 37 \qquad \textit{The solution is possibly 37}$$

Check:

$$\sqrt{37-1} + 1 = 7$$

$$\sqrt{36} + 1 = 7$$

$$6 + 1 = 7$$

$$7 = 7$$

Therefore, $x = 37$ is the solution.

If two radicals occur in an equation:

1. **Isolate each radical, one at a time.**
2. **Square both sides.**
3. **Repeat until all radicals have been eliminated.**
4. **Check for "extraneous solutions."**

$$\sqrt{x-5} + \sqrt{x} = 5$$

$$\sqrt{x-5} = 5 - \sqrt{x} \qquad \textit{Isolate one radical}$$

$$(\sqrt{x-5})^2 = (5 - \sqrt{x})^2 \qquad \textit{Square both members}$$

$$x - 5 = 25 - 10\sqrt{x} + x$$

$$-30 = -10\sqrt{x}$$

$$3 = \sqrt{x} \qquad \textit{Isolate second radical}$$

$$(3)^2 = (\sqrt{x})^2 \qquad \textit{Square both members}$$

$$9 = x \qquad \textit{The solution is possibly 9}$$

Check:

$$\sqrt{9-5} + \sqrt{9} = 5$$

$$\sqrt{4} + \sqrt{9} = 5$$

$$2 + 3 = 5$$

$$5 = 5$$

Therefore, $x = 9$ is the solution.

Consider

$$\sqrt{x+4} = 2x - 7$$

$$x + 4 = 4x^2 - 28x + 49 \qquad \textit{Square both sides}$$

$$4x^2 - 29x + 45 = 0$$

$$x = 5 \quad \text{or} \quad x = \frac{9}{4} \qquad \textit{Possible solutions found by quadratic formula}$$

Check:

$$\sqrt{5+4} = 2(5) - 7 \qquad \sqrt{\dfrac{9}{4} + 4} = 2\left(\dfrac{9}{4}\right) - 7$$

$$3 = 3 \qquad\qquad\qquad \dfrac{5}{2} = -\dfrac{5}{2} \text{ false}$$

Therefore, the solution is $x = 5$.

EXAMPLES

a. $\sqrt{5 - x} = -3$

No real solution because $\sqrt{5 - x}$ cannot be negative.

b. $\sqrt{3 - x} + x = 6$

$\sqrt{3 - x} = 6 - x$

$(\sqrt{3 - x})^2 = (6 - x)^2$

$3 - x = 36 - 12x + x^2$

$x^2 - 11x + 33 = 0$

$a = 1, b = -11, c = 33$

$x = \dfrac{-(-11) \pm \sqrt{(-11)^2 - 4(1)(33)}}{2(1)}$

$x = \dfrac{11 \pm \sqrt{121 - 132}}{2}$

$x = \dfrac{11 \pm \sqrt{-11}}{2}$

No real solution because $\sqrt{-11}$ is not real.

c. $\sqrt{x + 3} - \sqrt{x - 5} = 2$

$\sqrt{x + 3} = 2 + \sqrt{x - 5}$

$(\sqrt{x + 3})^2 = (2 + \sqrt{x - 5})^2$

$x + 3 = 4 + 4\sqrt{x - 5} + x - 5$

$4 = 4\sqrt{x - 5}$

$1 = \sqrt{x - 5}$

$(1)^2 = (\sqrt{x - 5})^2$

$1 = x - 5$

$x = 6$

Check: $\sqrt{x + 3} - \sqrt{x - 5} = 2$

$\sqrt{6 + 3} - \sqrt{6 - 5} = 2$

$\sqrt{9} - \sqrt{1} = 2$

$3 - 1 = 2$

$2 = 2$

Therefore, $x = 6$ is the solution

APPLICATION SOLUTION

To find the pressure, substitute the known values in the formula.

Formula: $G = 26.8d^2\sqrt{p}$

$G = 680,\ d = 3.5$

Substitute: $680 = 26.8(3.5)^2\sqrt{p}$

Solve: $\dfrac{680}{26.8(3.5)^2} = \sqrt{p}$ *Isolate the radical*

$2.071 \approx \sqrt{p}$ *Simplify*

$4.29 \approx p$ *Square both sides*

Answer: The pressure (to the nearest tenth) is 4.3 psi.

WARM UPS

Solve:

1. $\sqrt{x} = 5$ **2.** $\sqrt{a} = 3$ **3.** $\sqrt{s} = 8$

4. $\sqrt{t} = 12$ **5.** $\sqrt{x-1} = 2$ **6.** $\sqrt{b+2} = 4$

7. $\sqrt{x+3} = -2$ **8.** $\sqrt{x-6} = -3$ **9.** $\sqrt{x+2} = 3$

10. $\sqrt{x+1} = 10$ **11.** $\sqrt{x} + 2 = 3$ **12.** $\sqrt{x} + 1 = 10$

Exercises 10.4

ANSWERS

1. _____

2. _____

3. _____

4. _____

5. _____

6. _____

7. _____

8. _____

9. _____

10. _____

11. _____

12. _____

13. _____

14. _____

15. _____

16. _____

17. _____

18. _____

19. _____

20. _____

21. _____

22. _____

23. _____

24. _____

A

Solve:

1. $\sqrt{x} = 8$

2. $\sqrt{x+9} = 3$

3. $\sqrt{x-8} = -5$

4. $\sqrt{9-x} = 0$

5. $\sqrt{x} + 2 = 6$

6. $\sqrt{x} - 6 = -2$

B

7. $\dfrac{6}{\sqrt{x}} = 1$

8. $\dfrac{1}{\sqrt{x}} = 3$

9. $\sqrt{x+6} + 10 = 3$

10. $\sqrt{x+8} = \sqrt{-x}$

11. $\sqrt{x-5} = x - 5$

12. $\sqrt{x+8} + 4 = x$

13. $13 + \sqrt{x+7} = x$

14. $6 - \sqrt{x-9} = 3$

C

15. $\sqrt{x} - \sqrt{x+2} = 6$

16. $\sqrt{x+3} + \sqrt{x-2} = 10$

17. $\sqrt{x} + \sqrt{x+2} = 3$

18. $\sqrt{2x} + 3 = \sqrt{x-7}$

19. $\sqrt{x-5} - \sqrt{x-29} = 4$

20. $\sqrt{2x-2} = x - 5$

21. $\dfrac{2}{\sqrt{x+3}} + 5 = 0$

22. $\dfrac{1}{\sqrt{x-5}} - \dfrac{2}{\sqrt{x+10}} = 0$

23. $\sqrt{x^2 - 5x} = 2\sqrt{6}$

24. $\sqrt{x} - \dfrac{3}{\sqrt{x}} = 2$

ANSWERS
25.
26.
27.
28.
29.
30.
31.
32.
33.
34.
35.

D

25. Use the formula in the application to find the pressure of water from a fire hydrant outlet that is 4.5 inches in diameter and that discharges at 900 gallons per minute (round to nearest tenth).

26. Use the formula in the application to find the pressure of water from a fire hydrant outlet that is 3.75 inches in diameter and that discharges at 1100 gallons per minute (round to nearest tenth).

27. Find the stroke of a 4-cylinder, 180-in.^{3} displacement engine if the bore of each cylinder is 3.5 inches. (To the nearest hundredth.) The formula is

$$\text{Bore} = \sqrt{\frac{1}{.7854} \cdot \left(\frac{\text{Displacement}}{\text{\# of cylinders}}\right) \cdot \frac{1}{\text{Stroke}}}$$

28. At what height above sea level must a citizens band antenna be placed if it is to transmit a signal 30 miles? Distance traveled D is given by $D = \sqrt{h(h + 7800)}$ where h is height, and h and D are in miles.

29. The sum of the perimeters of two squares is 128 feet. One square has an area that is nine times the area of the other square. What is the area of each square?

30. The interest rate (r) (compounded annually) needed to have P dollars grow to A dollars at the end of two years is given by $r = \sqrt{\dfrac{A}{P}} - 1$. Find the value of A if $P = 2000$ and $r = .1$.

E   *Maintain Your Skills*

Write the following without fractions, using negative exponents:

31. $\dfrac{1}{x^2}$

32. $\dfrac{1}{(x + y)^2}$

33. $\dfrac{a^2}{b^3}$

Write the following using positive exponents only:

34. $a^2 b^{-1}$

35. $\dfrac{1}{x^{-3}}$

518

Chapter 10
POST-TEST

Solve:

1. **(Obj. 115)** $x^2 - 3x - 5 = 0$

2. **(Obj. 117)** $\sqrt{x + 39} + \sqrt{x + 7} = 8$

3. **(Obj. 114)** $4x^2 - 8 = 16$

4. **(Obj. 116)** $x^2 - 6x - 40 = 0$

5. **(Obj. 117)** $\dfrac{1}{\sqrt{x-1}} + 3 = 2$

6. **(Obj. 116)** $\dfrac{x}{x+3} + \dfrac{2}{x-6} = \dfrac{3x}{(x+3)(x-6)}$

7. **(Obj. 117)** $\sqrt{x + 5} - 6 = 13$

8. **(Obj. 114)** $x^2 - 100 = 0$

9. **(Obj. 115)** $2x^2 + 7x = 1$

10. **(Obj. 115)** $x^2 - 5x + 11 = 0$

11. **(Obj. 115)** A rectangular plot of ground has its length equal to 6 m less than twice its width. If the area is 108 m², what are the dimensions of the plot?

12. **(Obj. 117)** The height (S) of a rocket fired from ground level with an initial velocity of 320 ft/sec at any time (t) is given by the formula

$$S = 320t - 16t^2$$

(S in feet and t in seconds). In how many seconds will the rocket reach a height of 1600 ft?

ANSWERS TO WARM UPS (10.4)   **1.** $x = 25$ **2.** $a = 9$ **3.** $s = 64$ **4.** $t = 144$ **5.** $x = 5$

6. $b = 14$ **7.** no real solution   **8.** no real solution   **9.** $x = 7$ **10.** $x = 99$

11. $x = 1$ **12.** $x = 81$

11

Graphing

1. _____

The problems in the following pre-test are a sample of the material in the chapter. You may already know how to work some of these. If so, this will allow you to spend less time on those parts. As a result, you will have more time to give to the sections that you find more difficult or that are new to you. The answers are in the back of the text.

1. (Obj. 118) Locate the following points on the coordinate system: $A\,(-5,\,-5)$, $B\,(0,\,4)$, $C\,(-3,\,2)$, $D\,(4,\,-3)$, and $E\,(-2,\,0)$.

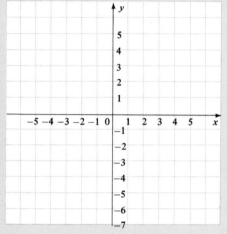

2. A _____

B _____

2. (Obj. 119) Identify the coordinates of the points on the following graph.

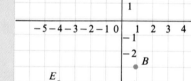

C _____

D _____

3. (Obj. 120) Draw the graph of $3y - 5x + 15 = 0$.

E _____

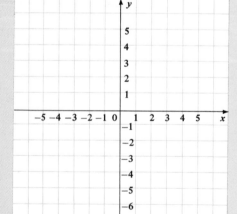

3. _____

522

4. (Obj. 121) Write the coordinates of the x-intercept of the line represented by the equation $2x - 7y - 28 = 0$.

5. (Obj. 122) Write the coordinates of the y-intercept of the line represented by $2x - 7y - 28 = 0$.

6. (Obj. 123) Write the slope of the line that passes through the points $(7, -4)$ and $(-3, 1)$.

7. (Obj. 124) Write the slope of the line represented by the equation $5x - 3y = 18$.

8. (Obj. 125) Find the distance between the points $(-5, -3)$ and $(-8, 1)$.

9. (Obj. 126) Draw the graph of the line through $(-3, 0)$ that has slope $m = \dfrac{2}{3}$.

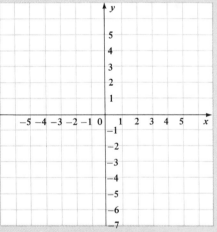

10. (Obj. 127) Solve the following system of equations by graphing.

$$\begin{cases} 3x - y = 3 \\ x + y = 5 \end{cases}$$

11. (Obj. 129) Solve the following system of equations by addition.

$$\begin{cases} 3x - 2y = 8 \\ 4x + 3y = 5 \end{cases}$$

12. (Obj. 128) Chang has $6000 in two savings accounts. One account pays 8% simple interest and the other pays 6% simple interest. If Chang received $410 in interest annually, how much money is invested in each account?

11.1 Rectangular Coordinate System

118. Given the coordinates for a point, locate that point on a rectangular coordinate system.

119. Find the coordinates of a point, given its graph.

APPLICATION

What are the coordinates of Mount St. Helens, Washington, which are indicated on the map?

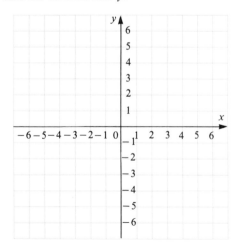

VOCABULARY

When two number lines are set at right angles to each other with the intersection at 0 (the *origin*), the resulting figure is called a *rectangular coordinate system.* The lines (*axes*) divide the plane into four parts called *quadrants.* The horizontal axis is called the *x-axis* and the vertical axis the *y-axis.* Points on the plane are identified by *ordered pairs* of numbers (x, y), called *coordinates* of the point. The *x-value,* the first member of the ordered pair, is called the *abscissa.* The *y-value,* the second member, is called the *ordinate.* A *graph* of a set of ordered pairs is the set of points corresponding to the ordered pairs.

HOW AND WHY

To find the graph of $(6, 3)$ we will first need to construct the rectangular coordinate system. Draw a horizontal number line and label it.

Now draw a vertical number line intersecting the first one at 0. Label the horizontal line x and the vertical one y.

To plot the point (6, 3), proceed 6 units to the right (positive) on the *x*-axis and 3 units up (positive) on the *y*-axis. Complete the rectangle, and the corner opposite the origin is the graph of (6, 3).

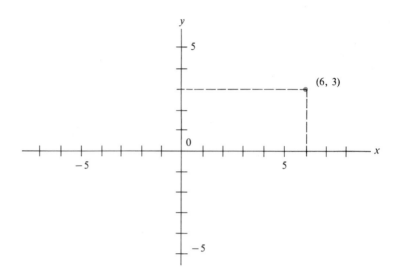

Note that the point could have been found by proceeding right 6 units and then up 3 units.

To find the coordinates of a point when its graph is given, read the *x* and *y* values off the axes.

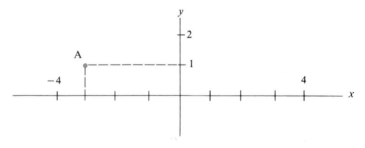

The point *A* has −3 for the *x*-value and 1 for the *y*-value, so the coordinates are (−3, 1).

EXAMPLES

a. Plot the following points: $A(3, -2)$, $B(-5, 3)$, $C(-4, -1)$, $D(1.5, 5)$, and $E(3, -4.5)$.

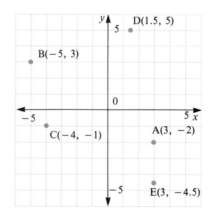

b. Identify the coordinates of the points on the following graph.

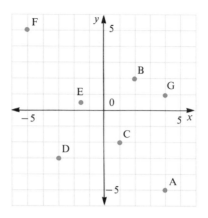

The coordinates are $A(4, -5)$, $B(2, 2)$, $C(1, -2)$, $D(-3, -3)$, $E(-1.5, .5)$, $F(-5, 5)$, and $G(4, 1)$.

APPLICATION SOLUTION

We return to the map of Washington and see that the horizontal coordinate is 3, while the vertical coordinate is F. Therefore, the coordinates of Mount St. Helens on this map are $(3, F)$.

WARM UPS

Plot the following points on the rectangular coordinate system given below.

1. $(3, 2)$ **2.** $(5, 1)$ **3.** $(-2, 5)$

4. $(-1, 3)$ **5.** $(-4, -2)$ **6.** $(3, -7)$

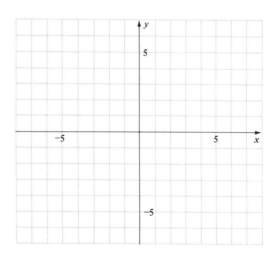

Name each of the following points using ordered pairs.

POINT

7. A

8. E

9. C

10. D

11. F

12. B

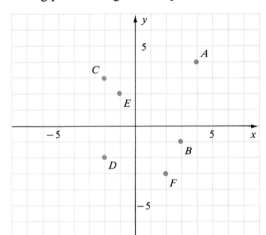

NAME _____ CLASS _____

NAME _____ CLASS _____

NAME _____ CLASS _____

ANSWERS

Exercises 11.1

A

Construct a rectangular coordinate system and plot the following points.

1. $(3, 4)$ **2.** $(-3, -2)$ **3.** $(-1, -4)$

4. $(-6, 2)$ **5.** $(-5, 4)$ **6.** $(4, -2)$

B

Name each of the following points using ordered pairs.

POINT

7. *A*

8. *B*

9. *C*

10. *D*

11. *E*

12. *F*

13. *G*

14. *H*

7. _____

8. _____

9. _____

10. _____

11. _____

12. _____

13. _____

14. _____

C

Construct a rectangular coordinate system and plot the following points.

15. $(2.5, 1)$ **16.** $\left(4\frac{1}{2}, -1\right)$ **17.** $(-1.5, -2)$

18. $(-3, 3.5)$ **19.** $(0, 5)$

529

ANSWERS

20. _____

21. _____

22. _____

23. _____

24. _____

25. _____

26. _____

27. _____

28. _____

29. _____

30. _____

31. _____

Name each of the following points using ordered pairs.

POINT

20. *A*

21. *B*

22. *C*

23. *D*

24. *E*

D

25. On the map of Washington in the application, what are the coordinates of Yakima?

26. On the map of Washington in the application, what are the coordinates of Seattle?

E   *Maintain Your Skills*

Write the following in scientific notation:

27. 38,200,000   28. .000000235   29. .0185

Write the following in place value notation:

30. 5.86×10^{-5} 31. 6.87×10^{11}

ANSWERS TO WARM UPS (11.1)   **1–6.**

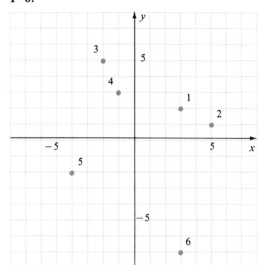

7. $(4, 4)$ 8. $(-1, 2)$

9. $(-2, 3)$ 10. $(-2, -2)$

11. $(2, -3)$ 12. $(3, -1)$

11.2 Graphs of Linear Equations

OBJECTIVE

120. Draw the graph of a linear equation by plotting and connecting several points.

APPLICATION

A company can depreciate the value of a small electronic control unit for tax purposes. One procedure is to use "straight-line depreciation." The procedure can be written in equation form

$$y = -20x + 250$$

where y represents the value of the unit and x represents the number of years the unit is in operation.

One of these units costs $250 and has a scrap value of $50 at the end of 10 years. Graph the equation and list the coordinates that show the value of the unit each year.

VOCABULARY

A *linear equation* is one that can be written in the form $ax + by + c = 0$ with a and b not both zero. The graph of a linear equation is a straight line.

HOW AND WHY

Consider the linear equation $3x + 2y - 6 = 0$. Its graph can be found by setting up a table of values. To make a table of values, assign any value for x and solve for y.

$x = 0$	$x = 3$
$3(0) + 2y - 6 = 0$	$3(3) + 2y - 6 = 0$
$2y = 6$	$9 + 2y - 6 = 0$
$y = 3$	$2y = -3$
	$y = -1.5$

$$x = -2 \qquad\qquad x = 4$$
$$3(-2) + 2y - 6 = 0 \qquad 3(4) + 2y - 6 = 0$$
$$-6 + 2y - 6 = 0 \qquad\qquad 2y = -6$$
$$2y = 12 \qquad\qquad\qquad y = -3$$
$$y = 6$$

x	y	(x, y)
0	3	$(0, 3)$
3	-1.5	$(3, -1.5)$
-2	6	$(-2, 6)$
4	-3	$(4, -3)$

Now construct a rectangular coordinate system, plot the points in the table, and draw the line joining the points.

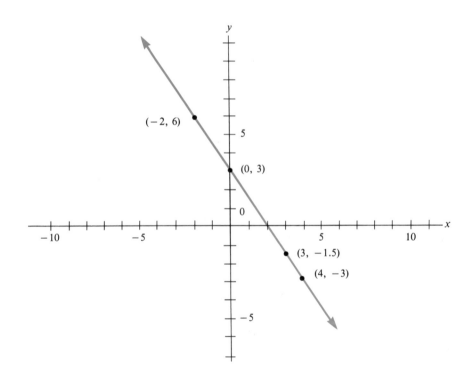

The significance of the graph lies in the fact that every ordered pair of numbers that makes the equation true represents a point that lies on the line. Every point on the line has coordinates that make the equation true.

To draw a graph of a line, given its equation,

1. **Determine three or more points by assigning values to x and solving for y.**
2. **Plot the points on a rectangular coordinate system.**
3. **Draw the line joining the points.**

EXAMPLES

Draw the graph of $x = 5y + 10$.

x	y
10	0
0	-2
15	1
5	-1

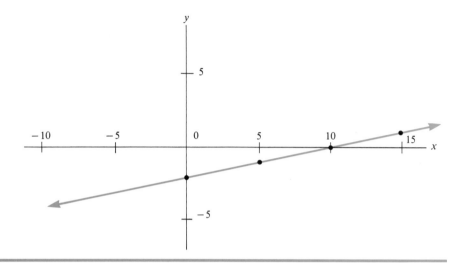

APPLICATION SOLUTION

When we draw the graph of

$y = -20x + 250$

we will be able to find the value of the electronic unit at any future time (up to 10 years) by looking at the x- and y-coordinates. Make a table of values:

x	y
0	250	$y = -20(0) + 250 = 250$
4	170	$y = -20(4) + 250 = 170$
8	90	$y = -20(8) + 250 = 90$

Plot the points with these coordinates, and draw the line joining these points. (The graph stops when $x = 10$ because the unit is scrapped after 10 years.) To make it possible to draw the graph on one page, we use the following scale: On the x-axis, one unit = 1 year; on the y-axis, one unit = 10 dollars.

Continued on next page.

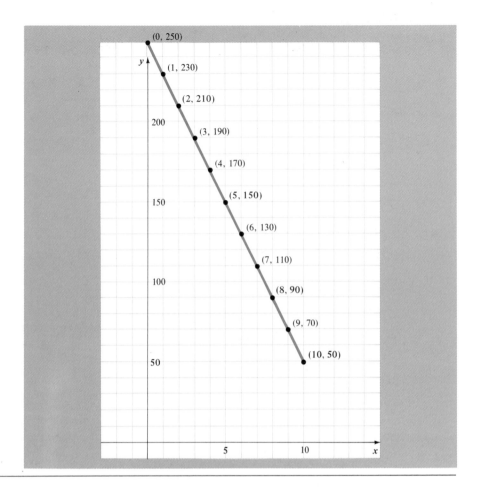

WARM UPS

Complete the following table of values.

1–4. $y = x + 1$

x	y
7	
3	
−9	
	0

5–8. $y = \frac{1}{2}x + 1$

x	y
0	
	0
−4	
	−4

9–12. $y = -2x + 4$

x	y
3	
5	
	8
	10

Exercises 11.2

ANSWERS

1. _____

2. _____

3. _____

4. _____

5. _____

6. _____

7. _____

8. _____

9. _____

10. _____

11. _____

12. _____

13. _____

14. _____

15. _____

16. _____

17. _____

18. _____

19. _____

20. _____

21. _____

22. _____

23. _____

24. _____

A

Make a table of values and graph each of the following.

1. $y = \frac{1}{2}x$ **2.** $y = 2x - 5$ **3.** $y = -x$

4. $y = -3x + 2$ **5.** $y + x = 6$ **6.** $2y + 2x = 5$

B

7. $5x - 7y = 10$ **8.** $6x + 5y = 15$ **9.** $3x = 4y + 12$

10. $7y - x = 14$ **11.** $-2x + y = 4$ **12.** $2x + y = 2$

13. $2x + 3y = 6$ **14.** $3x + 4y = 12$

C

15. $9x - 5y = 45$ **16.** $5x - 7y = 10$

17. $4x - 5y = 18$ **18.** $3x = 9y + 18$

19. $6x + 5y = 15$ **20.** $5x = 4y + 20$

21. $.2x + .5y = 1$ **22.** $.1x + .2y = 1$

23. $\frac{x}{4} - \frac{y}{7} = \frac{1}{28}$ **24.** $\frac{x}{3} - \frac{y}{5} = 2$

535

D

25. A business purchases an automobile for $8000 and will use straight-line depreciation over a period of 5 years to a value of $500. This can be described as $y = -1500x + 8000$, where y is the value of the automobile and x is the number of years from 0 to 5. Graph this equation.

26. A building purchased for $45,000 is depreciated over a 12.5-year period to no value. The formula for this is $y = 45000 - 3600x$, where y is the value of the building and x is the number of years from 0 to 12. Graph this equation. (*Hint:* Let one unit on the y-axis represent $10,000.)

27. If twice the length (x) of a rectangle is added to three times the width (y), the sum is 37 (i.e., $2x + 3y = 37$). The graph of the equation will give the length and width of all such rectangles. (Disregard coordinates with negative values.) Draw the graph.

28. If three times the length (x) of a rectangle is added to four times the width (y), the sum is 24 (i.e., $3x + 4y = 24$). The graph of the equation will give the length and width of all such rectangles. (Disregard coordinates with negative values.) Draw the graph.

29. The cost of renting a circular saw is $5 for the first day and $3 for each additional day the saw is kept. This can be expressed by the equation $y = 3x + 2$, where y is the total rental fee and x is the number of days the saw is kept. Draw the graph of this equation.

30. The cost of renting a car from Rent-A-Lemon is $15 plus 20¢ per mile. This can be expressed by the equation $y = .2x + 15$, where y is the rental fee and x is the number of miles driven. Draw the graph of the equation.

E   *Maintain Your Skills*

Solve:

31. $x^2 = 169$

32. $y^2 = -12$

33. $5a^2 = a^2 + 12$

34. $2x^2 + 12 = x^2 + 12$

35. $4x^2 - 3x - 4 = 2x^2 - 3x + 28$

ANSWERS TO WARM UPS (11.1)

1–4.

x	y
7	8
3	4
−9	−8
−1	0

5–8.

x	y
0	1
−2	0
−4	−1
−10	−4

9–12.

x	y
3	−2
5	−6
−2	8
−3	10

11.3 Properties of Lines

OBJECTIVES

121. Find the *x*-intercept given the equation of a line.
122. Find the *y*-intercept given the equation of a line.
123. Find the slope of a line given two points on the line.
124. Find the slope of a line given its equation.

APPLICATION

The roof of a house rises 3 ft for every 5 ft of run. $5y = 3x$ is the equation that will determine the amount of rise y that corresponds to a run x. The slope of the line determined by the equation is called the pitch of the roof. Find the pitch of the roof.

VOCABULARY

The *x and y intercepts* are the points where a line crosses (intercepts) the respective axes. The description of the slant of the line is called the *slope*.

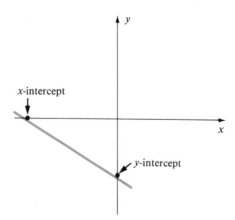

The slope is expressed as a ratio of the vertical change to the horizontal change.

HOW AND WHY

A look at the coordinate system will verify the fact that every point on the *y*-axis has its abscissa (*x*-value) zero and every point on the *x*-axis has its ordinate (*y*-value) zero.

To find the *y*-intercept, substitute 0 for *x* in the equation and solve for *y*.
To find the *x*-intercept, substitute 0 for *y* in the equation and solve for x.

Consider

$$3x + 4y = 12$$

If $x = 0$, $3(0) + 4y = 12$

$$4y = 12$$

$$y = 3$$

Therefore, the y-intercept is at $(0, 3)$.

If $y = 0$,

$$3x + 4(0) = 12$$

$$3x = 12$$

$$x = 4$$

Therefore, the x-intercept is at $(4, 0)$.

The slope of a line is the ratio of the change in the vertical to the change in the horizontal between any two points on the line. The slope is constant regardless of the two points chosen.

Consider

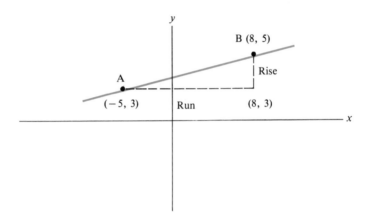

Notice that if you start at point A and proceed to point B, you can proceed horizontally (run) until below point B and then vertically (rise) until you reach B itself. The slope is then the ratio $\dfrac{\text{rise}}{\text{run}}$. Slope $= m = \dfrac{\text{rise}}{\text{run}}$.

The rise is from $y = 3$ to $y = 5$ and is given by $5 - 3 = 2$. The run is from $x = -5$ to $x = 8$ and is given by $8 - (-5) = 13$. Therefore,

$$m = \frac{5 - 3}{8 - (-5)} = \frac{2}{13}$$

In general, if two points, $A(x_1, y_1)$ and $B(x_2, y_2)$, are given, then the slope of the line through them is

$$\text{Slope} = m = \frac{y_2 - y_1}{x_2 - x_1} \quad \text{or} \quad \frac{y_1 - y_2}{x_1 - x_2}$$

The slope of a line can also be found by solving the equation for y.

When the equation is written in the form $y = mx + b$, the coefficient of x is the slope and $(0, b)$ is the y-intercept.

Consider

$$3y + 2x = 8$$

$$3y = -2x + 8$$

$$y = -\frac{2}{3}x + \frac{8}{3}$$

So $m = -\frac{2}{3}$ and the y-intercept is $\left(0, \frac{8}{3}\right)$.

EXAMPLES

a. Find the slope of the line passing through the points $(6, 4)$ and $(-5, 13)$.

$$m = \frac{y_2 - y_1}{x_2 - x_1} = \frac{13 - 4}{-5 - 6} = \frac{9}{-11} = -\frac{9}{11}$$

b. Find the slope of the line represented by the equation $x - y = 6$. Identify two points

x	y
0	-6
6	0

$$m = \frac{0 - (-6)}{6 - 0} = \frac{6}{6} = 1$$

c. $3x - 2y = 18$

$$-2y = -3x + 18$$

$$y = \frac{3}{2}x - 9$$

Therefore, the slope is $\frac{3}{2}$ and the y-intercept is $(0, -9)$

d. $6x + 5y = 11$

$$5y = -6x + 11$$

$$y = -\frac{6}{5}x + \frac{11}{5}$$

Therefore, the slope is $-\frac{6}{5}$ and the y-intercept is $\left(0, \frac{11}{5}\right)$.

APPLICATION SOLUTION

Now we can find the pitch of the roof (slope of the line) in the application.

$$5y = 3x$$

$$y = \frac{3}{5}x$$

So the pitch (slope) is $\frac{3}{5}$.

WARM UPS

Find the *x*-intercept and the *y*-intercept of the graph of each of the following equations:

1. $5x + 2y = 10$ **2.** $5x - 6y = 30$

3. $2x - 3y = -6$ **4.** $x - 7y = -14$

Find the slope of the line containing the given pair of points:

5. $(4, 3)(2, 1)$ **6.** $(-2, 1), (-3, 2)$

7. $(3, -8), (3, 5)$ **8.** $(11, -6), (-1, -6)$

Find the slope of the line and the *y*-intercept of the graph of each of the following equations:

9. $2y = 6x + 12$ **10.** $2y = -2x - 12$

11. $3y = 2x + 15$ **12.** $3y = x - 4$

Exercises 11.3

ANSWERS

1. _____

2. _____

3. _____

4. _____

5. _____

6. _____

7. _____

8. _____

9. _____

10. _____

11. _____

12. _____

13. _____

14. _____

15. _____

16. _____

17. _____

A

Find the x-intercept and the y-intercept of the graph of each of the following equations:

1. $3x + y = 9$ **2.** $8x - 3y = 24$

Find the slope of the line containing the given pair of points:

3. $(4, 7), (8, 10)$ **4.** $(-11, 3), (7, 0)$

Find the slope of the line and the y-intercept of the graph of each of the following equations:

5. $y = -\dfrac{1}{2}x + 6$ **6.** $y = \dfrac{3}{8}x + \dfrac{5}{8}$

B

Find the x-intercept and the y-intercept of the graph of each of the following equations:

7. $2x + 5y = 11$ **8.** $3x + 7y = 18$ **9.** $6x - 7y = 12$

Find the slope of the line containing the given points:

10. $(6, -5), (-3, -2)$ **11.** $(-3, -6), (-11, 12)$

12. $(4, -6), (-5, -3)$

Find the slope of the line and the y-intercept of the graph of each of the following equations:

13. $-2x + 5y = 8$ **14.** $4x - 5y = -12$

C

Find the x-intercept and the y-intercept of the graph of each of the following equations:

15. $\dfrac{1}{3}x + \dfrac{1}{2}y = -6$ **16.** $\dfrac{1}{2}x + \dfrac{1}{5}y = -3$

17. $-\dfrac{1}{7}x + \dfrac{1}{3}y = \dfrac{1}{2}$

ANSWERS

18. _____

19. _____

20. _____

21. _____

22. _____

23. _____

24. _____

25. _____

26. _____

27. _____

28. _____

29. _____

30. _____

31. _____

32. _____

Find the slope of the line containing the given points for each of the following:

18. $(5.6, -3.2), (4.1, 3.7)$ **19.** $(4.2, 2), (2.5, -3.1)$

20. $\left(\dfrac{2}{3}, -\dfrac{1}{2}\right), \left(\dfrac{1}{6}, \dfrac{3}{4}\right)$

Find the slope of the line and the y-intercept of the graph of each of the following equations:

21. $\dfrac{1}{2}x - \dfrac{1}{3}y = -4$ **22.** $\dfrac{2}{3}x - \dfrac{1}{2}y = 8$

23. $4x + \dfrac{3}{5}y = -\dfrac{1}{2}$ **24.** $\dfrac{1}{2}x + \dfrac{4}{7}y = -\dfrac{2}{3}$

D

25. A part of the road from Reno to Lake Tahoe, Nevada, has a 6% grade. The equation $100y = 6x$ describes the grade or slope. What is the slope of the road?

26. A home in Aspen, Colorado, has a steep roof. For every four feet of run it has three feet of rise. What is the slope of the line representing the roof?

27. The U-Drive-It auto leasing firm uses the following formula to compute the cost (c) of driving an auto x miles:

$c = .17x + 10$

What is the cost per mile (slope of line)? What is the cost if the car is not driven during a day (c-value of the c-intercept)?

E   *Maintain Your Skills*

Solve, using the quadratic formula:

28. $x^2 - 12x + 32 = 0$ **29.** $x^2 - 5x + 12 = 0$

30. $5x^2 - 7x - 12 = 0$ **31.** $3x^2 - 5x - 6 = 0$

32. $12x^2 + 24x - 36 = 0$

ANSWERS TO WARM UPS (11.3)   **1.** (2, 0) and (0, 5)   **2.** (6, 0) and (0, −5)   **3.** (−3, 0) and (0, 2)

4. (−14, 0) and (0, 2)   **5.** $m = 1$ **6.** $m = -1$

7. No slope   **8.** $m = 0$ **9.** $m = 3$, (0, 6)

10. $m = -1$, (0, −6)   **11.** $m = \dfrac{2}{3}$, (0, 5)   **12.** $m = \dfrac{1}{3}$, $\left(0, \dfrac{-4}{3}\right)$

11.4 Distance Between Two Points

OBJECTIVE

125. Find the distance between two points on a graph, given their coordinates.

APPLICATION

A triangle has vertices at $A\,(0, 0)$, $B\,(10, 0)$, and $C\,(6, 8)$. Find the lengths of its sides.

VOCABULARY

No new vocabulary.

HOW AND WHY

The distance between two points on a graph is given by

$$d = \sqrt{(x_2 - x_1)^2 + (y_2 - y_1)^2}$$

where (x_1, y_1) and (x_2, y_2) are the coordinates of the points A and B.

This formula can be developed using the theorem of Pythagoras and the following illustration.

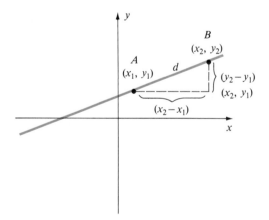

EXAMPLES

a. Find the distance between (3, 2) and (5, −1). Let $(x_1, y_1) = (3, 2)$ and $(x_2, y_2) = (5, -1)$.

$$d = \sqrt{(5 - 3)^2 + (-1 - 2)^2}$$

$$d = \sqrt{(2)^2 + (-3)^2} = \sqrt{4 + 9} = \sqrt{13}$$

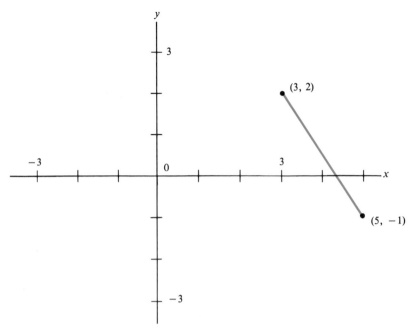

b. Find the distance between $(-3, -7)$ and $(-4, 3)$. Let $(x_1, y_1) = (-3, -7)$ and $(x_2, y_2) = (-4, 3)$.

$$d = \sqrt{[-4 - (-3)]^2 + [3 - (-7)]^2}$$

$$d = \sqrt{(-1)^2 + (10)^2}$$

$$d = \sqrt{1 + 100}$$

$$d = \sqrt{101} \approx 10.05$$

APPLICATION SOLUTION

First plot the points and draw the triangle.

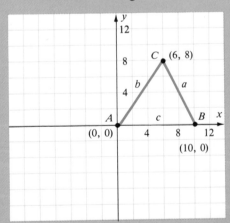

Now use the distance formula to find the lengths of the sides.

$$a = \sqrt{(10 - 6)^2 + (0 - 8)^2} \qquad b = \sqrt{(6 - 0)^2 + (8 - 0)^2}$$

$$= \sqrt{16 + 64} \qquad\qquad\qquad = \sqrt{36 + 64}$$

$$= \sqrt{80} \qquad\qquad\qquad\quad = \sqrt{100}$$

$$= 4\sqrt{5} \qquad\qquad\qquad\quad = 10$$

Continued on next page.

$$c = \sqrt{(10 - 0)^2 + (0 - 0)^2}$$
$$= \sqrt{100}$$
$$= 10$$

So the lengths of the sides are 10, 10, and $4\sqrt{5}$.

WARM UPS

Find the distance between each of the following pairs of points.

1. (0, 3)(4, 0)   2. (5, 7)(2, 3)

3. (9, 7)(13, 10)   4. (−2, −2)(1, 2)

5. (8, 0)(0, 6)   6. (5, 1)(−1, 9)

7. (3, 1)(9, 9)   8. (−3, −7)(5, −1)

9. $(\sqrt{3}, 1)(0, 1)$ 10. $(-\sqrt{8}, 1)(0, 0)$

11. (5, 0)(0, 12)   12. (1, 0)(0, 1)

Exercises 11.4

ANSWERS

1. _____
2. _____
3. _____
4. _____
5. _____
6. _____
7. _____
8. _____
9. _____
10. _____
11. _____
12. _____
13. _____
14. _____
15. _____
16. _____
17. _____
18. _____
19. _____
20. _____
21. _____
22. _____
23. _____
24. _____

A

Find the distance between each of the following pairs of points.

1. $(6, 5)(9, 9)$ 　　　　　**2.** $(-6, 3)(0, -5)$ 　　　　　**3.** $(3, 4)(-2, 16)$

4. $(-2, -3)(-5, 1)$ 　　　**5.** $(\sqrt{5}, 0)(0, 2)$ 　　　　**6.** $(\sqrt{7}, \sqrt{2})(0, 0)$

B

7. $(3, 4)(2, -1)$ 　　　　　**8.** $(-6, 3)(3, 9)$

9. $(5, -7)(6, -9)$ 　　　　**10.** $(-3, -5)(-2, 7)$

11. $(6, 3)(-6, -3)$ 　　　　**12.** $(-5, -3)(-6, 8)$

13. $(-7, 11)(11, -7)$ 　　　**14.** $(11, 11)(-7, -7)$

C

15. $(21, -2)(12, 1)$ 　　　　**16.** $(-8, 43)(-2, 39)$

17. $(53, 61)(46, 53)$ 　　　**18.** $(-17, 15)(-27, 19)$

19. $(92, -10)(80, -2)$

Find the distance between each of the following pairs of points to the nearest tenth.

20. $(-17, 5)(-8, 13)$ 　　　**21.** $(42, 15)(27, 31)$

22. $(72, 18)(65, -12)$ 　　　**23.** $(-52, -21)(-78, -36)$

24. $(5, -50)(21, 50)$

ANSWERS

25.

26.

27.

28.

29.

30.

31.

32.

33.

D

25. Find the length and width of the rectangle whose vertices are at the points $(1, 7)$, $(5, 3)$ $(10, 8)$, and $(6, 12)$.

26. Find the length of the sides of the triangle whose vertices are at $(1, 1)$, $(9, -5)$, and $(3, 3)$.

27. Find the perimeter of the quadrilateral with vertices at $(2, 1)$, $(4, 4)$, $(9, 4)$, and $(4, -2)$.

28. Find the area of the rectangle with vertices at $(8, 0)$, $(4, -6)$, $(1, -4)$, and $(5, 2)$.

E   *Maintain Your Skills*

Solve:

29. $12x^2 - 8x = 5$

30. $(x - 2)(x + 3) = 2x^2 - 12$

31. $4(x - 2)^2 = (x + 1)^2 + 5$

32. $2 + \dfrac{3}{x} + \dfrac{2}{x^2} = 0$

33. $\dfrac{10}{x - 2} = x + 1$

11.5 Graphing Lines Using Slope and Intercept

OBJECTIVE	**126. Draw the graph of a line, given its equation, using the slope and the y-intercept.**
APPLICATION	A truck driver averages 80 km/hr. The formula for the distance driven is $d = 80t$, where d represents distance (in km) and t represents time (in hr). Draw a graph showing the relationship between the distance and time. (Let each unit on the x-axis represent 1 hour and each unit on the y-axis represent 100 km.)
VOCABULARY	No new vocabulary.

HOW AND WHY   Consider the equation $2x - 3y = 6$. First write it in the slope-intercept form $y = mx + b$.

$$2x - 3y = 6$$
$$-3y = -2x + 6$$
$$y = \frac{2}{3}x - 2$$

Therefore, $m = \dfrac{2}{3}$ and the y-intercept is at $(0, -2)$. We know the line crosses the y-axis at $(0, -2)$, and for every 3 units of run there is a corresponding 2 units of rise. We use this to draw the graph.

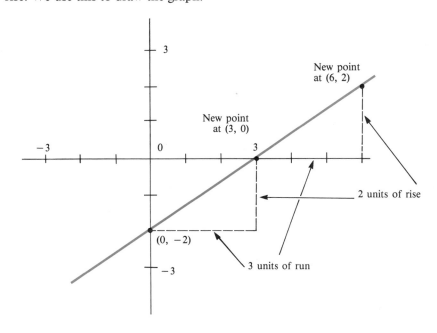

By sketching the run and rise we are able to obtain as many new points as we wish.

> **To draw the graph of a line using the slope and the *y*-intercept,**
>
> 1. **Write the equation in the form $y = mx + b$.**
> 2. **Plot the *y*-intercept $(0, b)$.**
> 3. **Using the slope *m*, starting at the *y*-intercept, locate an additional point.**
> 4. **Draw the line joining the points.**

EXAMPLE

Draw the graph $4x + 5y = 10$.

$$4x + 5y = 10$$

$$5y = -4x + 10$$

$$y = -\frac{4}{5}x + 2$$

$m = -\dfrac{4}{5}$, *y*-intercept at $(0, 2)$.

Note: Since the slope is negative, the rise will be down 4 (that is, a drop).

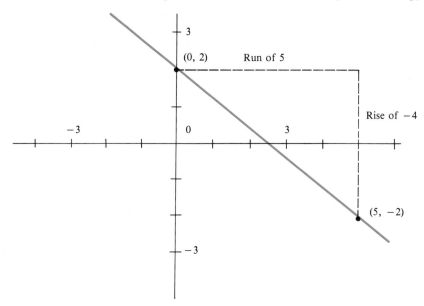

When using a negative slope, we will consider the rise to be negative and the run positive.

APPLICATION SOLUTION

> To draw the graph showing the relationship between distance and time for the trucker, first find the slope and the *y*-intercept.
>
> $d = 80t$ or
>
> $d = 80t + 0$
>
> $m = 80$, *y*-intercept $(0, 0)$
>
> To draw the graph, let each unit on the *t*-axis represent 1 hour and each unit on the *d*-axis represent 100 km.

Continued on next page.

Since the slope is $80 = \dfrac{80}{1}$, we know that for every 1 unit of run we have 80 units of rise. If we start at the y-intercept, we can use this to sketch the graph (Figure 11.1).

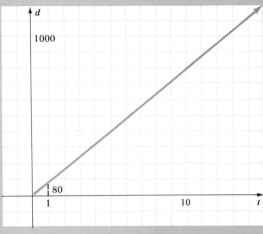

Figure 11.1

Exercises 11.5

A

Draw the graph of the line given its slope and y-intercept.

1. $m = \dfrac{2}{3}$, $(0, 5)$

2. $m = -\dfrac{3}{4}$, $(0, -1)$

3. $m = 3$, $(0, -5)$

4. $m = -2$, $(0, 2)$

5. $m = \dfrac{5}{7}$, $(0, 0)$

6. $m = -\dfrac{7}{9}$, $(0, 8)$

B

Draw the graph of each of the following equations using the slope and the y-intercept.

7. $4x + 5y = 20$

8. $7x - 3y = 21$

9. $2y = x - 4$

10. $3x = 4y + 5$

11. $5x - 6y = 3$

12. $2x + 3y = 9$

13. $7x - 3y = 18$

14. $2x - 3y = 6$

C

15. $-x + \dfrac{1}{2}y = 1$

16. $\dfrac{1}{2}x + \dfrac{2}{3}y = 3$

17. $-\dfrac{1}{5}x + \dfrac{1}{2}y = -1$

18. $-\dfrac{1}{3}x - y = -2$

19. $\dfrac{1}{4}x - \dfrac{3}{8}y = \dfrac{3}{4}$

20. $-\dfrac{1}{5}x - \dfrac{1}{3}y = \dfrac{1}{3}$

21. $.4x - .2y = 1$

22. $.3x + .5y = .1$

23. $.25x + .75y = 1$

24. $.5x - .25y = .25$

D

25. A tourist averages 50 mph as he drives coast to coast. The formula for distance (d) in terms of time (t) is $d = 50t$. Draw the graph.

26. The relationship between simple interest (I) at 10% and principal (P) is given by $I = .10P$. Draw the graph.

27. The cost of setting up a production line for transistor radios is $10,000. The additional cost to manufacture a radio is $5. The cost ($y$) of producing x radios is given by the equation

$$y = 10,000 + 5x$$

Graph the equation for 0 to 1000 radios. (*Hint:* Use 1 unit equal to $5000 on the y-axis and 1 unit equal to 100 radios on the x-axis.)

28. The profit on selling radios is the mark-up per radio less the cost of marketing (advertising, shipping, and so on). If the mark-up on each radio is $3 and the total cost of marketing 150 radios is $150, the profit equation is

$$y = 3x - 150$$

where y represents profit and x represents the number of radios sold. Draw the graph for 0 to 150 radios.

E   *Maintain Your Skills*

Solve:

29. $\sqrt{x + 2} = 12$

30. $\sqrt{x + 5} = \sqrt{2x + 3}$

31. $4x = \sqrt{x - 5}$

32. $\sqrt{2x + 3} = \sqrt{3x + 4}$

33. $\sqrt{5x - 9} + \sqrt{3x + 4} = 0$

11.6 Systems of Equations: Solving by Graphing

OBJECTIVE	**127. Solve a system of linear equations in two variables by graphing.**

APPLICATION	An electronics firm makes color and black-and-white television sets in two different plants. In one day the Deerberg plant can turn out 10 color sets and 50 black-and-white sets. The Elkland plant can produce 50 color sets and 10 black-and-white sets in a day. To fill an order for 100 color sets and 260 black-and-white sets, how many days at each plant are needed?

VOCABULARY	A set of equations such as $$\begin{cases} x + y = 4 \\ x + 2y = 6 \end{cases}$$ that contains more than one variable is called a *system of equations*. If the graphs of the two linear equations intersect (that is, have a point in common), the coordinates of the point are called the *solution of the system* (sometimes called the *simultaneous solution*).

HOW AND WHY	If two linear equations are graphed on the same set of axes, one of three possibilities occur.

First Possibility

The graphs of both equations are the same. Such a system is called a *dependent system*.

A pair of such equations is

$$\begin{cases} 3x - y = 2 & \textit{Equation of } \ell_1 \\ 6x - 2y = 4 & \textit{Equation of } \ell_2 \end{cases}$$

These equations are graphed in Figure 11.2.

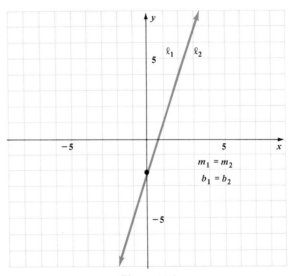

Figure 11.2

Since all points on the lines are in common, we say that the system has an unlimited number of solutions.

Second Possibility

The graphs of the equations are parallel lines. Such a system is called an *independent inconsistent system*. The system has no solution because there is no point of intersection.

A pair of such equations is

$$\begin{cases} 3x - 7y = 9 & \textit{Equation of } \ell_1 \\ 3x - 7y = 2 & \textit{Equation of } \ell_2 \end{cases}$$

The system can be written

$$\begin{cases} y = \dfrac{3}{7}x - \dfrac{9}{7} & \textit{Equation of } \ell_1 \\ y = \dfrac{3}{7}x - \dfrac{2}{7} & \textit{Equation of } \ell_2 \end{cases}$$

These equations are graphed in Figure 11.3.

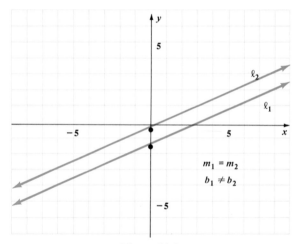

Figure 11.3

Third Possibility

The graphs of the equations are lines that intersect in exactly one point. The coordinates of the point of intersection are called the solution of the system. Such a system is called an *independent consistent system*.

A pair of such equations is

$$\begin{cases} 3x - 2y = 6 & \textit{Equation of } \ell_1 \\ 3x + 2y = 6 & \textit{Equation of } \ell_2 \end{cases}$$

These equations are graphed in Figure 11.4. The solution is (2, 0).

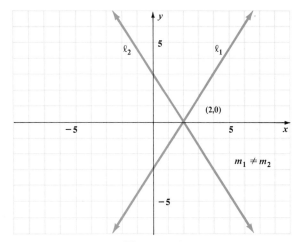

Figure 11.4

These three possibilities can be identified from their equations by comparing the slopes and y-intercepts.

TYPE OF SYSTEM	SLOPES OF LINES	y-INTERCEPTS	COMMON SOLUTIONS (POINTS)
Dependent	Same	Same	Unlimited
Independent inconsistent	Same	Different	No solution
Independent consistent	Different	Same or different	One solution

To solve a system of equations in two variables by graphing,

1. **Write each equation in the form $y = mx + b$.**
2. **Draw the graph of the system.**
3. **Write the solution(s).**
   a. **Unlimited (dependent system)**
   b. **No solution (independent inconsistent)**
   c. **Ordered pair of common solution (independent consistent)**

EXAMPLES

a. Solve the system:

$$\begin{cases} 2x - y - 1 = 0 \\ 8x - 4y - 8 = 0 \end{cases}$$

Write each equation in the form $y = mx + b$.

$2x - y - 1 = 0$ $8x - 4y - 8 = 0$

$-y = -2x + 1$ $-4y = -8x + 8$

$y = 2x - 1$ $y = 2x - 2$

Since the slopes are the same and the y-intercepts are different, the system has no solution. The graphs are parallel lines, as shown in the following graph:

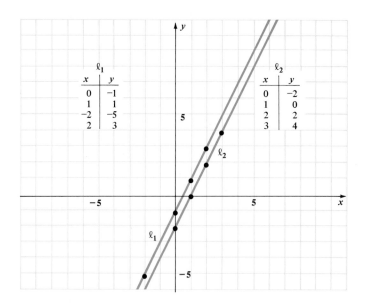

b. Solve the system:

$$\begin{cases} 2x + y - 8 = 0 \\ x - 2y - 9 = 0 \end{cases}$$

Write each equation in the form $y = mx + b$

$$2x + y - 8 = 0 \qquad x - 2y - 9 = 0$$

$$y = -2x + 8 \qquad -2y = -x + 9$$

$$y = \frac{1}{2}x - \frac{9}{2}$$

Graph the system.

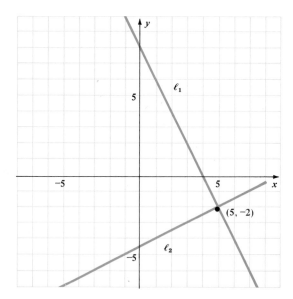

The solution is $(5, -2)$.

c. Solve the system:

$$\begin{cases} 4x - 5y - 10 = 0 \\ x + 3y + 9 = 0 \end{cases}$$

Write each equation in the form $y = mx + b$.

$$4x - 5y - 10 = 0 \qquad x + 3y + 9 = 0$$
$$-5y = -4x + 10 \qquad 3y = -x - 9$$
$$y = \frac{4}{5}x - 2 \qquad y = -\frac{1}{3}x - 3$$

Graph the system.

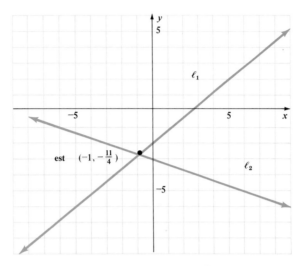

Our estimate of the solution is $\left(-1, -\frac{11}{4}\right)$. $\left[$ The exact solution is $\left(-\frac{15}{17}, -\frac{46}{17}\right).\right]$ A method of finding the exact solution is given in the next section.

APPLICATION SOLUTION

The following table is a summary of the information in the problem:

	Deerberg	Elkland	Order
Color Sets	10	50	100
B & W Sets	50	10	260

Simpler word form:

$$\begin{pmatrix} \text{number of color} \\ \text{sets manufactured} \\ \text{at Deerberg} \end{pmatrix} + \begin{pmatrix} \text{number of color} \\ \text{sets manufactured} \\ \text{at Elkland} \end{pmatrix} = \begin{pmatrix} \text{total} \\ \text{color} \\ \text{sets} \end{pmatrix}$$

$$\begin{pmatrix} \text{number of B \& W} \\ \text{sets manufactured} \\ \text{at Deerberg} \end{pmatrix} + \begin{pmatrix} \text{number of B \& W} \\ \text{sets manufactured} \\ \text{at Elkland} \end{pmatrix} = \begin{pmatrix} \text{total} \\ \text{B \& W} \\ \text{sets} \end{pmatrix}$$

Select variables:

If we let x represent the number of days the Deerberg plant operates and y represent the number of days the Elkland plant operates, we can add to the table.

Continued on next page.

	Deerberg (1 Day)	Elkland (1 Day)	Deerberg (x Days)	Elkland (y Days)	Order
Color Sets	10	50	$10x$	$50y$	100
B & W Sets	50	10	$50x$	$10y$	260

Translate to algebra:

With the information in the preceding table we can write the two equations:

$$\begin{cases} 10x + 50y = 100 \\ 50x + 10y = 260 \end{cases}$$

Solve:

$$\begin{cases} x + 5y = 10 & (\ell_1) \\ 5x + y = 26 & (\ell_2) \end{cases} \qquad \textit{Divide both sides by 10}$$

ℓ_1

x	y
0	2
10	0
5	1

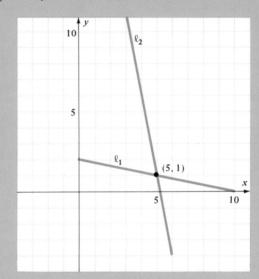

ℓ_2

x	y
5	1
4	6
6	−4

Answer:

The two lines intersect at the point (5, 1), which is the common solution. Therefore, if the Deerberg plant is run 5 days and the Elkland plant 1 day, the order can be filled.

Exercises 11.6

A

Solve by graphing:

1. $\begin{cases} x + y = 5 \\ x - y = 1 \end{cases}$

2. $\begin{cases} x + y = 12 \\ x - y = 2 \end{cases}$

3. $\begin{cases} x + y = 8 \\ 2x + y = 12 \end{cases}$

4. $\begin{cases} x - y = 3 \\ x + 2y = 12 \end{cases}$

5. $\begin{cases} 7x - y = 7 \\ x - y = 1 \end{cases}$

6. $\begin{cases} 2x + y = 8 \\ 2x - y = -4 \end{cases}$

B

7. $\begin{cases} y = x + 11 \\ y = -x + 3 \end{cases}$

8. $\begin{cases} y = -3x + 2 \\ y = 5x - 6 \end{cases}$

9. $\begin{cases} 2x + 4y = 2 \\ 3x + 6y = 3 \end{cases}$

10. $\begin{cases} x - y = 2 \\ 3y - 3x = -6 \end{cases}$

11. $\begin{cases} 3x + 4y = 6 \\ x - 2y = 2 \end{cases}$

12. $\begin{cases} 3x - y = 5 \\ x - y = 5 \end{cases}$

13. $\begin{cases} x + 3y = 15 \\ 2x + y = 10 \end{cases}$

14. $\begin{cases} 3y - 5x = 23 \\ 4x + y = 2 \end{cases}$

C

15. $\begin{cases} 2x - y = 3 \\ y - 2x = 4 \end{cases}$

16. $\begin{cases} 6x + 3y = -3 \\ 4x + 2y = 2 \end{cases}$

17. $\begin{cases} 4x + 3y = 11 \\ 2x - y = 3 \end{cases}$

18. $\begin{cases} 5x + 4y = -3 \\ 3x - y = 5 \end{cases}$

Solve by graphing; estimate the solution to the system to the nearest half unit.

19. $\begin{cases} 2x + 4y = 9 \\ x + y = 1 \end{cases}$

20. $\begin{cases} x - 3y = 15 \\ 5x + 3y = 6 \end{cases}$

21. $\begin{cases} x + y = 3 \\ 2x - y = 5 \end{cases}$

22. $\begin{cases} 2x + 5y = 2 \\ 4x - 5y = 13 \end{cases}$

23. $\begin{cases} x + y = 8 \\ 7x - 2y = 9 \end{cases}$

24. $\begin{cases} x - y = 10 \\ 4x + y = 7 \end{cases}$

1. _____
2. _____
3. _____
4. _____
5. _____
6. _____
7. _____
8. _____
9. _____
10. _____
11. _____
12. _____
13. _____
14. _____
15. _____
16. _____
17. _____
18. _____
19. _____
20. _____
21. _____
22. _____
23. _____
24. _____

ANSWERS

25. _____

26. _____

27. _____

28. _____

29. _____

30. _____

31. _____

32. _____

D

25. The sum of two numbers is 12. The difference of the same numbers is 2. Write a system of equations for the two numbers and solve by graphing.

26. Find the number of days for each of the plants in the application to fill an order for 300 color sets and 300 black-and-white sets.

27. Two cross-country runners leave the same starting point one hour apart. The first to leave runs at an average speed of 5 mph and the second at an average speed of 7 mph. How far will they have run when the second runner overtakes the first? The table summarizes what we know.

	Rate	Time	Distance
First	5	1 hour longer	Same distance for both runners
Second	7		

If we let t represent the time for the *second* runner and d represent the distance of *each* runner, we can fill in the table.

	Rate	Time	Distance
First	5	$t + 1$	d
Second	7	t	d

Now, since $d = rt$, we have the system:

$$\begin{cases} d = 5(t + 1) \\ d = 7t \end{cases}$$

Construct a rectangular coordinate system and find the solution by graphing.

E   *Maintain Your Skills*

Solve:

28. $\dfrac{\frac{1}{4}}{\frac{2}{3}} = \dfrac{x}{5}$

29. $\dfrac{2.4}{7.2} = \dfrac{3.5}{x}$

30. $\dfrac{1}{x} = \dfrac{2\frac{2}{5}}{5}$

31. $\dfrac{12 \text{ ft}}{\$4.50} = \dfrac{x \text{ ft}}{\$22.50}$

32. $\dfrac{5 \text{ yd}}{x \text{ ft}} = \dfrac{3 \text{ ft}}{4 \text{ in}}$

562

11.7 Systems of Equations: Solving by Substitution

OBJECTIVE	**128. Solve a system of linear equations in two variables by substitution.**
APPLICATION	A landscaper wants 600 pounds of grass seed that is 45% bluegrass. He has two mixtures available. One contains 40% bluegrass and the other 70% bluegrass. How many pounds of each can he use?
VOCABULARY	No new words.

HOW AND WHY

Systems of equations can also be solved by substitution. This method is recommended when one of the variables has a coefficient of 1.

Consider the system

$$\begin{cases} 3x - y + 8 = 0 & (1) \\ x + 2y - 9 = 0 & (2) \end{cases}$$

We can solve the first equation for y by adding y to each side.

$$3x - y + 8 = 0$$
$$3x + 8 = y$$

If we substitute this value for y in the second equation, the variable y is eliminated and we can solve for x. (Substitution is permitted because we are solving for common values of x and y.)

$$x + 2\;y - 9 = 0 \qquad (2)$$
$$x + 2\;(3x + 8) - 9 = 0 \qquad \textit{Substitute}$$
$$x + 6x + 16 - 9 = 0 \qquad \textit{Solve}$$
$$7x = -7$$
$$x = -1$$

Since $y = 3x + 8$, we can solve for y.

$$y = 3x + 8$$
$$y = 3(-1) + 8 = 5$$

Check:

$$3x - y + 8 = 0 \qquad\qquad x + 2y - 9 = 0$$
$$3(-1) - (5) + 8 = 0 \qquad (-1) + 2(5) - 9 = 0$$
$$-3 - 5 + 8 = 0 \qquad\qquad -1 + 10 - 9 = 0$$
$$0 = 0 \qquad\qquad\qquad\quad 0 = 0$$

The solution is $(-1, 5)$.

> To solve a system of linear equations in two variables by substitution,
>
> 1. Solve either equation for x (or y).
> 2. Eliminate x (or y) by substituting this expression for x (or y) in the other equation.
> 3. Solve.
> 4. Use this value of y (or x) to find the value of x (or y).
> 5. Check the value of x and y in both equations.
> 6. Write the solution as the ordered pair. (x, y).

During Step 3, if both x and y are eliminated, the new equation is true (identity) or false (contradition), which tells us the following about the system:

NEW EQUATION	TYPE OF SYSTEM	SOLUTIONS
Contradiction	Independent inconsistent	No solution
Identity	Dependent	Unlimited solutions

EXAMPLES

a. Solve $\begin{cases} x + 3y = 25 & (1) \\ 2x + y = 15 & (2) \end{cases}$

Step 1. Solve the first equation (1) for x.

$$x = 25 - 3y$$

Step 2. Substitute $(25 - 3y)$ for x in the second equation (2).

$$2(25 - 3y) + y = 15$$

Step 3. Solve for y.

$$50 - 6y + y = 15$$
$$50 - 5y = 15$$
$$-5y = -35$$
$$y = 7$$

Step 4. Find x (from the first equation).

$$x = 25 - 3y = 25 - 3(7)$$
$$= 25 - 21 = 4$$

Step 5. *Check.*

$$x + 3y = 25 \quad (1) \qquad\qquad 2x + y = 15 \quad (2)$$
$$4 + 3 \cdot 7 = 25 \qquad\qquad\qquad 2 \cdot 4 + 7 = 15$$
$$4 + 21 = 25 \qquad\qquad\qquad\quad 8 + 7 = 15$$
$$25 = 25 \qquad\qquad\qquad\qquad 15 = 15$$

Step 6. The solution is $(4, 7)$.

b. Solve $\begin{cases} x + y = 3 & (1) \\ 2x - y = 5 & (2) \end{cases}$

Step 1. Solve (1) for y.

$$y = 3 - x$$

Step 2. Substitute $y = 3 - x$ in (2).

$$2x - (3 - x) = 5$$

Step 3. Solve.

$$2x - 3 + x = 5$$
$$3x - 3 = 5$$
$$3x = 8$$
$$x = \frac{8}{3}$$

Step 4. Find y.

$$y = 3 - x = 3 - \frac{8}{3}$$
$$y = \frac{9}{3} - \frac{8}{3}$$
$$y = \frac{1}{3}$$

Step 5. *Check.*

$x + y = 3$ (1)	$2x - y = 5$ (2)
$\dfrac{8}{3} + \dfrac{1}{3} = 3$	$2 \cdot \dfrac{8}{3} - \dfrac{1}{3} = 5$
$\dfrac{9}{3} = 3$	$\dfrac{16}{3} - \dfrac{1}{3} = 5$
$3 = 3$	$\dfrac{15}{3} = 5$
	$5 = 5$

Step 6. The solution is $\left(\dfrac{8}{3}, \dfrac{1}{3} \right)$.

c. Solve $\begin{cases} x = 2y + 3 & (1) \\ 2x - 4y = 7 & (2) \end{cases}$

Step 1. Already done.

$$x = 2y + 3$$

Step 2.

$$2(2y + 3) - 4y = 7$$

Step 3.

$$4y + 6 - 4y = 7$$
$$6 = 7$$

Since this equation is false (contradiction), we can skip to Step 6.

Step 6. No solution.

APPLICATION SOLUTION

To solve the application, we summarize our information in the following table:

	First Mix	Second Mix	Desired Mix
% of Bluegrass (in decimal form)	.40	.70	.45
Number of Pounds	?	?	600

Simpler word form: Total number of pounds = 600

$$\begin{pmatrix} \text{amount of} \\ \text{bluegrass in} \\ \text{first mix} \end{pmatrix} + \begin{pmatrix} \text{amount of} \\ \text{bluegrass in} \\ \text{second mix} \end{pmatrix} = \begin{pmatrix} \text{total} \\ \text{amount of} \\ \text{bluegrass} \end{pmatrix}$$

Assign variables: Let x be the number of pounds of bluegrass in the first mix and y be the number of pounds of bluegrass in the second mix.

	First Mix	Second Mix	Desired Mix
% of Bluegrass (in decimal form)	.40	.70	.45
Number of Pounds	x	y	600
Number of Pounds of Bluegrass Seed	.40x	.70y	.45(600) = 270

Translate to algebra:
$$\begin{cases} x + y = 600 & (1) \\ .40x + .70y = 270 & (2) \end{cases}$$

Solve:

Step 1.

$y = 600 - x$ (1)   *Solve* (1) *for* y

Step 2.

$4x + 7y = 2700$ (2)   *Multiply* (2) *by 10*

$4x + 7(600 - x) = 2700$ *Substitute* $600 - x$ *for* y

Step 3.

$4x + 4200 - 7x = 2700$

$-3x + 4200 = 2700$

$-3x = -1500$

$x = 500$

Step 4.   Substitute $x = 500$ in (1).

$y = 600 - 500$

$y = 100$

Continued on next page.

Step 5. *Check:*

$$x + y = 600 \quad (1)$$
$$500 + 100 = 600$$
$$600 = 600$$

$$.40x + .70y = 270 \quad (2)$$
$$.40(500) + .70(100) = 270$$
$$200 + 70 = 270$$
$$270 = 270$$

Step 6.

Answer: The solution to the system is (500, 100), which means that by using 500 pounds of 40% mix and 100 pounds of 70% mix, the landscaper can get 600 pounds of 45% bluegrass.

Exercises 11.7

A

Solve by substitution:

1. $\begin{cases} y = x - 5 \\ x + y = 3 \end{cases}$ **2.** $\begin{cases} y = x + 6 \\ x + y = 10 \end{cases}$

3. $\begin{cases} x = y + 1 \\ 6x - y = 6 \end{cases}$ **4.** $\begin{cases} x = y - 4 \\ 2y - 3x = 12 \end{cases}$

5. $\begin{cases} x + y = 11 \\ 2x - 3y = 12 \end{cases}$ **6.** $\begin{cases} x + y = -3 \\ 7x + 2y = 4 \end{cases}$

B

7. $\begin{cases} x - y = 1 \\ 3x + 4y = 17 \end{cases}$ **8.** $\begin{cases} x + y = -1 \\ 5x - 2y = -19 \end{cases}$

9. $\begin{cases} x + y = -8 \\ -3x + 2y = 9 \end{cases}$ **10.** $\begin{cases} 3x + 4y = -29 \\ x - y = 2 \end{cases}$

11. $\begin{cases} 4x + y = 2 \\ 5x + 3y = 20 \end{cases}$ **12.** $\begin{cases} x + 5y = 14 \\ 2x - y = -5 \end{cases}$

13. $\begin{cases} 2a + 3b = 1 \\ a - 4b = 6 \end{cases}$ **14.** $\begin{cases} 2p - q = -15 \\ p + 2q = 5 \end{cases}$

C

15. $\begin{cases} y = \frac{1}{2}x + 4 \\ 2x + 5y = 11 \end{cases}$ **16.** $\begin{cases} x = \frac{2}{3}y - 8 \\ 8x - y = 1 \end{cases}$

17. $\begin{cases} x + 2y = \frac{3}{4} \\ 2x + y = -\frac{3}{8} \end{cases}$ **18.** $\begin{cases} x - 2y = \frac{3}{2} \\ 2x + y = \frac{9}{8} \end{cases}$

19. $\begin{cases} 6x - y = 7 \\ 2x + 3y = 5 \end{cases}$ **20.** $\begin{cases} 2x + 3y = -4 \\ 3x + y = 6 \end{cases}$

21. $\begin{cases} x - 2y = 3 \\ 3x - 6y = 9 \end{cases}$ **22.** $\begin{cases} y = 2x + 5 \\ 6x - 3y = 1 \end{cases}$

23. $\begin{cases} 5x + y = 7 \\ 3x - 2y = 8 \end{cases}$ **24.** $\begin{cases} x - 4y = 9 \\ 2x + 3y = 9 \end{cases}$

1. _____
2. _____
3. _____
4. _____
5. _____
6. _____
7. _____
8. _____
9. _____
10. _____
11. _____
12. _____
13. _____
14. _____
15. _____
16. _____
17. _____
18. _____
19. _____
20. _____
21. _____
22. _____
23. _____
24. _____

25.

26.

27.

28.

29.

30.

31.

32.

33.

34.

35.

36.

37.

D

25. If the landscaper in the application needs 600 pounds of grass seed that is 55% bluegrass for a different soil, how many pounds of each of his mixtures should be used?

26. Ellie needs some change for her weekend garage sale. She has enough nickels and wants three times as many quarters as dimes. If she takes $20.40 to the bank, how many quarters and how many dimes can she get?

27. Joe Cool invested $18,000, part at 12% and part at 15% interest. If the return on each investment was the same, find the amount invested at each rate.

28. Melba Sharp and Jose Remus both work for the Sweeter-Than-Sweet Ice Cream Company. Melba earns $8 per hour and Jose earns $6.50 per hour. The owner paid them a total of $388 for a combined total of 50 hours worked. How much did each earn?

29. The cost of driving Ms. Chinn's Cadillac is 32¢ per mile. The cost of driving her Datsun pickup is 22¢ per mile. During the month of December the cars were driven a total of 660 miles at a cost of $186.20. How many miles was each car driven during December?

30. For the grand opening of the new Opera House, 2600 tickets were sold. Some of the tickets sold for $25 each, and the rest sold for $15 each. How many tickets of each price were sold if the sales totaled $46,000?

31. Cindy Sloe borrowed money at 18% and 20%. If the annual interest payment is $396, and she borrowed a total of $2100, how much did she borrow at each rate?

32. Alpine Community College pays its administrators an average of $2100 per month and its faculty an average of $1950 per month. If the monthly payroll for the 120 employees in the two categories is $237,000, how many administrators and how many faculty members does Alpine employ?

E   *Maintain Your Skills*

33. $S - M = C$. Find S if $M = 48$ and $C = -12$.

34. $V = dwh$. Find d if $w = 20$, $h = 16$, and $V = 64000$.

35. $I = \dfrac{E - e}{R}$. Solve for e.

36. $A = p(1 + r)^n$. Find A if $p = 10000$, $r = .09$, and $n = 3$.

37. $F = \dfrac{kmM}{r^2}$. Solve for r.

11.8 Systems of Equations: Solving by Addition

OBJECTIVE	**129.** Solve a system of linear equations in two variables by addition.

APPLICATION

A tank contains a salt solution that is 10% salt by weight. A second tank contains a solution that is 50% salt by weight. How many gallons of each solution should be mixed to make 100 gallons of a 20% salt solution?*

VOCABULARY

No new words.

HOW AND WHY

If we are given a system of equations in which none of the coefficients of x and y is 1, the following method is recommended:

$$\begin{cases} 2x + 3y = 5 & (1) \\ 3x + 4y = 7 & (2) \end{cases}$$

Step 1. Multiply both sides of the first equation by 3 and both sides of the second equation by -2. We do this so that the coefficients of x in the system are opposites.

$$\begin{matrix} 3 \\ -2 \end{matrix} \begin{cases} 2x + 3y = 5 \\ 3x + 4y = 7 \end{cases} \text{ or } \begin{cases} 6x + 9y = 15 \\ -6x - 8y = -14 \end{cases} \quad \begin{matrix} \textit{Multiply (1) by 3} \\ \textit{Multiply (2) by } -2 \end{matrix}$$

Step 2. Add the two equations so that the terms containing x are eliminated.

$$\begin{array}{r} 6x + 9y = 15 \\ -6x - 8y = -14 \\ \hline y = 1 \end{array}$$

Step 3.

Step 4. Substitute $y = 1$ in (1) and solve for x.

$$2x + 3 \cdot 1 = 5$$
$$2x = 2$$
$$x = 1$$

Step 5. Check in (2).

$$3(1) + 4(1) = 7$$
$$3 + 4 = 7$$
$$7 = 7$$

Step 6. The solution is (1, 1).

To solve a system of equations in two variables by addition when each equation is written in the form

$Ax + By = C$

1. Multiply one or both of the equations by factors so that the coefficients of y (or x) are opposites.
2. Add the two equations so that y (or x) is eliminated.
3. Solve the resulting equation for x (or y).
4. Find the value of the other variable by substitution in one of the original equations.
5. Check the answer in the original equation not used in Step 4.
6. Write the solution as an ordered pair.

* Courtesy of Exxon Co.

If both x and y are eliminated in Step 2 and the new equation is true (identity) or false (contradiction), it tells us the following about the system:

RESULT OF ADDING EQUATIONS	TYPE OF SYSTEM	SOLUTIONS
Contradiction	Independent inconsistent	No solution
Identity	Dependent	Unlimited solutions

EXAMPLES

a. Solve by addition:

$$\begin{cases} 3x + 2y = -8 & (1) \\ 2x - 3y = -6 & (2) \end{cases}$$

Step 1. Choose the factors 2 and -3 so the coefficients of x will be opposites.

$$\begin{matrix} 2 \\ -3 \end{matrix} \begin{cases} 3x + 2y = -8 & (1) \\ 2x - 3y = -6 & (2) \end{cases} \text{ or } \begin{cases} 6x + 4y = -16 \\ -6x + 9y = 18 \end{cases}$$

Step 2. Add the two equations.

$$13y = 2$$

Step 3. Solve.

$$y = \frac{2}{13}$$

Step 4. Now substitute for y into $3x + 2y = -8$ to find x.

$$3x + 2\left(\frac{2}{13}\right) = -8$$

$$3x + \frac{4}{13} = -8$$

$$39x + 4 = -104$$

$$39x = -108$$

$$x = -\frac{108}{39} = -\frac{36}{13}$$

Step 5. Check: Substitute $x = -\frac{36}{13}$ and $y = \frac{2}{13}$ in (2).

$$2\left(-\frac{36}{13}\right) - 3\left(\frac{2}{13}\right) = -6$$

$$-\frac{72}{13} - \frac{6}{13} = -6$$

$$-\frac{78}{13} = -6$$

$$-6 = -6$$

Step 6. The solution is $\left(-\frac{36}{13}, \frac{2}{13}\right)$.

b. Solve by addition:

$$\begin{cases} 9x - 4y - 36 = 0 & (1) \\ 18x - 8y - 11 = 0 & (2) \end{cases}$$

Write each equation in the form $Ax + By = C$.

$$\begin{cases} 9x - 4y = 36 \\ 18x - 8y = 11 \end{cases}$$

Step 1. Choose factors -2 and 1 (to eliminate x).

$$\begin{array}{c} -2 \\ 1 \end{array} \begin{cases} 9x - 4y = 36 \\ 18x - 8y = 11 \end{cases} \quad \text{or} \quad \begin{cases} -18x + 8y = -72 \\ 18x - 8y = 11 \end{cases}$$

Step 2. Add to eliminate x.

$$0 = -61$$

Since both variables were eliminated and $0 = -61$ is a contradiction, the system has no solution.

APPLICATION SOLUTION

To solve the application, we do the following:

Simpler word form:

$$\begin{pmatrix} \text{number of gallons} \\ \text{of 10\% solution} \end{pmatrix} + \begin{pmatrix} \text{number of gallons} \\ \text{of 50\% solution} \end{pmatrix} = 100 \text{ gallons}$$

$$\begin{pmatrix} \text{number of gallons} \\ \text{of salt in 10\%} \\ \text{solution} \end{pmatrix} + \begin{pmatrix} \text{number of gallons} \\ \text{of salt in 50\%} \\ \text{solution} \end{pmatrix} = \begin{pmatrix} \text{number of gallons} \\ \text{of salt in final} \\ \text{20\% solution} \end{pmatrix}$$

Select variables:

If we let a represent the number of gallons from the first tank (10% salt solution) and b represent the number of gallons from the second tank (50% salt solution), we can organize the information in a table:

	Number of Gallons from A	Number of Gallons from B	Total
Solution	a	b	100
Salt	$.10a$	$.50b$	$.20(100) = 20$

Translate to algebra:

The total number of gallons of solution is 100, so $a + b = 100$.
The total amount of salt is 20, so

$$.10a + .50b = 20$$

The system of equations is

$$\begin{cases} a + b = 100 & (1) \\ .10a + .50b = 20 & (2) \end{cases}$$

Continued on next page.

Solve:

Step 1.

$$1 \begin{cases} a + b = 100 \\ .10a + .50b = 20 \end{cases} \quad \text{or} \quad \begin{cases} a + b = 100 \\ -a - 5b = -200 \end{cases}$$
-10

Step 2. *Add.*

$$\begin{array}{r} a + b = 100 \\ -a - 5b = -200 \\ \hline -4b = -100 \end{array}$$

Step 3. Solve for b.

$$-4b = -100$$

$$b = 25$$

Step 4. Substitute $b = 25$ in (1).

$$a + 25 = 100$$

$$a = 75$$

Step 5. *Check:*

$$.10(75) + .50(25) = 20$$

$$7.5 + 12.5 = 20$$

$$20 = 20$$

Step 6. The solution of the system is (75, 25).

Answer:

To make the mixture, 75 gallons are needed from the first tank and 25 gallons from the second tank.

Exercises 11.8

A

Solve by addition:

1. $\begin{cases} x + y = 4 \\ x - y = 2 \end{cases}$

2. $\begin{cases} x + y = -1 \\ x - y = 3 \end{cases}$

3. $\begin{cases} 3x - 2y = 10 \\ 3x + 2y = 14 \end{cases}$

4. $\begin{cases} 2x + 3y = 6 \\ 2x - 3y = 6 \end{cases}$

5. $\begin{cases} 2x - y = 3 \\ -2x + y = 2 \end{cases}$

6. $\begin{cases} x + 5y = 4 \\ -x - 5y = -4 \end{cases}$

B

7. $\begin{cases} x + y = 12 \\ x - y = 2 \end{cases}$

8. $\begin{cases} x + 2y = 5 \\ 2x - y = 0 \end{cases}$

9. $\begin{cases} x + 3y = 10 \\ x - y = -2 \end{cases}$

10. $\begin{cases} 3x + y = -8 \\ x + y = -4 \end{cases}$

11. $\begin{cases} x + 2y = 15 \\ x - y = 0 \end{cases}$

12. $\begin{cases} x + 3y = 5 \\ x + y = 1 \end{cases}$

13. $\begin{cases} 2x + y = 5 \\ 4x + y = 6 \end{cases}$

14. $\begin{cases} x + 4y = 2 \\ x - 8y = -7 \end{cases}$

ANSWERS

1. _____

2. _____

3. _____

4. _____

5. _____

6. _____

7. _____

8. _____

9. _____

10. _____

11. _____

12. _____

13. _____

14. _____

C

15. $\begin{cases} x + y = 35 \\ x - 5y = 17 \end{cases}$

16. $\begin{cases} 5x + 2y = -2 \\ x - y = 8 \end{cases}$

17. $\begin{cases} x + 5y + 1 = 0 \\ 2x + 7y - 1 = 0 \end{cases}$

18. $\begin{cases} 2x + y - 1 = 0 \\ 6x + 5y - 13 = 0 \end{cases}$

19. $\begin{cases} \frac{3}{4}x + y - 6 = 0 \\ 2x + y - 11 = 0 \end{cases}$

20. $\begin{cases} 3x + y = 7 \\ x - \frac{5}{2}y = -\frac{1}{2} \end{cases}$

21. $\begin{cases} 3x = 5y + 10 \\ 7x - 3y - 14 = 0 \end{cases}$

22. $\begin{cases} 7x - 9 = 13y \\ 3y + 7 = 9x \end{cases}$

23. $\begin{cases} x = 3 + 4y \\ 8y = 6x - 15 \end{cases}$

24. $\begin{cases} 6x + 3y - 9 = 0 \\ 6y + 12x = 18 \end{cases}$

D

25. How many ounces each of 85% pure gold and 70% pure gold are needed to make up 15 ounces of 75% pure gold?

26. Greg has 80 coins, all nickels and dimes. The value of the coins is $6.75. Write a system of equations and find how many nickels and how many dimes he has.

27. The Buy Me I'm Beautiful furniture company produces chairs at a cost of $110 each and sofas at a cost of $322 each. If the cost of production of these two items last month was $12,450 and 65 items were produced, how many sofas and how many chairs were made?

28. The Grow It Then Mow It seed company sells rye grass seed for $4.80 per pound and bluegrass seed for $5.40 per pound. They decide to market 500 pounds of a mixture of the two seeds. If the price of the mixture is $5.25 per pound, how many pounds of each seed did they use?

29. Mr. Tall Texan owns two oil wells. One well pumps oil that sells for $36 a barrel and the other pumps oil that sells for $42 a barrel. If Mr. Texan pumped 3600 barrels of oil last year for a sale of $136,200, how many barrels were pumped from each well?

30. Marlene Swift sold a record 210 subscriptions during the magazine sale at P.S. 76. The subscriptions were for either $27 per year or $48 for two years. If her sales amounted to $6930, how many one-year and how many two-year subscriptions did Marlene sell?

31. Janet Hardsell makes a 3.5% commission for all sales she makes while in the store and a 6.6% commission for all sales she makes in the field. Last month her sales totaled $45,500. How much did she sell in the store and in the field if her total commission was $2600?

32. A cereal company intends to add enough dried fruit to each box of cereal so that each box will contain 21 grams of protein and 338 grams of carbohydrates. The approximate food values are as follows:

	Protein	Carbohydrates
Dried fruit	2%	75%
Cereal	10%	85%

How much cereal and how much dried fruit should each box contain?

33. Ida's mother is on a diabetic diet. The doctor's prescription for her diet includes 99 g of carbohydrate, 91 g of protein, and 30 g of fat. Her breakfast and lunch for Saturday have already provided 66 g of carbohydrate, 61 g of protein, and 18 g of fat. The bread, vegetables, and dessert at dinner will provide 33 g of carbohydrate and 9 g of protein. The approximate food values of liver and bacon are (in grams per 100 grams of weight)

	Carbohydrate	Protein	Fat
Liver	0	25	8
Bacon	0	20	40

How many grams of bacon and how many grams of liver should Ida fix for her mother's dinner to balance her diet for the day? (Round all weights to the nearest gram.)

ANSWERS

28. _____

29. _____

30. _____

31. _____

32. _____

33. _____

34. _____

35. _____

36. _____

37. _____

38. _____

E   *Maintain Your Skills*

Do the indicated operations:

34. $\dfrac{4x - 6y}{4xy} \div \dfrac{8x - 12y}{16y^2}$

35. $\dfrac{x}{y} - \dfrac{x + y}{x} - \dfrac{y}{x + y}$

36. $\dfrac{\dfrac{x - 1}{x}}{\dfrac{1}{x} + 1}$

37. $\left(\dfrac{x^2 + 4x + 4}{x - 2}\right)\left(\dfrac{x^2 - 6x + 8}{x^2 + 6x + 8}\right)$

Solve:

38. $\dfrac{4}{x + 1} - \dfrac{1}{x} = 0$

1. _____

1. (Obj. 120) Draw the graph of $2y + 3x - 12 = 0$.

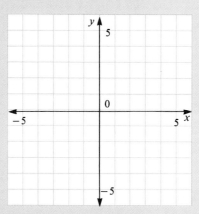

2. _____

2. (Obj. 125) Find the distance between the points $(-3, -2)$ and $(-1, 3)$.

3. (Obj. 122) Write the coordinates of the y-intercept of the line represented by $x = 2y + 8$.

4. (Obj. 123) Write the slope of the line that passes through the points $(-3, -2)$ and $(-1, 3)$.

3. _____

5. (Obj. 118) Locate the following points on the coordinate system: $A\ (-6, 4)$, $B\ (0, -3)$, $C\ (-1, -5)$, $D\ (-5, 0)$, and $E\ (5, 1)$.

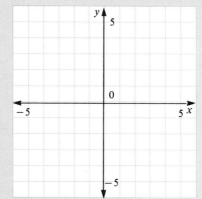

4. _____

6. (Obj. 127) Solve the following sytem of equations by graphing.

$$\begin{cases} 4x + 3y = 6 \\ 2x - y = -2 \end{cases}$$

5. _____

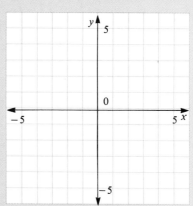

6. _____

7. _____

8. _____

9. _____

10. _____

11. A _____

    B _____

    C _____

    D _____

    E _____

12. _____

7. (Obj. 121) Write the coordinates of the *x*-intercept of the line represented by the equation $3x - 4y - 12 = 0$.

8. (Obj. 129) Solve the following system of equations by addition.

$$\begin{cases} 5s + 4t = -2 \\ -2s + 5t = 3 \end{cases}$$

9. (Obj. 126) Draw the graph of the line through $(4, 0)$ that has slope $m = -\dfrac{1}{2}$.

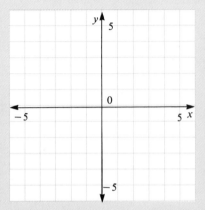

10. (Obj. 124) Write the slope of the line represented by the equation $-2y = 3x + 6$.

11. (Obj. 119) Identify the coordinates of the points on the graph below.

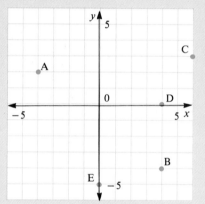

12. (Obj. 128) The voltage across a resistor is equal to the product of the current times the resistance. When a current of 8 amperes passes through a resistor and 5 amperes passes through another resistor, the sum of the voltages is 180 volts. If the sum of the two resistors is 30 ohms, what is the value of each resistor?

580

12

Trigonometry

1. _____

2. _____

3. _____

4. _____

5. _____

6. _____

7. _____

8. _____

Chapter 12
PRE-TEST

The problems in the following pre-test are a sample of the material in the chapter. You may already know how to work some of these. If so, this will allow you to spend less time on those parts. As a result, you will have more time to give to the sections that you find more difficult or that are new to you. The answers are in the back of the text.

1. **(Obj. 130)** In $\triangle XYZ$, $X = 43°$ and $Z = 19°$. What is the measure of Y?

2. **(Obj. 130)** In $\triangle RST$, $R = 35.4°$ and $T = 12.8°$. What is the measure of S?

3. **(Obj. 131)** In $\triangle DEF$, find the exact values of sin F, cos F, and tan F.

4. **(Obj. 131)** In $\triangle UVW$, find the exact values of sin U, cos U, and tan U.

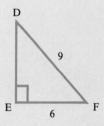

5. **(Obj. 132)** Find the exact value of cos 60°.

6. **(Obj. 133)** Find the approximate value of tan 34.7° (to the nearest ten-thousandth).

7. **(Obj. 134)** If sin $T = 0.2255$, find T to the nearest tenth of a degree.

8. **(Obj. 135)** In $\triangle FGH$, $f = 7$, $g = 13$, and $H = 90°$. Find h, F, and G to the nearest tenth.

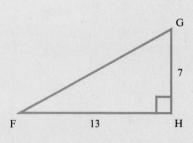

ANSWERS

9. _____

10. _____

11. _____

12. _____

13. _____

14. _____

15. _____

16. _____

17. _____

18. _____

19. _____

20. _____

9. **(Obj. 135)** In $\triangle PQR$, $q = 13$, $P = 90°$, and $R = 32.6°$. Find p, r, and Q to the nearest tenth.

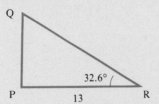

10. **(Obj. 136)** Find the value of sin 146° correct to four decimal places.

11. **(Obj. 136)** Find the value of cos 146° correct to four decimal places.

12. **(Obj. 136)** Find the value of tan 97.8° correct to four decimal places.

13. **(Obj. 137)** If cos $Y = -0.3333$, find Y to the nearest tenth of a degree.

14. **(Obj. 137)** If sin $L = 0.2299$, find two values of L to the nearest tenth of a degree.

15. **(Obj. 138)** Solve $\triangle DEF$ given $d = 22$, $e = 18$, and $f = 12$. Find the angles to the nearest tenth of a degree.

16. **(Obj. 139)** Solve $\triangle LMN$ given $L = 40°$, $n = 16$, and $m = 10$. Find all measures to the nearest tenth.

17. **(Obj. 140)** Solve $\triangle XYZ$ given $x = 15$, $z = 24$, and $Z = 52°$. Find all measures to the nearest tenth.

18. **(Obj. 141)** Solve $\triangle ABC$ given $B = 40°$, $C = 55°$, and $c = 14$. Find all measures to the nearest tenth.

19. **(Obj. 135)** Rene measures the angle of elevation of the top of a building that is on the opposite bank of a river. The angle measures 37.3°. If she knows that the building is 27 metres high, how wide is the river to the nearest metre?

20. **(Obj. 139)** Points X and Y are on opposite sides of a hill. Point A is located 409 metres from X and 322 metres from Y. The angle at A measures 61.8°. To the nearest metre, what would be the length of a tunnel from X to Y, through the hill?

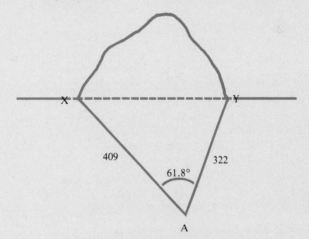

12.1 Measurement of Angles

OBJECTIVE

130. Find the degree measure of the third angle in a triangle, given two of the angles.

APPLICATION

A stairway makes an angle of 42 degrees with the downstairs floor. What angle does it make with the wall?

VOCABULARY

The word *line* is used here to mean a straight line that has no end in either direction. Any two points on the line can be used to name the line.

Line *MN* or line *MP* or line *NP* are possible names.
A *half line* contains all the points on one side of a point on a line.
A *ray* is a half line plus its endpoint.

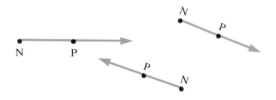

The above rays are read "ray *NP*" (not "ray *PN*").
An *angle* is formed when two rays start from a common point.

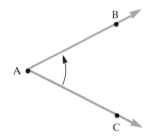

Ray *AB* and ray *AC* form angle *A* (or angle *CAB*). *A* is used to designate the angle or the measure of the angle.
A *degree* is a commonly used measure of angles.
Angles are classified according to their measures. *Obtuse angles* have measures between 90° and 180°; a *right angle* measures 90°; and *acute angles* have measures between 0° and 90°.
Minutes and *seconds* are smaller units of measure for an angle.

360 degrees = 1 revolution

60 minutes = 1 degree

60 seconds = 1 minute

If C = 36 degrees, 20 minutes, 30 seconds, we can write $C = 36° \ 20' \ 30''$.
In this text, angle measures of less than a degree will be expressed to the nearest tenth of a degree. So, $C \approx 36.3°$.

The angles in a triangle always have measures whose sum is 180°.

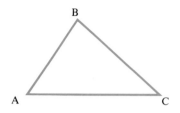

In triangle ABC ($\triangle ABC$) we have

$$A + B + C = 180°$$

HOW AND WHY

> **If the measure of two angles of a triangle is known, the measure of the third can be found by subtracting the first two measures from 180°.**

Consider $\triangle RST$.

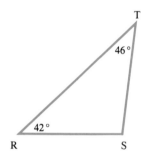

$R = 42°$

$T = 46°$

$S = ?$

$$R + S + T = 180°$$
$$42° + S + 46° = 180°$$
$$S = 180° - 88°$$
$$S = 92°$$

EXAMPLES

a.

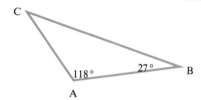

In $\triangle ABC$,

$A = 118°$

$B = 27°$

$C = ?$

$$A + B + C = 180°$$
$$118° + 27° + C = 180°$$
$$C = 180° - 145°$$
$$C = 35°$$

b.

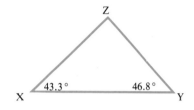

In $\triangle XYZ$,

$X = 43.3°$

$Y = 46.8°$

$Z = ?$

$$X + Y + Z = 180°$$
$$43.3° + 46.8° + Z = 180°$$
$$Z = 180° - 90.1°$$
$$Z = 89.9°$$

APPLICATION SOLUTION

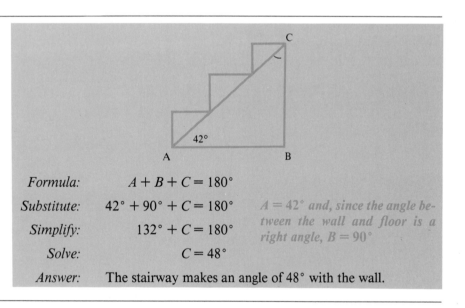

Formula: $A + B + C = 180°$

Substitute: $42° + 90° + C = 180°$ *A = 42° and, since the angle be-*
                                          tween the wall and floor is a
Simplify: $132° + C = 180°$ *right angle, B = 90°*

Solve: $C = 48°$

Answer: The stairway makes an angle of 48° with the wall.

WARM UPS

In each triangle, find the measure of the third angle.

1. In $\triangle ABC$, $A = 30°$, $B = 60°$

2. In $\triangle DEF$, $D = 45°$, $E = 45°$

3. In $\triangle MNP$, $P = 60°$, $M = 40°$

4. In $\triangle LMN$, $L = 90°$, $M = 80°$

5. In $\triangle PDQ$, $D = 40°$, $Q = 50°$

6. In $\triangle XYZ$, $Z = 20°$, $Y = 50°$

7. In $\triangle RST$, $R = 10°$, $S = 70°$

8. In $\triangle JKL$, $K = 25°$, $J = 25°$

9. In $\triangle ABC$, $C = 30°$, $B = 30°$

10. In $\triangle WXY$, $Y = 110°$, $X = 30°$

11. In $\triangle DEF$, $D = 130°$, $F = 40°$

12. In $\triangle PQR$, $P = 100°$, $R = 75°$

Exercises 12.1

A

In each triangle, find the measure of the third angle.

1. In $\triangle GHJ$, $J = 40°$, $H = 40°$

2. In $\triangle MNP$, $P = 20°$, $N = 70°$

3. In $\triangle ABC$, $C = 30°$, $A = 40°$

4. In $\triangle PQR$, $P = 100°$, $R = 20°$

5. In $\triangle RST$, $S = 110°$, $R = 10°$

6. In $\triangle PQR$, $Q = 140°$, $P = 20°$

B

In each triangle, find the measure of the third angle.

7. In $\triangle ABC$, $A = 23°$, $B = 88°$

8. In $\triangle DEF$, $D = 41°$, $F = 41°$

9. In $\triangle XYZ$, $Y = 50°$, $Z = 101°$

10. In $\triangle RST$, $R = 110°$, $S = 65°$

11. In $\triangle GHJ$, $J = 142°$, $G = 27°$

12. In $\triangle DEF$, $F = 107°$, $D = 32°$

13. In $\triangle XYZ$, $X = 117°$, $Z = 19°$

14. In $\triangle ARK$, $K = 78°$, $R = 67°$

C

In each triangle, find the measure of the third angle.

15. In $\triangle PQR$, $P = 35.6°$, $R = 37.3°$

16. In $\triangle GHJ$, $H = 57.5°$, $J = 62.3°$

17. In $\triangle LMN$, $M = 87.4°$, $N = 86.2°$

18. In $\triangle ACE$, $A = 75.9°$, $E = 75.3°$

19. In $\triangle TUV$, $V = 19.9°$, $T = 88.1°$

20. In $\triangle ART$, $R = 78.7°$, $A = 83.5°$

21. In $\triangle WXY$, $Y = 47.1°$, $X = 38.2°$

22. In $\triangle RST$, $T = 55.5°$, $R = 64°$

23. In $\triangle ABC$, $C = 7.3°$, $A = 102.8°$

24. In $\triangle TUV$, $V = 114.1°$, $U = 27.8°$

ANSWERS

1. _____
2. _____
3. _____
4. _____
5. _____
6. _____
7. _____
8. _____
9. _____
10. _____
11. _____
12. _____
13. _____
14. _____
15. _____
16. _____
17. _____
18. _____
19. _____
20. _____
21. _____
22. _____
23. _____
24. _____

D

25. A stairway makes an angle of 39 degrees with the downstairs floor. What angle does it make with the wall?

26. A stairway makes an angle of 47 degrees with the downstairs floor. What angle does it make with the wall?

27. A rocket carrying a shuttle takes off vertically. An observer sees the rocket at 38 degrees elevation. An astronaut would see the observer at what angle?

28. A rocket carrying a shuttle takes off vertically. An observer sees the rocket at 56.9 degrees elevation. An astronaut would see the observer at what angle?

29. A beam of light is projected vertically to make a spot of light on the clouds. A short distance away, a person facing the light measures the angle of her line of sight to the spot of light as 82 degrees. What angle does her line of sight make with the vertical beam of light?

30. A beam of light is projected vertically to make a spot of light on the clouds. A short distance away, a person facing the light measures the angle of his line of sight to the spot of light as 73.3 degrees. What angle does his line of sight make with the vertical beam of light?

E   *Maintain Your Skills*

Name each of the following points using ordered pairs.

31. *A*

32. *B*

33. *C*

34. *D*

35. *E*

12.2 Trigonometric Ratios: Part I

OBJECTIVE

131. Given any two sides of a right triangle, find the sine, cosine, and/or tangent of either acute angle.

APPLICATION

A road rises 9 feet for every 100 feet of horizontal distance. The grade of the road is the tangent of the angle expressed as a percent. What is the grade of the road?

VOCABULARY

A *right triangle* is any triangle that has one angle whose measure is 90°.
In the figure below, triangle ABC is a right triangle with right angle C ($C = 90°$).
Triangle ADE is also a right triangle with right angle D.

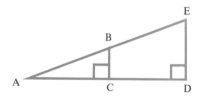

These right triangles are *similar triangles.* From geometry we know that similar triangles have equal corresponding angles and proportional corresponding sides. Therefore,

$B = E$ *Corresponding angles*

$C = D$ *Both are 90° angles*

$\dfrac{\overline{BC}}{\overline{AC}} = \dfrac{\overline{ED}}{\overline{AD}}$ *$\overline{BC}$ represents the measure of the side of the triangle opposite angle A*

$\dfrac{\overline{BC}}{\overline{AB}} = \dfrac{\overline{ED}}{\overline{AE}}$

$\dfrac{\overline{AC}}{\overline{AB}} = \dfrac{\overline{AD}}{\overline{AE}}$

Since these ratios are always equal (and very useful), they have special names.
The *sine* of angle $A = \sin A = \dfrac{\overline{BC}}{\overline{AB}} \left(\text{or } \dfrac{\overline{ED}}{\overline{AE}} \right)$.

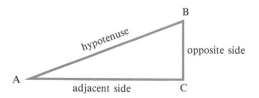

We can say

$$\sin A = \frac{\text{opposite}}{\text{hypotenuse}} = \frac{\overline{BC}}{\overline{AB}}$$

$$\text{cosine } A = \cos A = \frac{\text{adjacent}}{\text{hypotenuse}} = \frac{\overline{AC}}{\overline{AB}}$$

$$\text{tangent } A = \tan A = \frac{\text{opposite}}{\text{adjacent}} = \frac{\overline{BC}}{\overline{AC}}$$

These three ratios are called *trigonometric ratios.*

If we had chosen angle B, the opposite side would have been $\overline{AC}$ and the adjacent side would have been $\overline{BC}$, so that

$$\sin B = \frac{\text{opp}}{\text{hyp}} = \frac{\overline{AC}}{\overline{AB}}$$

$$\cos B = \frac{\text{adj}}{\text{hyp}} = \frac{\overline{BC}}{\overline{AB}}$$

$$\tan B = \frac{\text{opp}}{\text{adj}} = \frac{\overline{AC}}{\overline{BC}}$$

It is traditional to label the angles of a triangle using capital letters and the sides opposite the angles with the corresponding lower case letters. For example,

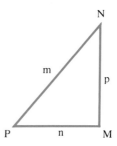

so $\overline{PM} = n$

$\overline{MN} = p$

$\overline{PN} = m$

HOW AND WHY

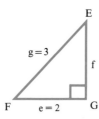

In this triangle $g = \overline{EF} = 3$ and $e = \overline{FG} = 2$; therefore,

$$f^2 + e^2 = g^2$$

$$f^2 + 4 = 9$$

$$f^2 = 5$$

$$f = \sqrt{5}$$

To find the trigonometric ratios of an acute angle of a right triangle, substitute into the following formulas.

$$\sin A = \frac{\text{opp}}{\text{hyp}}, \qquad \cos A = \frac{\text{adj}}{\text{hyp}}, \qquad \tan A = \frac{\text{opp}}{\text{adj}}$$

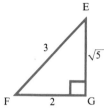

$$\sin F = \frac{\text{opp}}{\text{hyp}} = \frac{\sqrt{5}}{3} \qquad \sin E = \frac{\text{opp}}{\text{hyp}} = \frac{2}{3}$$

$$\cos F = \frac{\text{adj}}{\text{hyp}} = \frac{2}{3} \qquad \cos E = \frac{\text{adj}}{\text{hyp}} = \frac{\sqrt{5}}{3}$$

$$\tan F = \frac{\text{opp}}{\text{adj}} = \frac{\sqrt{5}}{2} \qquad \tan E = \frac{\text{opp}}{\text{adj}} = \frac{2}{\sqrt{5}} = \frac{2\sqrt{5}}{5}$$

EXAMPLES

a. In $\triangle RST$ find $\sin S$, $\cos S$, and $\tan S$.

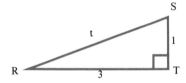

$$s^2 + r^2 = t^2$$
$$9 + 1 = t^2$$
$$10 = t^2$$
$$\sqrt{10} = t$$

and so

$$\sin S = \frac{3}{\sqrt{10}} = \frac{3\sqrt{10}}{10}$$

$$\cos S = \frac{1}{\sqrt{10}} = \frac{\sqrt{10}}{10}$$

$$\tan S = \frac{3}{1} = 3$$

b. In $\triangle UVW$ find $\sin V$, $\cos V$, and $\tan V$.

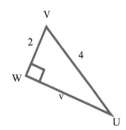

Here $w = 4$ and $u = 2$, so

$$v^2 + u^2 = w^2$$
$$v^2 + 4 = 16$$
$$v^2 = 12$$
$$v = \sqrt{12} = 2\sqrt{3}$$

Then

$$\sin V = \frac{2\sqrt{3}}{4} = \frac{\sqrt{3}}{2}$$

$$\cos V = \frac{2}{4} = \frac{1}{2}$$

$$\tan V = \frac{2\sqrt{3}}{2} = \sqrt{3}$$

APPLICATION SOLUTION

Formula:	$\tan A = \dfrac{\text{opposite}}{\text{adjacent}}$
Substitute:	$\tan A = \dfrac{9}{100}$ *The opposite side has length 9 and the adjacent side has length 100*
Simplify:	$\tan A = .09 = 9\%$
Answer:	The grade of the road is 9%.

WARM UPS

In each triangle, find $\sin A$, $\cos A$, and $\tan A$.

1. In $\triangle ABC$, $a = 6$, $c = 10$, and $C = 90°$

2. In $\triangle ABC$, $a = 8$, $c = 10$, and $C = 90°$

3. In $\triangle ABC$, $a = 4$, $c = 5$, and $C = 90°$

4. In $\triangle ABC$, $a = 3$, $c = 5$, and $C = 90°$

5. In $\triangle ABC$, $a = 12$, $c = 20$, and $C = 90°$

6. In $\triangle ABC$, $a = 20$, $c = 25$, and $C = 90°$

In each triangle, find $\sin T$, $\cos T$, and $\tan T$.

7. In $\triangle RST$, $r = 13$, $s = 12$, $R = 90°$

8. In $\triangle RST$, $r = 13$, $s = 5$, $R = 90°$

9. In $\triangle RST$, $r = 15$, $s = 9$, $R = 90°$

10. In $\triangle RST$, $r = 15$, $s = 12$, $R = 90°$

11. In $\triangle RST$, $r = 20$, $s = 16$, $R = 90°$

12. In $\triangle RST$, $r = 25$, $s = 15$, $R = 90°$

Exercises 12.2

A

In each triangle, find sin A, cos A, and tan A.

1. In $\triangle ABC$, $c = 10$, $b = 6$, $C = 90°$

2. In $\triangle ABC$, $c = 20$, $b = 12$, $C = 90°$

3. In $\triangle ABC$, $c = 13$, $b = 12$, $C = 90°$

4. In $\triangle ABC$, $c = 15$, $b = 12$, $C = 90°$

5. In $\triangle ABC$, $c = 26$, $b = 10$, $C = 90°$

6. In $\triangle ABC$, $c = 26$, $b = 24$, $C = 90°$

B

In each triangle, find sin T, cos T, and tan T.

7. In $\triangle RST$, $r = 1$, $t = 1$, $S = 90°$

8. In $\triangle RST$, $r = 2$, $t = 2$, $S = 90°$

9. In $\triangle RST$, $r = 3$, $t = 3$, $S = 90°$

10. In $\triangle RST$, $r = 1$, $t = 3$, $S = 90°$

11. In $\triangle RST$, $r = 2$, $t = 3$, $S = 90°$

12. In $\triangle RST$, $r = 2$, $t = 4$, $S = 90°$

13. In $\triangle RST$, $r = 3$, $t = 5$, $S = 90°$

14. In $\triangle RST$, $r = 5$, $t = 4$, $S = 90°$

C

In each triangle, find sin D, cos D, and tan D.

15. In $\triangle DEF$, $d = 3$, $e = 1$, $F = 90°$

16. In $\triangle DEF$, $d = 1$, $e = 3$, $F = 90°$

17. In $\triangle DEF$, $f = 3$, $e = 1$, $F = 90°$

18. In $\triangle DEF$, $f = 3$, $e = 2$, $F = 90°$

1. _____
2. _____
3. _____
4. _____
5. _____
6. _____
7. _____
8. _____
9. _____
10. _____
11. _____
12. _____
13. _____
14. _____
15. _____
16. _____
17. _____
18. _____

ANSWERS

19.

20.

21.

22.

23.

24.

25.

26.

27.

28.

29.

30.

31.

32.

33.

34.

35.

19. In $\triangle DEF$, $d = 2$, $e = 3$, $F = 90°$

20. In $\triangle DEF$, $d = 3$, $e = 2$, $F = 90°$

21. In $\triangle DEF$, $f = 1$, $e = 3$, $E = 90°$

22. In $\triangle DEF$, $d = 3$, $f = 5$, $E = 90°$

23. In $\triangle DEF$, $f = 6$, $e = 3$, $F = 90°$

24. In $\triangle DEF$, $d = 3$, $e = 6$, $E = 90°$

D

25. A mountain highway rises 4.2 metres for every 100 metres of horizontal distance. The grade is the tangent of the angle expressed as a percent. What is the grade of the highway?

26. A fire road rises 4.3 yards for every 50 yards of horizontal distance. The grade is the tangent of the angle expressed as a percent. What is the grade of the road?

27. A ski slope rises 2.55 feet for every 25 feet of horizontal distance. What is the grade of the slope?

28. A ski jump rises 3.2 feet for every 20 feet of horizontal distance. What is the grade of the slope?

E   *Maintain Your Skills*

Make a table of values and graph each of the following.

29. $y = \dfrac{2x}{3}$

30. $y = x + 3$

31. $y = \dfrac{-3x}{4} - 2$

32. $2x + 3y = 12$

33. $5x + 3y = -12$

596

ANSWERS TO WARM UPS (12.2)

1. $\sin A = \dfrac{3}{5}$ $\cos A = \dfrac{4}{5}$ $\tan A = \dfrac{3}{4}$

2. $\sin A = \dfrac{4}{5}$ $\cos A = \dfrac{3}{5}$ $\tan A = \dfrac{4}{3}$

3. $\sin A = \dfrac{4}{5}$ $\cos A = \dfrac{3}{5}$ $\tan A = \dfrac{4}{3}$

4. $\sin A = \dfrac{3}{5}$ $\cos A = \dfrac{4}{5}$ $\tan A = \dfrac{3}{4}$

5. $\sin A = \dfrac{3}{5}$ $\cos A = \dfrac{4}{5}$ $\tan A = \dfrac{3}{4}$

6. $\sin A = \dfrac{4}{5}$ $\cos A = \dfrac{3}{5}$ $\tan A = \dfrac{4}{3}$

7. $\sin T = \dfrac{5}{13}$ $\cos T = \dfrac{12}{13}$ $\tan T = \dfrac{5}{12}$

8. $\sin T = \dfrac{12}{13}$ $\cos T = \dfrac{5}{13}$ $\tan T = \dfrac{12}{5}$

9. $\sin T = \dfrac{4}{5}$ $\cos T = \dfrac{3}{5}$ $\tan T = \dfrac{4}{3}$

10. $\sin T = \dfrac{3}{5}$ $\cos T = \dfrac{4}{5}$ $\tan T = \dfrac{3}{4}$

11. $\sin T = \dfrac{3}{5}$ $\cos T = \dfrac{4}{5}$ $\tan T = \dfrac{3}{4}$

12. $\sin T = \dfrac{4}{5}$ $\cos T = \dfrac{3}{5}$ $\tan T = \dfrac{4}{3}$

12.3 Trigonometric Ratios: Part II

OBJECTIVES

132. Find the exact value of the sine, cosine, and tangent of an angle of $30°, 45°$, and $60°$.
133. Find the approximate value of the sine, cosine, and tangent of any acute angle from a table (see table at end of chapter) or a calculator with keys for trigonometric ratios.
134. Find the acute angle, to the nearest tenth of a degree, given the value of the sine, cosine, or tangent. Use a table or calculator.

APPLICATION

A tuned circuit in Greg's CB receiver has a reactance of 153.29 ohms. Find the impedance (to the nearest hundredth) of the circuit if the phase angle is $52.7°$. The formula for impedance is

$$Z_T = \frac{X}{\sin \theta}$$

VOCABULARY

A *30°–60° right triangle* is shown below.

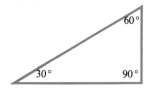

A *45° right triangle* has two angles and two sides equal (isosceles) as shown.

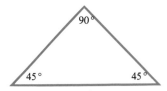

From geometry we know that the lengths of the sides are multiples of the following lengths.

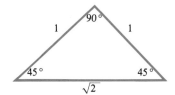

HOW AND WHY

The two special triangles shown above are used in many applications, so it is useful to memorize the exact values of the trigonometric ratios of these angles.

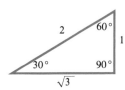

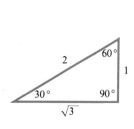

sin 30°	$\dfrac{\text{opp}}{\text{hyp}}$	$\dfrac{1}{2}$
sin 45°	$\dfrac{\text{opp}}{\text{hyp}}$	$\dfrac{1}{\sqrt{2}}$ or $\dfrac{\sqrt{2}}{2}$
sin 60°	$\dfrac{\text{opp}}{\text{hyp}}$	$\dfrac{\sqrt{3}}{2}$

cos 30°	$\dfrac{\text{adj}}{\text{hyp}}$	$\dfrac{\sqrt{3}}{2}$
cos 45°	$\dfrac{\text{adj}}{\text{hyp}}$	$\dfrac{1}{\sqrt{2}}$ or $\dfrac{\sqrt{2}}{2}$
cos 60°	$\dfrac{\text{adj}}{\text{hyp}}$	$\dfrac{1}{2}$

tan 30°	$\dfrac{\text{opp}}{\text{adj}}$	$\dfrac{1}{\sqrt{3}}$ or $\dfrac{\sqrt{3}}{3}$
tan 45°	$\dfrac{\text{opp}}{\text{adj}}$	$\dfrac{1}{1}$ or 1
tan 60°	$\dfrac{\text{opp}}{\text{adj}}$	$\dfrac{\sqrt{3}}{1}$ or $\sqrt{3}$

To find the three ratios of any other acute angle, unless the sides of the triangle are given, we must use a table of values or a calculator. Tables usually show 4 or 5 decimal places and calculators may display 7, 8, or 9 decimal places. (A four-place table is available at the end of this chapter.)

The low cost of scientific calculators (which have keys labeled "sin", "cos", and "tan") is making the use of tables obsolete. The examples below show how the calculator is used to find these values.

When the sine, cosine, or tangent of an angle is known, the table is read in reverse (from inside to out) to find the angle. The table is limited to finding angles to the nearest tenth of a degree. A scientific calculator has keys labeled $\sin^{-1}$, $\cos^{-1}$, and $\tan^{-1}$ that are used to find angles. A key marked $\boxed{\text{INV}}$ or $\boxed{\text{2nd}}$, which is pushed before the above keys, is found on most of these calculators. Consult your calculator manual.

EXAMPLES

a. $\sin 74° \approx 0.9613$ (four-place table)

CALCULATOR	DISPLAY

First, be sure your calculator is in "degree mode". Check the manual for your calculator.

$\boxed{74}$ 74.

$\boxed{\sin}$ 0.9612617

$\sin 74° \approx 0.9613$ (rounded to four decimal places)

b. $\cos 45° \approx 0.7071$ (four-place table)

CALCULATOR	DISPLAY

"degree mode"

$\boxed{45}$ 45.

$\boxed{\cos}$ 0.7071068

$\cos 45° \approx 0.7071$ (rounded to four decimal places)

c. $\tan 17.5° \approx 0.3153$ (four-place table)

CALCULATOR	DISPLAY

"degree mode"

$\boxed{17.5}$ 17.5

$\boxed{\tan}$ 0.3152988

$\tan 17.5° \approx 0.3153$ (rounded to four decimal places)

d. $\cos 66.4°$

Table: For angles larger than 45°, look in the right-hand column and read upward from the bottom. For cos, look at the labels at the bottom of the table, not the top.

$\cos 66.4° \approx 0.4003$ (four-place table)

e. $\sin 37° \approx 0.6018$ (four-place table or rounded calculator value)

f. $\cos 22.8° \approx 0.9219$ (four-place table or rounded calculator value)

g. tan 80.6°

CALCULATOR	DISPLAY
"degree mode"	
80.6	80.6
tan	6.0405103

tan 80.6° ≈ 6.0405   (rounded to four decimal places)

h. sin 43.2° ≈ 0.6845   (four-place table or rounded calculator value)

i. sin A = 0.3535

   Table: Read down the column labeled "sin" to .3535. The angle is read from the first column.

$A \approx 20.7°$

CALCULATOR	DISPLAY
"degree mode"	
.3535	0.3535
INV or 2nd	0.3535
sin or sin⁻¹	20.701541

$A \approx 20.7°$ (rounded to nearest tenth)

j. cos B = 0.8321

$B \approx 33.7°$	(0.8320 is closest entry in table)
$B \approx 33.684933°$	(calculator)
$\approx 33.7°$	(rounded to nearest tenth)

k. tan C = 1.5166

   Table: Reading down the column labeled "tan", we come to an end (at 0.8541). Now reverse, to read up the "tan" column from the bottom of the table (at 1.1708). Continuing up, find 1.5166 and read the angle at the right.

$C \approx 56.6°$

CALCULATOR	DISPLAY
"degree mode"	
1.5166	1.5166
INV or 2nd	1.5166
tan or tan⁻¹	56.600354

$C \approx 56.6°$ (rounded to nearest tenth)

APPLICATION SOLUTION

Formula: $Z_T = \dfrac{X}{\sin \theta}$,   where $X = 153.29$, $\theta = 52.7°$

Substitute: $Z_T = \dfrac{153.29}{\sin 52.7°}$

Simplify: $Z_T \approx \dfrac{153.29}{.7955}$

$Z_T \approx 192.70$

Answer: So the impedance is 192.70 ohms.

WARM UPS

Use the four-place trigonometric tables or a calculator to find each value correct to four decimal places.

1. $\sin 21°$	**2.** $\cos 21°$	**3.** $\tan 21°$
4. $\sin 36°$	**5.** $\cos 36°$	**6.** $\tan 36°$
7. $\sin 45°$	**8.** $\cos 45°$	**9.** $\tan 45°$
10. $\sin 56°$	**11.** $\cos 56°$	**12.** $\tan 56°$

Exercises 12.3

A

Use the four-place trigonometric tables or a calculator to find each value correct to four decimal places.

1. sin 7°   **2.** cos 7°   **3.** tan 7°

4. sin 82°   **5.** cos 82°   **6.** tan 82°

B

Use the four-place trigonometric tables or a calculator to find each value correct to four decimal places.

7. sin 37.3°   **8.** cos 37.3°   **9.** tan 37.3°

10. sin 10.2°   **11.** cos 10.2°   **12.** tan 10.2°

13. sin 46.2°   **14.** tan 8.4°

C

Use the four-place trigonometric tables or a calculator to find the measure of each acute angle correct to the nearest tenth of a degree.

15. cos $B = 0.6211$ **16.** tan $Y = 0.6297$

17. sin $B = 0.3338$ **18.** tan $A = 3.7214$

19. cos $C = 0.4217$ **20.** sin $M = 0.9146$

21. sin $A = 0.9061$ **22.** cos $B = 0.9061$

23. tan $C = 0.9061$ **24.** tan $N = 1.7000$

1. _____
2. _____
3. _____
4. _____
5. _____
6. _____
7. _____
8. _____
9. _____
10. _____
11. _____
12. _____
13. _____
14. _____
15. _____
16. _____
17. _____
18. _____
19. _____
20. _____
21. _____
22. _____
23. _____
24. _____

ANSWERS

25. _____

26. _____

27. _____

28. _____

29. _____

30. _____

31. _____

32. _____

33. _____

D

25. A tuned circuit has a reactance of 128.63 ohms. Use the formula in the application to find the impedance (to the nearest hundredth) of the circuit if the phase angle is 49.6 degrees.

26. A tuned circuit has a reactance of 173.05 ohms. Use the formula in the application to find the impedance (to the nearest hundredth) of the circuit if the phase angle is 53.8 degrees.

27. If the tread of a stairway is 20 cm and the rise is 16.5 cm, what angle, to the nearest tenth of a degree, does the stairway make with the floor?

28. If the tread of a stairway is 7 inches and the rise is 5 inches, what angle, to the nearest tenth of a degree, does the stairway make with the floor?

E   *Maintain Your Skills*

Find the slope of the line that joins each of the following pairs of points:

29. $(-5, 2), (4, 7)$ **30.** $(5, 9), (8, 2)$ **31.** $(2, -4), (-8, 7)$

Find the slope of the line and its y-intercept:

32. $4x - 5y = -8$ **33.** $4y = 6x$

12.4 Right Triangle Solutions

OBJECTIVE

135. Solve a right triangle.

APPLICATION

What angle does a rafter make with the horizontal if it has a rise of 4.5 feet in a horizontal run of 13.2 feet, to the nearest tenth of a degree?

VOCABULARY

To *solve* a right triangle is to find the measures of the sides and angles not already given.

HOW AND WHY

If three parts of a right triangle are known, including the right angle and at least one side, the measures of the remaining angles and sides can be found. To find them we can use any of the following formulas.

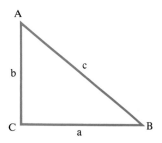

$$a^2 + b^2 = c^2 \qquad A + B + C = 180°$$

$$\sin A = \frac{a}{c} \qquad \cos A = \frac{b}{c} \qquad \tan A = \frac{a}{b}$$

$$\sin B = \frac{b}{c} \qquad \cos B = \frac{a}{c} \qquad \tan B = \frac{b}{a}$$

In the use of tables or calculators for trigonometric ratios, it is understood that the values are approximations. The examples illustrate how the formulas may be applied.

EXAMPLES

a.

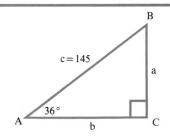

Given: $c = 145$

$A = 36°$

$C = 90°$

Find: a, b, B

To find B	To find a	To find b
$A + B + C = 180°$	$\sin A = \dfrac{a}{c}$	$\cos A = \dfrac{b}{c}$
$36° + B + 90° = 180°$		
$B = 54°$	$\sin 36° = \dfrac{a}{145}$	$\cos 36° = \dfrac{b}{145}$
	$0.5878 \approx \dfrac{a}{145}$	$0.8090 \approx \dfrac{b}{145}$
	$85.2 \approx a$	$117.3 \approx b$

b.

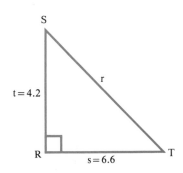

Given: $R = 90°$

$t = 4.2$

$s = 6.6$

Find: r, S, T

To find r	To find S	To find T
$t^2 + s^2 = r^2$	$\tan S = \dfrac{s}{t}$	$S + R + T = 180°$
$17.64 + 43.56 = r^2$		$57.5° + 90° + T \approx 180°$
$7.8 \approx r$	$\tan S = \dfrac{6.6}{4.2} \approx 1.5714$	$T \approx 32.5°$
	$S \approx 57.5°$	

c. A tunnel is to be dug down at an angle of 11.3° from a level surface. What is the vertical distance (to the nearest tenth of a metre) between two points that are 500 metres apart along the tunnel? First sketch a figure and label it.

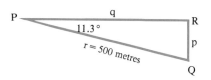

$\overline{PQ} = r = 500$ metres

$P = 11.3°$

Find p

$\sin P = \dfrac{p}{r}$ so $\sin 11.3° = \dfrac{p}{500}$

$0.1959 \approx \dfrac{p}{500}$

$98.0 \approx p$

So the vertical distance is approximately 98 metres.

APPLICATION SOLUTION

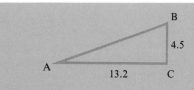

Formula: $\sin A = \dfrac{4.5}{13.2} \approx 0.3409$

$A \approx 19.9°$

Answer: So the rafter makes an angle of 19.9° with the horizontal.

Exercises 12.4

A

Find lengths to the nearest tenth of a unit and angles to the nearest tenth of a degree.

1. Given: $e = 6$

$E = 30°$

$D = 90°$

Find: d, f, F

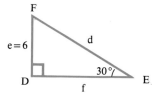

2. Given: $s = 4$

$S = 45°$

$T = 90°$

Find: r, t, R

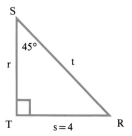

3. Given: $a = 8$

$B = 60°$

$C = 90°$

Find: b, c, A

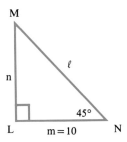

4. Given: $m = 10$

$N = 45°$

$L = 90°$

Find: ℓ, n, M

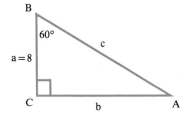

5. Given: $y = 1$

$X = 30°$

$Z = 90°$

Find: x, z, Y

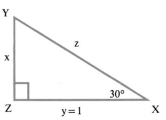

6. Given: $r = 2$

$R = 60°$

$S = 90°$

Find: s, t, T

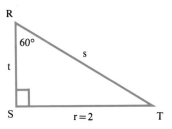

ANSWERS

7. _____

8. _____

9. _____

10. _____

11. _____

12. _____

13. _____

14. _____

15. _____

B

7. Given: $r = 10.1$
$Q = 45°$
$P = 90°$
Find: p, q, R

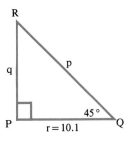

8. Given: $b = 7$
$a = 10$
$A = 90°$
Find: c, B, C

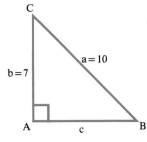

9. Given: $a = 37.4$
$b = 25$
$C = 90°$
Find: c, A, B

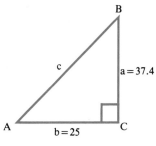

10. Given: $s = 77.3$
$R = 21.1°$
$S = 90°$
Find: r, t, T

11. Given: $x = 305.0$
$Z = 41.1°$
$Y = 90°$
Find: y, z, X

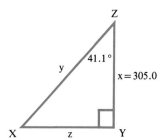

C

12. Given: $e = 6.5, f = 3.7, D = 90°$
Find: d, E, F

13. Given: $s = 4, r = 12.2, T = 90°$
Find: t, S, R

14. Given: $a = 8.8, c = 14.2, C = 90°$
Find: b, B, A

15. Given: $m = 22.7, n = 31.6, L = 90°$
Find: ℓ, N, M

16. Given: $y = 40$, $x = 31.4$, $Z = 90°$

   Find: z, X, Y

16. _____

D

17. What angle does a rafter make with the horizontal if it has a rise of 5.8 feet in a horizontal run of 14.6 feet, to the nearest tenth of a degree?

17. _____

18. What angle does a rafter make with the horizontal if it has a rise of 3.8 feet in a horizontal run of 10.7 feet, to the nearest tenth of a degree?

18. _____

19. A drain pipe is to be laid down in a ditch at angle of 5.7° from a level surface. The pipe is 121 feet long. How much lower is one end of the pipe than the other, to the nearest tenth of a foot?

19. _____

20. A tunnel is to be dug down at an angle of 17.5° from a level surface. What is the vertical distance (to the nearest tenth of a foot) between two points that are 300 feet apart along the tunnel?

20. _____

21. When the shadow of an office building is 25 feet long, the angle of elevation of the sun is 80°. How tall is the building, to the nearest foot?

80°

25

21. _____

22. _____

22. When the shadow of a tree is 31 feet long, the angle of elevation of the sun is 62°. How tall is the tree, to the nearest foot?

23. _____

23. A guy wire stretches from the top of a 50-foot pole to the ground, and makes a 63° angle with the ground. Find the length of the guy wire, to the nearest foot.

24. A supporting cable stretches from the top of a 37-foot antenna, and makes a 43° angle with the ground. Find the length of the cable, to the nearest foot.

25. _____

25. A pilot sights a wrecked aircraft at an angle of depression of 35°. What is the distance, to the nearest ten feet, from the pilot to the wreck site if the pilot is flying at 3500 feet? (*Hint:* the angle of depression is measured from the horizontal flight path to the wreck.)

25. _____

26. A search plane sights a lost hiker at an angle of depression of 23°. What is the distance, to the nearest ten feet, from the search plane to the hiker if the pilot is flying at 2800 feet? (*Hint:* the angle of depression is measured from the horizontal flight path to the hiker.)

26. _____

611

27. The truss of a bridge is shown below. What is the height of the truss?

28. The truss of a bridge is shown below. What is the height of the truss?

29. A light plane flies 55 miles east from C and then flies 40 miles north. What is the bearing, to the nearest degree, of the plane from C, to the nearest tenth of a degree? (Angle $C = ?$)

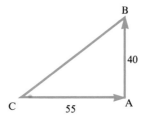

30. A small passenger plane flies 237 miles east from C and then flies 107 miles north. What is the bearing, to the nearest degree, of the plane from C, to the nearest tenth of a degree? (Angle $C = ?$)

31. A power line tower is known to be 70 feet high and is located on the bank of a river. If the angle from the horizontal measured from the opposite bank to the top of the tower is 39°, how wide is the river to the nearest foot?

32. A paratrooper jumps from a height of 3200 feet. Because of prevailing winds, the angle his path makes with the ground is 78°. How far from the point below his plane (at the time he jumped) does he land?

33. A sky diver jumps from a height of 2700 feet. Because of prevailing winds, the angle her path makes with the ground is 82°. How far from the point below her plane (at the time she jumped) does she land?

E   *Maintain Your Skills*

Find the distance between each of the following pairs of points:

34. (8, 2), (−3, 5)   **35.** (−8, −2), (5, 7)   **36.** (4, 4), (8, −6)

37. (0, 0), (8, 12)   **38.** (9, 9), (−9, −9)

12.5 Trigonometric Ratios of Obtuse Angles

OBJECTIVE

136. Find the approximate values of the sine, cosine, and tangent of an obtuse angle, using a table or a scientific calculator.

APPLICATION

The major application is in the solution of triangles.

VOCABULARY

No new vocabulary.

HOW AND WHY

The trigonometric ratio of an obtuse angle can be approximated by using the tables for acute angles and the following formulas.

> **If A is an obtuse angle,**
>
> $\sin A = \sin (180 - A)$
>
> $\cos A = -\cos (180 - A)$
>
> $\tan A = -\tan (180 - A)$

These formulas, as well as those for larger angles, are developed in a formal treatment of trigonometry.

EXAMPLES

a. Using the tables and the formulas of this section:
Find the sine, cosine, and tangent of 120°.

$\sin 120° = \sin (180° - 120°) = \sin 60° \approx 0.8660$

$\cos 120° = -\cos (180° - 120°) = -\cos 60° = -0.5000$

$\tan 120° = -\tan (180° - 120°) = -\tan 60° \approx -1.7321$

b. With a calculator, the formulas are not necessary.
Find the sine, cosine, and tangent of 148°.

CALCULATOR	DISPLAY
148	148.
sin	0.5299193
148	148.
cos	−0.8480481
148	148.
tan	−0.6248694

So, to four decimal places:

$\sin 148° \approx 0.5299$

$\cos 148° \approx -0.8480$

$\tan 148° \approx -0.6249$

c. Find the sine, cosine, and tangent of 165°.

$\sin 165° = \sin 15° \approx 0.2588$

$\cos 165° = -\cos 15° \approx -0.9659$

$\tan 165° = -\tan 15° \approx -0.2679$

WARM UPS

Find the value of each trigonometric ratio correct to four decimal places.

1. sin 150°	**2.** cos 150°	**3.** tan 150°
4. sin 135°	**5.** cos 135°	**6.** tan 135°
7. sin 100°	**8.** cos 100°	**9.** tan 100°
10. sin 170°	**11.** cos 170°	**12.** tan 170°

Exercises 12.5

ANSWERS

1. _____
2. _____
3. _____
4. _____
5. _____
6. _____
7. _____
8. _____
9. _____
10. _____
11. _____
12. _____
13. _____
14. _____
15. _____
16. _____
17. _____
18. _____
19. _____
20. _____
21. _____
22. _____
23. _____
24. _____
25. _____
26. _____
27. _____
28. _____
29. _____

A

Find the value of each trigonometric ratio correct to four decimal places.

1. $\sin 127°$ **2.** $\cos 127°$ **3.** $\tan 127°$

4. $\sin 115°$ **5.** $\cos 115°$ **6.** $\tan 115°$

B

7. $\sin 98°$ **8.** $\cos 98°$ **9.** $\tan 98°$

10. $\sin 104°$ **11.** $\cos 104°$ **12.** $\tan 104°$

13. $\sin 95°$ **14.** $\cos 95°$

C

15. $\sin 121.7°$ **16.** $\cos 121.7°$ **17.** $\tan 121.7°$

18. $\sin 118.6°$ **19.** $\cos 118.6°$ **20.** $\tan 118.6°$

21. $\tan 108.5°$ **22.** $\cos 111.1°$ **23.** $\sin 152.9°$

24. $\tan 140.8°$

E   *Maintain Your Skills*

Draw the graph of each line, given the slope and the y-intercept:

25. $m = -2, (0, -2)$ **26.** $m = 1, (0, 5)$ **27.** $m = \frac{2}{3}, (0, -1)$

Draw the graph of each of the following equations, using the slope and the y-intercept:

28. $8x - 3y = 6$ **29.** $5x + 2y = 8$

ANSWERS TO WARM UPS (12.5)	**1.** 0.5000	**2.** −0.8660	**3.** −0.5774	**4.** 0.7071	**5.** −0.7071
	6. −1.0000	**7.** 0.9848	**8.** −0.1736	**9.** −5.6713	**10.** 0.1736
	11. −0.9848	**12.** −0.1763			

12.6 Obtuse Angles from Trigonometric Ratios

OBJECTIVE

137. Find the measure of an obtuse angle, given the value of its sine, cosine, or tangent.

APPLICATION

The major application is in the solution of triangles.

VOCABULARY

No new vocabulary.

HOW AND WHY

To find the obtuse angle given one of its trigonometric ratios, we use the absolute value of the ratio to find an acute angle from the table. This angle is then subtracted from 180°.

> **If A is an obtuse angle and**
>
> **if $\sin A = w$, then $A = 180° −$ (acute angle whose sin $= |w|$)**
> **if $\cos A = x$, then $A = 180° −$ (acute angle whose cos $= |x|$)**
> **if $\tan A = y$, then $A = 180° −$ (acute angle whose tan $= |y|$)**

A is identified as being obtuse if the cosine is negative or the tangent is negative.

Find the obtuse angle whose tangent is −0.9753. That is, find A if $\tan A = -0.9753$.

$A = 180° −$ (acute angle whose tan $= |-0.9753|$)

$A = 180° −$ (acute angle whose tan $= 0.9753$)

$A \approx 180° − 44.3°$

$A \approx 135.7°$

EXAMPLES

a. Using the table:

Find A if $\cos A = -0.3656$

A is obtuse since the cosine is negative.

$A = 180° −$ (acute angle whose cos $= |-0.3656|$)

$A \approx 180° − 68.6°$

$A \approx 111.4°$

With a calculator:

CALCULATOR	DISPLAY
"degree mode"	
.3656	0.3656
+/− or CHS	−0.3656
INV or 2nd	−0.3656
cos or cos⁻¹	111.44451

$A \approx 111.4°$ (rounded to nearest tenth)

b. Find A if sin $A = 0.3567$

A can be either acute or obtuse.

$A = $ acute angle whose sin $= 0.3567$ or $A = 180° - $ (acute angle whose sin $= 0.3567$)

$A \approx 20.9°$ or $A \approx 180° - 20.9°$

$A \approx 159.1°$

c. Using the table:

Find B if tan $B = -0.5631$

B is obtuse since the tangent is negative.

$B = 180° - $ (acute angle whose tan $= |-0.5631|$)

$B \approx 180° - 29.4°$

$B \approx 150.6°$

With a calculator:

We still need the formula because the calculator displays the opposite of the acute angle. (To understand why requires a full course in trigonometry.)

CALCULATOR	DISPLAY
"degree mode"	
.5631	0.5631
+/− or CHS	−0.5631
INV or 2nd	−0.5631
tan or tan⁻¹	−29.383861
+	−29.383861
180	180.
=	150.61614

$B \approx 150.6°$ (rounded to nearest tenth)

d. Find C if cos $C = -0.7695$

The angle is obtuse since the cosine is negative.

$C = 180° - $ (acute angle whose cos $= |-0.7695|$)

$C \approx 180° - 39.7°$

$C \approx 140.3°$

Exercises 12.6

ANSWERS

1. _____

2. _____

3. _____

4. _____

5. _____

6. _____

7. _____

8. _____

9. _____

10. _____

11. _____

12. _____

13. _____

14. _____

15. _____

16. _____

17. _____

18. _____

19. _____

20. _____

21. _____

22. _____

23. _____

24. _____

25. _____

26. _____

27. _____

28. _____

29. _____

A

Find the obtuse angle to the nearest tenth of a degree.

1. $\sin A = 0.6123$ **2.** $\cos C = -0.6123$

3. $\tan P = -0.6123$ **4.** $\sin R = 0.7799$

5. $\cos T = -0.7799$ **6.** $\tan P = -0.7799$

B

7. $\sin A = 0.3621$ **8.** $\cos B = -0.3621$

9. $\tan C = -0.3621$ **10.** $\sin X = 0.9123$

11. $\cos Z = -0.9123$ **12.** $\tan Y = -0.9123$

13. $\sin W = 0.3456$ **14.** $\cos Q = -0.3456$

C

15. $\cos M = -0.0982$ **16.** $\tan N = -1.6234$

17. $\cos P = -0.9200$ **18.** $\tan A = -1.7111$

19. $\cos B = -0.1111$ **20.** $\tan C = -2.2461$

21. $\sin T = 0.6534$ **22.** $\cos S = -0.1098$

23. $\tan C = -0.9820$ **24.** $\cos V = -0.4682$

E *Maintain Your Skills*

Solve the following systems by graphing:

25. $\begin{cases} 4x - 3y = 15 \\ 3x + y = 8 \end{cases}$ **26.** $\begin{cases} 2x - y = 8 \\ 8x - 4y = 32 \end{cases}$

27. $\begin{cases} 5x + 3y = 12 \\ 10x + 6y = 18 \end{cases}$ **28.** $\begin{cases} 3x + y = 4 \\ x - y = 0 \end{cases}$

29. $\begin{cases} y = 3x - 2 \\ y = -x + 6 \end{cases}$

619

12.7 The Law of Cosines

OBJECTIVES

138. Solve an oblique triangle, given all three sides and no angles.
139. Solve an oblique triangle, given two sides and the angle formed by them.

APPLICATION

A piece of aluminum 8 inches wide is to be formed into an eaves trough with a 5-inch opening at the top. What is the angle formed by the sides of the trough?

VOCABULARY

An *oblique triangle* is any triangle that does not contain a right angle.
The *law of cosines* is an expression stating the relationship between the cosines of the angles of a triangle and the sides. Given the triangle *ABC*,

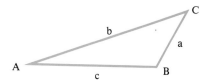

the law of cosines is stated as follows:

$$a^2 = b^2 + c^2 - 2bc \cos A \quad \text{or} \quad \cos A = \frac{b^2 + c^2 - a^2}{2bc}$$

$$b^2 = a^2 + c^2 - 2ac \cos B \quad \text{or} \quad \cos B = \frac{a^2 + c^2 - b^2}{2ac}$$

$$c^2 = a^2 + b^2 - 2ab \cos C \quad \text{or} \quad \cos C = \frac{a^2 + b^2 - c^2}{2ab}$$

HOW AND WHY

As with any formula, if all of the unknowns are given except for one, it can be found by substituting the known ones in the formula. This follows for each of the forms of the law of cosines. Given triangle *ABC* with $a = 21$, $b = 32$, and $c = 15$, find the size of the angles.

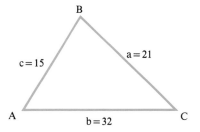

Since all sides are given, any one of the angles can be found by the law of cosines. First find A:

$$a^2 = b^2 + c^2 - 2bc \cos A$$

$$2bc \cos A = b^2 + c^2 - a^2$$

$$\cos A = \frac{b^2 + c^2 - a^2}{2bc}$$

$$\cos A = \frac{(32)^2 + (15)^2 - (21)^2}{2(32)(15)}$$

$$\cos A = \frac{1024 + 225 - 441}{960} = \frac{808}{960}$$

$$\cos A \approx 0.8417$$

$$A \approx 32.7°$$

Now find B:

$$b^2 = a^2 + c^2 - 2ac \cos B$$

$$2ac \cos B = a^2 + c^2 - b^2$$

$$\cos B = \frac{a^2 + c^2 - b^2}{2ac}$$

$$\cos B = \frac{(21)^2 + (15)^2 - (32)^2}{2(21)(15)}$$

$$\cos B = \frac{441 + 225 - 1024}{630} = \frac{-358}{630}$$

$$\cos B \approx -0.5683$$

$$B \approx 124.6°$$

Since the sum of the angles is $180°$,

$$C \approx 180° - 124.6° - 32.7°$$

$$C \approx 22.7°$$

The angles of the triangle are $124.6°$, $32.7°$, and $22.7°$.

EXAMPLES

a. Solve the triangle ABC, given $a = 13.7$, $b = 12.5$, and $C = 49.7°$.

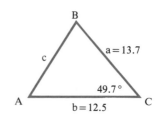

Side c can be found by using the law of cosines:

$c^2 = a^2 + b^2 - 2ab \cos C$

$c^2 = (13.7)^2 + (12.5)^2 - 2(13.7)(12.5)(\cos 49.7°)$

$c^2 \approx (13.7)^2 + (12.5)^2 - 2(13.7)(12.5)(0.6468)$

$c^2 \approx 187.69 + 156.25 - 221.53$

$c^2 \approx 122.41$

$c \approx 11.1$ (nearest tenth)

Angle B can also be found using the law of cosines:

$\cos B = \dfrac{a^2 + c^2 - b^2}{2ac}$

$\cos B \approx \dfrac{(13.7)^2 + (11.1)^2 - (12.5)^2}{2(13.7)(11.1)}$

$\cos B \approx \dfrac{187.69 + 123.21 - 156.25}{304.14} = \dfrac{154.65}{304.14}$

$\cos B \approx 0.5085$

$B \approx 59.4°$

The remaining angle must bring the total to 180°; therefore,

$A \approx 180° - 59.4° - 49.7°$

$A \approx 70.9°$

The solution of the triangle is $A \approx 70.9°$, $B \approx 59.4°$, $C = 49.7°$, $a = 13.7$, $b = 12.5$, and $c \approx 11.1$.

b. The distance across a lake can be found by measuring the distance from a point to each end of the lake as well as the angle these lines form.

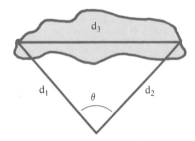

Find the distance across the lake if $d_1 = 2$ mi, $d_2 = 2.6$ mi, and $\theta = 30°$. Using the law of cosines we have

$(d_3)^2 = (d_1)^2 + (d_2)^2 - 2(d_1)(d_2) \cos \theta$

$(d_3)^2 = (2)^2 + (2.6)^2 - 2(2)(2.6) \cos 30°$

$(d_3)^2 \approx 4 + 6.76 - 10.4\,(0.8660)$

$(d_3)^2 \approx 10.76 - 9.01$

$(d_3)^2 \approx 1.75$

$d_3 \approx 1.3$ (nearest tenth)

Therefore, the lake is approximately 1.3 miles across.

APPLICATION SOLUTION

First, we can label the figure as follows.

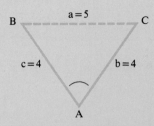

Formula: $\cos A = \dfrac{b^2 + c^2 - a^2}{2bc}$ where $b = 4$, $c = 4$, and $a = 5$

Substitute: $\cos A = \dfrac{16 + 16 - 25}{2(4)(4)}$

Simplify: $\cos A = \dfrac{7}{32} = .21875$

$A \approx 77.4°$ (nearest tenth)

So the trough angle is about 77.4°.

Exercises 12.7

In each of the following problems, find lengths correct to the nearest tenth of a unit and angles to the nearest tenth of a degree, unless otherwise stated.

1. A piece of aluminum 10 inches wide is to be formed into an eaves trough with a 4-inch opening at the top. What is the angle formed by the sides of the trough? (See application.)

2. A piece of aluminum 11 inches wide is to be formed into an eaves trough with a 4.5-inch opening at the top. What is the angle formed by the sides of the trough?

3. Two cars leave the same place, traveling on different roads that form a 75° angle with each other. How far apart are they at the end of 1 hour if one averages 55 mph and the other 60 mph?

4. Two cars leave the same place, traveling on different roads that form an 83° angle with each other. How far apart are they at the end of 2 hours if one averages 35 mph and the other 40 mph?

5. The distance across a lake is found by sighting both ends from a point away from the lake. One end is 120 feet and the other 150 feet from the point, and the angle between the lines of sight is 70°. What is the distance across the lake, to the nearest foot?

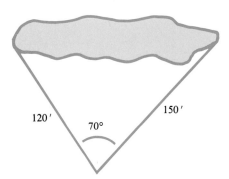

120′   70°   150′

6. The distance across a reservoir is found by sighting both ends from a point away from the reservoir. One end is 165 feet and the other 215 feet from the point, and the angle between the lines of sight is 57°. What is the distance across the reservoir, to the nearest foot?

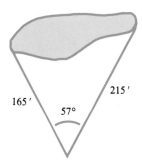

165′   57°   215′

7. A bridge from point A to point C is to be built to replace the detour around the lake. What is the length of the bridge, given the measurements in the figure below, to the nearest foot?

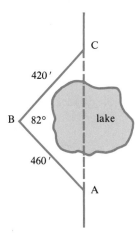

8. A bridge from point X to point Y is to be built to replace the detour around the washout. What is the length of the bridge, given the measurements in the figure below, to the nearest foot?

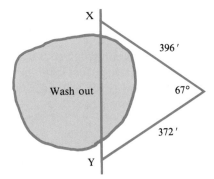

9. Point B is 28 miles north of point A, and point C is in an easterly direction 35 miles from point B and 45 miles from point A. (See figure.) What are the angles at A and B (to the nearest degree)?

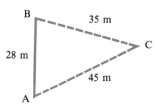

10. Point B is 50 kilometres north of point A, point C is in an easterly direction 72 kilometres from point B and 85 kilometres from point A. (See figure.) What are the angles at A and B (to the nearest degree)?

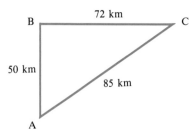

11. A ship travels a course directly west for 4 hours at a speed of 20 knots (nautical miles per hour). For the next four hours the ship travels at 18 knots in a direction

626

that is 25° west of north. How far is the ship from its starting point, to the nearest nautical mile?

12. A cargo vessel travels a course directly south for 3 hours at a speed of 17 knots (nautical miles per hour). For the next five hours the cargo vessel travels at 14 knots in a direction that is 55° west of south. How far is the cargo vessel from its starting point, to the nearest nautical mile?

13. A surveyor wants to find the length of a lake. He places three stakes at points *A*, *B*, and *C*. He measures the distance from *A* to *B* as 920 feet and the distance from *A* to *C* as 530 feet, both to the nearest ten feet. The angle at *A* is measured to be 130°. What is the length of the lake to the nearest 10 feet?

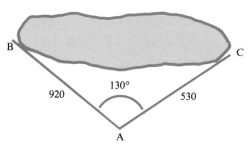

14. A surveyor wants to find the length of a swampy pasture. She places three stakes at points *A*, *B*, and *C*. She measures the distance from *A* to *B* as 1240 feet and the distance from *A* to *C* as 760 feet, both to the nearest ten feet. The angle at *A* is measured to be 110°. What is the length of the pasture to the nearest 10 feet?

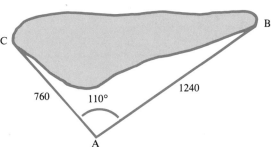

15. In physics, when two forces act on an object at the same time, the object will not move in the direction of either force, but rather in a direction between the forces along a line called the resultant. The resultant is pictured by the diagonal of a parallelogram whose adjacent sides represent the magnitudes of the two forces. See the figure below.

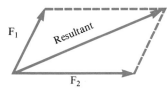

Two forces act on an object so that the angle between them is 57.6°. If one force has a magnitude of 78 pounds and the other a magnitude of 92 pounds, find the magnitude of the resultant force. (*Hint:* Solve triangle *RST*, where the distance from *S* to *T* is 92 and the distance from *T* to *R* is 78. *T* = 180° − 57.6° = 122.4°. Find the distance from *R* to *S* to the nearest pound.)

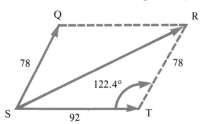

ANSWERS

16. _____

17. _____

18. _____

19. _____

20. _____

21. _____

22. _____

23. _____

16. Two forces act on an object so that the angle between them is 39.3°. If one force has a magnitude of 39 kg and the other a magnitude of 47 kg, find the magnitude of the resultant force. (*Hint:* See exercise 15.)

17. When two forces act on an object with magnitudes of 200 pounds and 375 pounds, respectively, the resultant force has a magnitude of 480 pounds. Find the angle between the two forces to the nearest tenth of a degree. (*Hint:* First find angle *B* and then subtract from 180°.)

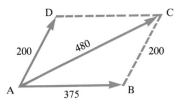

18. When two forces act on an object with magnitudes of 130 pounds and 90 pounds, respectively, the resultant force has a magnitude of 210 pounds. Find the angle between the two forces to the nearest tenth of a degree. (*Hint:* First find angle *B* and then subtract from 180°.)

E   *Maintain Your Skills*

Solve the following systems using substitution:

19. $\begin{cases} 4x + 5y = -8 \\ y = x + 2 \end{cases}$

20. $\begin{cases} 5m - 2n = 24 \\ m - 2n = 8 \end{cases}$

21. $\begin{cases} 3x - y = 4 \\ 6x + 9y = -25 \end{cases}$

22. $\begin{cases} y = \dfrac{2x}{3} - 4 \\ 6x - 5y = 28 \end{cases}$

23. $\begin{cases} 4x + 9y = 71 \\ 2x + y = 4 \end{cases}$

12.8 The Law of Sines

OBJECTIVES

140. Solve an oblique triangle, given any two sides and the angle opposite one of them.
141. Solve an oblique triangle, given any two angles and the side opposite one of them.

APPLICATION

The angle formed by the rafters of a house at the ridge is 80°, and the rafters form equal angles with the width of the house. If the rafters are 16 feet in length with a 2-foot overhang, what is the width of the house to the nearest foot?

VOCABULARY

The law of sines is a rule giving the relationship between the sides of a triangle and the sines of the angles.
Given triangle *ABC*,

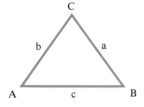

the law of sines is

$$\frac{a}{\sin A} = \frac{b}{\sin B} = \frac{c}{\sin C}$$

HOW AND WHY

The law of sines is a statement of equality among three ratios. If three parts of any pair of the ratios are known, the fourth part can be found. To identify which parts are known, sketch the triangle and label the parts.

Solve the triangle given $A = 75°$, $a = 10$, and $c = 7$. (Solving the triangle means finding B, C, and b.) First sketch the triangle.

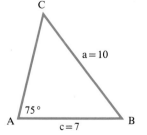

$a = 10$
$b = ?$
$c = 7$
$A = 75°$
$B = ?$
$C = ?$

Since A, a, and c are given, we use this portion of the law of sines:

$$\frac{a}{\sin A} = \frac{c}{\sin C}$$

$$\frac{10}{\sin 75°} = \frac{7}{\sin C}$$

$$\sin C = \frac{7(\sin 75°)}{10}$$

$$\sin C \approx \frac{(7)(0.9659)}{10}$$

$$\sin C \approx .6761$$

$$C \approx 42.5°$$

Now two angles are known and we can find the third.

$$A + B + C = 180°$$

$$75° + B + 42.5° \approx 180°$$

$$B \approx 62.5°$$

To find b use the following ratios:

$$\frac{a}{\sin A} = \frac{b}{\sin B}$$

$$\frac{10}{\sin 75°} \approx \frac{b}{\sin 62.5°}$$

$$b \approx \frac{10(\sin 62.5°)}{\sin 75°}$$

$$b \approx \frac{10(0.8870)}{0.9659}$$

$$b \approx 9.2 \text{ (nearest tenth)}$$

The solution of the triangle is $A = 75°$, $B \approx 62.5°$, $C \approx 42.5°$, $a = 10$, $b \approx 9.2$, and $c = 7$.

If side a had been equal to 6.8, there would have been two possible solutions.

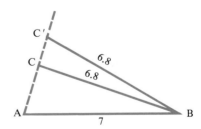

This is known as the "ambiguous case." In this text the problem will dictate which solution is desired. A formal treatment of the ambiguous case is a subject of trigonometry.

EXAMPLES   **a.** Solve the triangle given $A = 133.5°$, $B = 36.4°$, and $b = 42$.

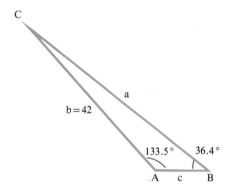

Since two angles are given, the third can be found.

$$A + B + C = 180°$$

$$133.5° + 36.4° + C = 180°$$

$$C = 10.1°$$

Side a can be found using

$$\frac{a}{\sin A} = \frac{b}{\sin B}$$

$$\frac{a}{\sin 133.5°} = \frac{42}{\sin 36.4°}$$

$$a = \frac{42 \,(\sin 133.5°)}{\sin 36.4°}$$

$$\sin 133.5° = \sin (180° - 133.5°)$$

$$= \sin 46.5°$$

$$a = \frac{42 \,(\sin 46.5°)}{\sin 36.4°}$$

$$a \approx \frac{42 \,(0.7254)}{0.5934}$$

$$a \approx 51.3 \text{ (nearest tenth)}$$

Side c can be found using

$$\frac{b}{\sin B} = \frac{c}{\sin C}$$

$$\frac{42}{\sin 36.4°} = \frac{c}{\sin 10.1°}$$

$$c = \frac{42 \,(\sin 10.1°)}{\sin 36.4°}$$

$$c \approx \frac{42 \,(0.1754)}{0.5934}$$

$$c \approx 12.4 \text{ (nearest tenth)}$$

The solution of the triangle is $A = 133.5°$, $B = 36.4°$, $C = 10.1°$, $a \approx 51.3$, $b = 42$, and $c \approx 12.4$.

b. The diagonal of a parallelogram makes angles of 32° and 41.5° with the sides. If the diagonal is 20 inches long, what are the lengths of the sides of the parallelogram?

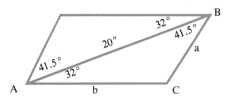

In △*ABC* we see that $A = 32°$, $B = 41.5°$, and $c = 20''$. Since two angles are known, the third can be found.

$$C = 180° - 32° - 41.5°$$

$$C = 106.5°$$

Side *b* can be found by using

$$\frac{b}{\sin B} = \frac{c}{\sin C}$$

$$b = \frac{c(\sin B)}{\sin C}$$

$$b = \frac{20 (\sin 41.5°)}{\sin 106.5°}$$

$$b = \frac{20 (\sin 41.5°)}{\sin 73.5°}$$

$$b \approx \frac{20 (0.6626)}{0.9588}$$

$$b \approx 13.8 \text{ (to the nearest tenth)}$$

Side *a* can be found by using

$$\frac{a}{\sin A} = \frac{c}{\sin C}$$

$$a = \frac{c(\sin A)}{\sin C}$$

$$a = \frac{20 (\sin 32°)}{\sin 106.5°}$$

$$a \approx \frac{20 (0.5299)}{0.9588}$$

$$a \approx 11.1 \text{ (to the nearest tenth)}$$

Therefore, to the nearest tenth the lengths of the sides of the parallelogram are 13.8″ and 11.1″.

APPLICATION SOLUTION

First, we draw a figure as follows.

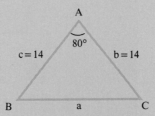

Since there is a 2-foot overhang on each of the rafters, we know that sides b and c are both 14 feet. Angle A is given as 80°. The angles at the width of the house are equal and their sum is $180° - 80° = 100°$. So each angle measures 50°.

Formula: $\dfrac{a}{\sin A} = \dfrac{b}{\sin B}$ where $A = 80°$, $b = 14$, and $B = 50°$

Substitute: $\dfrac{a}{\sin 80°} = \dfrac{14}{\sin 50°}$

Simplify: $a = \dfrac{14(\sin 80°)}{\sin 50°} \approx 17.998$

Answer: So the width is approximately 18 feet.

Exercises 12.8

1. The angle formed by the rafters of a house at the ridge is 130°, and the rafters form equal angles with the width of the house. If the rafters are 18 feet in length with a 1-foot overhang, what is the width of the house to the nearest foot?

2. The angle formed by the rafters of a house at the ridge is 152°, and the rafters form equal angles with the width of the house. If the rafters are 15 feet in length with a 2-foot overhang, what is the width of the house to the nearest foot?

3. An observer at point A is 600 yards directly west of an observer at point B. They are both looking at point C, which is located 55° east of north from A and 65° west of north from B. Which observer is closer to point C and by how many yards (to the nearest yard)?

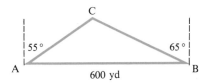

4. An observer at point B is 500 metres directly east of an observer at point A. They are both looking at point C, which is located 72° east of north from A and 54° west of north from B. Which observer is closer to point C and by how many metres (to the nearest metre)?

5. Two observers who are 500 m apart observe an airplane from opposite sides at the same time. One measures the angle of elevation to be 46.3°, while the other calculates it to be 58.6°. What is the height of the airplane (to the nearest ten metres)?

6. Two observers who are 850 m apart observe an airplane from opposite sides at the same time. One measures the angle of elevation to be 62.4°, while the other calculates it to be 77.5°. What is the height of the airplane (to the nearest ten metres)?

7. Two searchlights spot a plane in the air. If the light beams make angles of 47° and 58° with the ground, and the lights are 5000 feet apart, find the height of the plane to the nearest 100 feet. (*Hint:* First find the distance from B to C.)

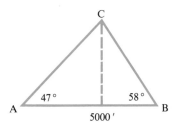

8. Two searchlights spot a plane in the air. If the light beams make angles of 53° and 68° with the ground, and the lights are 7200 metres apart, find the height of the plane to the nearest 100 metres. (*Hint:* See Exercise 7.)

9. Two forces (F_1 and F_2) are to be applied to an object at R to produce a resultant force with magnitude 165 newtons. The magnitude of F_1 is 93 newtons. If the

angle between F_1 and F_2 is 65°, find the magnitude of F_2 to the nearest newton. (_Hint:_ $U = 180° - 28° - 37°$. Find the distance from T to U in triangle TUR.)

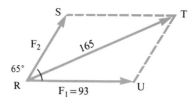

10. Two forces (F_1 and F_2) are to be applied to an object at R to produce a resultant force with magnitude 247 newtons. The magnitude of F_1 is 122 newtons. If the angle between F_1 and F_2 is 52°, find the magnitude of F_2 to the nearest newton. (_Hint:_ See Exercise 9.)

11. An isosceles triangle has equal sides of length 20′ and equal angles of 50.3°. Find the length of the third side to the nearest foot.

12. An isosceles triangle has equal sides of length 35 m and equal angles of 39.7°. Find the length of the third side to the nearest metre.

13. A 400-foot bridge spans a canyon in central Oregon. In order to calculate the depth of the canyon, engineers measured the angle from the bridge to the deepest point in the canyon from both ends of the bridge. What is the depth (d) of the canyon to the nearest foot?

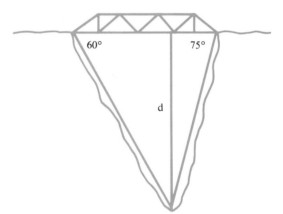

14. A 550-foot bridge spans a river bed in Texas. In order to calculate the depth of the river bed, engineers measured the angle from the bridge to the deepest point in the river bed from both ends of the bridge. What is the depth (d) of the river bed to the nearest foot?

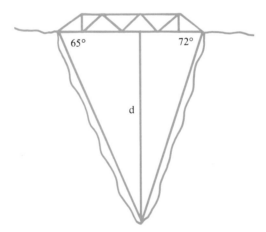

15. Two forces, F_1 and F_2, are applied to an object at point P to produce a resultant force. F_1 has magnitude 55 pounds and F_2 has magnitude 30 pounds. If the angle between F_1 and the resultant is to be $25°$, find the angle between the two forces to the nearest degree and the magnitude of the resultant to the nearest pound. (*Hint:* Angle $WPT = 25°$. Find angle PTW; then $W = 180° -$ angle $PTW - 25°$, and angle $WPM = 180° - W$.)

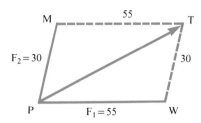

16. Two forces, F_1 and F_2, are applied to an object at point P to produce a resultant force. F_1 has magnitude 78 pounds and F_2 has magnitude 44 pounds. If the angle between F_1 and the resultant is to be $15°$, find the angle between the two forces to the nearest degree and the magnitude of the resultant to the nearest pound. (*Hint:* See Exercise 15.)

ANSWERS

16. _____

17. _____

18. _____

19. _____

20. _____

21. _____

E   *Maintain Your Skills*

Solve the following systems by addition:

17. $\begin{cases} 5a - 3b = -30 \\ 3a - 5b = -34 \end{cases}$

18. $\begin{cases} 7x - 2y = -4 \\ 3x - 4y = 14 \end{cases}$

19. $\begin{cases} 5a + 2b = 16 \\ 3a - 4b = 20 \end{cases}$

20. $\begin{cases} 2x + 3y = 2 \\ 4x - 6y = -2 \end{cases}$

21. $\begin{cases} 7x - 3y = 25 \\ 5x - 2y = 18 \end{cases}$

Chapter 12
POST-TEST

1. **(Obj. 132)** Find the exact value of sin 45°.

2. **(Obj. 137)** If sin R = 0.5650, find two values of R to the nearest tenth of a degree.

3. **(Obj. 130)** In $\triangle DEF$, E = 16.4° and F = 122°. What is the measure of D?

4. **(Obj. 139)** Solve $\triangle ABC$ given a = 5, b = 7, and C = 48°. Find all measures to the nearest tenth.

5. **(Obj. 135)** In $\triangle DEF$, d = 12, e = 9, and F = 90°. Find f, D, and E to the nearest tenth.

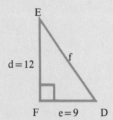

6. **(Obj. 140)** Solve $\triangle ABC$ given A = 63°, a = 15, and c = 9. Find all measures to the nearest tenth.

7. **(Obj. 130)** In $\triangle ABC$, A = 42° and B = 36°. What is the measure of C?

8. **(Obj. 137)** If tan Q = −5.2422, find Q to the nearest tenth of a degree.

9. **(Obj. 136)** Find the approximate value of cos 160.4°.

10. **(Obj. 131)** In $\triangle ABC$ find sin A, cos A, and tan A.

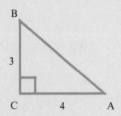

11. **(Obj. 131)** In $\triangle XYZ$ find the exact values of sin Y, cos Y, and tan Y.

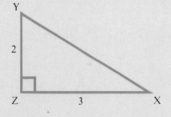

12. **(Obj. 133)** Find the approximate value of cos 32.7°.

13. **(Obj. 135)** In $\triangle ABC$, a = 8, C = 90°, and A = 40°. Find c, b, and B to the nearest tenth.

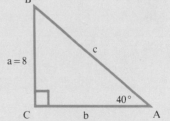

639

14. (Obj. 136) Find the approximate value of sin 136°.

15. (Obj. 137) Find the approximate value of tan 136°.

16. (Obj. 138) Solve $\triangle ABC$ given $a = 6$, $b = 8$, and $c = 12$. Find angles to the nearest tenth of a degree.

17. (Obj. 134) If cos $P = 0.6691$, find P to the nearest tenth of a degree.

18. (Obj. 141) Solve $\triangle ABC$ given $A = 28°$, $B = 124°$, and $c = 48$. Find all measures to the nearest tenth.

19. (Obj. 139) A fishing boat leaves its home port in a due west direction, traveling to a lightship that is 47 miles out to sea. The boat then turns to the right 125° and travels to a buoy that is known to be 22 miles from the lightship. How far is the boat from its home port? (To the nearest mile.)

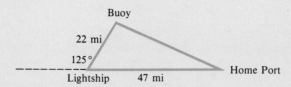

20. (Obj. 141) The base of one face of a pyramid is 350′ and the angles the edges make with the base are both 75°. What is the length of the edge of the pyramid? (To the nearest ten feet.)

Table of Trigonometric Functions

Deg.	Sin	Tan	*Cot	Cos		Deg.	Sin	Tan	Cot	Cos	
0.0	0.00000	0.00000	∞	1.0000	**90.0**	**6.0**	0.10453	0.10510	9.514	0.9945	**84.0**
.1	.00175	.00175	573.0	1.0000	89.9	.1	.10626	.10687	9.357	.9943	83.9
.2	.00349	.00349	286.5	1.0000	.8	.2	.10800	.10863	9.205	.9942	.8
.3	.00524	.00524	191.0	1.0000	.7	.3	.10973	.11040	9.058	.9940	.7
.4	.00698	.00698	143.24	1.0000	.6	.4	.11147	.11217	8.915	.9938	.6
.5	.00873	.00873	114.59	1.0000	.5	.5	.11320	.11394	8.777	.9936	.5
.6	.01047	.01047	95.49	0.9999	.4	.6	.11494	.11570	8.643	.9934	.4
.7	.01222	.01222	81.85	.9999	.3	.7	.11667	.11747	8.513	.9932	.3
.8	.01396	.01396	71.62	.9999	.2	.8	.11840	.11924	8.386	.9930	.2
.9	.01571	.01571	63.66	.9999	89.1	.9	.12014	.12101	8.264	.9928	83.1
1.0	0.01745	0.01746	57.29	0.9998	**89.0**	**7.0**	0.12187	0.12278	8.144	0.9925	**83.0**
.1	.01920	.01920	52.08	.9998	88.9	.1	.12360	.12456	8.028	.9923	82.9
.2	.02094	.02095	47.74	.9998	.8	.2	.12533	.12633	7.916	.9921	.8
.3	.02269	.02269	44.07	.9997	.7	.3	.12706	.12810	7.806	.9919	.7
.4	.02443	.02444	40.92	.9997	.6	.4	.12880	.12988	7.700	.9917	.6
.5	.02618	.02619	38.19	.9997	.5	.5	.13053	.13165	7.596	.9914	.5
.6	.02792	.02793	35.80	.9996	.4	.6	.13226	.13343	7.495	.9912	.4
.7	.02967	.02968	33.69	.9996	.3	.7	.13399	.13521	7.396	.9910	.3
.8	.03141	.03143	31.82	.9995	.2	.8	.13572	.13698	7.300	.9907	.2
.9	.03316	.03317	30.14	.9995	88.1	.9	.13744	.13876	7.207	.9905	82.1
2.0	0.03490	0.03492	28.64	0.9994	**88.0**	**8.0**	0.13917	0.14054	7.115	0.9903	**82.0**
.1	.03664	.03667	27.27	.9993	87.9	.1	.14090	.14232	7.026	.9900	81.9
.2	.03839	.03842	26.03	.9993	.8	.2	.14263	.14410	6.940	.9898	.8
.3	.04013	.04016	24.90	.9992	.7	.3	.14436	.14588	6.855	.9895	.7
.4	.04188	.04191	23.86	.9991	.6	.4	.14608	.14767	6.772	.9893	.6
.5	.04362	.04366	22.90	.9990	.5	.5	.14781	.14945	6.691	.9890	.5
.6	.04536	.04541	22.02	.9990	.4	.6	.14954	.15124	6.612	.9888	.4
.7	.04711	.04716	21.20	.9989	.3	.7	.15126	.15302	6.535	.9885	.3
.8	.04885	.04891	20.45	.9988	.2	.8	.15299	.15481	6.460	.9882	.2
.9	.05059	.05066	19.74	.9987	87.1	.9	.15471	.15660	6.386	.9880	81.1
3.0	0.05234	0.05241	19.081	0.9986	**87.0**	**9.0**	0.15643	0.15838	6.314	0.9877	**81.0**
.1	.05408	.05416	18.464	.9985	86.9	.1	.15816	.16017	6.243	.9874	80.9
.2	.05582	.05591	17.886	.9984	.8	.2	.15988	.16196	6.174	.9871	.8
.3	.05756	.05766	17.343	.9983	.7	.3	.16160	.16376	6.107	.9869	.7
.4	.05931	.05941	16.832	.9982	.6	.4	.16333	.16555	6.041	.9866	.6
.5	.06105	.06116	16.350	.9981	.5	.5	.16505	.16734	5.976	.9863	.5
.6	.06279	.06291	15.895	.9980	.4	.6	.16677	.16914	5.912	.9860	.4
.7	.06453	.06467	15.464	.9979	.3	.7	.16849	.17093	5.850	.9857	.3
.8	.06627	.06642	15.056	.9978	.2	.8	.17021	.17273	5.789	.9854	.2
.9	.06802	.06817	14.669	.9977	86.1	.9	.17193	.17453	5.730	.9851	80.1
4.0	0.06976	0.06993	14.301	0.9976	**86.0**	**10.0**	0.1736	0.1763	5.671	0.9848	**80.0**
.1	.07150	.07168	13.951	.9974	85.9	.1	.1754	.1781	5.614	.9845	79.9
.2	.07324	.07344	13.617	.9973	.8	.2	.1771	.1799	5.558	.9842	.8
.3	.07498	.07519	13.300	.9972	.7	.3	.1788	.1817	5.503	.9839	.7
.4	.07672	.07695	12.996	.9971	.6	.4	.1805	.1835	5.449	.9836	.6
.5	.07846	.07870	12.706	.9969	.5	.5	.1822	.1853	5.396	.9833	.5
.6	.08020	.08046	12.429	.9968	.4	.6	.1840	.1871	5.343	.9829	.4
.7	.08194	.08221	12.163	.9966	.3	.7	.1857	.1890	5.292	.9826	.3
.8	.08368	.08397	11.909	.9965	.2	.8	.1874	.1908	5.242	.9823	.2
.9	.08542	.08573	11.664	.9963	85.1	.9	.1891	.1926	5.193	.9820	79.1
5.0	0.08716	0.08749	11.430	0.9962	**85.0**	**11.0**	0.1908	0.1944	5.145	0.9816	**79.0**
.1	.08889	.08925	11.205	.9960	84.9	.1	.1925	.1962	5.097	.9813	78.9
.2	.09063	.09101	10.988	.9959	.8	.2	.1942	.1980	5.050	.9810	.8
.3	.09237	.09277	10.780	.9957	.7	.3	.1959	.1998	5.005	.9806	.7
.4	.09411	.09453	10.579	.9956	.6	.4	.1977	.2016	4.959	.9803	.6
.5	.09585	.09629	10.385	.9954	.5	.5	.1994	.2035	4.915	.9799	.5
.6	.09758	.09805	10.199	.9952	.4	.6	.2011	.2053	4.872	.9796	.4
.7	.09932	.09981	10.019	.9951	.3	.7	.2028	.2071	4.829	.9792	.3
.8	.10106	.10158	9.845	.9949	.2	.8	.2045	.2089	4.787	.9789	.2
.9	.10279	.10334	9.677	.9947	84.1	.9	.2062	.2107	4.745	.9785	78.1
6.0	0.10453	0.10510	9.514	0.9945	**84.0**	**12.0**	0.2079	0.2126	4.705	0.9781	**78.0**
	Cos	Cot	*Tan	Sin	Deg.		Cos	Cot	Tan	Sin	Deg.

*Interpolation in this section of the table is inaccurate.

(From Flanders, H., and Price, J.J.: Trigonometry. New York, Academic Press, 1975, pp. 222–225.)

Deg.	Sin	Tan	Cot	Cos		Deg.	Sin	Tan	Cot	Cos	
12.0	0.2079	0.2126	4.705	0.9781	**78.0**	**18.0**	0.3090	0.3249	3.078	0.9511	**72.0**
.1	.2096	.2144	4.665	.9778	77.9	.1	.3107	.3269	3.060	.9505	71.9
.2	.2113	.2162	4.625	.9774	.8	.2	.3123	.3288	3.042	.9500	.8
.3	.2130	.2180	4.586	.9770	.7	.3	.3140	.3307	3.024	.9494	.7
.4	.2147	.2199	4.548	.9767	.6	.4	.3156	.3327	3.006	.9489	.6
.5	.2164	.2217	4.511	.9763	.5	.5	.3173	.3346	2.989	.9483	.5
.6	.2181	.2235	4.474	.9759	.4	.6	.3190	.3365	2.971	.9478	.4
.7	.2198	.2254	4.437	.9755	.3	.7	.3206	.3385	2.954	.9472	.3
.8	.2215	.2272	4.402	.9751	.2	.8	.3223	.3404	2.937	.9466	.2
.9	.2233	.2290	4.366	.9748	77.1	.9	.3239	.3424	2.921	.9461	71.1
13.0	0.2250	0.2309	4.331	0.9744	**77.0**	**19.0**	0.3256	0.3443	2.904	0.9455	**71.0**
.1	.2267	.2327	4.297	.9740	76.9	.1	.3272	.3463	2.888	.9449	70.9
.2	.2284	.2345	4.264	.9736	.8	.2	.3289	.3482	2.872	.9444	.8
.3	.2300	.2364	4.230	.9732	.7	.3	.3305	.3502	2.856	.9438	.7
.4	.2317	.2382	4.198	.9728	.6	.4	.3322	.3522	2.840	.9432	.6
.5	.2334	.2401	4.165	.9724	.5	.5	.3338	.3541	2.824	.9426	.5
.6	.2351	.2419	4.134	.9720	.4	.6	.3355	.3561	2.808	.9421	.4
.7	.2368	.2438	4.102	.9715	.3	.7	.3371	.3581	2.793	.9415	.3
.8	.2385	.2456	4.071	.9711	.2	.8	.3387	.3600	2.778	.9409	.2
.9	.2402	.2475	4.041	.9707	76.1	.9	.3404	.3620	2.762	.9403	70.1
14.0	0.2419	0.2493	4.011	0.9703	**76.0**	**20.0**	0.3420	0.3640	2.747	0.9397	**70.0**
.1	.2436	.2512	3.981	.9699	75.9	.1	.3437	.3659	2.733	.9391	69.9
.2	.2453	.2530	3.952	.9694	.8	.2	.3453	.3679	2.718	.9385	.8
.3	.2470	.2549	3.923	.9690	.7	.3	.3469	.3699	2.703	.9379	.7
.4	.2487	.2568	3.895	.9686	.6	.4	.3486	.3719	2.689	.9373	.6
.5	.2504	.2586	3.867	.9681	.5	.5	.3502	.3739	2.675	.9367	.5
.6	.2521	.2605	3.839	.9677	.4	.6	.3518	.3759	2.660	.9361	.4
.7	.2538	.2623	3.812	.9673	.3	.7	.3535	.3779	2.646	.9354	.3
.8	.2554	.2642	3.785	.9668	.2	.8	.3551	.3799	2.633	.9348	.2
.9	.2571	.2661	3.758	.9664	75.1	.9	.3567	.3819	2.619	.9342	69.1
15.0	0.2588	0.2679	3.732	0.9659	**75.0**	**21.0**	0.3584	0.3839	2.605	0.9336	**69.0**
.1	.2605	.2698	3.706	.9655	74.9	.1	.3600	.3859	2.592	.9330	68.9
.2	.2622	.2717	3.681	.9650	.8	.2	.3616	.3879	2.578	.9323	.8
.3	.2639	.2736	3.655	.9646	.7	.3	.3633	.3899	2.565	.9317	.7
.4	.2656	.2754	3.630	.9641	.6	.4	.3649	.3919	2.552	.9311	.6
.5	.2672	.2773	3.606	.9636	.5	.5	.3665	.3939	2.539	.9304	.5
.6	.2689	.2792	3.582	.9632	.4	.6	.3681	.3959	2.526	.9298	.4
.7	.2706	.2811	3.558	.9627	.3	.7	.3697	.3979	2.513	.9291	.3
.8	.2723	.2830	3.534	.9622	.2	.8	.3714	.4000	2.500	.9285	.2
.9	.2740	.2849	3.511	.9617	74.1	.9	.3730	.4020	2.488	.9278	68.1
16.0	0.2756	0.2867	3.487	0.9613	**74.0**	**22.0**	0.3746	0.4040	2.475	0.9272	**68.0**
.1	.2773	.2886	3.465	.9608	73.9	.1	.3762	.4061	2.463	.9265	67.9
.2	.2790	.2905	3.442	.9603	.8	.2	.3778	.4081	2.450	.9259	.8
.3	.2807	.2924	3.420	.9598	.7	.3	.3795	.4101	2.438	.9252	.7
.4	.2823	.2943	3.398	.9593	.6	.4	.3811	.4122	2.426	.9245	.6
.5	.2840	.2962	3.376	.9588	.5	.5	.3827	.4142	2.414	.9239	.5
.6	.2857	.2981	3.354	.9583	.4	.6	.3843	.4163	2.402	.9232	.4
.7	.2874	.3000	3.333	.9578	.3	.7	.3859	.4183	2.391	.9225	.3
.8	.2890	.3019	3.312	.9573	.2	.8	.3875	.4204	2.379	.9219	.2
.9	.2907	.3038	3.291	.9568	73.1	.9	.3891	.4224	2.367	.9212	67.1
17.0	0.2924	0.3057	3.271	0.9563	**73.0**	**23.0**	0.3907	0.4245	2.356	0.9205	**67.0**
.1	.2940	.3076	3.251	.9558	72.9	.1	.3923	.4265	2.344	.9198	66.9
.2	.2957	.3096	3.230	.9553	.8	.2	.3939	.4286	2.333	.9191	.8
.3	.2974	.3115	3.211	.9548	.7	.3	.3955	.4307	2.322	.9184	.7
.4	.2990	.3134	3.191	.9542	.6	.4	.3971	.4327	2.311	.9178	.6
.5	.3007	.3153	3.172	.9537	.5	.5	.3987	.4348	2.300	.9171	.5
.6	.3024	.3172	3.152	.9532	.4	.6	.4003	.4369	2.289	.9164	.4
.7	.3040	.3191	3.133	.9527	.3	.7	.4019	.4390	2.278	.9157	.3
.8	.3057	.3211	3.115	.9521	.2	.8	.4035	.4411	2.267	.9150	.2
.9	.3074	.3230	3.096	.9516	72.1	.9	.4051	.4431	2.257	.9143	66.1
18.0	0.3090	0.3249	3.078	0.9511	**72.0**	**24.0**	0.4067	0.4452	2.246	0.9135	**66.0**
	Cos	Cot	Tan	Sin	Deg.		Cos	Cot	Tan	Sin	Deg.

Deg.	Sin	Tan	Cot	Cos	
24.0	0.4067	0.4452	2.246	0.9135	**66.0**
.1	.4083	.4473	2.236	.9128	65.9
.2	.4099	.4494	2.225	.9121	.8
.3	.4115	.4515	2.215	.9114	.7
.4	.4131	.4536	2.204	.9107	.6
.5	.4147	.4557	2.194	.9100	.5
.6	.4163	.4578	2.184	.9092	.4
.7	.4179	.4599	2.174	.9085	.3
.8	.4195	.4621	2.164	.9078	.2
.9	.4210	.4642	2.154	.9070	65.1
25.0	0.4226	0.4663	2.145	0.9063	**65.0**
.1	.4242	.4684	2.135	.9056	64.9
.2	.4258	.4706	2.125	.9048	.8
.3	.4274	.4727	2.116	.9041	.7
.4	.4289	.4748	2.106	.9033	.6
.5	.4305	.4770	2.097	.9026	.5
.6	.4321	.4791	2.087	.9018	.4
.7	.4337	.4813	2.078	.9011	.3
.8	.4352	.4834	2.069	.9003	.2
.9	.4368	.4856	2.059	.8996	64.1
26.0	0.4384	0.4877	2.050	0.8988	**64.0**
1	.4399	.4899	2.041	.8980	63.9
.2	.4415	.4921	2.032	.8973	.8
.3	.4431	.4942	2.023	.8965	.7
.4	.4446	.4964	2.014	.8957	.6
.5	.4462	.4986	2.006	.8949	.5
.6	.4478	.5008	1.997	.8942	.4
.7	.4493	5029	1.988	.8934	.3
.8	.4509	.5051	1.980	.8926	.2
.9	.4524	.5073	1.971	.8918	63.1
27.0	0.4540	0.5095	1.963	0.8910	**63.0**
.1	.4555	.5117	1.954	.8902	62.9
.2	.4571	.5139	1.946	.8894	.8
.3	.4586	5161	1.937	.8886	.7
.4	.4602	.5184	1.929	.8878	.6
.5	.4617	.5206	1 921	.8870	.5
.6	.4633	.5228	1.913	.8862	.4
.7	.4648	.5250	1.905	.8854	.3
.8	.4664	.5272	1.897	.8846	.2
.9	.4679	.5295	1.889	.8838	62.1
28.0	0.4695	0.5317	1.881	0.8829	**62.0**
.1	.4710	.5340	1.873	.8821	61.9
.2	.4726	.5362	1.865	.8813	.8
.3	.4741	.5384	1.857	.8805	.7
.4	.4756	.5407	1.849	.8796	.6
.5	.4772	.5430	1.842	.8788	.5
.6	.4787	.5452	1.834	.8780	.4
.7	.4802	.5475	1.827	.8771	.3
.8	.4818	.5498	1.819	.8763	.2
.9	.4833	.5520	1.811	.8755	61.1
29.0	0.4848	0.5543	1.804	0.8746	**61.0**
.1	.4863	.5566	1.797	.8738	60.9
.2	.4879	.5589	1.789	.8729	.8
.3	.4894	.5612	1.782	.8721	.7
.4	.4909	.5635	1.775	.8712	.6
.5	.4924	.5658	1.767	.8704	.5
.6	.4939	.5681	1.760	.8695	.4
.7	.4955	.5704	1.753	.8686	.3
.8	.4970	.5727	1.746	.8678	.2
.9	.4985	.5750	1.739	.8669	60.1
30.0	0.5000	0.5774	1.732	0.8660	**60.0**
	Cos	Cot	Tan	Sin	Deg.

Deg.	Sin	Tan	Cot	Cos	
30.0	0.5000	0.5774	1.7321	0.8660	**60.0**
.1	.5015	.5797	1.7251	.8652	59.9
.2	.5030	.5820	1.7182	.8643	.8
.3	.5045	.5844	1.7113	.8634	.7
.4	.5060	.5867	1.7045	.8625	.6
.5	.5075	.5890	1.6977	.8616	.5
.6	.5090	.5914	1.6909	.8607	.4
.7	.5105	.5938	1.6842	.8599	.3
.8	.5120	.5961	1.6775	.8590	.2
.9	.5135	.5985	1.6709	.8581	59.1
31.0	0.5150	0.6009	1.6643	0.8572	**59.0**
.1	.5165	.6032	1.6577	.8563	58.9
.2	.5180	.6056	1.6512	.8554	.8
.3	.5195	.6080	1.6447	.8545	.7
.4	.5210	.6104	1.6383	.8536	.6
.5	.5225	.6128	1.6319	.8526	.5
.6	.5240	.6152	1.6255	.8517	.4
.7	.5255	.6176	1.6191	.8508	.3
.8	.5270	.6200	1.6128	.8499	.2
.9	.5284	.6224	1.6066	.8190	58.1
32.0	0.5299	0.6249	1.6003	0.8480	**58.0**
.1	.5314	.6273	1.5941	.8471	57.9
.2	.5329	.6297	1.5880	.8462	.8
.3	.5344	.6322	1.5818	.8453	.7
.4	.5358	.6346	1.5757	.8443	.6
.5	.5373	.6371	1.5697	.8434	.5
.6	.5388	.6395	1.5637	.8425	.4
.7	.5402	.6420	1.5577	.8415	.3
.8	.5417	.6445	1.5517	.8406	.2
.9	.5432	.6469	1.5458	.8396	57.1
33.0	0.5446	0.6494	1.5399	0.8387	**57.0**
.1	.5461	.6519	1.5340	.8377	56.9
.2	.5476	.6544	1.5282	.8368	.8
.3	.5490	.6569	1.5224	.8358	.7
.4	.5505	.6594	1.5166	.8348	.6
.5	.5519	.6619	1.5108	.8339	.5
.6	.5534	.6644	1.5051	.8329	.4
.7	.5548	.6669	1.4994	.8320	.3
.8	.5563	.6694	1.4938	.8310	.2
.9	.5577	.6720	1.4882	.8300	56.1
34.0	0.5592	0.6745	1.4826	0.8290	**56.0**
.1	.5606	.6771	1.4770	.8281	55.9
.2	.5621	.6796	1.4715	.8271	.8
.3	.5635	.6822	1.4659	.8261	.7
.4	.5650	.6847	1.4605	.8251	.6
.5	.5664	.6873	1.4550	.8241	.5
.6	.5678	.6899	1.4496	.8231	.4
.7	.5693	.6924	1.4442	.8221	.3
.8	.5707	.6950	1.4388	.8211	.2
.9	.5721	.6976	1.4335	.8202	55.1
35.0	0.5736	0.7002	1.4281	0.8192	**55.0**
.1	.5750	.7028	1.4229	.8181	54.9
.2	.5764	.7054	1.4176	.8171	.8
.3	.5779	.7080	1.4124	.8161	.7
.4	.5793	.7107	1.4071	.8151	.6
.5	.5807	.7133	1.4019	.8141	.5
.6	.5821	.7159	1.3968	.8131	.4
.7	.5835	.7186	1.3916	.8121	.3
.8	.5850	.7212	1.3865	.8111	.2
.9	.5864	.7239	1.3814	.8100	54.1
36.0	0.5878	0.7265	1.3764	0.8090	**54.0**
	Cos	Cot	Tan	Sin	Deg.

Deg.	Sin	Tan	Cot	Cos		Deg.	Sin	Tan	Cot	Cos	
36.0	0.5878	0.7265	1.3764	0.8090	**54.0**	**40.5**	0.6494	0.8541	1.1708	0.7604	**49.5**
.1	.5892	.7292	1.3713	.8080	53.9	.6	.6508	.8571	1.1667	.7593	.4
.2	.5906	.7319	1.3663	.8070	.8	.7	.6521	.8601	1.1626	.7581	.3
.3	.5920	.7346	1.3613	.8059	.7	.8	.6534	.8632	1.1585	.7570	.2
.4	.5934	.7373	1.3564	.8049	.6	.9	.6547	.8662	1.1544	.7559	49.1
.5	.5948	.7400	1.3514	.8039	.5	**41.0**	0.6561	0.8693	1.1504	0.7547	**49.0**
.6	.5962	.7427	1.3465	.8028	.4	.1	.6574	.8724	1.1463	.7536	48.9
.7	.5976	.7454	1.3416	.8018	.3	.2	.6587	.8754	1.1423	.7524	.8
.8	.5990	.7481	1.3367	.8007	.2	.3	.6600	.8785	1.1383	.7513	.7
.9	.6004	.7508	1.3319	.7997	53.1	.4	.6613	.8816	1.1343	.7501	.6
37.0	0.6018	0.7536	1.3270	0.7986	**53.0**	.5	.6626	.8847	1.1303	.7490	.5
.1	.6032	.7563	1.3222	.7976	52.9	.6	.6639	.8878	1.1263	.7478	.4
.2	.6046	.7590	1.3175	.7965	.8	.7	.6652	.8910	1.1224	.7466	.3
.3	.6060	.7618	1.3127	.7955	.7	.8	.6665	.8941	1.1184	.7455	.2
.4	.6074	.7646	1.3079	.7944	.6	.9	.6678	.8972	1.1145	.7443	48.1
.5	.6088	.7673	1.3032	.7934	.5	**42.0**	0.6691	0.9004	1.1106	0.7431	**48.0**
.6	.6101	.7701	1.2985	.7923	.4	.1	.6704	.9036	1.1067	.7420	47.9
.7	.6115	.7729	1.2938	.7912	.3	.2	.6717	.9067	1.1028	.7408	.8
.8	.6129	.7757	1.2892	.7902	.2	.3	.6730	.9099	1.0990	.7396	.7
.9	.6143	.7785	1.2846	.7891	52.1	.4	.6743	.9131	1.0951	.7385	.6
38.0	0.6157	0.7813	1.2799	0.7880	**52.0**	.5	.6756	.9163	1.0913	.7373	.5
.1	.6170	.7841	1.2753	.7869	51.9	.6	.6769	.9195	1.0875	.7361	.4
.2	.6184	.7869	1.2708	.7859	.8	.7	.6782	.9228	1.0837	.7349	.3
.3	.6198	.7898	1.2662	.7848	.7	.8	.6794	.9260	1.0799	.7337	.2
.4	.6211	.7926	1.2617	.7837	.6	.9	.6807	.9293	1.0761	.7325	47.1
.5	.6225	.7954	1.2572	.7826	.5	**43.0**	0.6820	0.9325	1.0724	0.7314	**47.0**
.6	.6239	.7983	1.2527	.7815	.4	.1	.6833	.9358	1.0686	.7302	46.9
.7	.6252	.8012	1.2482	.7804	.3	.2	.6845	.9391	1.0649	.7290	.8
.8	.6266	.8040	1.2437	.7793	.2	.3	.6858	.9424	1.0612	.7278	.7
.9	.6280	.8069	1.2393	.7782	51.1	.4	.6871	.9457	1.0575	.7266	.6
39.0	0.6293	0.8098	1.2349	0.7771	**51.0**	.5	.6884	.9490	1.0538	.7254	.5
.1	.6307	.8127	1.2305	.7760	50.9	.6	.6896	.9523	1.0501	.7242	.4
.2	.6320	.8156	1.2261	.7749	.8	.7	.6909	.9556	1.0464	.7230	.3
.3	.6334	.8185	1.2218	.7738	.7	.8	.6921	.9590	1.0428	.7218	.2
.4	.6347	.8214	1.2174	.7727	.6	.9	.6934	.9623	1.0392	.7206	46.1
.5	.6361	.8243	1.2131	.7716	.5	**44.0**	0.6947	0.9657	1.0355	0.7193	**46.0**
.6	.6374	.8273	1.2088	.7705	.4	.1	.6959	.9691	1.0319	.7181	45.9
.7	.6388	.8302	1.2045	.7694	.3	.2	.6972	.9725	1.0283	.7169	.8
.8	.6401	.8332	1.2002	.7683	.2	.3	.6984	.9759	1.0247	.7157	.7
.9	.6414	.8361	1.1960	.7672	50.1	.4	.6997	.9793	1.0212	.7145	.6
40.0	0.6428	0.8391	1.1918	0.7660	**50.0**	.5	.7009	.9827	1.0176	.7133	.5
.1	.6441	.8421	1.1875	.7649	49.9	.6	.7022	.9861	1.0141	.7120	.4
.2	.6455	.8451	1.1833	.7638	.8	.7	.7034	.9896	1.0105	.7108	.3
.3	.6468	.8481	1.1792	.7627	.7	.8	.7046	.9930	1.0070	.7096	.2
.4	.6481	.8511	1.1750	.7615	.6	.9	.7059	.9965	1.0035	.7083	45.1
40.5	0.6494	0.8541	1.1708	0.7604	**49.5**	**45.0**	0.7071	1.0000	1.0000	0.7071	**45.0**
	Cos	Cot	Tan	Sin	Deg.		Cos	Cot	Tan	Sin	Deg.

CHAPTER 1

Pre-test

1. 743 **2.** 7,769 **3.** 1,663 **4.** 2,232

5. 7,789 **6.** 326 **7.** 17,589 **8.** 366,788

9. 306 **10.** 333 **11.** 16 R 7 **12.** 2,744

13. 44 **14.** 88 **15.** 14 **16.** 1, 7, 53, 371

17. prime **18.** 5 · 7 · 11 **19.** 3 · 3 · 3 · 19

20. 56 **21.** 180 **22.** 6 shelves **23.** 180

24. 37 stations **25.** 432 cans

Section 1.1

1. five hundred forty **3.** five thousand forty-two

5. 3,006 **7.** two thousand five hundred two

9. twenty-five thousand two hundred **11.** 243

13. 243,000

15. seventy-eight million eighty-seven thousand seven hundred eighty

17. fifty million fifty thousand five hundred

19. five billion seven hundred twenty-two million forty-four thousand nine hundred ten

21. 406,242,713 **23.** 6,000,606 **25.** $78,063

27. seven hundred eighty-five miles **29.** 4,000,000,000

Section 1.2

1. 33 **3.** 159 **5.** 28 **7.** 309 **9.** 4,339

11. 93 **13.** 1,186 **15.** 18,208 **17.** 2,279

19. 29,118 **21.** 21,786 **23.** 16,199

25. $5,928 **27.** 55,412 tickets **29.** $3,584

Section 1.3

1. 488 **3.** 0 **5.** 11 R 1 **7.** 774 **9.** 18,798

11. 62 **13.** 38 R 19 **15.** 271,404

17. 13,683,250 **19.** 632,553 **21.** 160

23. 309 R 3 **25.** 108 cases; 3 extra cans

27. $1,638 **29.** 408 cans

Section 1.4

1. 14 **3.** 16 **5.** 16 **7.** 10 **9.** 1 **11.** 144

13. 100,000 **15.** 243 **17.** 1,000,000 **19.** 256

21. 100,000,000 **23.** 18 **25.** $400,000,000

27. 25,500,000,000,000 miles

29. 2,000,000,000,000 cells

Section 1.5

1. 16 **3.** 1 **5.** 80 **7.** 70 **9.** 6 **11.** 219

13. 40 **15.** 1,694 **17.** 4,500 **19.** 60 **21.** 29

23. 175 **25.** $453 more **27.** $7,300 **29.** $245

31. 800 ohms

Section 1.6

1. 1, 5, 25 **3.** 1, 5, 7, 35

5. $1 \times 40, 2 \times 20, 4 \times 10, 5 \times 8$

7. 1, 2, 3, 4, 6, 9, 12, 18, 36

9. 1, 2, 3, 4, 6, 8, 12, 16, 24, 32, 48, 96

11. $1 \times 100, 2 \times 50, 4 \times 25, 5 \times 20, 10 \times 10$

13. $1 \times 96, 2 \times 48, 3 \times 32, 4 \times 24, 6 \times 16, 8 \times 12$

15. 1, 2, 4, 8, 16, 32, 64, 128

17. 1, 5, 7, 35, 49, 245

19. 1, 2, 3, 4, 12, 37, 111, 148, 222, 444

21. $1 \times 847, 7 \times 121, 11 \times 77$

23. $1 \times 1,311, 3 \times 437, 19 \times 69, 23 \times 57$

25.
    1 program of 150 minutes
150 programs of 1 minute
    2 programs of 75 minutes
 75 programs of 2 minutes
    3 programs of 50 minutes
 50 programs of 3 minutes
    5 programs of 30 minutes

30 programs of   5 minutes
6 programs of 25 minutes
25 programs of   6 minutes
10 programs of 15 minutes
15 programs of 10 minutes

27.   1 program of 90 minutes
90 programs of   1 minute
2 programs of 45 minutes
45 programs of   2 minutes
3 programs of 30 minutes
30 programs of   3 minutes
5 programs of 18 minutes
18 programs of   5 minutes
6 programs of 15 minutes
15 programs of   6 minutes
9 programs of 10 minutes
10 programs of   9 minutes

Section 1.7

1. composite   **3.** composite   **5.** composite

7. composite   **9.** prime   **11.** prime

13. composite   **15.** prime   **17.** prime

19. composite   **21.** prime   **23.** prime

Section 1.8

1. 2^3 **3.** prime   **5.** $2 \cdot 7$ **7.** $2 \cdot 3 \cdot 5$

9. $2^4 \cdot 3$ **11.** 2^6 **13.** prime   **15.** prime

17. $2^3 \cdot 3 \cdot 5$ **19.** $2 \cdot 3^2 \cdot 11$ **21.** $2 \cdot 3^2 \cdot 23$

23. prime

Section 1.9

1. 15   **3.** 20   **5.** 24   **7.** 30   **9.** 180

11. 72   **13.** 300   **15.** 180   **17.** 480   **19.** 720

21. 1,440   **23.** 6,370   **25.** 48   **27.** 108   **29.** 60

CHAPTER 2

Pre-test

1. $\frac{3}{7}$ **2.** $\frac{1}{15}$ **3.** $\frac{1}{2}$ **4.** $21\frac{35}{81}$ **5.** $\frac{3}{20}$ **6.** $\frac{15}{44}$

7. $\frac{23}{26}$ **8.** $\frac{1}{27}$ **9.** 9   **10.** $\frac{1}{6}, \frac{7}{36}, \frac{1}{4}$

11. $\frac{29}{54}, \frac{5}{9}, \frac{11}{18}, \frac{2}{3}$ **12.** $\frac{8}{21}$ **13.** $\frac{41}{30}$ or $1\frac{11}{30}$

14. $19\frac{13}{36}$ **15.** $7\frac{17}{24}$ **16.** $\frac{11}{18}$ **17.** $11\frac{5}{6}$

18. $26\frac{2}{7}$ **19.** $8\frac{5}{42}$ **20.** $\frac{3}{7}$ **21.** $\frac{8}{35}$

22. $147\frac{7}{16}$ lb   **23.** $37\frac{3}{8}$ ft   **24.** \$5.83

25. 44 bags with $\frac{4}{9}$ lb left over

Section 2.1

1. $\frac{4}{7}$ **3.** $2\frac{1}{8}$ **5.** $\frac{15}{4}$ **7.** $\frac{6}{8}$ **9.** $8\frac{7}{10}$

11. $8\frac{13}{15}$ **13.** $\frac{43}{8}$ **15.** $\frac{5}{3}$ **17.** $65\frac{1}{3}$ **19.** $15\frac{4}{5}$

21. $\frac{5}{1}$ **23.** $\frac{50}{3}$ **25.** 7 ft   **27.** 195 sections

29. 9   **31.** 139 parts   **33.** 525   **35.** 966

Section 2.2

1. $\frac{13}{15}$ **3.** $\frac{1}{4}$ **5.** $\frac{7}{8}$ **7.** $\frac{8}{5}$ **9.** $\frac{45}{32}$ **11.** $\frac{5}{7}$

13. $\frac{11}{12}$ **15.** $\frac{2}{3}$ **17.** $\frac{49}{51}$ **19.** $\frac{20}{33}$ **21.** $\frac{3}{5}$

23. $\frac{2}{11}$ **25.** $\frac{5}{8}$ inch   **27.** $\frac{25}{3}$ fbm   **29.** $\frac{1}{4}$

31. 181,162   **33.** 817,587   **35.** 2,001

Section 2.3

1. $\frac{12}{7}$ **3.** $\frac{3}{14}$ **5.** $\frac{8}{5}$ **7.** $\frac{1}{3}$ **9.** 3   **11.** $3\frac{13}{14}$

13. $\frac{1}{3}$ **15.** $\frac{16}{35}$ **17.** $\frac{3}{5}$ **19.** $\frac{7}{8}$ **21.** 31

23. $\frac{3}{8}$ **25.** 220 miles   **27.** 18 drawings

29. \$20,250 less   **31.** $40\frac{25}{27}$ lb/in²   **33.** 25

35. 1,000

Section 2.4

1. 35   **3.** $\frac{6}{7}$ **5.** 4   **7.** $1\frac{1}{3}$ **9.** $\frac{1}{2}$ **11.** $18\frac{1}{3}$

13. $\frac{9}{14}$ **15.** $\frac{2}{3}$ **17.** $6\frac{2}{69}$ **19.** $5\frac{13}{28}$ **21.** $\frac{41}{87}$

23. $\frac{136}{165}$ **25.** $5\frac{2}{5}$ turns   **27.** $3\frac{1}{2}$ omelets

29. $4\frac{1}{6}$ in   **31.** $4\frac{1}{2}$ in or $\frac{1}{8}$ yd   **33.** 15   **35.** 10

37. 18

Section 2.5

1. 20 **3.** 200 **5.** 24 **7.** 51 **9.** 66

11. 336 **13.** 875 **15.** 40 **17.** 221 **19.** 119

21. 187 **23.** $\frac{21}{28}$ **25.** $\frac{25}{40}, \frac{28}{40}, \frac{30}{40}$

27. $\frac{45}{72}, \frac{56}{72}, \frac{60}{72}$

29. 1, 2, 3, 4, 6, 12, 13, 26, 39, 52, 78, 156

31. 1(210), 2(105), 3(70), 5(42), 6(32), 7(30), 10(21), 14(15)

Section 2.6

1. $\frac{3}{4}$ **3.** $\frac{5}{6}$ **5.** $\frac{2}{9}, \frac{1}{3}, \frac{4}{9}$ **7.** $\frac{5}{6}$ **9.** $\frac{7}{2}$

11. $\frac{2}{5}, \frac{4}{7}, \frac{2}{3}$ **13.** $\frac{5}{8}, \frac{7}{10}, \frac{3}{4}$ **15.** $\frac{9}{10}$ **17.** $1\frac{3}{4}$

19. $\frac{11}{24}, \frac{17}{36}, \frac{35}{72}$ **21.** $\frac{6}{14}, \frac{13}{28}, \frac{17}{35}$

23. $\frac{11}{30}, \frac{17}{45}, \frac{7}{18}, \frac{2}{5}$ **25.** Pedro's

27. $\frac{3}{32}, \frac{1}{8}, \frac{1}{4}, \frac{5}{16}, \frac{3}{8}, \frac{1}{2}, \frac{9}{16}$

29. $\frac{1}{2}$ gallon (largest), $\frac{1}{4}$ gallon (smallest)

31. composite **33.** composite **35.** prime

Section 2.7

1. $\frac{2}{3}$ **3.** $\frac{1}{3}$ **5.** $\frac{25}{24}$ **7.** $\frac{3}{5}$ **9.** $\frac{47}{60}$ **11.** $\frac{7}{5}$

13. $\frac{61}{48}$ **15.** $\frac{37}{42}$ **17.** $\frac{229}{240}$ **19.** $\frac{59}{195}$ **21.** $\frac{173}{150}$

23. $\frac{131}{240}$ **25.** $\frac{113}{48}$ ft or $2\frac{17}{48}$ ft **27.** $\frac{5}{16}$ of a point

29. $\frac{41}{24}$ ohm or $1\frac{17}{24}$ ohm **31.** 1,898

33. 1,568,336 **35.** 1

Section 2.8

1. $5\frac{11}{12}$ **3.** $6\frac{2}{9}$ **5.** 11 **7.** $26\frac{3}{10}$ **9.** $720\frac{5}{24}$

11. $86\frac{11}{30}$ **13.** $719\frac{31}{36}$ **15.** $235\frac{7}{11}$ **17.** $113\frac{1}{8}$

19. $13\frac{1}{25}$ **21.** $14\frac{13}{66}$ **23.** $141\frac{13}{36}$

25. $76\frac{3}{4}$ hours **27.** $80\frac{9}{16}$ lb **29.** $21\frac{5}{16}$ in

31. $3 \cdot 5 \cdot 23$ **33.** $3^3 \cdot 5^2$ **35.** $2^3 \cdot 3 \cdot 19$

Section 2.9

1. $\frac{5}{9}$ **3.** $\frac{3}{5}$ **5.** $\frac{7}{45}$ **7.** $\frac{1}{3}$ **9.** $\frac{7}{16}$ **11.** $\frac{3}{14}$

13. $\frac{11}{30}$ **15.** $\frac{21}{200}$ **17.** $\frac{47}{140}$ **19.** $\frac{7}{30}$ **21.** $\frac{1}{72}$

23. $\frac{119}{225}$ **25.** $\frac{9}{8}$ in or $1\frac{1}{8}$ in **27.** $\frac{5}{12}$ cup

29. $\frac{13}{60}$ of the loan **31.** 24 **33.** 900 **35.** 1,140

Section 2.10

1. $2\frac{1}{3}$ **3.** $4\frac{1}{2}$ **5.** $1\frac{1}{2}$ **7.** $3\frac{5}{8}$ **9.** $\frac{3}{4}$

11. $118\frac{1}{6}$ **13.** $3\frac{13}{36}$ **15.** $115\frac{1}{9}$ **17.** $28\frac{1}{3}$

19. $1\frac{17}{36}$ **21.** $9\frac{3}{14}$ **23.** $28\frac{19}{72}$ **25.** $9\frac{5}{36}$ gal

27. $1\frac{5}{6}$ ft **29.** $10\frac{19}{32}$ in **31.** $103\frac{1}{8}$ volts

33. $2\frac{1}{12}$ **35.** $2\frac{5}{22}$ **37.** $\frac{413}{12}$

Section 2.11

1. $\frac{2}{7}$ **3.** $\frac{1}{7}$ **5.** $\frac{1}{6}$ **7.** $\frac{8}{15}$ **9.** $\frac{1}{2}$ **11.** $\frac{4}{7}$

13. $\frac{1}{3}$ **15.** $13\frac{1}{2}$ **17.** $\frac{3}{5}$ **19.** $1\frac{4}{9}$ **21.** $1\frac{8}{9}$

23. 1 **25.** $11\frac{1}{8}$ in **27.** $6\frac{12}{23}$ ohms

29. $\frac{11}{32}$ of the shop **33.** $2\frac{16}{27}$ or $\frac{70}{27}$

35. $1\frac{89}{351}$ or $\frac{440}{351}$ **37.** $\frac{2}{15}$

CHAPTER 3

Pre-test

1. eighty-one and eighty-one ten-thousandths

2. .067, .0678, .068, .07 **3.** 36.50 **4.** 32.91

5. 28.1228 **6.** .305 **7.** 18.699 **8.** .208

9. 56.196 **10.** 0.01287 **11.** 29.9052

12. 62.44 **13.** 1.006 **14.** 112.3 **15.** 1.90

16. .8875   **17.** 1.118   **18.** 7.63   **19.** 9.539

20. .971   **21.** $41.61   **22.** 4 hours and 20 minutes

23. 9.9 hours   **24.** 1.57 in   **25.** $704.69

Section 3.1

1. four tenths   **3.** three and twenty-six hundredths

5. .11   **7.** five and four hundredths

9. five hundred and four tenths   **11.** .15

13. 5000.05

15. eighteen and two hundred five ten-thousandths

17. three hundred and forty-five thousandths

19. ten and one thousand eleven ten-thousandths

21. 1000.005   **23.** .231

25. Seventy-three and eighty-five hundredths

27. 3.6×5.125 **29.** 189   **31.** 64   **33.** $\frac{96}{128}$

Section 3.2

1. .11, .12, .14   **3.** .5, .503, .510

5. .019, .02, .021, .022   **7.** 0.692, 0.748, 0.75

9. 6.139, 6.14, 6.141   **11.** 7.19, 7.21, 7.27, 7.3

13. .0097, .0709, .079, .0907

15. .86, .899, .9, .903, .91

17. .1159, .116, .1163, .117

19. .072, .0729, .073, .073001, .073015

21. .88759, .88799, .888, .8881

23. 8.2975, 8.3401, 8.3599, 8.36   **25.** Trepid

27. 98.35 cents   **29.** $\frac{16}{25}$ **31.** $\frac{3}{4}, \frac{19}{25}, \frac{7}{8}$ **33.** $\frac{2}{3}, \frac{4}{5}, \frac{7}{8}$

Section 3.3

1. 2.7, 2.65, 2.653   **3.** 12.3, 12.30, 12.302

5. 10.0, 9.99, 9.989   **7.** 53.3, 53.31, 53.313

9. 2.2, 2.18, 2.179   **11.** 0.8, 0.79, 0.793

13. 0.8, 0.79, 0.789   **15.** 7000, 6900, 6920

17. 4000, 4000, 3970   **19.** 1000, 1000, 970

21. 37, 36.6, 36.59   **23.** 5, 5.4, 5.40   **25.** $484

27. .50 in   **29.** $1,370   **31.** 37 mpg   **33.** $\frac{1}{2}$

35. $\frac{1}{3}$

Section 3.4

1. 1.29   **3.** 8.835   **5.** 2.07   **7.** 30.1099

9. 872.1691   **11.** .948   **13.** 2.76   **15.** 37.4165

17. 58.6   **19.** 170.5   **21.** 25.585   **23.** 74.88

25. 276.69 tons, 34.73 tons   **27.** 2.7 cc

29. 19.7 miles   **31.** Yes   **33.** $\frac{1}{12}$ **35.** $21\frac{3}{4}$

37. $\frac{35}{72}$

Section 3.5

1. .14   **3.** 5.6   **5.** 2.16   **7.** 224.9   **9.** 4.2488

11. 5.576   **13.** .1328   **15.** 7.742   **17.** 31.329

19. 3.4524   **21.** .70824   **23.** 852.066

25. $16.72   **27.** $135.22   **29.** $38.85

31. $388.51   **33.** $2\frac{1}{2}$ **35.** $1\frac{1}{2}$ **37.** 4

Section 3.6

1. 1.1   **3.** .0004   **5.** 1.13   **7.** 1.31   **9.** .018

11. .052   **13.** 6.307   **15.** .2   **17.** 5.945

19. 2.424   **21.** 7.79   **23.** 6.92   **25.** 56.3 gal

27. $4.17   **29.** 48 mpg   **31.** 586 ft   **33.** $\frac{53}{24}$

35. $\frac{253}{120}$ **37.** $\frac{599}{360}$

Section 3.7

1. .65   **3.** .792   **5.** 3.625   **7.** .07375

9. 3.348   **11.** .3   **13.** 1.0   **15.** .29   **17.** 5.30

19. .85   **21.** .778   **23.** .818   **25.** .03125

27. .781 in   **29.** Win .49; Lose .51   **31.** .005

33. $10\frac{1}{8}$ **35.** $13\frac{1}{2}$ **37.** $16\frac{1}{9}$

Section 3.8

1. .01   **3.** 24.8   **5.** .9675   **7.** 7.765   **9.** .36

11. .013   **13.** 3.954   **15.** 38.559   **17.** .1742

19. 3.78   **21.** 3.9   **23.** 11.936   **25.** $1,094.16

27. $488.84   **29.** $696.65   **31.** $12\frac{7}{24}$ **33.** $8\frac{5}{6}$

35. $4\frac{1}{4}$

CHAPTER 4

Pre-test

1. 28.5 ft   **2.** 20 pt   **3.** 67 oz   **4.** 9 gal 2 qt

5. 1 yd 2 ft 6 in   **6.** 7.844 kg   **7.** 3,277,000 cm

8. 3.475 ℓ **9.** 130 g 3 cg   **10.** 14 kℓ 910 ℓ

11. $10.50/hr   **12.** 23.6 m/sec   **13.** 733 ft/sec

14. $V = 150.72$ **15.** $S = 7250$ **16.** 75.36 m

17. 287.88 m   **18.** 45.5 ft   **19.** 254.34 yd

20. 7.85 m²   **21.** 16.935 yd²   **22.** 137.2 m³

23. 78125 m³   **24.** 533.8 ft²   **25.** 1169.4 ft²

Section 4.1

1. 60 in   **3.** 16 qt   **5.** 6 hr 5 min   **7.** 14 qt

9. 67 oz   **11.** 4,260 lb   **13.** 6 hr 13 min 16 sec

15. 269 ft   **17.** 276 in   **19.** 8 ft 9 in

21. 17 gal 3 qt 1 pt   **23.** 3 yd 1 ft

25. 1,027 lb 8 oz   **27.** 385 miles   **29.** $84

31. two and five ten-thousandths
33. three thousand, eight hundred seventy-six ten-
millionths   **35.** 0.004335

Section 4.2

1. 5,000 ℓ **3.** 1,300 m   **5.** 7 g 1 dg

7. 1,300 ℓ **9.** .245 ℓ **11.** 86 dg   **13.** 101 mℓ

15. 422 m   **17.** 700 m   **19.** 7,150 m

21. 7,013 ℓ **23.** 11,850 m   **25.** 14.22 m

27. .75 mg   **29.** 11.75 mℓ

31. .0099, 0.0572, 0.485, 2.345

33. 0.24, 2.004, 2.04, 2.4

35. 1.2233, 1.2234, 1.2244, 1.2345

Section 4.3

1. 3.5 ft   **3.** $4\frac{1}{3}$ yd   **5.** 3 tons   **7.** 336 sec

9. 7.528 km   **11.** 63,360 in   **13.** 30 mph

15. 211 in   **17.** 1.25 hr   **19.** 1,500 lb/in

21. 40 m/sec   **23.** .354 m²   **25.** 10.5 lb

27. .012 ℓ **29.** $246.40   **31.** 32.96 gal

33. 0.235   **35.** 12.898   **37.** 1,234,600

Section 4.4

1. 49   **3.** 94   **5.** 38   **7.** 40   **9.** 40

11. 112   **13.** 40   **15.** .2   **17.** $\frac{121}{7}$

19. $3\frac{4}{37}$ or $\frac{115}{37}$ **21.** $A = 464$ **23.** $R = .2$

25. 15,700 sq ft   **27.** $19.50   **29.** $1.56

31. 52.921   **33.** 151.6727   **35.** 120.736

Section 4.5

1. 52 cm   **3.** 628 in   **5.** 57 in   **7.** 50.24 ft

9. 36 in   **11.** 43 cm   **13.** 49.7 ft   **15.** 103 cm

17. 41.68 in   **19.** 60.84 in   **21.** 132.48 in

23. 70.84 in   **25.** $34.84   **27.** 360 ft

29. 1.57 in   **31.** 108 min or 1 hr 48 min

33. 10.966   **35.** 12.13   **37.** 2.1318

Section 4.6

1. 314 cm²   **3.** 6 in²   **5.** 4 ft²   **7.** 96 in²

9. 243 ft²   **11.** 3.36 m²   **13.** 452.16 in²

15. 452.6 ft²   **17.** 295.4426 ft²   **19.** 466.56 ft²

21. 1,890.69 in²   **23.** 207.255 cm²   **25.** $250.67

27. $111.91   **29.** $208.80   **31.** $47.36

33. 0.463448   **35.** 30.14238

Section 4.7

1. 30,000 ft²   **3.** 255 ft²   **5.** 2,395 mm²

7. 418.08 in²   **9.** 562.5 cm²   **11.** 81.64 cm²

13. 12.86 ft²   **15.** 150 m²   **17.** 64.94 ft²

19. 49.60 m²   **21.** 50 m²   **23.** 85.12 ft²

25. $236.18   **27.** 10 lb   **29.** $4,296.93

31. 18.92   **33.** 2.05   **35.** 2.7

Section 4.8

1. 525 m³   **3.** 1,582.56 in³   **5.** 125 in³

7. 3,140 cm³   **9.** 267.9 ft³   **11.** 468 cm³

13. .2 m³   **15.** 6,104.16 in³   **17.** 1,387.5 ft³

19. 10,889.5 mm³   **21.** 1,191.936 ft³

23. 362.984 m³   **25.** 75.6 tons   **27.** 69.08 gal

29. 20 loads   **31.** .96   **33.** 0.2   **35.** 0.464

Section 4.9

1. 15,000 cm²   **3.** 2,813.44 in²   **5.** 376 in²

7. 1,858.56 in²   **9.** 138.788 m²   **11.** 100.48 in²

13. 96 in²   **15.** 3.84 ft²   **17.** 2,512 in² or $17\frac{4}{9}$ ft²

19. 144 ft² or 16 yd²   **21.** 603 in²   **23.** 163.28 in²

25. 879.2 ft²   **27.** 577.76 ft²   **29.** $144.38

31. 7.326   **33.** 3.37   **35.** 50.9

CHAPTER 5

Pre-test

1. (a) -32, (b) 32   **2.** -5 **3.** 0.83   **4.** 0

5. -127 **6.** -82 **7.** 231   **8.** $\frac{9}{8}$

9. -14.83 **10.** 640   **11.** 96   **12.** $-\frac{2}{9}$

13. -14 **14.** 35   **15.** -15 **16.** $-\frac{25}{16}$

17. -169 **18.** 248   **19.** -66 **20.** 113

21. $-\frac{16}{3}$ **22.** $-\frac{25}{6}$ **23.** $-31°$ **24.** -3.4 yd

25. $-20°$C

Section 5.1

1. 3   **3.** 2.1   **5.** $\frac{1}{6}$ **7.** 31   **9.** $\frac{2}{3}$ **11.** 7

13. .035   **15.** $-3\frac{1}{8}$ **17.** -0.23 **19.** $4\frac{15}{16}$

21. 21.75   **23.** 4.5   **25.** $-2\frac{7}{8}$ points   **27.** 12°C

29. A.D. 1875 or $+1875$ **31.** 60 in   **33.** 300 in

35. 29 min 58 sec

Section 5.2

1. -13 **3.** 8   **5.** -8 **7.** 9   **9.** 7

11. -185 **13.** -11.4 **15.** $-\frac{1}{6}$ **17.** 2.7

19. -27 **21.** -23.09 **23.** $1\frac{8}{9}$

25. -50 pounds or 50 pounds lighter

27. $6,458 profit   **29.** 5,466 books   **31.** .35 ℓ

33. 1.9 km   **35.** 4 kℓ 8 hℓ

Section 5.3

1. -7 **3.** -13 **5.** -27 **7.** 41   **9.** -9

11. -90 **13.** -5 **15.** 1   **17.** $-\frac{9}{40}$ **19.** 9.81

21. -4.3 **23.** $\frac{67}{70}$ **25.** 63 degrees   **27.** $378.99

29. 20,602 ft   **31.** 6.25 ft or $6\frac{1}{4}$ ft   **33.** 42,915 ft

35. 1,000 $\frac{lb}{in}$

Section 5.4

1. 10   **3.** 0   **5.** -95 **7.** -108 **9.** -60

11. 0.048   **13.** $\frac{3}{8}$ **15.** 30.24   **17.** -4

19. $-\frac{10}{9}$ **21.** 110   **23.** $-\frac{27}{35}$ **25.** $-10°$C

27. -12.5 lb   **29.** -33.96 points   **31.** 79

33. 1,225   **35.** 1,576

Section 5.5

1. -14 **3.** 29   **5.** 21   **7.** -1.01 **9.** -310

11. -50 **13.** $-.008125$ **15.** -1 **17.** 4

19. 4   **21.** 5   **23.** $-\frac{4}{31}$ **25.** $-4°$

27. $-$$69.67   **29.** $-$$30.61   **31.** 14 ft

33. 62.8 in   **35.** 30 ft

Section 5.6

1. -7 **3.** -16 **5.** -5 **7.** 13   **9.** -1.875

11. -39 **13.** -33 **15.** 131   **17.** -84

19. -157 **21.** 117   **23.** -135 **25.** No

27. 13.8°C   **29.** Yes   **31.** 144 cm²

33. 108 cm²   **35.** 1,962.5 cm²

CHAPTER 6

Pre-test

1. $x = 11$ **2.** $x = 5 - 3y$ **3.** y^{15} **4.** z^6

5. 19,683   **6.** $21y$ **7.** $4a^2bc^2$

8. $2a + 3b - 2c$ **9.** $15x + 6y + 28$

10. $x = -12$ **11.** $a = -22$ **12.** $x^2 - 3x - 28$

13. $6x^2 + 19x + 15$ **14.** $x = 17$ **15.** $x = 2$

16. $-a + 3c - 2b$ **17.** $2x + 3$

18. $6ac(2 - b + 3bc)$ **19.** $(x - 8)(x + 7)$

20. $(3x + 2)(4x - 1)$ **21.** $(5x - 8)(5x + 8)$

22. $x = 25, 1$ **23.** $x = 8, 7$ **24.** 15

25. 15 m $\times$ 32 m

Section 6.1

1. $x = 6$ **3.** $x = 30$ **5.** $x = -4$

7. $x = \dfrac{9}{7}$ or $x = 1\dfrac{2}{7}$ **9.** $x = 5$ **11.** $b = -3$

13. $x = \dfrac{4 - y}{2}$ **15.** $y = 1$ **17.** $x = 0$ **19.** $b = 1$

21. $x = 3$ **23.** $\ell = \dfrac{P - 2w}{2}$ **25.** $60°C$

27. 8 cm   **29.** 27 ft/sec   **31.** 14,826 yd²

33. 523.3 cm³   **35.** 48,000 ft³

Section 6.2

1. 64   **3.** 4   **5.** y^{10} **7.** 5,184   **9.** $\dfrac{16}{81}$

11. x^{11} **13.** y^{12} **15.** x^3 **17.** $8a^3b^6$

19. $27a^{12}b^6x^9$ **21.** $81a^4b^4x^4y^4$ **23.** x^{25}

25. $\dfrac{4}{3}\pi r$ **27.** $V = w^4$ **29.** $V = \pi r^5$

31. 60,000 cm²   **33.** 48 m²   **35.** 169.56 ft²

Section 6.3

1. $9a$ **3.** $-2a$ **5.** $-2ab$ **7.** $-2y$ **9.** $-7st$

11. $5ab + 4a$ **13.** $-48a^2b^2$ **15.** $-\dfrac{8}{3}xy$

17. $17x^3$ **19.** $25z^3$ **21.** $a = 2$ **23.** $x = 20$

25. \$15,000   **27.** 4 m   **29.** 5   **31.** $\dfrac{3}{4}$ **33.** 4

Section 6.4

1. $x - 6$ **3.** $6x + 14$ **5.** $2a - 2$ **7.** $2a + 4w$
9. $-x^2 + 2x - 10$

11. $-10a^2 - b^2 - 8ac$ **13.** $-.3bc - 4.3ad$

15. $13x - 7y$ **17.** $2x + \dfrac{7}{4}y$ **19.** $x + 15$

21. $x - 8y + 2$ **23.** $x = -2.75$

25. $C = \dfrac{26}{5}n^2 - 44n + 70$ **27.** $5x$

29. Minh 816, Cindy 4,084   **31.** $d = 208t$

33. -128 **35.** $-12\dfrac{1}{6}$

Section 6.5

1. $r_1 = 260$, $r_2 = 410$, $r_3 = 560$, $r_4 = 710$

3. \$5.45, \$9.30   **5.** 187, 64

7. $L = 310$ ft, $W = 135$ ft   **9.** First test 995 seeds, second test 990 seeds, third test 1,015 seeds

11. 4.75 in, 5.25 in, 5.75 in, 6.25 in, 6.75 in, 7.25 in

13. antimony: 1.6 lb; copper: .8 lb; tin: 9.6 lb

15. 1.5 hr or 1 hr 30 min   **17.** 13 days

19. 75 of each   **21.** 24 of each kind

23. 64,000 residents   **25.** 12.5 oz   **27.** -173

29. -42.88

Section 6.6

1. $6a - 3$ **3.** $-6y + 2$ **5.** $-15a^2 + 6ab$

7. $6x^2 + 10x - 56$ **9.** $-2r^2 + 11rs - 12s^2$

11. $x^2 + 10x + 25$ **13.** $12x^2 + 11x - 15$

15. $9y^2 - 16$ **17.** $-12x^2 + 26xy + 9wx - 15wy - 10y^2$

19. $8a^2 - 16a^2b - 2ab + 24ab^2 - 15b^2$

21. $x = -\dfrac{24}{11}$ or $x = -2\dfrac{2}{11}$ **23.** $12x - 32y + 2$

25. 3 hr   **27.** 7:24 p.m.   **29.** -384 **31.** 18

33. 1

Section 6.7

1. $x^2 - 2x - 3$ **3.** $a - b$ **5.** $2bc - 4abc^2$

7. $xy + 2y - 3$ **9.** $2c + 3abc^2 + bc^2$

11. $\dfrac{7}{12}y^2 + \dfrac{2}{3}yz + y - \dfrac{5}{4}y^2z$ **13.** $x + 2$

15. $x - 2$ **17.** $2x - 5$ **19.** $2a + 3b$

21. $x^2 + 2x + 1$ **23.** $x^2 - x + 1$

25. $\dfrac{A}{2} = \ell w + h\ell + hw$ **27.** No   **29.** Yes

31. 15   **33.** 4   **35.** 30.8

Section 6.8

1. $y(6x - 5)$ **3.** $x(2x - 3)$ **5.** $xz(15y - 7w)$

7. $\pi h(r^2 - R^2)$ **9.** $2(2ab + 4bc - 3cd)$

11. $\pi r(rh + r + 2)$ **13.** Cannot be factored

15. $x^2y^2(1 + xy - x^2y^2)$ **17.** $x^2y^2(8x + 9y - 7)$

19. $12x(2x^3 - x^2 - 3x - 4)$

21. $6a^2b(4ab^2 + 3c^2 + 5c)$

23. $4xy(xy + 2 - 3x^2 - 4y)$ **25.** $x = 2$

27. $y = -2$ **29.** $y = 4 - 3x$

Section 6.9

1. $(x + 6)(x - 7)$ **3.** $(x - 10)(x + 8)$

5. $(a - 4)(a - 1)$ **7.** $b(x - 10)(x + 2)$

9. $(2y - 5)(y + 1)$ **11.** $(3a - 4)(a - 2)$

13. $(3x - y)(x - 3y)$ **15.** Cannot be factored

17. $(5x - 4y)(5x - 4y)$ or $(5x - 4y)^2$

19. Cannot be factored   **21.** $(2x - 1)(6x + 5)$

23. $(3x + 2)(5x - 3)$ **25.** $(2x + 1)(3x + 2)$

27. $(3x + 2)(5x + 3)$ **29.** Cannot be factored

31. x^{22} **33.** $16a^{10}b^4c^6$ **35.** $\dfrac{x^6}{y^{12}}$

Section 6.10

1. $(a + 4)(a - 4)$ **3.** $(2x + 3)(2x - 3)$

5. $(a + 8)(a - 8)$ **7.** $(3a + 8b)(3a - 8b)$

9. $(5x + 6y)(5x - 6y)$ **11.** $(12a + 11b)(12a - 11b)$

13. $(20x + 3)(20x - 3)$ **15.** $3(4a + 5)(4a - 5)$

17. $9(x + 2)(x - 2)$ **19.** $4a(3a + 4)(3a - 4)$

21. $(5b + 4)(5b - 4)$ **23.** $14a(a + 2b)(a - 2b)$

25. 53 and 47   **27.** 135 and 125   **29.** $8x^2 + 7x$

31. $x = 1$ **33.** $a = c - 3b$

Section 6.11

1. $x = 7$ or $x = 8$ **3.** $x = 9$ or $x = 4$

5. $x = 0$ or $x = 1$ or $x = 3$ **7.** $x = 9$ or $x = -1$

9. $b = -5$ or $b = 2$

11. $x = 3$ or $x = -3$

13. $x = -24$ or $x = -6$

15. $x = -\dfrac{2}{3}$ or $x = -3$ **17.** $x = 3$ or $x = \dfrac{1}{3}$

19. $x = -\dfrac{1}{2}$ **21.** $x = -3$ or $x = 0$

23. $x = -7$ or $x = -\dfrac{5}{2}$ **25.** 9 sec   **27.** 7, 9

29. 8 ft by 11 ft   **31.** $8x + 4y$ **33.** $x = -6$

35. $b = \dfrac{3 - 2a}{4}$

CHAPTER 7

Pre-test

1. $\dfrac{56x^2 - 21x}{45}$ **2.** $\dfrac{119p^3}{45q^2}$ **3.** $\dfrac{n - 7}{3(n - 4)}$ **4.** $\dfrac{3x}{4y^2}$

5. $x^2 - 6x - 27$ **6.** $180ab^3c$ **7.** $\dfrac{5c(a + 1)}{a - 1}$

8. $\dfrac{4np}{m^2}$ **9.** $\dfrac{1}{y - 5}$ **10.** $30a^2b^3c$

11. $135(a + 2b)$ **12.** $\dfrac{5}{2y}$ **13.** $\dfrac{a^2 - a - 9}{(a - 3)(a - 4)}$

14. $\dfrac{5}{x + 5}$ **15.** $\dfrac{y^2 + 15}{(y - 5)(y + 3)}$ **16.** $\dfrac{2a + 5b}{a - 2b}$

17. $\dfrac{1 - x}{1 + x}$ **18.** $a = \dfrac{44}{5}$ **19.** $a = -\dfrac{31}{5}$

20. $h = 10$ **21.** $b = 20$ **22.** $e = E - IR$

23. $a = \dfrac{S - 2bh}{2b + 2h}$ **24.** 24 days

25. 50 mph, 60 mph

Section 7.1

1. $\dfrac{5s}{7t}$ **3.** $\dfrac{15mn}{2p^2}$ **5.** $-\dfrac{21a^2b^2}{2c^3}$ **7.** $\dfrac{-2a - 2b}{m^2 - mn}$

9. $\dfrac{3c^2 - 5c}{p^2 - 4p}$ **11.** $-\dfrac{152ax}{195by}$ **13.** $\dfrac{45a^2b + 27ab^2}{3ax + cx}$

15. $\dfrac{yz + y^2}{z^2 - yz}$ **17.** $\dfrac{y^2 - 4y - 21}{6t^2 - t - 2}$ **19.** $-\dfrac{77x^2}{150y^3}$

21. $\dfrac{3x^2 + 9x - 12}{2x^3 + 2x^2}$ **23.** $\dfrac{c^3 - 36c}{4a^4 + 8a^3}$

25. $x^2 + 4x - 96$ **27.** $4x^3 + x + 15$

29. $15x^3 - 11x^2 - 36x + 32$

Section 7.2

1. $2a$ **3.** $\dfrac{a}{2}$ **5.** $14xy$ **7.** $\dfrac{2xy^2}{3}$ **9.** $\dfrac{1}{2}$

11. $10x^2$ **13.** $4x^2 + 6xy + 2y^2$ **15.** $\dfrac{4c^4}{3ab^3}$

17. $\dfrac{x}{x + 4}$ **19.** $\dfrac{x + 3}{x + 8}$ **21.** $-27mx^2y$

23. $12x^2 - 13x - 35$ **25.** $2x^3 + 3x^2 - x + 4$

27. $4x + 7$ **29.** $x^2 + 7x - 5$

Section 7.3

1. $\dfrac{8}{5y}$ **3.** $\dfrac{7}{4}$ **5.** $\dfrac{11}{x - 3}$ **7.** $\dfrac{x + 4}{x - 4}$ **9.** $\dfrac{6a}{5}$

11. $\dfrac{xy + 2x}{a + y}$ **13.** $\dfrac{x + 2}{x + 5}$ **15.** $\dfrac{x^2 + 2x + 3}{(x + 1)(x + 2)}$

17. $\dfrac{8x + 8y}{3}$ **19.** $\dfrac{bc^2}{x + 2}$ **21.** $\dfrac{x - 4}{x - 5}$

23. $\dfrac{2a + 1}{4(a - 1)(a + 4)}$ **25.** $2\dfrac{2}{5}$ days   **27.** $3\dfrac{9}{37}$ ohms

29. $12(3x + 4y)$ **31.** $12xy(4xy - 6y^2 + 3x^2)$

33. Cannot be factored

Section 7.4

1. $\dfrac{x}{3}$ **3.** $\dfrac{4xy}{3}$ **5.** $\dfrac{x}{y}$ **7.** $\dfrac{x-3}{2}$ **9.** $\dfrac{2}{3}$

11. $\dfrac{8a}{5b}$ **13.** $\dfrac{5b^2}{9a}$ **15.** $\dfrac{2a^2 + 4a}{3a - 6}$

17. $\dfrac{(a+b)(x-y)}{(x+y)(a-b)}$ **19.** $\dfrac{4x}{a}$ **21.** $\dfrac{2a-1}{x-4y}$

23. $\dfrac{(b-4)(b-6)(b+1)}{(b+2)(b+2)(b-2)}$ **25.** $(x-13)(x+12)$

27. $(a+7)(4a-9)$ **29.** Cannot be factored

Section 7.5

1. $8a^2b^2$ **3.** $24a^2b^2$ **5.** $(x+y)(x-y)$

7. $7(2a+3)(2a-3)$ **9.** $(x+y)^2$

11. $(b+c)(b-c)$ **13.** $(y+5)(y+2)(y+1)$

15. $(x-5)(x+5)(x+2)$ **17.** $(x+3)^2(x-3)$

19. $72x^3(x-1)$ **21.** $(x+7)(x+3)(x+2)$

23. $36a^3b(c+1)(c-1)$ **25.** $60x^2y^3$

27. $(x+3)(x-3)^2$ **29.** $(4y+7z)(4y-7z)$

31. $(11ab+12c)(11ab-12c)$ **33.** $3a(4b+5c)(4b-5c)$

Section 7.6

1. $\dfrac{7a}{3}$ **3.** $\dfrac{a}{2}$ **5.** $\dfrac{2a}{5b}$ **7.** $\dfrac{3x+7}{10}$ **9.** $\dfrac{4x+xy}{3y}$

11. $\dfrac{3xy + 5y^2 + 8x^2}{x^2y^2}$ **13.** $\dfrac{35a+3}{5a}$

15. $\dfrac{x^2 + 3x + 1}{x}$ **17.** $\dfrac{9x+33}{x^2+4x-5}$

19. $\dfrac{x^2 + y^2 + y}{x^2 - y^2}$ **21.** $\dfrac{x^2 + 3x + 1}{x}$

23. $\dfrac{2x^2 + 7x - 7}{(x+4)(x-5)(x+5)}$ **25.** $x = -4$ or $x = 5$

27. $x = -\dfrac{3}{2}$ or $x = 5$ **29.** $x = \dfrac{-7}{3}$ or $x = 5$

Section 7.7

1. $\dfrac{x}{2}$ **3.** $\dfrac{2x-3}{y-1}$ **5.** $\dfrac{1}{3a}$ **7.** $\dfrac{3b-8a}{12ab}$

9. $\dfrac{5x-2z}{10y}$ **11.** $\dfrac{3-35a}{5a}$ **13.** $\dfrac{x^2 - 3x - 1}{x}$

15. $\dfrac{-3a - 11b}{a^2 - b^2}$ **17.** $\dfrac{y+4}{y^2 + y}$ **19.** $\dfrac{-2ab - b^2}{a^2 + ab}$

21. $\dfrac{1 - x^2 + 3x}{x}$ **23.** $\dfrac{-2x - 11}{(x+1)(x+4)(x-4)}$

25. $\dfrac{15w}{xyz}$ **27.** $\dfrac{4x^2 - 16}{15x^3}$ **29.** $\dfrac{-7y^3 + 7yz^2}{4a^2b - 4b^3}$

Section 7.8

1. $\dfrac{5}{12}$ **3.** $\dfrac{2x}{3}$ **5.** $\dfrac{17}{6}$ **7.** $\dfrac{a^2c + ab}{c}$

9. $\dfrac{wx + wy}{xy}$ **11.** $\dfrac{ab+1}{ab-1}$ **13.** $-\dfrac{1}{y}$

15. $\dfrac{x^2z + x^2y}{y^2z + xyz^2}$ **17.** $\dfrac{10x^2 + 4}{5x}$ **19.** $\dfrac{20x + 8}{x^3}$

21. $\dfrac{x+1}{x - x^2}$ **23.** $-\dfrac{3}{a}$ **25.** $\dfrac{5}{x+3}$ **27.** $\dfrac{x}{x+2}$

29. $x^2 + 6x + 9$

Section 7.9

1. $x = 2$ **3.** $x = 1$ **5.** $x = \dfrac{4}{3}$ **7.** $a = 6$

9. $a = -\dfrac{1}{6}$ **11.** $x = -10$ **13.** $a = 2$

15. $x = 0$ **17.** $x = 4$ **19.** $b = -\dfrac{1}{7}$

21. $a = -22$ **23.** $x = -\dfrac{3}{10}$ **25.** $r_1 = 20$, $r_2 = 5$

27. $342\dfrac{6}{7}$ mph, $267\dfrac{6}{7}$ mph   **29.** $1\dfrac{1}{5}$ hr

31. 12 days   **33.** $\dfrac{3x}{8}$ **35.** $2x - 6$ **37.** 1

Section 7.10

1. $P = 21$ **3.** $g = 16$ **5.** $\ell = 20.5$ **7.** $h = 8$

9. $d \approx 21$ **11.** $w = 1.6$ **13.** $d \approx 20$

15. $g = \dfrac{S}{t - \dfrac{1}{2}t^2}$ **17.** $a = \dfrac{S - bh}{b + h}$ **19.** $f = \dfrac{ts}{t+s}$

21. $N = \dfrac{k}{T-S}$ **23.** $m = \dfrac{y-b}{x}$ **25.** $S = \dfrac{5252H}{T}$

27. 8 years   **29.** $24x^2$ **31.** $a^2 - b^2$

33. $(x+1)(x+2)(x+3)$

CHAPTER 8

Pre-test

1. $\dfrac{1}{8}$ **2.** True   **3.** False   **4.** 90   **5.** $x = 20$

6. 20 lb   **7.** \$4.38   **8.** \$78.96   **9.** \$3,625

10. 4 hrs   **11.** $27\frac{3}{4}$% or 27.75%   **12.** 350%

13. 3.25   **14.** .002   **15.** 75%   **16.** 255.6%

17. $2\frac{3}{4}$ **18.** $\frac{83}{600}$ **19.** 75   **20.** 45%

21. 107.9   **22.** 80%   **23.** 75%   **24.** 240%

25. $8.64

Section 8.1

1. $\frac{8\text{ people}}{11\text{ chairs}}$ **3.** $\frac{1}{2}$ **5.** $\frac{4}{5}$ **7.** True   **9.** False

11. False   **13.** $\frac{6}{7}$ **15.** $\frac{75\text{ miles}}{1\text{ hour}}$

17. $\frac{1.3\text{ television sets}}{1\text{ house}}$ **19.** $\frac{27}{2}$ **21.** True

23. True   **25.** a. $\frac{3}{4}$, b. $\frac{3}{7}$ **27.** No

29. $\frac{97.6\text{ people}}{1\text{ sq mile}}$ **31.** $\frac{4a-b}{6}$ **33.** $\frac{14x}{9y}$

35. $\frac{x^2-2x+1}{x}$

Section 8.2

1. $x=16$ **3.** $y=3\frac{1}{3}$ **5.** $w=22.5$

7. $R=150$ **9.** $x=12$ **11.** $y=7.5$

13. $w=1.5$ **15.** $y=.01$

17. $R=5\frac{1}{2}$ or $R=5.5$ **19.** $x=24$

21. $y=2\frac{13}{16}$ or $y=2.8125$ **23.** $A=2.17$

25. $x=234.375$ or $x=234\frac{3}{8}$ **27.** $z=14.57$

29. $b=.47$ **31.** a

33. $\frac{-2x-12y}{(x+y)(x-y)}$ or $\frac{-2x-12y}{x^2-y^2}$

35. $\frac{-x^2+2x+1}{x}$

Section 8.3

1. 3   **3.** x **5.** $\frac{3}{x}=\frac{5}{10}$ **7.** .4 dram or $\frac{2}{5}$ dram

9. 57 rows   **11.** $26\frac{2}{3}$ lb   **13.** $1,215   **15.** $136

17. 63.4 gal   **19.** 487 points   **21.** 1,200 miles

23. 5 jobs   **25.** 48 boys

27. $40,045; $16,018; $16,018   **29.** 5.4 yd

31. $\frac{cf}{de}$ **33.** $\frac{1}{x+3}$ **35.** $\frac{x-3}{x+4}$

Section 8.4

1. 490 lb   **3.** $4\frac{1}{6}$ hr or 4 hr 10 min   **5.** $247.25

7. 237.6 lb   **9.** $26\frac{2}{3}$ lb   **11.** 20 hours   **13.** $\frac{44}{15}$

15. $\frac{xy-1}{xy+1}$ **17.** $-\frac{1}{y}$

Section 8.5

1. 27%   **3.** 160%   **5.** 160%   **7.** 48%

9. $10\frac{3}{5}$%   **11.** $77\frac{7}{9}$%   **13.** 515%   **15.** $12\frac{3}{5}$%

17. $63\frac{7}{11}$%   **19.** $91\frac{2}{3}$%   **21.** 8, 8%

23. $22\frac{2}{9}, 22\frac{2}{9}$%   **25.** 10   **27.** 63%   **29.** 6%

31. 7%   **33.** $x=\frac{1}{2}$ **35.** $x=\frac{7}{12}$ **37.** $x=\frac{10}{3}$

Section 8.6

1. 36%   **3.** 8%   **5.** 160%   **7.** 21.4%

9. 700%   **11.** 1321%   **13.** 127%   **15.** 2.56%

17. 575%   **19.** 10.25%   **21.** $3\frac{1}{2}$%   **23.** $27\frac{2}{3}$%

25. 3%   **27.** 85%   **29.** 37.5%   **31.** $r=.075$

33. $d=25$ **35.** $m=\frac{Fr^2}{kM}$

Section 8.7

1. .14   **3.** .923   **5.** .03   **7.** .16   **9.** .82

11. .0215   **13.** 5.63   **15.** .537   **17.** .00125

19. 3.147   **21.** $.35\frac{1}{6}$ or $.3516\frac{2}{3}$ **23.** $.00\frac{1}{4}$ or .0025

25. .05   **27.** .20 or .2   **29.** False   **31.** True

33. False

Section 8.8

1. 7%   **3.** 22%   **5.** 85%   **7.** $66\frac{2}{3}$%

9. $187\frac{1}{2}$%   **11.** $5\frac{1}{4}$%   **13.** $8\frac{1}{2}$%   **15.** 33.3%

17. 1.3%   **19.** 38.1%   **21.** 368.8%   **23.** 101.3%

25. 60%   **27.** 57.1%   **29.** 1.7%   **31.** $x = 1.25$

33. $x = 24$ **35.** $x = 8.1$

Section 8.9

1. $\dfrac{1}{20}$ **3.** $1\dfrac{1}{4}$ **5.** $\dfrac{7}{10}$ **7.** $\dfrac{14}{25}$ **9.** $1\dfrac{3}{25}$

11. $\dfrac{3}{400}$ **13.** $\dfrac{23}{250}$ **15.** $\dfrac{33}{200}$ **17.** $\dfrac{1}{300}$ **19.** $\dfrac{1}{9}$

21. $\dfrac{1}{40}$ **23.** $\dfrac{117}{700}$ **25.** $\dfrac{1}{8}$ **27.** $\dfrac{9}{50}$ **29.** $\dfrac{9}{100}$

31. $2.50   **33.** 2,100   **35.** 400 miles

Section 8.10

Percent	Fraction	Decimal
10%		.1
	$\dfrac{3}{10}$	.3
75%	$\dfrac{3}{4}$	
90%		.9
	$1\dfrac{9}{20}$	1.45
37.5% or $37\dfrac{1}{2}$%		.375
.1%	$\dfrac{1}{1000}$	
100%	$\dfrac{1}{1}$ or 1	
225%		2.25
80%	$\dfrac{4}{5}$	
	$\dfrac{11}{200}$	.055
87.5% or $87\dfrac{1}{2}$%	$\dfrac{7}{8}$	
	$\dfrac{1}{200}$	.005
60%	$\dfrac{3}{5}$	
	$\dfrac{5}{8}$	.625

Section 8.10 *continued*

Percent	Fraction	Decimal
	$\dfrac{3}{400}$	.0075
$66\dfrac{2}{3}$%		$.66\dfrac{2}{3}$
	$\dfrac{1}{4}$	.25
20%	$\dfrac{1}{5}$	
	$\dfrac{2}{5}$	.4
	$\dfrac{1}{3}$	$.33\dfrac{1}{3}$
12.5% or $12\dfrac{1}{2}$%	$\dfrac{1}{8}$	
70%		.7

1. $k = 128,000$ **3.** $3\dfrac{1}{3}$ hrs.   **5.** $108

Section 8.11

1. 50%   **3.** 40   **5.** 36   **7.** 26.8 or $26\dfrac{4}{5}$

9. $133\dfrac{1}{3}$%   **11.** 25   **13.** 5.58 or $5\dfrac{29}{50}$

15. .0224   **17.** 128.6   **19.** 156.7%   **21.** 18.3

23. $8.30\dfrac{1}{3}$ **25.** 85%   **27.** .03%   **29.** 1850%

Section 8.12

1. 25%   **3.** $171.90   **5.** $900   **7.** 87%

9. 28 questions   **11.** 15%   **13.** 62.5% or $62\dfrac{1}{2}$%

15. $1.457   **17.** 96¢ per dozen   **19.** 4,761

21. $160   **23.** 7.2%   **25.** 59%   **27.** 217%

29. 24%   **31.** 70%

CHAPTER 9

Pre-test

1. 13   **2.** $\dfrac{1}{2}$ **3.** $c = 6.5$ **4.** $\sqrt{143}$

5. $3\sqrt{7} + \sqrt{35}$ **6.** -5 **7.** $6\sqrt{7}$ **8.** $3y\sqrt{13x}$

9. $3b\sqrt[3]{2a^2}$ **10.** $\dfrac{\sqrt{70}}{10}$ **11.** $\dfrac{5\sqrt{14}}{7}$

12. $\dfrac{(3\sqrt{11}+3\sqrt{7})}{4}$ **13.** $8\sqrt{3}$ **14.** $8\sqrt{2}$ **15.** $2\sqrt[3]{3}$

16. $6^{1/2}a^{5/2}$ **17.** $x\sqrt[4]{5y^3}$ **18.** $6x^{-2}-y^{-3}$

19. $\dfrac{a^3b}{6}$ **20.** 3.9×10^{-5} **21.** 271,000,000

22. $2\sqrt{161}$ ft ≈ 25.4 ft   **23.** 6.7×10^7 miles

24. $(.32A)^{1/2}$ **25.** $I=k/d^2$

Section 9.1

1. 10   **3.** -5 **5.** -2 **7.** 11   **9.** $\dfrac{11}{13}$

11. $\dfrac{2}{3}$ **13.** 3   **15.** 2.83   **17.** 4.90   **19.** 14.14

21. 29.87   **23.** 3.26   **25.** 15 ft   **27.** $\dfrac{264}{49}$ or $5\dfrac{19}{49}$

29. 10   **31.** 15.0 in   **33.** 86.7%   **35.** 84.4%

Section 9.2

1. $c=5$ **3.** $a=5$ **5.** $c=15$ **7.** $c=7.5$

9. $b=15$ **11.** $a=36$ **13.** $c=30$

15. $c=2.24$ **17.** $c=3.61$ **19.** $a=11.18$

21. $c=11.70$ **23.** $b=10.54$ **25.** 13.93 ft

27. 14.42 ft   **29.** 61.0 ft   **31.** 127.3 ft   **33.** $1\dfrac{12}{25}$

35. $\dfrac{7}{200}$

Section 9.3

1. $\sqrt{21}$ **3.** 10   **5.** $\sqrt{42}$ **7.** 3   **9.** 12

11. $\sqrt{abc}$ **13.** 14   **15.** $\sqrt{15}+2\sqrt{3}$

17. $3\sqrt{6}+\sqrt{30}$ **19.** $6-\sqrt{30}$ **21.** $\sqrt{30}+\sqrt{105}$

23. -4 **25.** 2.4 ft   **27.** 13.4 in²

	Fraction	Decimal	Percent
29.	$\dfrac{7}{8}$	.875	87.5%
31.	$1\dfrac{9}{25}$	1.36	136%

Section 9.4

1. $4\sqrt{2}$ **3.** $2\sqrt{10}$ **5.** $\sqrt{26}$ **7.** $4c$ **9.** $4\sqrt{5}$

11. $2\sqrt{21}$ **13.** $2\sqrt[3]{7}$ **15.** $4\sqrt{11}$ **17.** $7\sqrt{5}$

19. $15\sqrt{5}$ **21.** $5w\sqrt{6}$ **23.** $3\sqrt[3]{3}$

25. $30\sqrt{5}\approx67.1$ ft   **27.** $60\sqrt{3}$ volts   **29.** 28.08

31. 43.75   **33.** 380

Section 9.5

1. $\dfrac{\sqrt{5}}{5}$ **3.** $\dfrac{\sqrt{65}}{13}$ **5.** $\dfrac{\sqrt{42}}{6}$ **7.** $\dfrac{\sqrt{6}}{2}$ **9.** $\dfrac{\sqrt{3}}{2}$

11. $\dfrac{\sqrt{6}}{3}$ **13.** $\dfrac{2\sqrt{5}}{5}$ **15.** $\dfrac{6\sqrt{5}}{5}$ **17.** $\dfrac{\sqrt{3}}{9}$

19. $\dfrac{3\sqrt{15}}{10}$ **21.** $\dfrac{11\sqrt{6}-11\sqrt{3}}{3}$ **23.** $\dfrac{8+2\sqrt{3}}{13}$

25. $\dfrac{\pi}{2}\approx1.57$ sec   **27.** \$5,000   **29.** 9,812 students

31. 18%

Section 9.6

1. $9\sqrt{5}$ **3.** $4\sqrt{7}$ **5.** $2\sqrt{2}$ **7.** $6\sqrt{5}$ **9.** $5\sqrt{2}$

11. $4\sqrt{6}$ **13.** $\sqrt{6x}$ **15.** $7\sqrt{2}$ **17.** $3\sqrt{2}$

19. $13\sqrt{2}-8\sqrt{3}$ **21.** $-5\sqrt{7}$ **23.** $-21\sqrt{3}$

25. $140\sqrt{6}$ volts   **27.** 32   **29.** 5   **31.** 5.3

Section 9.7

1. $7^{1/2}$ **3.** $x^{1/2}$ **5.** $\sqrt[3]{12}$ **7.** $57^{1/2}$ **9.** $(ab)^{1/2}$

11. $\sqrt{7b}$ **13.** $\sqrt[3]{xy}$ **15.** $2x^{2/3}$ **17.** $7x^{3/2}$

19. $5^{1/4}x^{3/4}$ **21.** $4x\sqrt{2x}$ **23.** $2\sqrt[3]{2x^2}$ **25.** $x=64$

27. $x=13$ **29.** 13   **31.** 18.33   **33.** No

Section 9.8

1. $\dfrac{1}{t}$ **3.** $\dfrac{1}{3x^2}$ **5.** $2b^{-3}$ **7.** $\dfrac{1}{ab^2}$ **9.** $\dfrac{x^2}{3}$

11. $7x^{-2}$ **13.** $4(m+n)^{-1}$ **15.** $\dfrac{2}{x}$ **17.** x^3

19. $\dfrac{1}{(x+y)^{2/3}}$ **21.** $x^{-2}y^{-2}$ **23.** $(a+b)^{-1/2}$

25. $z=\dfrac{k}{d^2}$ **27.** 10   **29.** 6   **31.** $\sqrt{30}-\sqrt{10}$

Section 9.9

1. 3×10^4 **3.** 3×10^{-4} **5.** 710

7. 6.92×10^5 **9.** 3.48×10^{-3} **11.** .468

13. 32.9   **15.** 8.446×10^3 **17.** 7.76×10^{-1}

19. 6.7×10^{-5} **21.** 69,000,000   **23.** .000000503

25. 5.24×10^7 mi²   **27.** .00000002 cm

29. 172,000,000 km   **31.** \$186,000,000,000

33. 1.28×10^5 **35.** $4\sqrt{3}$ **37.** $3\sqrt[3]{3}$ **39.** $4x\sqrt{y}$

CHAPTER 10

Pre-test

1. $x = 3\sqrt{5}$ or $x = -3\sqrt{5}$

2. $x = 8$ or $x = -8$ 3. $x = 3$ or $x = 5$

4. $x = \dfrac{5}{2}$ or $x = 1$ 5. $x = \dfrac{5}{4}$ or $x = \dfrac{2}{3}$

6. $x = -\dfrac{1}{2}$ or $x = -2$ 7. $x = -2$

8. $x = 10$ 9. $x = 119$ 10. $x = -\dfrac{1}{3}$

11. 46 ft   12. $\dfrac{75 \pm \sqrt{921}}{16}$ sec ≈ 2.79 sec or 6.58 sec

Section 10.1

1. $x = \pm 5$ 3. $x = \pm 11$ 5. No real roots

7. $x = \pm\sqrt{5}$ 9. $x = \pm 2\sqrt{2}$ 11. $x = \pm 6$

13. $x = \pm 4$ 15. $x = \pm\sqrt{59}$ 17. No real roots

19. $x = \pm 4$ 21. $x = \pm\sqrt{38}$ 23. $x = \pm 4$

25. $D \approx .68$ ft   27. 9.5 sec   29. $10\sqrt{2} \approx 14.14$ ft

31. $\dfrac{5\sqrt{2}}{4}$ 33. $\dfrac{\sqrt{15}}{5}$ 35. $\dfrac{1 + \sqrt{3}}{-2}$

Section 10.2

1. $x = 6$ or $x = -3$ 3. $y = -7$ or $y = -5$

5. $x = -7$ or $x = -2$ 7. $x = -\dfrac{2}{3}$ or $x = \dfrac{3}{2}$

9. $x = -\dfrac{3}{5}$ or $x = \dfrac{7}{4}$ 11. $x = \dfrac{1}{4}$ or $x = \dfrac{1}{3}$

13. No real roots   15. $x = \dfrac{5 \pm \sqrt{65}}{2}$

17. $x = \dfrac{1 \pm \sqrt{57}}{2}$ 19. $x = \dfrac{3 \pm \sqrt{29}}{10}$

21. $x = \dfrac{-1 \pm \sqrt{222}}{17}$ 23. $x = \dfrac{5}{2}$ or $x = -\dfrac{4}{5}$

25. 4 in   27. 10%   29. 18.32 in by 21.32 in

31. $\sqrt{2}$ 33. $12\sqrt{3}$ 35. $\sqrt{2}$

Section 10.3

1. $x = -1$ or $x = 5$ 3. $x = -\dfrac{10}{3}$ or $x = 6$

5. $x = -\dfrac{1}{2}$ or $x = 0$ 7. $x = \dfrac{1 \pm \sqrt{109}}{6}$

9. $x = -4$ or $x = 5$ 11. $x = 5 \pm 2\sqrt{7}$

13. No real roots   15. $x = -\dfrac{5}{3}$ or $x = 1$

17. $x = \dfrac{-7 \pm \sqrt{57}}{4}$ 19. $x = 4 \pm 2\sqrt{21}$

21. $x = -6$ 23. $x = \dfrac{9 \pm \sqrt{161}}{2}$ 25. 2 ft

27. 12 cm by 18 cm   29. 17 cm by 20 cm

31. $7^{1/3}$ 33. $15^{1/2}$ 35. $a\sqrt[3]{b}$

Section 10.4

1. $x = 64$ 3. No real solution   5. $x = 16$

7. $x = 36$ 9. No real solution

11. $x = 6$ or $x = 5$ 13. $x = 18$

15. No real solution   17. $x = \dfrac{49}{36}$ 19. $x = 30$

21. No real solution   23. $x = -3$ or $x = 8$

25. $p \approx 2.8$ psi   27. 4.68 in   29. 64 sq ft, 576 sq ft

31. x^{-2} 33. $a^2 b^{-3}$ 35. x^3

CHAPTER 11

Pre-test

1.

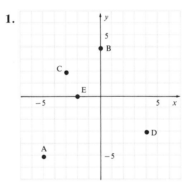

2. $A(4, 4)\ B(1, -3)\ C(-5, 3)\ D(0, -5)\ E(-4, -4)$

3.

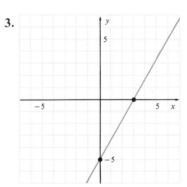

4. $(14, 0)$ 5. $(0, -4)$ 6. $m = -\dfrac{1}{2}$

7. $m = \dfrac{5}{3}$ 8. 5

9.

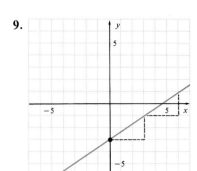

10.

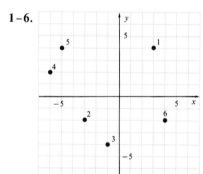

11. $(2, -1)$ **12.** $2,500 at 8%, $3,500 at 6%

Section 11.1

1–6.

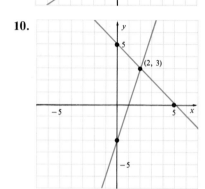

7. $(3, 4)$ **8.** $(4, -2)$ **9.** $(2, -4)$ **10.** $(-4, -2)$

11. $(-1, 4)$ **12.** $(-1, -3)$ **13.** $(-3, 1)$ **14.** $(2, 0)$

15–19.

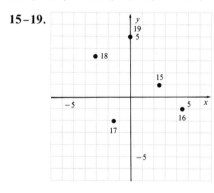

20. $(4, 1.5)$ **21.** $(-2, 3.5)$ **22.** $(-3, -.5)$

23. $(0, -2.5)$ **24.** $(4.5, -4)$ **25.** $(5, E)$

26. $(3, C)$ **27.** 3.82×10^7 **28.** 2.35×10^{-7}

29. 1.85×10^{-2} **30.** .000056   **31.** 687,000,000,000

Section 11.2

1.

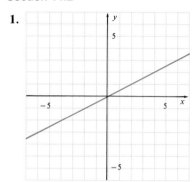

3.

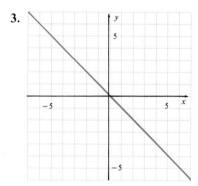

5.

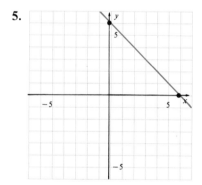

7.

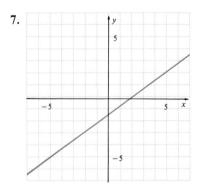

9.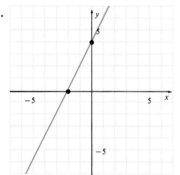

11.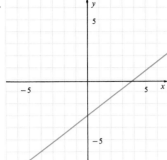

13.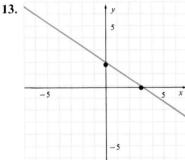

15.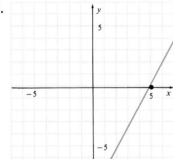

17.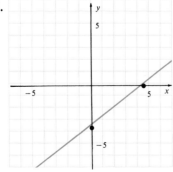

19.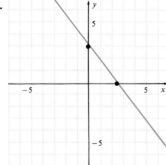

21.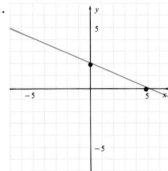

23.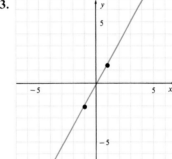

25.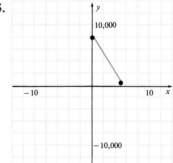

27.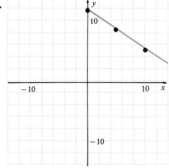

29.

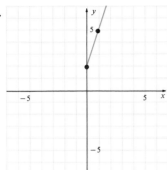

31. $x = \pm 13$ **33.** $a = \pm 3$ **35.** $x = \pm 4$

Section 11.3

1. $(3, 0)(0, 9)$ **3.** $m = \dfrac{3}{4}$ **5.** $m = -\dfrac{1}{2}, (0, 6)$

7. $\left(\dfrac{11}{2}, 0\right)\left(0, \dfrac{11}{5}\right)$ **9.** $(2, 0)\left(0, -\dfrac{12}{7}\right)$

11. $m = -\dfrac{9}{4}$ **13.** $m = \dfrac{2}{5}, \left(0, \dfrac{8}{5}\right)$

15. $(-18, 0)(0, -12)$ **17.** $\left(-\dfrac{7}{2}, 0\right)\left(0, \dfrac{3}{2}\right)$

19. $m = 3$ **21.** $m = \dfrac{3}{2}, (0, +12)$

23. $m = -\dfrac{20}{3}, \left(0, -\dfrac{5}{6}\right)$ **25.** $m = \dfrac{3}{50}$

27. 17¢, \$10   **29.** No real solution   **31.** $x = \dfrac{5 \pm \sqrt{97}}{6}$

Section 11.4

1. 5   **3.** 13   **5.** 3   **7.** $\sqrt{26}$ **9.** $\sqrt{5}$ **11.** $6\sqrt{5}$
13. $18\sqrt{2}$ **15.** $3\sqrt{10}$ **17.** $\sqrt{113}$ **19.** $4\sqrt{13}$
21. 21.9   **23.** 30.0   **25.** $\ell = 5\sqrt{2}, w = 4\sqrt{2}$
27. $5 + 2\sqrt{13} + \sqrt{61}$ **29.** $x = \dfrac{2 \pm \sqrt{19}}{6}$

31. $x = \dfrac{9 \pm \sqrt{51}}{3}$ **33.** $x = -3$ or $x = 4$

Section 11.5

1.

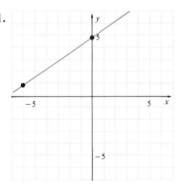

3.

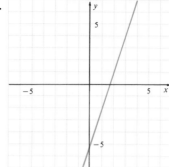

5.

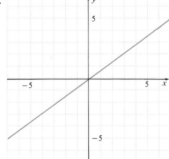

7.

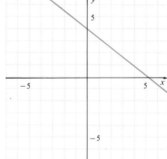

9.

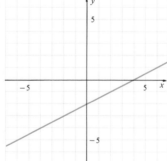

11.

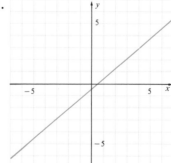

13.

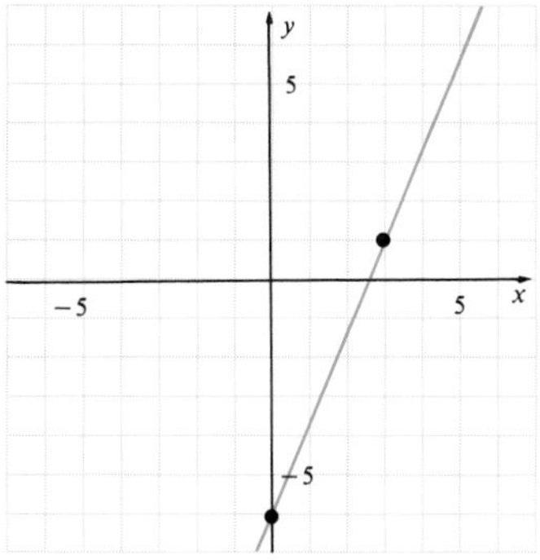

15.

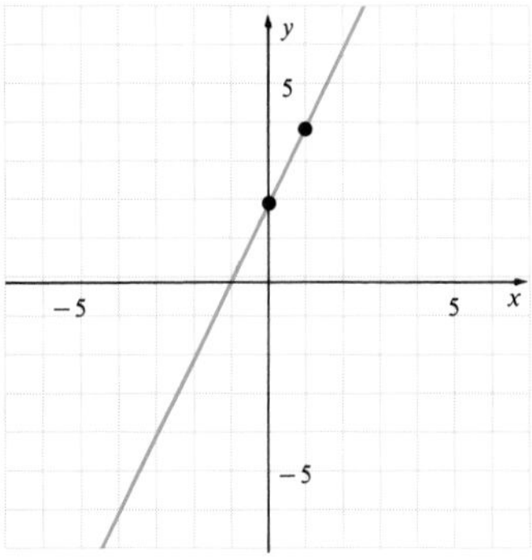

17.

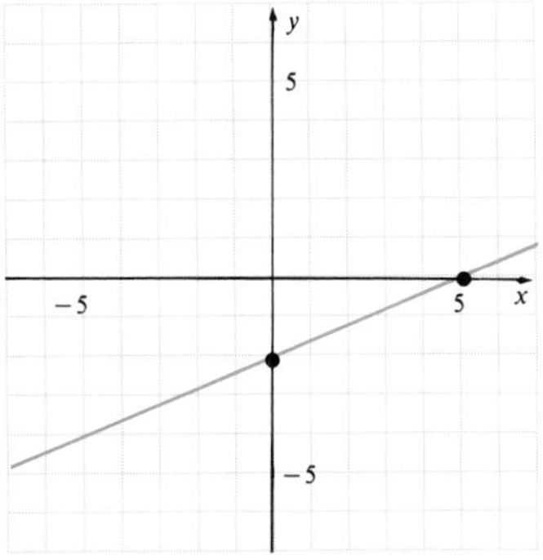

19.

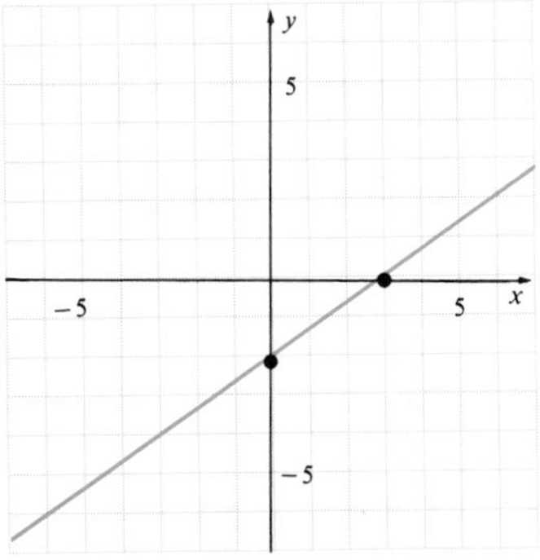

21.

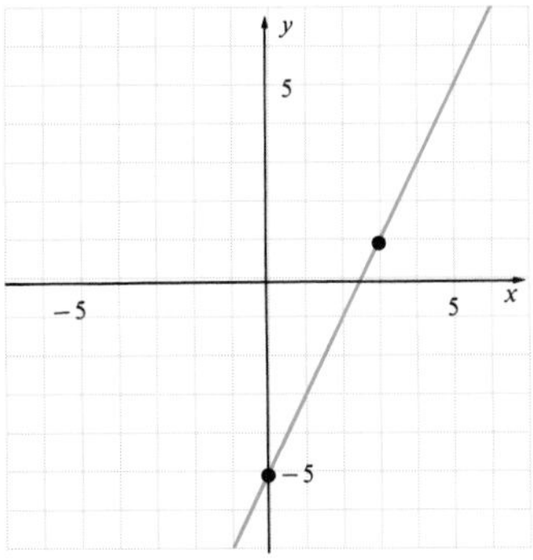

23.

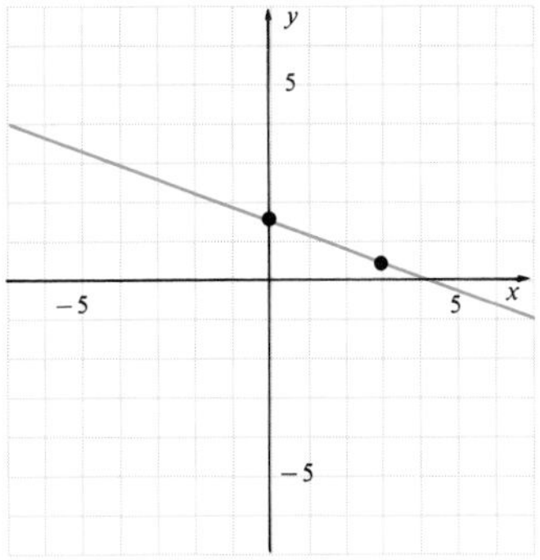

25.

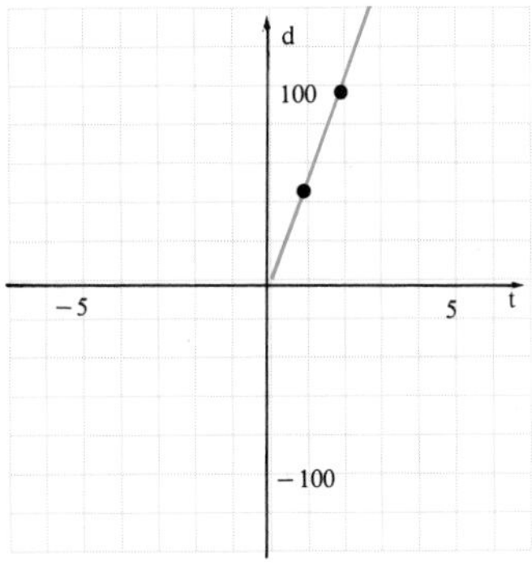

27.

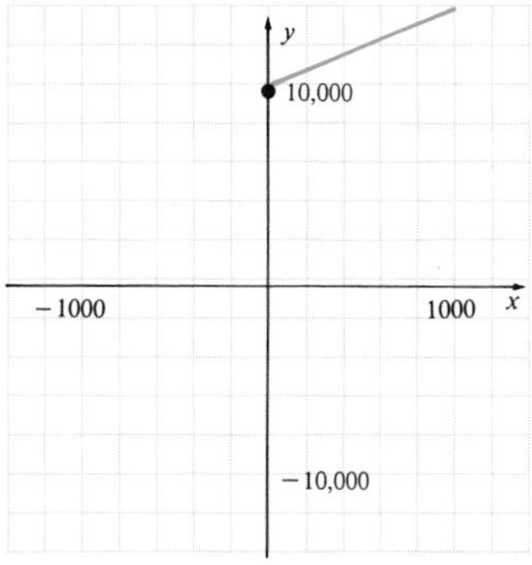

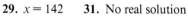

29. $x = 142$ 31. No real solution

33. No real solution

Section 11.6

1.

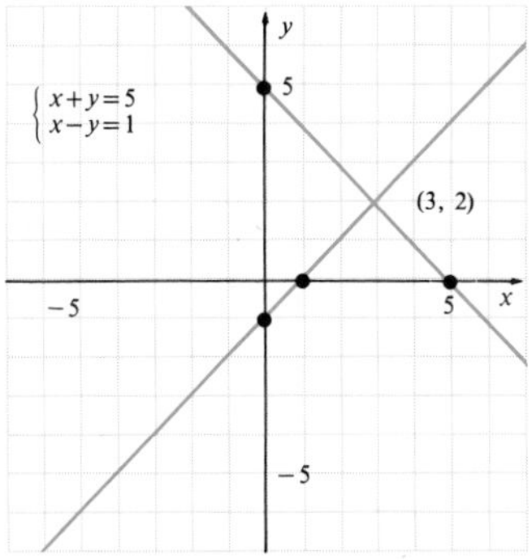

3.

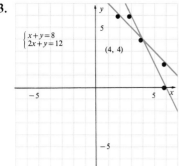

$\begin{cases} x+y=8 \\ 2x+y=12 \end{cases}$

(4, 4)

5.

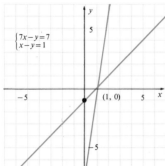

$\begin{cases} 7x-y=7 \\ x-y=1 \end{cases}$

(1, 0)

7.

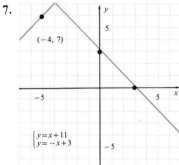

(−4, 7)

$\begin{cases} y=x+11 \\ y=-x+3 \end{cases}$

9.

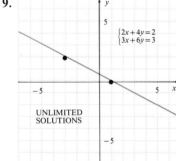

$\begin{cases} 2x+4y=2 \\ 3x+6y=3 \end{cases}$

UNLIMITED SOLUTIONS

11.

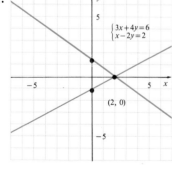

$\begin{cases} 3x+4y=6 \\ x-2y=2 \end{cases}$

(2, 0)

13.

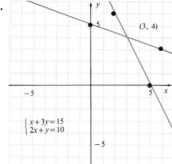

(3, 4)

$\begin{cases} x+3y=15 \\ 2x+y=10 \end{cases}$

15.

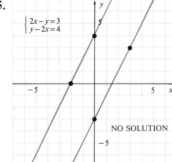

$\begin{cases} 2x-y=3 \\ y-2x=4 \end{cases}$

NO SOLUTION

17.

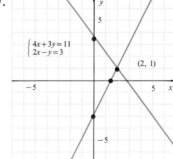

$\begin{cases} 4x+3y=11 \\ 2x-y=3 \end{cases}$

(2, 1)

19.

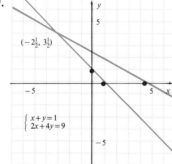

$\left(-2\tfrac{1}{2},\ 3\tfrac{1}{2}\right)$

$\begin{cases} x+y=1 \\ 2x+4y=9 \end{cases}$

21.

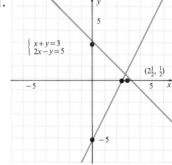

$\begin{cases} x+y=3 \\ 2x-y=5 \end{cases}$

$\left(2\tfrac{1}{2},\ \tfrac{1}{2}\right)$

23.

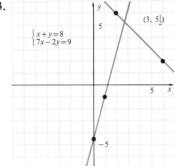

$$\begin{cases} x+y=8 \\ 7x-2y=9 \end{cases}$$

$(3, 5\frac{1}{2})$

25.

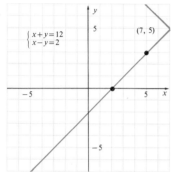

$$\begin{cases} x+y=12 \\ x-y=2 \end{cases}$$

$(7, 5)$

27.

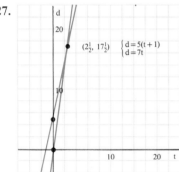

$(2\frac{1}{2}, 17\frac{1}{2})$ $\begin{cases} d=5(t+1) \\ d=7t \end{cases}$

They will both have run $17\frac{1}{2}$ miles $\left(\text{in } 2\frac{1}{2} \text{ hours}\right)$.

29. $x = 10.5$ **31.** $x = 60$ ft

Section 11.7

1. $(4, -1)$ **3.** $(1, 0)$ **5.** $(9, 2)$ **7.** $(3, 2)$

9. $(-5, -3)$ **11.** $(-2, 10)$ **13.** $(2, -1)$

15. $(-2, 3)$ **17.** $\left(-\frac{1}{2}, \frac{5}{8}\right)$ **19.** $\left(\frac{13}{10}, \frac{4}{5}\right)$

21. Unlimited solutions   **23.** $\left(\frac{22}{13}, -\frac{19}{13}\right)$

25. 300 lb of 40%, 300 lb of 70%

27. \$10,000 @ 12%, \$8,000 @ 15%

29. Cadillac, 410 mi; Datsun, 250 mi

31. \$1,200 @ 18%, \$900 @ 20%   **33.** $S = -60$

35. $e = E - IR$ **37.** $r = \pm\sqrt{\dfrac{kmM}{F}}$

Section 11.8

1. $(3, 1)$ **3.** $(4, 1)$ **5.** No solution   **7.** $(7, 5)$

9. $(1, 3)$ **11.** $(5, 5)$ **13.** $\left(\frac{1}{2}, 4\right)$ **15.** $(32, 3)$

17. $(4, -1)$ **19.** $(4, 3)$ **21.** $\left(\frac{20}{13}, -\frac{14}{13}\right)$

23. $\left(\frac{9}{4}, -\frac{3}{16}\right)$ **25.** 5 oz 85%; 10 oz 70%

27. 25 sofas, 40 chairs

29. 2,500 @ \$36/barrel, 1,100 @ \$42/barrel

31. 13,000 @ 3.5%, 32,500 @ 6.6%

33. 71 grams liver, 16 grams bacon

35. $\dfrac{x^3 - 3xy^2 - y^3}{xy(x+y)}$ **37.** $\dfrac{x^2 - 2x - 8}{x+4}$

CHAPTER 12

Pre-test

1. $118°$ **2.** $131.8°$

3. $\sin F = \dfrac{\sqrt{5}}{3}$, $\cos F = \dfrac{2}{3}$, $\tan F = \dfrac{\sqrt{5}}{2}$

4. $\sin U = \dfrac{5\sqrt{29}}{29}$, $\cos U = \dfrac{2\sqrt{29}}{29}$, $\tan U = \dfrac{5}{2}$

5. $\dfrac{1}{2}$ **6.** 0.6924   **7.** $13.0°$

8. $h \approx 14.8$, $F \approx 28.3°$, $G \approx 61.7°$

9. $p \approx 15.4$, $r \approx 8.3$, $Q = 57.4°$ **10.** 0.5592

11. -0.8920 **12.** -7.3002 **13.** $109.5°$

14. $13.3°$ or $166.7°$

15. $D \approx 92.1°$, $E \approx 54.8°$, $F \approx 33.0°$

16. $\ell \approx 10.5$, $M \approx 37.6°$, $N \approx 102.4°$

17. $y \approx 30.1$, $Y \approx 98.5°$, $X \approx 29.5°$

18. $a \approx 17.0$, $b \approx 11.0$, $A = 85°$ **19.** 35 m

20. 383 m

Section 12.1

1. $G = 100°$ **3.** $B = 110°$ **5.** $R = 60°$

7. $C = 69°$ **9.** $X = 29°$ **11.** $H = 11°$

13. $Y = 44°$ **15.** $Q = 107.1°$ **17.** $L = 6.4°$

19. $U = 72°$ **21.** $W = 94.7°$ **23.** $B = 69.9°$

25. $51°$ **27.** $52°$ **29.** $8°$ **31.** $(0, 0)$

33. $(-6, 5)$ **35.** $(4, -3)$

Section 12.2

1. $\sin A = \dfrac{4}{5}$, $\cos A = \dfrac{3}{5}$, $\tan A = \dfrac{4}{3}$

3. $\sin A = \dfrac{12}{13}$, $\cos A = \dfrac{5}{13}$, $\tan A = \dfrac{12}{5}$

5. $\sin A = \dfrac{12}{13}$, $\cos A = \dfrac{5}{13}$, $\tan A = \dfrac{12}{5}$

7. $\sin T = \dfrac{1}{\sqrt{2}}$, $\cos T = \dfrac{1}{\sqrt{2}}$, $\tan T = 1$

9. $\sin T = \dfrac{1}{\sqrt{2}}$, $\cos T = \dfrac{1}{\sqrt{2}}$, $\tan T = 1$

11. $\sin T = \dfrac{3}{\sqrt{13}}$, $\cos T = \dfrac{2}{\sqrt{13}}$, $\tan T = \dfrac{3}{2}$

13. $\sin T = \dfrac{5}{\sqrt{34}}$, $\cos T = \dfrac{3}{\sqrt{34}}$, $\tan T = \dfrac{5}{3}$

15. $\sin D = \dfrac{3}{\sqrt{10}}$, $\cos D = \dfrac{1}{\sqrt{10}}$, $\tan D = 3$

17. $\sin D = \dfrac{2\sqrt{2}}{3}$, $\cos D = \dfrac{1}{3}$, $\tan D = 2\sqrt{2}$

19. $\sin D = \dfrac{2}{\sqrt{13}}$, $\cos D = \dfrac{3}{\sqrt{13}}$, $\tan D = \dfrac{2}{3}$

21. $\sin D = \dfrac{2\sqrt{2}}{3}$, $\cos D = \dfrac{1}{3}$, $\tan D = 2\sqrt{2}$

23. $\sin D = \dfrac{\sqrt{3}}{2}$, $\cos D = \dfrac{1}{2}$, $\tan D = \sqrt{3}$ **25.** 4.2%

27. 10.2%

29.

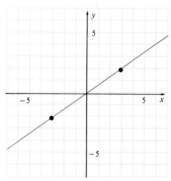

31.

33.

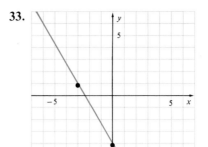

Section 12.3

1. .1219   **3.** .1228   **5.** .1392   **7.** .6060

9. .7618   **11.** .9842   **13.** .7218   **15.** $B = 51.6°$

17. $B = 19.5°$ **19.** $C = 65.1°$ **21.** $A = 65.0°$

23. $C = 42.2°$ **25.** $Z_T = 168.91$ ohms   **27.** 39.5°

29. $\dfrac{5}{9}$ **31.** $-\dfrac{11}{10}$ **33.** $\dfrac{3}{2}$, (0, 0)

Section 12.4

1. $d = 12$, $f \approx 10.4$, $F = 60°$

3. $b \approx 13.9$, $c = 16$, $A = 30°$

5. $x \approx 0.6$, $z \approx 1.2$, $Y = 60°$

7. $p \approx 14.3$, $q \approx 10.1$, $R = 45°$

9. $c \approx 45.0$, $A \approx 56.2°$, $B \approx 33.8°$

11. $y \approx 404.7$, $z \approx 266.1$, $X = 48.9°$

13. $t \approx 12.8$, $S \approx 18.2°$, $R \approx 71.8°$

15. $\ell \approx 38.9$, $N \approx 54.3°$, $M \approx 35.7°$ **17.** 21.7°

19. 12.0 ft   **21.** 142 ft   **23.** 56 ft   **25.** 6,100 ft

27. 53.2 ft   **29.** 36.0°   **31.** 86.4 ft   **33.** 379.5 ft

35. $5\sqrt{10}$ **37.** $4\sqrt{13}$

Section 12.5

1. 0.7986   **3.** -1.3270 **5.** -0.4226

7. 0.9903   **9.** -7.1154 **11.** -0.2419

13. 0.9962   **15.** 0.8508   **17.** -1.6191

19. -0.4787 **21.** -2.9887 **23.** 0.4555

25.

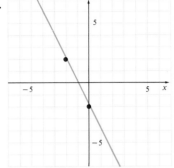

27.

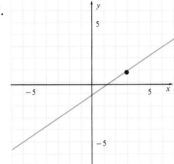

29.

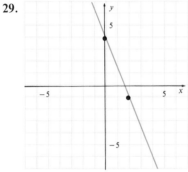

Section 12.6

  1. $A \approx 142.2°$ **3.** $P \approx 148.5°$ **5.** $T \approx 141.3°$

  7. $A \approx 158.8°$ **9.** $C \approx 160.1°$ **11.** $Z \approx 155.8°$

13. $W \approx 159.8°$ **15.** $M \approx 95.6°$ **17.** $P \approx 156.9°$

19. $B \approx 96.4°$ **21.** $T \approx 139.2°$ **23.** $C \approx 135.5°$

25.

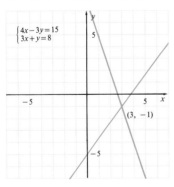

27.

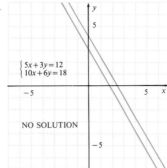

NO SOLUTION

29.

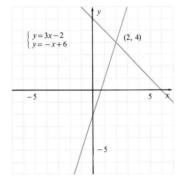

Section 12.7

  1. $47.2°$ **3.** 70.1 miles   **5.** 157 feet

  7. 578 feet   **9.** $B \approx 90°, A \approx 51°$ **11.** 128 knots

13. 1,320 feet   **15.** 149 pounds   **17.** $70.6°$

19. $(-2, 0)$ **21.** $\left(\dfrac{1}{3}, -3\right)$ **23.** $\left(-\dfrac{5}{2}, 9\right)$

Section 12.8

  1. 31 ft   **3.** A is closer by 105 yd   **5.** 320 m

  7. 3,200 ft   **9.** 103 newtons   **11.** 26 ft

13. 473 ft   **15.** $76°, R \approx 69$ lb   **17.** $(-3, 5)$

19. $(4, -2)$ **21.** $(4, 1)$

Index